				IIIA	IVA	VA	VIA	VIIA	0
									2 **He** 4.00260
				5 **B** 10.81	6 **C** 12.011	7 **N** 14.0067	8 **O** 15.9994	9 **F** 18.998403	10 **Ne** 20.179
	IB	IIB		13 **Al** 26.98154	14 **Si** 28.0855	15 **P** 30.97376	16 **S** 32.06	17 **Cl** 35.453	18 **Ar** 39.948
28 **Ni** 58.69	29 **Cu** 63.546	30 **Zn** 65.38	31 **Ga** 69.72	32 **Ge** 72.59	33 **As** 74.9216	34 **Se** 78.96	35 **Br** 79.904	36 **Kr** 83.80
46 **Pd** 106.42	47 **Ag** 107.868	48 **Cd** 112.41	49 **In** 114.82	50 **Sn** 118.69	51 **Sb** 121.75	52 **Te** 127.60	53 **I** 126.9045	54 **Xe** 131.29
78 **Pt** 195.08	79 **Au** 196.9665	80 **Hg** 200.59	81 **Tl** 204.383	82 **Pb** 207.2	83 **Bi** 208.9804	84 **Po** (209)	85 **At** (210)	86 **Rn** (222)

Metals ← → Nonmetals

63 **Eu** 151.96	64 **Gd** 157.25	65 **Tb** 158.9254	66 **Dy** 162.50	67 **Ho** 164.9304	68 **Er** 167.26	69 **Tm** 168.9342	70 **Yb** 173.04	71 **Lu** 174.967
95 **Am** (243)	96 **Cm** (247)	97 **Bk** (247)	98 **Cf** (251)	99 **Es** (252)	100 **Fm** (257)	101 **Md** (258)	102 **No** (259)	103 **Lr** (260)

Introductory Chemistry
for Health Professionals

Introductory Chemistry
for Health Professionals

Ken Liska
San Diego Mesa College

Lucy T. Pryde
Southwestern College, Chula Vista

Macmillan Publishing Company
New York

Collier Macmillan Publishers
London

Macmillan Publishing Company
866 Third Avenue, New York, New York 10022

Collier Macmillan Canada, Inc.

Library of Congress Cataloging in Publication Data

Liska, Ken
    Introductory chemistry for health professionals.

    Includes index.
    1. Chemistry.   I. Pryde, Lucy T.   II. Title.
[DNLM: 1. Chemistry   QD 31.2 L769i]
QD33.L757   1984   540   83-11321
ISBN 0-02-370980-4

Printing: 1 2 3 4 5 6 7 8   Year: 4 5 6 7 8 9 0 1 2

ISBN   0-02-370980-4

Preface

TO THE STUDENT

We wrote this book with a special kind of student in mind: the pre-health professional. We wrote it for those of you who are planning a career in nursing, nutrition, dietetics, dental hygiene, medical technology, physical therapy, x-ray technology, or any other of the related health professions. As we wrote each chapter, we asked ourselves this question: How do chemical principles relate to patient care and to clinical practice? We discovered dozens of modern applications of chemistry to practice, and you in turn will discover these as you read each chapter.

The text is written at a basic level with considerable descriptive material included. We have presented enough theory to set the stage for understanding applications important for the health professional. So that we would have more room for clinical applications, we have left out detailed discussions of reaction mechanisms, seldom-encountered chemical topics, many types of calculations, and much theoretical organic chemistry. In their place we have included topics which should prove fascinating and useful to you in your future career. Here's just one example: in Chapter 10, we tell the story of a woman who drank several liters of ethylene glycol (auto antifreeze) and was taken to a hospital. The emergency room personnel knew that ethylene glycol is oxidized in the body to oxalic acid, the real poisonous threat. We describe the chemistry involved in glycol metabolism and the treatment that saved this patient from serious injury.

Be sure to become acquainted with the important features of this book that will help you to learn from it. Each chapter starts with an introduction that will provide you with an overview of what the chapter is about. The sequence of topics has been carefully considered. Sample problems are provided to help you master the ideas presented. At the end of the chapter you will find a summary restating the key ideas of the chapter. Key terms and study questions complete the chapter. The answers to all of the odd-numbered

v

problems are in an appendix. You will also find a very large glossary of chemical, medical, and biological terms in an appendix.

In addition to this text, we have provided a study guide and a laboratory manual suited to your needs and interests. The student study guide contains many more sample problems, self-tests, and other study aids.

You should realize that studying chemistry may not be the same process for you as reading other types of books. Learning chemistry has to be a very interactive process. We encourage you to read each section several times, to underline, make notes, summarize material in your own words, and try all sample problems. Examine the figures and tables carefully to see what ideas are presented. Keep track of new terms as they arise so that you can build your chemical vocabulary gradually. Use the chapter-end questions and the study guide to test your own progress. Ask your teacher for other sources to read and study. You may find that computer-assisted programs are available at your school to help you learn basic skills in chemistry.

As you start your study of chemistry, we wish you all will find the joy of inquiry and the pride of success. It may not always be easy for you, but beginning to understand the chemical nature of our world is a goal well worth pursuing.

TO THE TEACHER

We have long taken the stance that in the area of education of pre-health professionals, we must relate the theory to what is happening in the clinic, on the hospital ward, or in consultation with a patient or his family. Our own feelings were reinforced by Lance Factor and Robert Kooser, who examined and criticized textbooks written for the allied health field. In a report entitled "Value Presuppositions in Science Textbooks—a Critical Bibliography", they wrote:

> It is possible, we believe, to argue that the standard chemical facts are actually irrelevant to the practice of allied health sciences. This notion of the irrelevance of basic chemistry . . . does suggest that what is and is not relevant to a practitioner in the allied health sciences has been established in the confines of chemistry department offices and not from the floor of an emergency room.

We have taken that statement seriously. This textbook for students of the allied health professions places as much emphasis on clinical applications as it does on chemistry itself. Ultimately, there must be decisions made on how to balance the necessary foundations of theory and mathematical reasoning with the clinical topics we feel are essential. Whatever decisions have been made to omit material have been based on the idea of gaining more room to give the students of the health professions all the clinical relevance they can possibly get in their early training.

This text includes chapters you will not find in most other allied health textbooks. Chapter 10 on Medicinal Chemistry, Chapter 14 on Steroids, Chapter 15 on Body Fluids, Chapter 16 on Biotransformations, and Chapter 17 on Diet and Body Chemistry all present related material in one place rather than placing it piecemeal in other chapters.

We have included many topics that bear directly on the modern clinical scene. For example, we have discussed transdermal drug delivery systems, synthetic polymers used in drug prostheses, methotrexate in cancer chemotherapy, interferon and the anti-viral state, current theories of atherosclerosis, clinical signs of electrolyte imbalance, diagnosis of pathology using enzymes, application of blood gas data, hyperbaric chambers, and herpetic infections. We discuss recombinant DNA techniques, hemodialysis, salt and the diet, inborn errors of metabolism, and modern scanning techniques such as PETT. We have used the language of the health professional and have included some highly original illustrations such as crystallographic evidence for a receptor site with the antimetabolite in place.

Finally, as you consider the effectiveness of this text as a tool for learning, you should know that it is part of an entire package of **text, laboratory manual, teachers guide,** and **student study guide.** All parts of the package have been prepared by the authors of this text.

ACKNOWLEDGEMENTS

We are indebted to many people for helping us to bring this project to fruition. Those who undertook major reviews of the manuscript include Stanley Mehlman (SUNY, Farmingdale), Michael E. Kurz (Illinois State University), Thomas Berke (Brookdale Community College), Theodore Richerzhagen (Walla Walla Community College), and Ronald Chriss (Western Connecticut State College). We thank these colleagues for their many thoughtful suggestions that helped to improve the manuscript.

Photographic assistance has been generously provided by Lucille Fisher (Editor, American Lung Association), Heniz Hoenecke, M.D. (Sharp Hospital, San Diego), Cortez F. Enloe, M.D. (Editor, Nutrition Today magazine), Dale Blumenthal (Picture Editor, National Institutes of Health), Martin Petersen, and Philip R. Pryde, PhD (San Diego State University). Additional technical assistance has been provided by Jonathan Apple II.

We would like to express our gratitude to all of our colleagues, friends, and families who offered encouragement and support in so many ways while we pursued this project. A special thanks is extended to Paula Liska and Philip Pryde whose patience and understanding are truly appreciated. Lastly, we wish to acknowledge with pleasure the assistance we have received from the editorial and production staff at Macmillan, especially from Dr. Gregory Payne, Mr. Thomas Vance, Mr. J. Edward Neve, and Ms. Holly Reid.

Contents

3 | Molecules and Compounds   46

4 | Quantifying Chemistry   74

11 | Biochemistry of the Amide Group: Peptides and Proteins
288

12 | Carbohydrates
316

13 | Fats and Other Lipids   352

14 | Steroids   379

15 | Body Fluids   399

16 | Biotransformation of Chemicals, Drugs, and Food   446

1

Fundamentals of Matter and Measurement

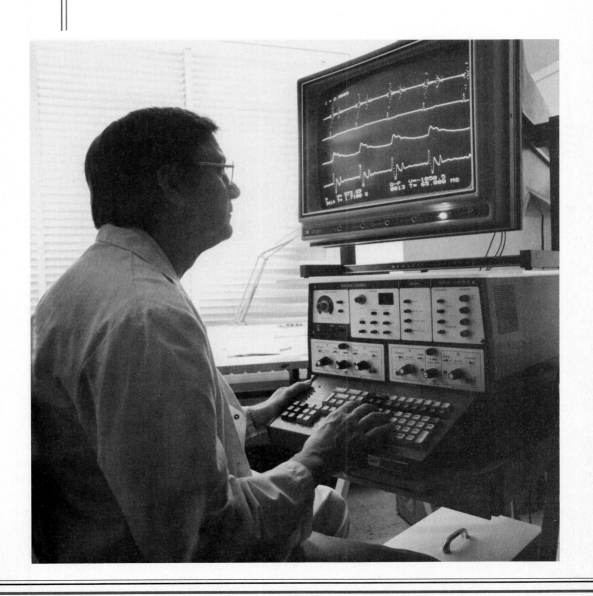

Your study of chemistry begins with an examination of matter and measurement. In this chapter, you will learn to classify matter and to recognize physical and chemical properties. You will also begin to learn what the practicing health professional already knows—that taking accurate measurements and recording meaningful data is a prime concern. This is especially true when recording blood pressure, weight gain or loss, body temperature, respiratory rate, head circumference, or daily urine output.

Speed has become important in data acquisition. You are receiving your education in the computer age, and the recording, storage, and retrieval of information has been vastly accelerated. The researcher in our photo is processing data describing a drug's effects on heart function. He can do this in one hour's time, a process that formerly required two days. The difference is modern instrumentation.

As you study this book and work your laboratory exercises, keep your mind alert to the acquisition, storage, and retrieval of data and to the modern apparatus that help the health professional accomplish them. (Photo courtesy of Eli Lilly and Co.)

INTRODUCTION

Welcome to the fascinating world of chemistry! Our goal in this first chapter is to introduce you to some of the important ideas of chemistry and to develop the means to express those ideas in precisely defined words and mathematical symbols. These skills will give you a working ability to communicate within the field of chemistry, and therefore allow you to relate basic chemistry to the world of the life sciences.

The field of **chemistry** is the study of matter and the changes it undergoes. Chemists deal with both the structure and behavior of matter. Both physical and chemical properties are used to help characterize a particular sample of matter. There are a limited number of elemental building blocks for matter but an amazing number of ways in which these basic units can be put together. The great variety of matter is what makes the study of life processes so challenging and interesting. We shall see many examples in this book of the ways in which important life-sustaining functions are dependent upon chemicals and their properties. Each of these examples serves as a reminder of the essentially practical nature of chemistry in helping us to understand the principles of nursing, nutrition, medical technology, inhalation therapy, and other allied health sciences.

MATTER

Because chemistry is the study of matter, we need to have a clear idea of the nature of matter in order to proceed. **Matter** is defined as anything that has mass and occupies space. That certainly covers a lot of territory, which helps to explain why this is such an important definition for the field of chemistry. We can begin to understand this definition by visualizing any object taking up its share of space. The exact meaning of the term *mass* may not be as familiar to you.

Mass and **weight** are terms that are often used interchangeably, but in fact they are not exactly the same thing. You are probably more familiar with the term **weight,** which is the measurement of the gravitational pull on an object. Weight is a force but mass is not. The **mass** of an object is the measurement of the object's tendency to stay at rest if it is stationary or its tendency to stay in motion if it is already moving. Mass can also be thought of as the quantity of matter present compared with a standard set of masses.

Mass is a more fundamental measurement than weight, for mass does not change as gravitational force changes. If you weigh 150 pounds (lb) on your bathroom scale at home, you would find quite a change in your weight on the surface of the moon with its gravitational pull being one-sixth that on the surface of the earth. (You would weigh 150 lb × ⅙, or just 25 lb!) However, your mass would not change with your travels. Mass is the more fundamental measurement for the amount of matter present.

Classification of Matter

Matter seems to be everywhere and has a wide variety of characteristics. It is soon clear that a system of classification of matter is needed. There are many different classification schemes, as you might imagine. For example, matter may be classified as living or nonliving. It may also be classified according to its state—solid, liquid, or gas. Some types of matter in nature can usefully be categorized as metallic or nonmetallic. Drugs might be categorized as having antibiotic properties or not.[1] Any characteristic of matter can be used as the basis of a classification scheme if it proves convenient and helpful for some particular purpose.

As we examine any number of samples of matter, we note that much of it is not uniform in composition. You may actually see individual particles or regions in a given sample that are different from their neighbors'. Such mixed composition indicates that the sample is not made of just one substance but rather is made up of two or more substances mixed together in such a way that each retains some of its individuality. This type of matter is referred to as **heterogeneous matter.** An example would be a glass of soda with ice cubes.

---

[1] See Appendix V for a classification of drugs based on pharmacological action.

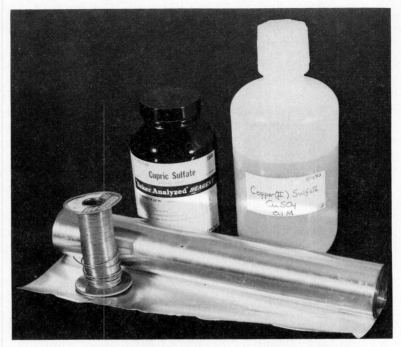

FIGURE 1.1   Systems of Classification. Copper can exist in many different forms. Here we show the element copper in both sheet and wire form. Copper can combine into the compound copper(II) sulfate (CuSO$_4$) which can then be dissolved in water to form a solution. How many different ways can the four items in this photograph be classified into groupings?

In contrast, many samples of matter will appear to be uniform in composition throughout the sample chosen for observation. This type of matter is called **homogeneous matter.** Closer examination may reveal that homogeneous matter is indeed made up of only one basic building block. This is the case for diamond, which is composed of pure carbon solid arranged in a crystalline structure. The homogeneous matter may, however, be a mixture, too, such as a true solution of one substance dissolved in another. If salt is totally dissolved in water, the resulting solution will be uniform in composition and yet composed of a mixture of substances. Any sample of this homogeneous mixture would have the same macroscopic (large scale) properties.

We can conclude that **mixtures** may be either homogeneous or heterogeneous in composition, but they do have the following similarities. They are all composed of two or more substances, present in indefinite proportions. They can be separated into their component parts by means of some physical process. For example, the salt and water mixture could be separated by gently boiling away the water. No chemical change takes place, only the physical separation of the components of the mixture. This does not always mean it will be easy to accomplish complete separation, but it is at least theoretically possible to do so by physical methods. There are many important biochemical mixtures. Homogeneous mixtures with particular interest to the health scientist are gastric juice, cerebrospinal spinal fluid (spinal tap fluid), intravenous (IV) solutions, and many types of medicines furnished in solution form. Some heterogeneous mixtures are blood and suspensions of drugs such as milk of magnesia or Calamine lotion.

Elements and Compounds

Some homogeneous substances fall into the classification of **pure substances.** These substances are not mixtures but rather are composed of just one characteristic material. Unlike homogeneous mixtures, pure substances have a constant, invariable composition by weight and a distinct set of characteristic properties. A pure substance cannot be separated by simple physical processes such as filtering or heating. Earlier we mentioned that the homogeneous mixture of salt and water could be separated by boiling away the water. Once that homogeneous mixture has been separated into its component pure substances, salt and water, further decomposition of either the salt or the water will not be possible unless chemical means are used. Reaction with another substance or the application of a strong electric current would alter the chemical nature of the separated pure substances.

There are many pure substances. In fact, the number is almost overwhelming, easily running into the millions. To make this system of classification more manageable, it can be further subdivided into **elements,** which cannot be decomposed by chemical means into anything simpler, and **compounds,** which are capable of being decomposed by chemical means into their simpler components, elements. A compound is simply a chemical combination of two or more different elements. Elements are the basic building blocks for all of the more complex forms of matter.

At the present time there are 106 known elements, with strong evidence having been presented for elements 107 and 109. You should take a look at the table of elements in the front of the book at this time. The most abundant elements in the body are listed for you in Table 1.1. Each element is listed by mass percent. This means that for every 100 lb of body weight, a person has 65 lb of oxygen either free or combined in the body. Although the listed elements are important, you will see in Chapter 2 that trace elements are also essential for the well-being of the life system.

Physical and Chemical Properties

A *property* is a quality or characteristic possessed by a substance. For example, the compound sugar is sweet, white, and crystalline. Charcoal is black, solid, and capable of being burned to release heat energy. Hydrogen sulfide is a

Table 1.1. Most Abundant Elements in the Body

Element	Percentage (by mass)
Oxygen	65.0
Carbon	18.0
Hydrogen	10.0
Nitrogen	3.0
Calcium	2.0
Phosphorus	1.0
Traces of other elements	1.0
	100.0

colorless gas with an odor typical of rotten eggs. Each pure substance has associated with it a unique set of properties that distinguish it from all other pure substances. Even when two different substances have some similar properties—both sugar and salt are comprised of white crystals—each will have some properties that distinguish it from the other. Therefore, a complete list of properties can be used to identify a substance. If all of the properties of two substances are identical, the chances are that they are samples of the same substance.

Properties used for identifying substances fall into two categories. The set of **physical properties** describes physical characteristics such as state, color, density, melting point, boiling point, volume, temperature, and mass. All of these physical properties can be observed or measured without changing the identity of the substance, although they may cause a change in the state of the substance. If you heat water to determine the boiling temperature, you still have the same chemical substance, water, just in its gaseous state. Such a change in the physical properties of a substance is called a **physical change.**

The other set of properties is called **chemical properties.** These are used to describe how a substance behaves when it changes its identity by interacting with another substance. We have already mentioned that charcoal burns, and you are familiar with the fact that iron rusts. Both of these are examples of chemical change. New substances with new properties are produced in each case. This is a characteristic of **chemical change.** When the charcoal has completely burned, you no longer have charcoal but rather carbon dioxide gas, a substance with a completely new set of properties.

Use these examples to check your understanding of the difference between physical and chemical change. When we inhale, oxygen gas enters our lungs and then diffuses into the blood stream. These are physical changes. The oxygen gas then chemically combines with the hemoglobin in the blood so that the oxygen may be carried throughout the body to tissues. Now the oxygen has undergone a chemical change. The oxyhemoglobin that results is a new substance with new physical and chemical properties.

FIGURE 1.2 Modern Materials in Medicine. **This is an example of the development of modern chemical materials useful in medical sciences. This is an ANGEL-CHIK™ Anti-Reflux Prosthesis. It is placed under the diaphragm and above the stomach. It can successfully be used in the body because the materials have been carefully developed so they are compatible with bodily tissues and fluids.** (Photo courtesy of American Heyer-Schulte Corporation, Goleta, CA.)

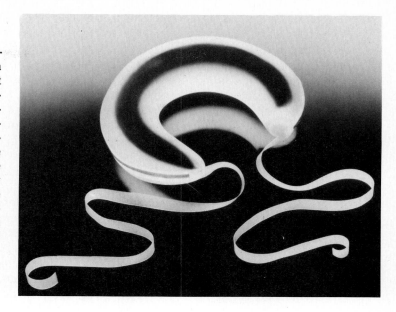

Properties also help to indicate the uses of different substances. For example, steel is strong and can be made resistant to rusting. This means it can be used as the structural material in bridges or in surgical pins for holding together mending bones. In fact, the specific properties of the steel can be finely adjusted by adding small amounts of elements such as nickel, tungsten, vanadium, and molybdenum to the primary metallic mixture (called an **alloy**) of iron, chromium, and manganese. The nonmetals carbon, sulfur, and phosphorus are also present in steel.

MEASUREMENT

You are continually making measurements of all kinds in your daily life. How much did you weigh this morning? Did you have two pieces of toast for breakfast? Is your friend older or younger than you are? Does your child or your patient have a fever? Did you obey the speed limit on the way to work?

These examples serve to illustrate several important points about the measurement process. The first is that the purpose of measurement is to extend your senses beyond qualitative judgments, to actually obtain a number that can then be used as a basis for further decision making. The second point is that all measurements are subject to a certain margin of error and it is important to identify the size of that error. For example, is your scale capable of detecting the loss or telling if you have lost 1 lb? Does the scale need to be corrected for temperature? For the fact that it does not read zero when nobody is on it?

The next point is that measurements are often interpreted as relative statements. They may be relative to a certain legal standard, such as the speed limit or the level of blood alcohol. Measurements may be compared to expected "normal" ranges of values for the blood level of sugar or cholesterol, body temperature, blood pressure, or daily urine output. (Appendix IV lists some of the expected normal values for blood and urine constituents.) Measurement may even be related to your own expectation, such as weight loss if you have been dieting. Health professionals know that "normal" values are subject to change as more data are accumulated on a specific condition or disease state. An example of this is the "safe" exposure limit to benzene in the workplace. The Environmental Protection Agency has drastically reduced this exposure limit in recent years.

The last point about measurement before we learn some specific units and methods of handling those units is to always remember to keep your common sense in gear. You recognize that it must be a mistake if someone directs you to add 3 cups of salt to one container of popcorn and you question what the directions really mean—perhaps the unit "cups" is incorrect or perhaps the quantity "container" has not been clearly communicated and the directions really call for making a 50-gallon (gal) drum of popcorn! On a more serious note, health professionals have the potential to make many mistakes in medication, often because of such things as a misplaced or a misread decimal point. The press has recently reported the case of a doctor who wrote an order for 1.0 milligram (mg) of a drug used for acute gout. The nurse did not observe the decimal point and administered 10 mg of the drug, alledgedly resulting in the death of the patient. Although an extreme case, this points out the need for careful communication and very careful reading of measurements.

Basic Units for Measurement

In everyday life, we tend to use many things without much thought, once those things are familiar to us. So it is with the English system of measurement. Most of us are accustomed to measuring length in feet, yards, or inches and to finding weight in pounds, time in seconds, and temperature in degrees **Fahrenheit.** On the other hand, many people in the United States are less familiar with the basic units of measurement used by the majority of the world and by most scientific disciplines—the International System of Units, commonly referred to as **SI.** (These initials stand for the French name, *Système Internationale d'Unités.*) SI is the modern successor to what used to just be referred to as the **metric system.** Both the metric system and the SI are systems of measurement using just a few base reference units, together with a standard set of prefixes. The SI is coming into increasing use in this country, too, and already many manufacturing companies and public agencies have switched to the International System of Units.

There are just a few so-called base units in the SI. These include the **meter** (m) to measure length, the **kilogram** (kg) to measure mass, the **second(s)** to measure time, and the **kelvin** (K) to measure temperature. The amount of a substance is measured in the unit mole (mol), the luminous intensity is measured in the candela (cd), and the electric current is measured in the ampere (A). All other SI units of measurement are derived from these base units.

If you are just starting to become familiar with SI reference units, it is useful to relate them to some of the common English units that you do know. Examine Figure 1.3 to help with this process.

The SI, like the metric system, is easy to use because the units of each type of measure are all related by the decimal system. For example, the standard unit of length is the meter, but for some things this is too large a measure. Consequently, a unit that is only 1/100 of a meter is used, called the centimeter (cm). The kilogram is also a rather large unit and often the unit **gram** (g) is used. If you have 1000 g of a substance, then you have a kilogram of that substance. The common paper clip can be helpful for quick approximation. One metal paper clip is approximately 1 cm wide and weighs about 1 g, or about 0.035 ounce (oz) in the English system. These useful common sense measures are shown in Figure 1.4.

All of the prefixes used in SI can be attached to any base unit in order to change its size. A commonly used unit of mass in the clinical laboratory would be milligram (mg); 1 mg is 1/1000 of a gram. (Note carefully the difference between *1000* times and *1/1000* times a base unit.) A full set of prefixes that we will find useful is given in Table 1.2.

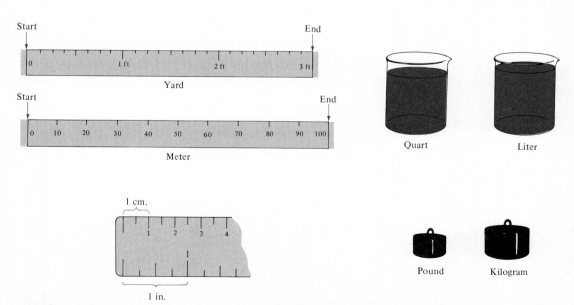

FIGURE 1.3   **The Relationship of Units in the English System to the SI.**

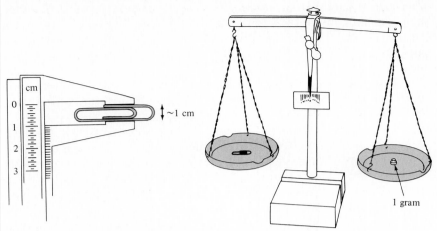

FIGURE 1.4   Common Sense Approximations of Length and Mass.

Conversion of Units

Since all of these prefixes differ by a power of ten, the conversion of one unit to another will always involve moving the decimal point. You may find that this needs some practice until you get used to the system. Once this is mastered, it will be a great deal easier to accomplish unit conversions in the SI than in the English system. Using powers of ten is mathematically much simpler than trying to sort out the English system with its conflicting conversion factors. For example, there are 16 oz in a pound if you are using customary (avoirdupois) measurement of weight, but 12 oz in a pint if you are using U.S. apothecaries' measurements.

The next few examples will show you how to set up and solve conversions within the SI. Observe carefully in these examples how the units are arranged.

Table 1.2.   Metric System Prefixes

Prefix	Abbreviation	Meaning
Nano-	n	1/1,000,000,000
Micro-	μ	1/1,000,000
Milli-	m	1/1000
Centi-	c	1/100
Deci-	d	1/10
Deka-	da	10
Hecto-	h	100
Kilo-	k	1000
Mega-	M	1,000,000
Giga-	G	1,000,000,000

The correct cancellation of units can be very helpful to you in many types of chemistry problems.

EXAMPLE 1.1

How many milligrams are there in 5.6 g?

Solution
Start by checking the definition of milli-. This prefix means ⅟₁₀₀₀. Therefore, there are 1000 mg in a gram because each milligram is ⅟₁₀₀₀ of a gram. Note that a milligram is a much *smaller* unit than a gram so that when you get your answer, it will require far *more* milligrams to represent the same mass as the given number of grams. Here is the mathematical solution.

$$5.6 \, g \times \frac{1000 \text{ mg}}{1 \, g} = 5600 \text{ mg}$$

Look once again at several features of this problem. The ratio 1000 mg to 1 g is an example of a **conversion factor.** It is a ratio that relates the size of one unit to another unit. Notice that the units in the conversion factor are arranged so that the cancellation of units with the given information produces the unit needed for the answer. If you had incorrectly multiplied the original number of grams by the ratio of 1 g to 1000 mg, the units would *not* have canceled. This means that whenever you carry out a conversion, you must both know the correct conversion factor to use *and* use it so that units cancel to yield the correct units for the answer you are seeking.

Also look once again at the size of the answer. It is reasonable that the answer be much larger than the given value because it will take many more of the smaller unit milligrams to represent the same mass as the given number of grams.

EXAMPLE 1.2

If a person weighs 55,000 g, what is the person's weight expressed in kilograms?

Solution
Start by checking the definition of kilo-. This prefix means 1000 times. The kilogram is a much *larger* unit than a gram so that when you get your answer, it will require far *fewer* kilograms to represent the same mass as the given number of grams. Here is the mathematical solution.

$$55,000 \, g \times \frac{1 \text{ kg}}{1000 \, g} = 55 \text{ kg}$$

Once again, check the proper cancellation of units and the relative size of the final answer.

■■■■■■■■ **EXAMPLE 1.3** ■■■■■■■■

If you have measured a distance to be 2.5 millimeters (mm), what will the distance be if expressed in centimeters?

Solution
Check the definitions and find that a millimeter is $\frac{1}{1000}$ of a meter and a centimeter is $\frac{1}{100}$ of a meter. A millimeter is therefore just one power of ten smaller than a centimeter and there are 10 mm in every centimeter. This, in turn, means that the value expressed in centimeters will be *smaller* than the same value expressed in millimeters because the unit centimeters is a larger unit. Here is the mathematical solution, done in two different but equivalent ways.

$$2.5 \text{ mm} \times \frac{1 \text{ m}}{1000 \text{ mm}} \times \frac{100 \text{ cm}}{1 \text{ m}} = 0.25 \text{ cm}$$

$$2.5 \text{ mm} \times \frac{1 \text{ cm}}{10 \text{ mm}} = 0.25 \text{ cm}$$

Scientific Notation

Often in chemistry we deal with awkward numbers. The actual number of atoms in a gram of iron, for example, is approximately 10,000 times a million times a million times a million, or 10,000,000,000,000,000,000,000. This is much more conveniently written as 1×10^{22} atoms. A human red blood cell has a diameter than is typically 8 millionths of a meter. This can be written 0.000008 m but is much more conveniently expressed as 8×10^{-6} m. Both very large and very small numbers are usually represented by scientists in this convenient format, called scientific notation.

Scientific notation expresses any number as a *product* of two factors. The first factor is between 1 and 10 and will have just one digit before the decimal point. The second factor is a power of ten. When these two factors are multiplied together, the number in its standard notation is obtained. Here are two more examples. The number expressed in scientific notation is shown on the left; the same number expressed in standard notation is shown on the right.

Three factors
of 10

$$5.6 \times 10^3 = 5.6 \times \overbrace{10 \times 10 \times 10} = 5600$$

$$2 \times 10^{-2} = 2 \times \underbrace{\frac{1}{10} \times \frac{1}{10}}_{\substack{\text{Two factors} \\ \text{of } \frac{1}{10}}} = 0.02$$

In these two examples, we can see that the presence of 10 raised to a positive power means that the number, when expressed in standard notation, will be a number larger than 1. A negative power of 10 in scientific notation means that the number, when expressed in standard notation, will be a number less than 1.

Exponential notation is a similar but more generalized notation in which the second factor is still a power of 10, but the first factor is not restricted to a value between 1 and 10. The number 298,000 can be correctly expressed as 298×10^3 in exponential notation, but to be in scientific notation must be expressed as 2.98×10^5.

If exponential and scientific notation are unfamiliar to you at this point, you will need to review some basic mathematical rules for handling exponents correctly. We suggest you use the Study Guide that accompanies this text. Your instructor will be able to recommend other sources for your review. It is important to have this skill because the use of scientific notation helps to simplify many problems. We will continue to use scientific notation wherever it is appropriate throughout this text.

Significant Figures

There is one more aspect of handling numbers that we need to consider. Any scientific measurement is reported with the number of digits appropriate to the design of the measuring device. If we used a very good balance, such as an analytical balance, we could find the mass of a small flask to be 96.4322 g. This number is expressed with six **significant figures,** the term used when a figure or digit actually has scientific meaning due to the measurement process. If we found the mass of the same flask using a different balance, such as a triple beam balance, we might only be able to read a mass of 96.43 g. This number has only four significant figures. In each case, the last significant figure is considered as the best estimate between measured divisions. As measurements are taken in the laboratory, we will practice matching the number of significant figures to the type of measuring device.

The number of significant figures, being dictated by the design of the measuring device, will not change if the measurement is expressed in some other unit. If 96.43 g is converted to milligrams, the measurement must still have four significant figures.

$$96.43 \, \cancel{g} \times \frac{1000 \, \text{mg}}{1 \, \cancel{g}} = 96,430 \, \text{mg}$$

The zero is not significant here as its only function is to indicate the position of the decimal point. To eliminate any possible misinterpretation, it is preferable to express such values using scientific notation rather than standard notation.

$$96.43 \text{ g} \times \frac{1 \times 10^3 \text{ mg}}{1 \text{ g}} = 9.643 \times 10^4 \text{ mg}$$

If the same measurement were converted to kilograms or any other mass unit, the value would once again retain four significant figures and be best expressed in scientific notation.

As we learn how to solve different types of problems, keep in mind that when numbers are converted to other units or combined in any calculation, we must still report all answers to reflect the capabilities of the original measuring devices used. It is not good practice either to eliminate digits with meaning or to report extra digits with no scientific meaning. (Calculators are very good at this last trick.) The proper handling of significant figures will continue to be noted in the text. Further practice will take place in the laboratory. The Study Guide contains more of the specific rules governing the correct use of significant figures when combining numbers.

EXAMPLE 1.4

Convert 5×10^2 cm to micrometers.

Solution

Check the definitions in Table 1.2 and find that 1 cm is $\frac{1}{100}$ of a meter, or 10^{-2} m. This means that there are 100 or 1×10^2 cm in a meter. A micrometer (μm) is $\frac{1}{1,000,000}$ of a meter, or 10^{-6} m. This means that there are 1,000,000 μm in a meter or 10^6 μm per meter. It also means that a centimeter and a micrometer are separated by four powers of ten. Since a micrometer is a smaller unit, the value expressed in micrometers must be *larger* than the original number of centimeters. Two ways to arrive at the solution are shown in the following.

$$5 \times 10^2 \text{ cm} \times \frac{1 \text{ m}}{1 \times 10^2 \text{ cm}} \times \frac{10^6 \text{ }\mu\text{m}}{1 \text{ m}} = 5 \times 10^6 \text{ }\mu\text{m}$$

$$5 \times 10^2 \text{ cm} \times \frac{10^4 \text{ }\mu\text{m}}{1 \text{ cm}} = 5 \times 10^6 \text{ }\mu\text{m}$$

Volume Measurements

You may have noticed that we have not yet talked about volume measurements. The seven base units of the SI do not include any volume units, so they must be derived from one of the base units. This derived unit is the cubic meter

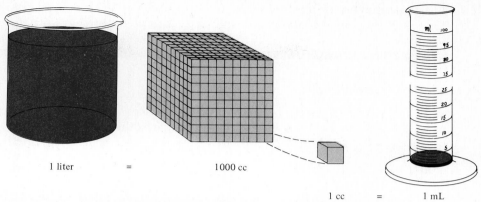

1 liter   =   1000 cc

1 cc   =   1 mL

FIGURE 1.5 Liquid Volume Relationships. **The volume of a cube 10 cm on each edge is 1000 cc, defined to be 1 L. One milliliter is the same as 1 cc.**

(m^3). If you think about this volume, a cube 1 m in each direction, you can appreciate that it is much too large for most laboratory work. The **cubic centimeter** (cm^3 or cc) and the cubic millimeter (mm^3) are more appropriately sized.

For liquid volume, it is still current practice, although not correct from a strict interpretation of SI, to use the **liter** (L) and its subunits. This is a holdover from the metric system and an extremely convenient one. Look back at Figure 1.3 and there you will see that 1 L is just slightly larger than a liquid U.S. quart. (Be careful here, this is not the same quart in Canada. The English system is not noted for its consistency.) Figure 1.5 shows how the various units of volume are related.

For clinical laboratory work, the deciliter (dL) and the microliter (μL) are very useful units, for many working volumes fall into these size ranges. The next example illustrates the use of volume units.

■■■■■■**EXAMPLE 1.5**■■■■■■

If the volume of a liver cell is in the range of 4×10^{-6} μL, what is that volume expressed in liters?

Solution
Check definitions to find that a microliter is $\frac{1}{1,000,000}$ of a liter or 10^{-6} L. This means there are 1,000,000 μL or 10^6 μL in a liter. Since the liter is the larger sized unit, it will take fewer of them to represent the same given volume.

$$4 \times 10^{-6} \, \mu\text{L} \times \frac{1 \, \text{L}}{1 \times 10^6 \, \mu\text{L}} = 4 \times 10^{-12} \, \text{L}$$

The measurement of liquid volume is very important to clinical work. Figure 1.6 shows you some of the most common pieces of equipment used for this purpose.

Temperature Measurement

The kelvin (K) is another SI unit that replaces an earlier metric system unit. The size of a degree kelvin is the same size as the **Celsius** degree, but the fixed points on which the temperature-measuring scale is based are not numerically

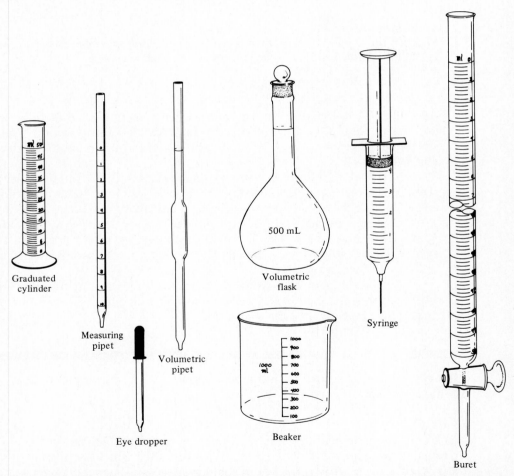

Graduated cylinder

Measuring pipet

Eye dropper

Volumetric pipet

500 mL

Volumetric flask

1000 ml

Beaker

Syringe

Buret

FIGURE 1.6 **Glassware Used for Liquid Measurement. Each of these pieces of glassware can be used for measuring liquid volume. The numbers marked on the sides of a beaker are not as accurate as numbers on a graduated cylinder or a measuring pipet. Volumetric pipets and flasks are designed for just one specific volume.**

the same. Note that the small degree abbreviation is not used with the kelvin unit as it is with the other temperature scale units. A comparison of three temperature scales is given in Figure 1.7.

To actually take temperature, the standard clinical thermometer is still in widespread use but is being replaced in many applications by digital thermometers. The newest development is a strip thermometer containing liquid crystals highly sensitive to temperature changes. (Liquid crystals are also used for many calculator displays.) You will most likely find many types of thermometers in actual use. In the next section you will practice conversions between these temperature scales.

Conversions Between Systems of Measurement

Although the SI or the metric system is used most commonly in chemistry, it is frequently helpful and often required to know how each measurement relates to a corresponding unit in the English system. Many patients in the United States will not be able to understand the meaning of the numbers at all if they are listed in SI, and so conversions between systems of measurement will continue to be needed. The process is very similar to that described earlier for conversion within the SI. You need first to consult a table that gives you the relationships among the units and then to set up a mathematical statement showing this conversion. Values other than powers of ten complicate this process slightly, but in this era of calculators, that does not seem to be a problem. Just remember to keep your common sense so that you will recognize when your calculator has accidentally received the wrong numerical information. Table 1.3 shows some useful comparisons between systems of measurement.

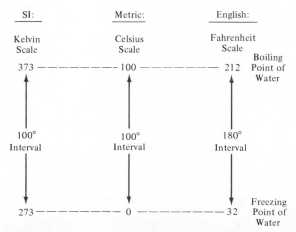

FIGURE 1.7   Temperature Scale Comparisons.

Table 1.3   Comparison of Units
Between Systems of Measurement

Mass and Weight
   1 kg = 2.20 lb
   1 lb = 454 g
   1 g = 0.0353 oz (avoirdupois)
   1 oz (avoirdupois) = 28.3 g

Length
   1 m = 39.4 in.
   1 in. = 2.54 cm
   1 km = 0.621 miles
   1 mile = 1.61 km

Volume
   1 L = 1.06 qt
   1 qt = 0.946 L
   1 fluid oz = 29.6 mL
   1 m³ = 264 gal

Temperature
   °C = K − 273.2
   K = °C + 273.2
   °F = 1.8°C + 32.0
   $°C = \dfrac{(°F - 32.0)}{1.8}$

Heat
   1 Joule (J) = 0.239 calories (cal)

━━━━━ **EXAMPLE 1.6** ━━━━━

"Normal" body temperature orally is 98.6°F. What is this value in degrees
Celsius?

Solution
Check Table 1.3 for the appropriate relationships.

$$°C = \frac{(°F - 32.0)}{1.8}$$

$$= \frac{(98.6 - 32.0)}{1.8}$$

$$= \frac{(66.6)}{1.8}$$

$$= 37.0$$

Note that in this problem you *first* subtract 32.0 from 98.6 within the
parentheses. Once you have the difference, you divide by 1.8.

EXAMPLE 1.7

A certain blood plasma must be stored at a temperature of 5.0°C. How can this be expressed in kelvin and in degrees Fahrenheit?

Solution
Check Table 1.3 for the appropriate relationships.

$$\begin{aligned} K &= {}^\circ C + 273.2 \\ &= 5.0 + 273.2 \\ &= 278.2 \end{aligned}$$

$$\begin{aligned} {}^\circ F &= 1.8{}^\circ C + 32.0 \\ &= 1.8(5.0) + 32.0 \\ &= 9.0 + 32.0 \\ &= 41.0 \end{aligned}$$

Therefore, a temperature of 5.0°F may be expressed as 41.0°F or as 278.2 K.

EXAMPLE 1.8

How many centimeters are there in 5.0 miles?

Solution
From Table 1.3 we find that each mile is equivalent to 1.61 km. Recall, or find from Table 1.2, that a kilometer is 1000 (1×10^3) times as large as a meter and that a centimeter is 1⁄100 (1×10^{-2}) of a meter. Therefore, it will take a larger number of centimeters to express the same distance that is given in miles.

$$5.0\ \cancel{\text{mile}} \times \frac{1.61\ \cancel{\text{km}}}{1\ \cancel{\text{mile}}} \times \frac{1 \times 10^3\ \cancel{\text{m}}}{1\ \cancel{\text{km}}} \times \frac{1 \times 10^2\ \text{cm}}{1\ \cancel{\text{m}}} = 8.0 \times 10^5\ \text{cm}$$

EXAMPLE 1.9

Express your weight in kilograms.

Solution
Check the tables for the appropriate conversion factors. Realize that the weight in kilograms will be smaller numerically since 1 kg is equivalent to 2.20 lb.

The numerical answer will depend on your actual weight, but to find that value the problem should be set up this way:

$$\text{weight in}\ \cancel{\text{lb}} \times \frac{1\ \text{kg}}{2.20\ \cancel{\text{lb}}} = \text{weight in kg}$$

For example, for a 115-lb person

$$115 \ \cancel{lb} \times \frac{1 \ kg}{2.20 \ \cancel{lb}} = 52.3 \ kg$$

━━━━━ **EXAMPLE 1.10** ━━━━━

How much is gasoline per gallon if it is selling for $0.30 a liter?

Solution
This requires changing liters to gallons so that the price can be reported in dollars per gallon. Table 1.3, a useful conversion factor is 1 L = 1.06 qt. (You could just as well use the fact that 1 qt is equivalent to 0.946 L; you could also go to some other table of conversion values and find a direct conversion of liters to gallons.)

$$\frac{\$0.30}{\cancel{L}} \times \frac{1 \ \cancel{L}}{1.06 \ \cancel{qt}} \times \frac{4 \ \cancel{qt}}{1 \ gal} = \frac{\$1.13}{gal}$$

━━━━━ **EXAMPLE 1.11** ━━━━━

The normal range for calcium in the blood is between 8.5 and 10.5 mg/dL. A test is performed and the patient is found to have 9.2 cg of calcium for every liter of blood. Is this patient within the normal range for this test?

Solution
You will have to convert centigrams to milligrams, and you will also have to convert deciliters to liters. Find the appropriate conversion factors in the tables.

$$\frac{9.2 \ \cancel{cg}}{\cancel{L}} \times \frac{10 \ mg}{\cancel{cg}} \times \frac{1 \ \cancel{L}}{10 \ dL} = \frac{9.2 \ mg}{dL}$$

This patient does fall within the normal tolerances expected for this test, as 9.2 is between 8.5 and 10.5 mg/dL.

The last example illustrates common types of units. Often the concentration of a component is expressed by reporting a certain mass present in a specified volume. In Chapter 7 we will learn a wide range of ways to express concentration.

In general, the ratio of mass of an object to its volume is known as **density**. Mathematically, it is defined in this way:

$$\text{density} = \frac{\text{mass}}{\text{volume}}$$

As with all numbers that come from measurements, the units must be reported clearly along with the numerical value. Density can be reported in a wide variety of units, such as grams per liter or milligrams per milliliter. As long as there is a ratio of mass to volume, the units are appropriate for reporting density.

EXAMPLE 1.12

Bone has a density of about 2.0 g/cc. What is the mass of a 45-cc bone?

Solution
Since density is mass per unit volume, the mass of this bone must be obtained by combining the given density with the given volume.

$$45 \text{ cc} \times \frac{2.0 \text{ g}}{\text{cc}} = 90. \text{ g}$$

SUMMARY

This chapter has provided an introduction into the world of chemistry. Chemists deal with the study of matter and the changes that matter can undergo. With over 5 million distinct chemical substances known, classification of matter is an essential step in trying to organize chemical knowledge. Physical and chemical properties help to determine the classification and possible utilization of substances.

Measurements are used to extend our senses and therefore to describe matter and energy quantitatively. The careful expression of measurements means using the appropriate number of significant figures to match the measuring device. Units must be expressed for any measurement. You should know the base units within the SI and know the commonly used prefixes to change the size of these units. We have practiced converting measurements to other units within the SI system, as well as converting between systems of measurement. Scientific notation is often helpful in expressing numbers that result from measurement or calculation.

KEY TERMS

Check your understanding of this chapter. Can you explain what is meant by each of these terms?

alloy	kelvin
Celsius	kilogram
chemical change	liter
chemical property	mass
chemistry	matter
compound	meter
conversion factor	metric system
cubic centimeter	mixture
cubic meter	physical change
density	physical property
element	pure substance
exponential notation	scientific notation
Fahrenheit	second
gram	SI
heterogeneous matter	significant figures
homogeneous matter	weight

STUDY QUESTIONS

1. Are the following classified as pure substances or as mixtures?
   a. Copper   b. Air
   c. Milk   d. Water
   e. Concrete   f. Wine
   g. Sulfur   h. Sugar solution
   i. Blood   j. Urine

2. Are each of these mixtures homogeneous or heterogeneous?
   a. Salt and pepper
   b. Alcohol and water
   c. Sulfur and iron
   d. Two different salts
   e. Milk of Magnesia suspension
   f. 5% dextrose in water (used for injection)

3. Can you suggest possible methods to separate the components of the mixtures in Question 2 (parts a–d)?

4. Here are some objects. Develop a classification system for these objects. Discuss your classification system with your classmates, for there are many possible answers.

   milk, newspaper, blouse, theater tickets, cat, book, candle, pencil, graduated cylinder, peanuts, coat, Valentine's day card, waste basket, picture frame, violin strings, running shoes, paper, eye glasses, house plant, spinach, penicillin

5. How could you separate the pure substance salt from a homogeneous mixture of salt and water?

6. A glass of soda with ice cubes is given as an example of heterogeneous matter. Will this example remain heterogeneous? Explain your reasoning.

7. What physical properties do salt and sugar have in common? How do they differ in physical properties?

8. Are each of the following classified as physical or chemical properties?
   a. Density   b. Ability to burn
   c. Explosiveness   d. Solubility
   e. Melting point   f. Corrosiveness

9. Classify each of the following as either physical or chemical changes.
   a. Freezing of water
   b. Rusting of iron
   c. Evaporation of dry ice
   d. Explosion of gun powder
   e. Tearing of paper
   f. Digestion of food
   g. Absorption of a vitamin from the intestinal tract
   h. Detoxification of a drug in the liver
   i. Synthesis of protein in muscle tissue

10. Classify each of the following as either physical or chemical changes.
    a. Boiling seawater
    b. Souring milk
    c. Burning gasoline
    d. Sugar dissolving in water
    e. Blood coagulating
    f. Spreading of a drop of ink in water
    g. Neutralizing battery acid with bicarbonate of soda
    h. Healing a broken bone
    i. Silver tarnishing

11. Why is it necessary to measure properties and not just depend on direct observation?

12. Name the base SI unit for measuring length. What is the base unit for mass? For temperature measurement?

13. Convert each of the following measurements to meters.
    a. 675 dm
    b. 92 dekameter (dam)
    c. 5,600,000 μm
    d. 800 mm
    e. 25 km

14. Convert each of the following measurements to meters. Express each answer in scientific notation.
    a. 780 cm
    b. 5×10^4 mm
    c. 4.7×10^2 cm
    d. 6×10^{-3} hm
    e. 3.55×10^4 μm

15. Convert each of the following to milliliters.
    a. 6.7 L
    b. 900 kL
    c. 6×10^{-4} L
    d. 0.0056 nL
    e. 0.045 L

16. Convert each of the following to milliliters. Express your answers in scientific notation.
    a. 1.6×10^2 L
    b. 444 μL
    c. 2.34×10^{-4} cL
    d. 560,000 dL
    e. 1.3×10^3 kL

17. Convert each of the following to kilograms.
    a. 750 g
    b. 5,000,000 dag
    c. 4×10^5 g
    d. 3.4×10^{-7} mg
    e. 7.124×10^4 cg

18. How many significant figures are there in each of these measurements?
    a. 24.69 g
    b. 0.00021 km
    c. 74,000 cm
    d. 4.1329 mg
    e. 12,000 m

19. Change each of the measurements in Question 18 into scientific notation. How many significant figures are now present in each case?

20. How many significant figures are present in each of these measurements?
    a. 2.35×10^3 mg
    b. 17.07823 g
    c. 24.3450 g
    d. 1.8×10^{-2} μm
    e. 2.020×10^1 mm

21. Which is larger: 200 m or 220 yd? Show calculations to support your judgment.

22. Which is a longer distance: 50 miles or 50 km? Show calculations to support your judgment.

23. You are traveling in Mexico, step on a scale, and discover that you weigh 90 kg. If your usual weight is 180 lb, should you start a diet or eat everything in sight?

24. Calculate the height of a 7-ft-tall basketball player in meters.

25. Many chemical properties are measured at standard temperature of 25°C. What temperature would this be in kelvin and in Fahrenheit?

26. If your "normal" body temperature is 99.0°F, what is this expressed on the Celsius temperature scale?

27. Make each of these conversions.
    a. 100°C to Fahrenheit
    b. 65°F to Celsius
    c. 100°F to Celsius
    d. 5°C to kelvin
    e. 50°F to kelvin

28. If the thermostat controlling your room's temperature is set at 68°F, what temperature will you set on your metric system thermostat to maintain the same temperature?

29. The doctor has directed that you call if the patient's temperature is above 40°C. You find that the patient has a temperature of 102°F. Should you call the doctor?

30. Make each of these conversions.
    a. 7.2 cg/L to kilograms per liter
    b. 42×10^2 mg/L to milligrams per deciliter
    c. 6.7 g/L to milligrams per milliliter
    d. 8.1 g/nL to milligrams per deciliter

31. A sample of urine has a density of 1.003 g/mL. What is its density expressed in milligrams per deciliter?

32. Concentrated hydrochloric acid has a density of 1.2 g/cc. How many cubic centimeters do you need to measure out if you want to be sure of having 20. g of concentrated hydrochloric acid?

33. Object A weighs 55 g and has volume of 15 cc. Object B weighs 295 g and has a volume of 120 cc. Which object is more dense? Explain.

2 Atoms and Elements

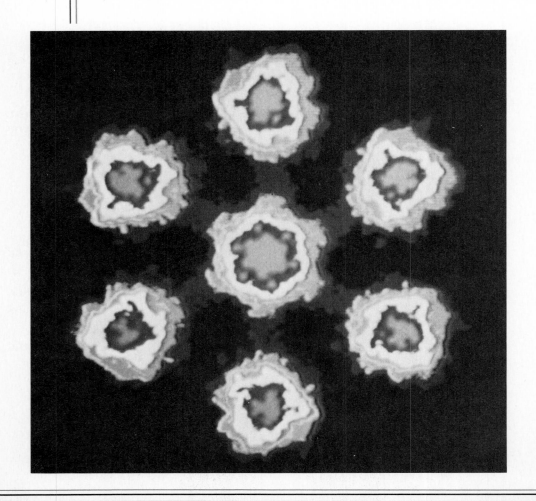

Throughout recorded history, mankind has attempted actually to see the smallest building block of matter. These pictures of uranium atoms, taken with a scanning electron microscope, date from 1976 and were among the first ever made of individual atoms. Thus, it is only in most recent history that the age-old quest to "see" atoms has actually been realized. One can only guess what advances will occur in the next decade in our understanding of the fundamental structure of all matter.

Many of the ideas presented in this chapter deal with the representation of the structure of matter by means of models. The true nature of these models has become most sophisticated and very mathematical in the time since the start of quantum theory. These complexities should not stop us from using the pictorial aspects of the theories, even if the advanced mathematics associated with quantum theory is well beyond the scope of this book. After all, we all employ electrical applicances every day, even if we do not have a full understanding of the full nature of electromagnetic field theory.

In this chapter you will see how to use the periodic table of elements to correlate the structure and properties of atoms. This table is surely one of the most remarkable organizing tools of modern chemistry and you will want to become very familiar with its organization. Within the table, we will identify certain elements that have particular relevance for health professionals. In particular, you will want to learn which elements are present in relative abundance in our bodies. You will also want to know which elements, even though present in small absolute amounts, are very important to our well-being. (Photo courtesy of M. Isaacson, Cornell University, and M. Ohtsuki, The University of Chicago.)

INTRODUCTION

Is it possible to use chemistry without an understanding of atoms? The answer to this question is definitely "yes." A technology of chemical practice existed for thousands of years without knowledge of atoms. Our modern understanding of the nature of matter, and therefore of the science behind the technology, dates to the beginning of the 1800s. John **Dalton** (1766–1844), an English scientist, proposed what we now call the **atomic theory.** By that time, a large amount of experimental evidence had accumulated that needed a rational explanation.

In this chapter we will consider some of Dalton's ideas on atoms before presenting our modern understanding of the structure of atoms. We will learn some appropriate symbolism to represent the subatomic structure of the atom.

One of the most important organizing concepts in all of chemistry is the periodic table. We will begin to understand the usefulness of the table by studying the basis for its organization. We will then correlate electronic structure with position on the periodic table. Some of the elements that are important for your health are the final topic in this chapter.

DALTON'S ATOMIC THEORY

Although Dalton truly did not originate the idea of atoms, he did revive many earlier thoughts about the particulate nature of matter and put them together into a theory that could be tested experimentally. Not all of these ideas are believed to be true today, for we know a great deal more about the atom than Dalton did. Still, Dalton's suggestions remain important stepping-stones in our understanding of matter. Here are some of Dalton's major ideas about atoms.

1. All matter consists of definite particles called atoms.
2. Atoms themselves are indestructible. In chemical reactions, they may carry out rearrangements but are not destroyed.
3. All atoms of one particular element are the same in mass and in all properties.
4. It therefore follows that atoms of different elements have different masses and different properties.
5. Atoms combine to form compounds in definite proportion by mass.

SUBATOMIC PARTICLES

The **atom** is defined as the smallest unit of an element that retains all the chemical properties of that element. However, the atom is not an indestructible unit as Dalton believed. Interpretation of modern evidence leads us to the conclusion that the atom is composed of smaller subparts, usually referred to as **subatomic particles.** The three major subatomic particles are the **electron,** the **proton,** and the **neutron.** The proton and the neutron are relatively heavy particles found in the center region of the atom called the **nucleus.** Only the relatively much lighter electron is outside the nucleus. The atom is essentially electrical in nature. The positive charges of the protons must be balanced by the negative charges of the electrons if the atom is to maintain electrical neutrality.

There are actually many other particles that have been identified within the atom, particularly in the event of the disintegration of an atom, but we will limit our consideration to just the three mentioned. The important features of these particles are listed in Table 2.1.

Nuclear Symbolism

All of the information about the subatomic composition of any particular atom can conveniently be represented by the use of **nuclear symbols.** To illustrate this chemical shorthand, consider the symbolism shown in the following for

Table 2.1. Subatomic Particles

	Particle		
	Electron	Proton	Neutron
Position in atom	Outside nucleus	Inside nucleus	Inside nucleus
Discoverer (year)	J.J. Thomson (1897)	Ernest Rutherford (1919)	James Chadwick (1932)
Electrical charge	Negative	Positive	Neutral
Mass			
Absolute (g)	9.110×10^{-28}	1.673×10^{-24}	1.675×10^{-24}
Relative, unified atomic mass units[a] (u)	0	1	1

[a] The **unified atomic mass unit** (u) is based on a particular type of carbon atom, carbon-12. This is assigned a mass of exactly 12 u making 1 u exactly one-twelfth the mass of carbon-12. One unified atomic mass unit is equal to $1.6605655 \times 10^{-27}$ kg.

On this scale, the mass of the proton is 1.00728 u and the mass of the neutron is 1.00867 u. Their masses differ starting in the third decimal place, and both are very closely equal to 1 unit. The exact mass of the electron, by comparison, is 5.484×10^{-4} u. This is so small in relation to the mass of the proton and neutron that the electron's mass is often given as "O" units. This does not mean it has no mass but just that its mass is much less than the mass of either of the other two subatomic particles.

a particular carbon atom. A great deal of information is packed into this compact representation. Here is the explanation.

The 14 is the **atomic mass number.** (This is often just called the **mass number**). Because the protons and neutrons are the only particles heavy enough to contribute significantly to the mass, this number is the sum of the protons and neutrons present in the atom. Because the atomic number informs you that six protons are present, there must be $14 - 6$ or eight neutrons present. Unified atomic mass units, or u, are understood for this value. It is traditional to round this mass number to the nearest whole number to use in this symbol.

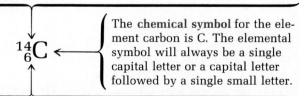

The **chemical symbol** for the element carbon is C. The elemental symbol will always be a single capital letter or a capital letter followed by a single small letter.

The **atomic number** gives the number of protons in the nucleus. If the atom is neutral, which you may assume in the absence of any other information, there must also be six electrons present in the atom.

Carbon always has an atomic number of 6. This fact means that the nuclear symbol $^{14}_{6}\text{C}$ may be written just as ^{14}C without any loss of clarity. You will see this same symbol further simplified as C-14 or carbon-14. These two forms are very easy to type or print because no subscripts or superscripts are used.

■■■■■■■■■■ **EXAMPLE 2.1** ■■■■■■■■■■■■■■■■■■■■■■■■■■

Explain the meaning of the symbol $^{35}_{17}Cl$.

Solution
The arrangement of the numbers in the symbol tells you which number is the atomic number and which is the mass number. The atomic number gives you the number of protons. It is also the number of electrons if the atom is assumed to be neutral. The mass number is a sum of the number of protons and the number of neutrons.

Chlorine has an atomic number of 17. There are seventeen protons in the nucleus and, if the atom is neutral, seventeen electrons surrounding the nucleus. The mass number is 35, so there must be $35 - 17 = 18$ neutrons also in the nucleus.

■■■■■■■■■■ **EXAMPLE 2.2** ■■■■■■■■■■■■■■■■■■■■■■■■■■

How does the subatomic structure of $^{14}_{6}C$ differ from $^{12}_{6}C$?

Solution
Both of these symbols represent carbon, but the mass numbers differ. Since the number of protons is the same, the number of neutrons must be different.

Each of these atoms has six protons in its nucleus. They are both carbon which is the element with atomic number 6. However, carbon-14 must have $14 - 6 = 8$ neutrons in its nucleus whereas carbon-12 must have $12 - 6 = 6$ neutrons in its nucleus. In both cases there will be six electrons if the atom is neutral.

ISOTOPES

The last example illustrates a key way in which Dalton's atomic theory has been updated. It is now known that atoms of the same element do *not* always have exactly the same number of neutrons and therefore do not have exactly the same mass. The number of neutrons does not vary widely for atoms of the same element, but there is more variation shown in atoms of the heavier elements than in atoms of the lighter elements. These differing atoms of the same element are called **isotopes.** Remember that isotopes are atoms of the *same* element that differ in mass because of the differing number of neutrons in their nuclei.

Because of the existence of isotopes, the atomic mass of an atom is not a fundamental identifying characteristic of that atom. Rather the atomic number is what determines the identity of the element. If, for example, a certain atom has 16 protons, then it is an atom of sulfur. Conversely if it is an atom of sulfur, it has an atomic number of 16. The same correspondence does not hold true for atomic mass number, for as the preceeding examples have shown, there are different types of atoms of the same element, each with a different mass. In the case of sulfur, there are four naturally occurring isotopes. These are

$$^{32}_{16}S \quad ^{33}_{16}S \quad ^{34}_{16}S \quad ^{36}_{16}S$$

Each of these isotopes of sulfur has the same atomic number, 16, and therefore has sixteen protons in the nucleus. However, there are sixteen, seventeen, eighteen and twenty neutrons present, respectively, in each isotope. Once again we can see that although the mass number can vary, the atomic number cannot if this atom is to retain its identity as sulfur.

Isotopes behave the same chemically but because their nuclei are different, they may be quite different in any property depending upon nuclear characteristics. One such property is the ability to decompose and spontaneously emit particles and energy from the nucleus. Both iodine-123 and iodine-131 are isotopes that spontaneously emit energy and particles that can be used for the detection and treatment of thyroid cancer. We will learn more about this type of isotope in Chapter 19.

EXAMPLE 2.3

Write a symbol for an isotope with ten protons and twelve neutrons.

Solution
Since the number of protons is the same as the atomic number, this is element 10. From the list of elements in the front of the text, find that element 10 is neon. The sum of the number of protons and neutrons is 22, which is the mass number.

The symbol is $^{22}_{10}Ne$.

EXAMPLE 2.4

Are the atoms $^{15}_{8}O$ and $^{15}_{7}N$ isotopes?

Solution
Check the definition of isotopes. These two atoms have the same mass number but not the same atomic number.

These two atoms are *not* isotopes because they are not of the same element. They do not have the same atomic number, so they do not fit the definition of isotopes.

Atomic Weight

Since elements occur in nature as a mixture of isotopes, a way is needed to reflect the natural distribution of isotopes for any element. You will notice on the Table of Elements inside the front cover of this book that what is listed is the atomic weight for each element. The concept of atomic weight involves both the relative natural abundance of each isotope and the mass of each isotope. Therefore the **atomic weight** of an element is defined as the average of all the naturally occurring isotopes of an element, taking into account their relative abundance. This process is illustrated in the next example.

EXAMPLE 2.5

Find the atomic weight of oxygen by using the following isotopic data.

Isotope	Mass (u)	Occurrence (%)
0-16	15.99	99.76
0-17	17.00	0.04
0-18	18.00	0.20

Solution
To find an average that depends both on the mass of the isotope and on its percentage of occurrence, you find the mass contribution from each isotope. This is accomplished by multiplying the mass times the percent, expressed as a decimal. Then the individual mass contributions are rounded so they can be summed to give the overall atomic weight for the element.

$$15.99 \times 0.9976 = 15.95$$
$$17.00 \times 0.0004 = 0.01$$
$$18.00 \times 0.0020 = \underline{0.04}$$
$$= 16.00 \text{ u}$$

When you double check the result in this problem against the Table of Elements, you find that the calculated atomic weight is a little different from the atomic weight listed. To get results that agree exactly with the table value, you would have to use better values for the masses of the individual isotopes. Oxygen-16, for example, has a known atomic mass of 15.99491 u, not 15.99 as we have used.

ELEMENTS AND THE PERIODIC TABLE

The front inside cover of this book contains the list of the known elements, together with their symbols and atomic weights. The back inside cover contains the **periodic table** which organizes the elements according to their properties.

Many different scientists shared the idea that the elements could be grouped so that their chemical and physical properties would fall into regularly repeating patterns. This observation is now referred to as the **periodic law.**

In 1869, Dimitri Mendeleev (1834–1907) published the most complete pattern of chemical families that had been arranged up to that time. The most remarkable thing about his pattern was that he found he had to leave some gaps in order to achieve the right placement for other elements. This daring step suggested the existence of new elements. The subsequent discovery of the ''missing'' elements helped to ensure the acceptance of Mendeleev's patterning system. Many changes have taken place and new elements added since Mendeleev's time, but the fundamental idea remains the same. Figure 2.1 gives a skeleton form of the modern periodic table, which we can use to identify certain features.

The horizontal rows on the periodic table are each called a **period.** A vertical column is referred to as a **group** or **family.** The set of groups that have an ''A'' designation are referred to as the **representative elements** and the set with a ''B'' designation are called the **transition elements.** Some special groups are labeled in the figure, including the IA group, which goes by the name of

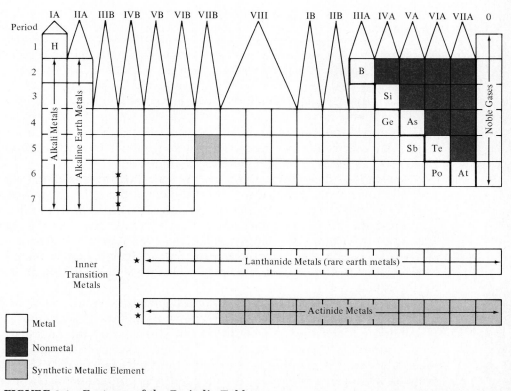

FIGURE 2.1 **Features of the Periodic Table.**

alkali metals, and the IIA group, which is called the **alkaline earth metals.** The **halogen family** occupies the VIIA position and the last group on the right is known as the **noble gases.** (This group is also called the rare gases. Earlier they were known as the inert gases, but as they are known to react, this name is no longer in use.)

The nonshaded areas on this table indicate the **metals** and the shaded areas, the **nonmetals.** The elements that fall on the diagonal zigzag line that separates these two regions are known as **semimetals,** earlier called the **metalloids.** These elements have properties that are intermediate between those of a typical metal and those of a typical nonmetal. Semimetals may behave as metals or nonmetals, depending on the substances with which they react. Some of the properties associated with the different regions of the periodic table are shown in Table 2.2.

Knowing the basic organization of the periodic table can be a great help in predicting the properties of the elements. We will see many examples of the usefulness of the table in the chapters ahead. If the organizing principle behind the periodic table is the fact there is a periodic occurrence of similar physical and chemical properties, then the question arises as to why this should be so. The answer lies in the electronic structure of the elements.

Electronic Structure

To understand the relationship between organization of the periodic table and electronic structure, we need first to have a more complete idea of the arrangement of the electrons within the atom. So far we know that electrons are subatomic particles located outside the nucleus and that they have a very small mass and carry a negative electric charge. Now we can further say that the electrons are located in certain orderly patterns in space, each associated with particular values of energy. These arrangements of electrons around the nucleus of an atom are called **electron configurations.**

The earliest description of the electron arrangements in atoms was given by Neils Bohr (1885–1962) who, in 1913, likened the electrons to planets. He envisioned the electrons circling the nucleus in a path, much as the planets go around the sun. This "solar system model" of electronic structure was an important contribution to our understanding of the atom, but like many models, it has since been replaced by ideas better able to explain observable properties of the elements. Bohr's idea of **energy levels** for electrons within atoms is still an important one.

The best current understanding of the electron's arrangement within the atom stems from the field of science known as **quantum mechanics.** Without a detailed mathematical analysis, we can still employ the conclusions that come from this type of treatment. The quantum mechanical view of the electrons within atoms does not try to pin down the exact paths of the electrons. Instead, it concerns itself with the probability of finding an electron within a

Table 2.2. General Properties Related to Position on the Periodic Table

Metals	Semimetals	Nonmetals	Noble gases
Solid (one liquid—mercury)	Some properties of metals, some of nonmetals	Generally gases; some solids, one liquid	Enter into very limited numbers of chemical reactions
Shiny		Not shiny	In general, nonreactive
Can be pounded into sheets		Cannot be pounded into sheets or drawn into wires	
Can be drawn into wires		Poor conductors of heat or electricity	
High electrical conductivity		Oxides, if soluble, form acidic solutions in water[a]	
High thermal conductivity		Tend to form negative ions[b]	
Oxides, if soluble, form alkaline solutions in water[a]			
Tend to form positive ions[b]			

Active Metals (Groups IA and IIA)	Transition Metals
Highly reactive chemically	Generally less active chemically
	Coinage metals of copper, silver, and gold are used because of their unreactive nature

→ Increasing Metallic Character →

↓ *Increasing Metallic Character* ↓

[a] To be discussed in Chapters 4 and 8.
[b] To be discussed in Chapter 3.

35

certain region of space around the nucleus. Instead of a "track" or a path for the electron, this model sees an "electron cloud" that represents the enclosed volume of space in which there is a specified (commonly 90 or 95 percent) probability of finding a particular electron. This probability region is called an **orbital** to differentiate it fron the **orbit** of the Bohr model's electron. The contrast between the Bohr model and the quantum mechanical model is shown schematically in Figure 2.2.

The electron cloud shown in Figure 2.2 is for a lowest energy electron closest to the nucleus. This energy level is numbered 1. (Using the older Bohr Theory model, the first energy level was called the K shell.) A second electron, if present, could be found in this same electron cloud. The maximum capacity of any orbital is just two electrons, so no additional electrons past two could be in this same probability region. If more electrons are present, they must be in higher energy orbitals that extend further than the first energy level from the nucleus. The energy levels are numbered sequentially out from the nucleus. (In the Bohr Theory model, the shells are lettered sequentially so that successive levels are K, L, M, N, . . .)

One of the most surprising results of the quantum mechanical view of electrons in atoms is that not all of the higher energy orbitals are spherical in shape. This is quite different from the view of electrons proposed in the Bohr Theory model. The relative sizes and shapes of all first and second energy level orbitals are shown in Figure 2.3. Note that the shape is given a letter designation, either s or p in this case, that goes with the energy level number. The number and letter together serve to describe two important aspects of that orbital, its energy and its shape. Note also that the first energy level only has an s orbital, but that the second energy level is large enough to have an s orbital and three different orientations of p orbitals.

If every time we wanted to describe the arrangement of electrons within

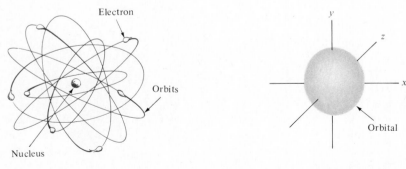

Bohr "Solar System" Model

Quantum Mechanical Electron
Density Model of an Electron
Probability Region

FIGURE 2.2 **Contrasting the Bohr Model with the Quantum Mechanical View of an Electron.**

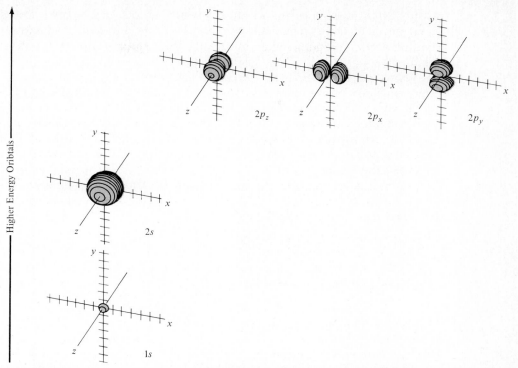

FIGURE 2.3   Energy Levels and Orbital Shapes. Rather than showing electron densities, these contour drawings are boundary-surface diagrams. They show the different possible orbital shapes for an electron of hydrogen. They are sketched from computer-generated diagrams and are drawn to scale. Each marker on the axis indicates an interval of 200 picometers. The most stable orbital for an electron of hydrogen is the 1s orbital. (Diagrams used with permission from W. G. Davies and J. W. Moore, Copyright 1975.)

an atom, we had to draw or sketch these orbital shapes, it would be very time-consuming and difficult. As is usual in chemistry, there is a shorthand system for representing the electron configuration of any atom. What this system does is to identify the energy level, the shape of the probability region, and how many electrons are found in that orbital shape.

For example, there are seven electrons in the element nitrogen. Two of these are found in the first energy level and five are in the second energy level. Those in the first energy level are arranged in the same shape shown in Figure 2.2, which is called the s shape. The five in the second energy level are distributed between the s shape and the p shape shown in Figure 2.3. All of this information can be represented by the following shorthand notation:

N $1s^2 2s^2 2p^3$

Notice that the sum of the electrons in the *second* energy level is 5 and the total of all of the superscripts is equal to 7, the number of electrons in the neutral atom. Table 2.3 gives you the electron configuration of the first twenty elements.

Now go back through Table 2.3 and try to find patterns within the electronic configurations that correspond to patterns within the periodic table. What you will find are similar electron configurations based on the position of the element in the periodic table. There is a very close relationship between electron arrangement and periodic table arrangement. Consider the elements in the same family, for example, the halogen family. Do you see the similarity between the electron configurations of fluorine and chlorine? They both have the same type of electron configuration in their outermost energy level.

Fluorine $1s^2 2s^2 2p^5$

Chlorine $1s^2 2s^2 2p^6 3s^2 3p^5$

The electrons in the outermost energy level are often called the *valence electrons* and in this case there are seven valence electrons for each element.

Table 2.3. Electron Configurations of the First Twenty Elements

Element	Electron configuration	Total electrons
Hydrogen	$1s^1$	1
Helium	$1s^2$	2
Lithium	$1s^2 2s^1$	3
Beryllium	$1s^2 2s^2$	4
Boron	$1s^2 2s^2 2p^1$	5
Carbon	$1s^2 2s^2 2p^2$	6
Nitrogen	$1s^2 2s^2 2p^3$	7
Oxygen	$1s^2 2s^2 2p^4$	8
Fluorine	$1s^2 2s^2 2p^5$	9
Neon	$1s^2 2s^2 2p^6$	10
Sodium	$1s^2 2s^2 2p^6 3s^1$	11
Magnesium	$1s^2 2s^2 2p^6 3s^2$	12
Aluminum	$1s^2 2s^2 2p^6 3s^2 3p^1$	13
Silicon	$1s^2 2s^2 2p^6 3s^2 3p^2$	14
Phosphorus	$1s^2 2s^2 2p^6 3s^2 3p^3$	15
Sulfur	$1s^2 2s^2 2p^6 3s^2 3p^4$	16
Chlorine	$1s^2 2s^2 2p^6 3s^2 3p^5$	17
Argon	$1s^2 2s^2 2p^6 3s^2 3p^6$	18
Potassium	$1s^2 2s^2 2p^6 3s^2 3p^6 4s^1$	19
Calcium	$1s^2 2s^2 2p^6 3s^2 3p^6 4s^2$	20

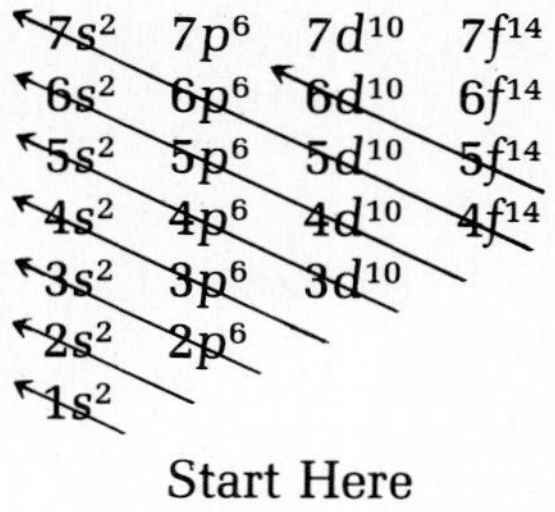

FIGURE 2.4 Order of Orbital Filling. **This scheme can be used to place electrons in their correct orbitals. Start with the bottom arrow that indicates filling the 1s orbital with a maximum of two electrons. Proceed to the 2s orbital, shown by the second arrow. The third arrow indicates the filling of a maximum of six electrons in the 2p orbitals, followed by a maximum of two electrons in the 3s orbital. The arrows shown are sufficient to account for the electrons in all known elements.**

$$7s^2 \quad 7p^6 \quad 7d^{10} \quad 7f^{14}$$
$$6s^2 \quad 6p^6 \quad 6d^{10} \quad 6f^{14}$$
$$5s^2 \quad 5p^6 \quad 5d^{10} \quad 5f^{14}$$
$$4s^2 \quad 4p^6 \quad 4d^{10} \quad 4f^{14}$$
$$3s^2 \quad 3p^6 \quad 3d^{10}$$
$$2s^2 \quad 2p^6$$
$$1s^2$$

Start Here

This similarity in electron configuration is exactly what is responsible for their similar chemical behavior and their placement in the same group in the periodic table. Check out this family similarity for the alkali metals and for the alkaline earth metals, too.

Table 2.3 also shows the regular way in which electrons are distributed. The first electrons enter the most stable position, the 1s energy level. This orbital is filled after two electrons have been added, so the next higher energy level must be used in the case of the last electron in lithium. When that orbital is also filled, the next available orbital has a p shape and is also at the second level. You will recall from Figure 2.3 that there are three p orbitals altogether and that is why the maximum capacity of the 2p orbitals is six electrons. Figure 2.4 is a memory device to help you remember the order in which electrons fill available energy levels. With this scheme you can predict the electron configurations of many other elements and not be limited to the twenty shown in Table 2.3. Figure 2.4 also shows you that each type of orbital has a certain frequency of occurrence. The s orbital occurs only once in each numerical energy level. The p orbital occurs three times in every energy level past the first level. Other types of orbitals include the d, occurring 5 times in every level past the second level. The f orbitals occur 7 times in every level past the third level.

EXAMPLE 2.6

Predict the electron configuration for aluminum, element 13.

Solution
Because aluminum has an atomic number of 13, it will have thirteen protons. If it is neutral, it will have thirteen electrons. To distribute the electrons, consult Figure 2.4. Remember to stop when you have used all thirteen electrons.

The configuration is $1s^2 2s^2 2p^6 3s^2 3p^1$.

When you are done, check the configuration you have written with that listed in Table 2.3. Also be sure to notice that there are three electrons in the third energy level; this is reasonable because aluminum is a Group IIIA element.

━━━■ **EXAMPLE 2.7** ■━━━

Predict the electron configuration for krypton (element 36).

Solution
The thirty-six electrons will be distributed by the same ground rules used in the previous example. Fill each orbital to capacity as you go and follow the direction of the arrows in Figure 2.4

The configuration is $1s^22s^22p^63s^23p^64s^23d^{10}4p^6$. This configuration represents the order of orbital filling for the electrons in krypton. There is a total of eight electrons in the fourth or outermost energy level, corresponding to an VIIIA group member. The group is alternatively called group 0.

Pay special attention to the noble gas family's electron configurations. Notice that in each case the orbitals present are completely full. The first three noble gases are

He $1s^2$
Ne $1s^22s^22p^6$
Ar $1s^22s^22p^63s^23p^6$

There is a certain degree of chemical stability associated with the outermost orbitals being completely filled. Other than helium, which is just too small and light to have more than two electrons, all the noble gases have electron configurations that include completely full s and p orbitals in their highest energy levels. Since the maximum capacity of the s orbital is two, and the maximum capacity of the three p orbitals is six, this means that each of the noble gases other than helium has eight electrons in its highest energy level. Therefore the noble gas elements do not need to gain or lose electrons in a chemical reaction in order to achieve the octet. A very powerful generalization about chemical behavior of all the elements is that they will try, through chemical interaction, to attain this same stable eight electrons in their highest energy level. This tendency is called the **octet rule.**

Not all elements are capable of actually achieving an octet of electrons. The lightest elements are too small to achieve a stable octet. They will try to have helium's stable two electron structure. Even for the heavier elements, it is usually more feasible for the representative elements actually to achieve an octet than for the transition elements to do so. In the next chapter we will study the tendency of atoms to combine and see how important the octet rule is in helping to understand chemical bonding.

ELEMENTS IMPORTANT FOR YOUR HEALTH

In Chapter 1, we listed the most abundant elements in your body. There are also many elements that are classified as **trace elements** by mass percentage and while they occur in only tiny amounts, they are still very important for proper functioning of the body. Many of these trace elements are thought to be essential but their exact method of functioning is not always known. Other trace elements have been identified as important, usually because the absence of that element is associated with a breakdown in some bodily function. Generally they occur in combined form as salts or other compounds. As we end this chapter on atoms and elements, you can use Table 2.4 to help focus on a few elements that will be of significance to your health studies. Locate each of the elements on the periodic table and learn their symbols.

It is important that the body maintain the correct ranges of concentrations of these trace elements. A lack of copper for example, can impair growth but it is a poison if taken in larger amounts than the trace seemingly required by the body. Selenium can be extremely toxic and zinc can contribute to heart disease, at concentrations above those required by the body for well-being. Another factor is the interaction that is possible among the elements. At least some effects of toxic trace elements result not so much from the properties of the element itself as from its interference with the normal function of other elements. Researchers have reported that excesses of vitamin C appear to

Table 2.4.   Trace Elements in the Body

Element	Reason considered essential
Chromium	Used in normal glucose metabolism and also seems to be needed for proper insulin action.
Cobalt	Needed to form vitamin B_{12}
Copper	Occurs in proteins and enzymes, needed for growth and to aid formation of red blood cells, tissue in both bone and brain
Fluorine	Essential for the formation of healthy teeth
Iodine	Essential to have balanced amount within the body for proper thyroid functioning
Iron	Needed to form hemoglobin and present in several enzymes
Manganese	Needed for normal nerve function, for enzymes, and for proper fat and carbohydrate metabolism. Involved in the proper reproduction of liver, kidney, eye, and bone cells
Molybdenum	Involved in an enzyme used for nucleic acid metabolism
Selenium	Essential for growth and health in some animal species and thought to be essential in humans too. Deficiency seems to be associated with incidence of stroke and heart disease
Zinc	Present in some enzymes and bone. Needed for normal growth patterns

enhance the biological availability of iron but depress that of copper and make selenium totally unavailable for use, even if present in the body.

SUMMARY

All matter consists of atoms. In chemical reactions, the atoms may be rearranged but they are not destroyed. Atoms themselves are composed of three principal subatomic particles. The proton and neutron are found in the nucleus; the electron is found outside the nucleus.

A nuclear symbol consists of the chemical symbol for the atom, the atomic number, and the atomic mass number. The atomic number tells you the number of protons in the nucleus; it is also the fundamental value that identifies the element present. If the atom is neutral, the number of electrons will match the number of protons. The difference between the atomic mass number and the atomic number gives you the number of neutrons present in a particular isotope. The atomic weight is a weighted average of all isotopes of that element.

The periodic table is organized with regular repetition of chemical and physical properties. Trends are observed in both vertical columns (the groups or families) and periods (the horizontal rows). The regular patterns are correlated with similar patterns of electronic structure. The modern quantum theory gives us a way to predict these patterns of electrons in atoms. Of all the elements on the periodic table, there are some that are recognized for their importance to your health, even though they occur only in trace amounts.

KEY TERMS

Check your understanding of this chapter. Can you explain what is meant by each of these terms?

alkali metal	metal
alkaline earth metal	metalloid
atom	neutron
atomic mass number	noble gas
atomic number	nonmetal
atomic weight	nuclear symbol
Dalton's atomic theory	nucleus
electron	octet rule
electron configuration	orbit
energy level	orbital
family	period
group	periodic law
halogen	periodic table
isotope	proton
mass number	quantum mechanics

representative elements
semimetal
subatomic particle
symbol

trace element
transition element
unified atomic mass unit
valence electrons

STUDY QUESTIONS

1. Which of Dalton's five ideas about atomic theory listed at the beginning of this chapter, are still believed true today and which have been modified?

2. One unified atomic mass unit corresponds to 1.66×10^{-24} g. Use this relationship to find the mass in grams of an atom of nitrogen-14.

3. Use the same relationship in Question 2 to express the mass of an atom of a neon isotope in unified atomic mass units, given that its absolute mass is 3.285×10^{-23} g.

4. Name the subatomic particles which are associated with the following characteristics.
   a. Heaviest mass
   b. Lightest mass
   c. Positive charge
   d. Negative charge
   e. Neutral
   f. Outside nucleus

5. List the subatomic particles found in each of these atoms. Assume that all atoms are electrically neutral. Which particles are in the nucleus?
   a. $^{31}_{15}P$ b. $^{40}_{20}Ca$
   c. $^{233}_{92}U$ d. $^{33}_{16}S$

6. a. Write a symbol to represent an atom with nine protons and ten neutrons. Assume that the atom is neutral.
   b. Write a symbol to represent an atom with ten protons and ten neutrons. Again assume that the atom is neutral.
   c. Compare the symbols in parts a and b. Do these two symbols represent isotopes? Explain.

7. Complete the following table.

Element	Atomic number	Mass number	Number of protons	Number of neutrons
$^{4}_{2}He$				
$^{56}_{26}Fe$				
$^{238}_{92}U$				
$^{75}_{33}As$				

8. The element hydrogen can be found in three different isotopes. $_{1}^{1}\text{H}$ ("normal" hydrogen), $_{1}^{2}\text{H}$ (deuterium), and $_{1}^{3}\text{H}$ (tritium). Answer the following questions.
   a. How do these isotopes differ from each other?
   b. The atomic weight of hydrogen is 1.00797 u. Without consulting any table giving the percentages of each isotope, what can you conclude just from the atomic weight as to which isotope must be most abundant?

9. Each line in this table represents a neutral atom of a different isotope. Complete the table.

Element	Atomic number	Mass number	Number of protons	Number of neutrons	Number of electrons
		90	38		38
C-13					
			13	14	13

10. Sulfur occurs as four natural isotopes. They are given here with their percentage occurrence and the atomic mass of each isotope.

    sulfur-32 (95.0%, 31.97 u)

    sulfur-33 (0.76%, 32.97 u)

    sulfur-34 (4.22%, 33.97 u)

    sulfur-36 (0.014%, 35.97 u)

    Use these data to calculate the atomic weight of the element sulfur. Compare your results with the weights listed on the periodic table.

11. Determine the atomic weight for the element neon. The three isotopes are as follows.

    neon-20 (90.92%, 19.99 u)

    neon-21 (0.26%, 20.99 u)

    neon-22 (8.82%, 21.99 u)

12. Classify each of the following elements as metal, nonmetal, semimetal, or noble gas.
    a. Oxygen   **b.** Gold
    c. Sulfur   **d.** Helium
    e. Iron   **f.** Lithium
    g. Antimony   **h.** Cadmium
    i. Bromine   **j.** Radium
    k. Cesium   **l.** Cerium

13. What elements make up the IVA subgroup? What similarity in electron configuration would you expect for all the members of this subgroup?

14. What elements make up the VIA subgroup? What similarity in electron configuration would you expect for all the members of this subgroup?

15. Which atom would you predict is larger: fluorine or chlorine? Why?

16. Sketch the shapes for the orbitals that would be used for each of the electrons in nitrogen.

17. How does the shape of a *p* orbital differ from what the Bohr theory would have predicted?

18. Are metals or nonmetals the more abundant type of element on the periodic table?

19. Sodium has an electron configuration of $1s^2 2s^2 2p^6 3s^1$. Explain the meaning of this symbolism.

20. Write the electron configuration of each of these elements. Consult Table 2.2 and Figure 2.4 for help.
    a. Oxygen
    b. Fluorine
    c. Argon
    d. Potassium

21. Write the electron configuration of each member of the IIIB subgroup. Use Figure 2.4 to help you.

22. Given that the electron configuration of a certain neutral atom is $1s^2 2s^2 2p^6 3s^2 3p^6 4s^2 3d^1$, answer the following questions.
    a. What element is represented?
    b. What group does this element belong to?

23. What is the maximum number of electrons associated with all orbitals possible within the second energy level? How does this number compare with the number of elements in the second period of the periodic table?

24. Is an *orbit* the same as an *orbital*? Explain.

25. Where will element 107 fall on the periodic table?

26. If an element does not make up a large percentage by weight of the body, does this mean it is unimportant? Explain.

27. Is an element such as arsenic always toxic? Use an appropriate reference suggested by your instructor to find the limits of toxicity for arsenic.

3 Molecules and Compounds

Our photo shows a visual inspection stage in the production of Roerig's Geocillin tablets. Geocillin is a semisynthetic penicillin—an antibiotic used in the control of disease-causing bacteria.

Each tablet contains 382 mg of Geocillin. Each molecule of Geocillin consists of atoms of C, H, O, N, and S, joined in a precise way. Chemists understand the type of bonding that holds these atoms together, the size and shape of the molecule, and its physical and chemical properties.

What we are saying here about Geocillin is true for any drug on the market, any fabric, any dye, any food additive, any insecticide, any biochemical, any industrial chemical. Atoms are the building blocks for all substances, and it is possible to study the types of bonding in substances and thereby to predict physical and chemical properties.

Scientists use their knowledge of the chemistry of molecules and compounds to create (synthesize) new substances that might have value in our society.

For the remainder of your study of chemistry, remember that if a substance is pure it can be represented by a formula; it is composed of atoms of one or more elements joined in a unique, specific way. (Photo courtesy of Pfizer, Inc.)

INTRODUCTION

Many common pure substances are not individual elements but rather consist of elements chemically combined into compounds. Although there are just over 100 elements, there are quite literally millions of compounds. Our major goal in this chapter is to understand why and how these compounds form. We will also describe the bonding within different types of compounds and see how the bonding types correlate with the typical properties of these compounds.

The next step is to develop some ideas that will enable you to predict the types of bonds that are present when elements join. This chapter will also help you to become more familiar with the formulas and names of the compounds that you will be meeting as you continue your health studies.

WHY DO COMPOUNDS FORM?

All **chemical bonding** can be described as a force between atoms. The nature of this force is not always the same. We will study two major types of bonding in this chapter: covalent and ionic. In both cases, the result is a combination of atoms such that each member of the combination now has a more stable electron configuration than any had individually. The ideal is for each atom to fill its valence electron orbitals and thus obey the octet rule. Sometimes this means that the individual atoms achieve a noble gas electron structure by becoming electrically charged rather than remaining neutral. In other cases, the octet rule can be followed only if the atoms remain neutral but share some of their electrons. In still other cases, the combining atoms cannot achieve an octet of valence electrons at all, but they still may be able to improve their electron configurations and therefore become more stable than the separated atoms would be.

The result of chemical bonding is the production of a compound. In Chapter 1, we learned that a **compound** contains two or more elements combined in a fixed ratio by mass. The simplest unit of a compound to retain the properties of the compound is called a **molecule;** the simplest unit of an element is the atom.

Covalent Bonding in Simple Molecules

Hydrogen is the simplest of all atoms. There is just a single proton and a single electron in a neutral atom of the simplest isotope of hydrogen. The electron outside the nucleus is attracted to the proton in the nucleus because of their opposite electrical charges. Now what happens if two hydrogen atoms approach each other? The electron of one hydrogen atom may also be attracted to the proton of the second atom, because of their opposite electrical charges. Of course, the same could be said for the electron of the second hydrogen being attracted to the proton of the first hydrogen. Countering this attraction is the fact that like electrical charges repel and so the electron in the first hydrogen atom will also be repelled by the electron in the second atom, as well as attracted to its nucleus. The two nuclei will also repel, as they are both positively charged. Which type of force wins this attraction-repulsion battle?

The fact is that the attractive forces win out over the repulsive forces. Hydrogen gas occurs as molecules, not atoms, under most common conditions of temperature and pressure. The molecules of hydrogen gas are known to be made up of two bonded atoms and can be represented as H_2. This is an example of a **diatomic molecule,** one in which two atoms join to form a single molecule.

It is believed that when these two atoms join, a new **molecular orbital,** encompassing both nuclei, is formed. A molecular orbital, like an atomic orbital, is a probability region showing, with 90 or 95 percent certainty, the volume of space in which the electrons will be found. Each molecular orbital holds two electrons, just as each atomic orbital does. This new probability region in the case of diatomic hydrogen is shown in Figure 3.1.

A pair of electrons has been used to form this new molecular orbital. These two electrons now have a greater probability of being found between the two nuclei than they do when they are just associated with one or the other nucleus. The two electrons in the covalent bond are said to be shared

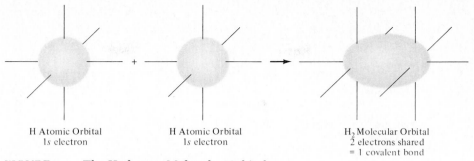

H Atomic Orbital
1s electron

H Atomic Orbital
1s electron

H_2 Molecular Orbital
2 electrons shared
= 1 covalent bond

FIGURE 3.1 The Hydrogen Molecular Orbital.

by both nuclei. When electrons are shared by two nuclei, the type of bond is called a **covalent bond.** There is now a fixed distance between the two nuclei, called the **covalent bond length.** This particular length is the most favorable for maximizing the attractive forces. If the nuclei are any farther apart, the attractive forces will be weaker. If the nuclei are any closer, the repulsion forces overshadow the forces holding the molecule together. By sharing electrons, each hydrogen atom has been able to have a share in two electrons, the total number of electrons in the nearest noble gas, helium. (Remember that hydrogen cannot obey the octet rule because it is too small.)

The pair of electrons itself is sometimes referred to as the chemical bond and can be represented in either of two convenient shorthand ways—as a pair of electron dots, or as a short line connecting the two chemical symbols.

H:H   or   H—H

The first representation is called either a **Lewis structure** or an **electron dot representation.** The second form is called either a simplified Lewis Structure, a line structure, or a **structural formula.** Both representations show the single covalent bond that results when the two electrons occupy the same molecular orbital, an orbital associated with both hydrogen nuclei. Notice that *one* line is used to represent *one pair* of electrons.

Other simple diatomic molecules such as F_2 and Cl_2 again show how electrons pair to form a covalent bond. All of the halogens have seven electrons in their outer energy level, and the unpaired electron is in a p shaped orbital. Because this orbital has a different shape from the s orbital used in forming diatomic hydrogen gas, the molecular orbital that results for the halogens differs from that of hydrogen, shown in Figure 3.1. The shape is given in Figure 3.2.

For simplicity in depicting these atoms and their bonds on a two-dimensional surface, Lewis diagrams (electron dot diagrams) are used more commonly than the molecular orbital representation. You will need to remind yourself that the arrangement of electrons in atoms is really three dimensional

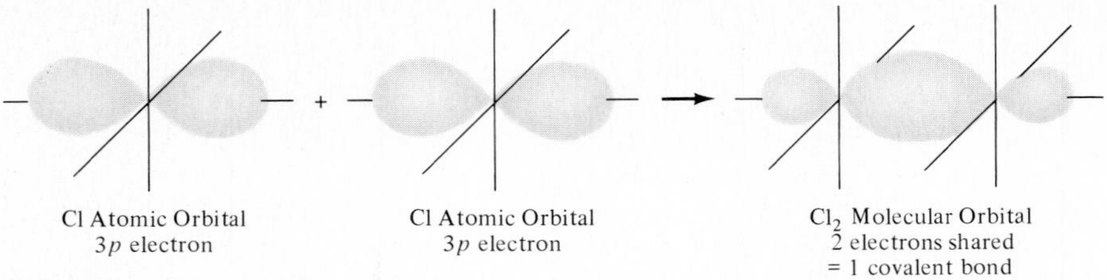

FIGURE 3.2   The Chlorine Molecular Orbital.

and that the pairs of electrons shown between the nuclei are found in a three-dimensional probability region that surrounds both nuclei, with the highest probability of finding the bonding electrons being between the atoms. Another difference between the molecular orbital representation and the Lewis structure is that only the bonding electrons are shown in the molecular orbital. It is common to show even the nonbonding electrons in the electron dot representation, although only those electrons in the outermost energy level. This practice makes it easy to count to see if the octet rule has been followed. Check out these conventions for the halogens in the following table.

███████ **EXAMPLE 3.1** ███████

Has bromine followed the octet rule in forming the diatomic molecule, Br_2?

Solution
Either consult Table 3.1 or use the electron configuration. You should realize in either case that bromine is a halogen family element (Group VIIA) and therefore has seven electrons in the highest energy level of each individual atom. If it now has eight valence electrons per atom, then it is obeying the octet rule.

Table 3.1.   Lewis Structures for the Halogen Family

		Lewis structures	
Name	*Formula*	*Electron dot*	*Line structure*
Fluorine (gas)	F_2	:F̈:F̈:	F—F
Chlorine (gas)	Cl_2	:C̈l:C̈l:	Cl—Cl
Bromine (liquid)	Br_2	:B̈r:B̈r:	Br—Br
Iodine (solid)	I_2	:Ï Ï:	I—I

In sharing a pair of electrons between the two nuclei, each atom is able to claim eight electrons in its outermost energy level. There are seven valence electrons in each bromine's electron configuration, and each bromine has a share in one electron from the other atom in the diatomic molecule. Therefore each bromine atom is following the octet rule. The nearest noble gas is Xe and it does have eight electrons in its outermost energy level, as do all the noble gases except He.

Double and Triple Bonds

We have seen that sharing electrons is an important concept for diatomic molecules such as hydrogen gas and the halogen family. There are some diatomic molecules that are not able to follow the octet rule by sharing just one pair of electrons. What happens, for example, in the case of oxygen gas? It is also known to form diatomic molecules, and yet there are only six electrons in the outermost energy level of each atom. Each oxygen atom would require two more electrons in order to fill to eight. Is this possible by sharing? Yes, it is, but only if *two* pairs of electrons or four electrons altogether are shared. A similar problem arises in the case of nitrogen, which is also known to form with diatomic molecules but has only five valence electrons per atom. A sharing of *three* electron pairs or six electrons is required this time. These Lewis structures are shown in Table 3.2.

Oxygen atoms share two pairs of electrons and now each atom has four of its own valence electrons and four shared electrons. The two shared pairs of electrons form what is referred to as a **double bond**. The nitrogen has two electrons of its own per atom and 6 electrons that it shares, forming the octet for each atom. The three bonded pairs are called a **triple bond**.

The covalent bond length reflects the fact that increased sharing of electrons causes the nuclei to pull closer together. Oxygen, nitrogen, and fluorine atoms are similar in atomic size as they all have valence electrons in the second energy level. The fluorine-to-fluorine single bond length is 0.142 nanometer (nm), but the oxygen-to-oxygen double bond length is 0.121 nm and the nitrogen-to-nitrogen triple bond length is 0.109 nm.

Table 3.2. Lewis Structures for Oxygen and Nitrogen Diatomic Molecules

| Name | Formula | *Lewis structures* | |
		Electron dot	Line structure
Oxygen	O_2	$:\ddot{O}::\ddot{O}:$	$O{=}O$
Nitrogen	N_2	$:N:::N:$	$N{\equiv}N$

Polar Covalent Compounds

All of the covalent compounds we have looked at so far have been formed by two atoms of the same element joining together. The pair of electrons that is shared is therefore guaranteed to the shared equally between the two nuclei. Such a bond is given the special name of a **nonpolar covalent bond.**

If two different atoms join in a covalent bond, they may still attain an electron configuration of the nearest noble gas. The difference in this case is that one of the atoms may have a greater tendency to pull the shared electrons closer to its nuclei, and, consequently, the electron pair is not equally shared and a **polar covalent bond** results. When hydrogen joins with chlorine to form hydrogen choloride, a polar covalent bond forms. The electron pair in this case is attracted more strongly to the chlorine than to the hydrogen atom. As a result, the molecule is partially negative ($\sigma-$) at the chlorine end and partially positive ($\sigma+$) at the hydrogen end. Each partial charge is less than a unit charge, and their sum is still zero.

The polarity (unequal electron sharing) within a bond can also be represented by an arrow pointing toward the partially negative end of the bond and a + sign on the tail indicating the partially positive region. Such a polar molecule is called a **dipole** because it has two electrical poles. In the presence of another dipole or an electrical field, the hydrogen chloride molecule will always turn so that the hydrogen atom will point toward the negative electrical pole and the chloride atom will point toward the positive pole. The electron cloud for polar hydrogen chloride is shown in Figure 3.3.

In order to predict if a bond will be polar, we need to have a way to quantify the relative electron attracting abilities of the elements. This can be done through the property called **electronegativity.** Electronegativity values have no units. The values are defined on a relative scale that describes the ability of an atom to attract bonded electrons to itself. The nonmetallic elements are the most electronegative. Fluorine, which is near the upper right-hand corner of the periodic table, is the most electronegative of all the elements. The metallic elements are the least electronegative. Francium, at the lower

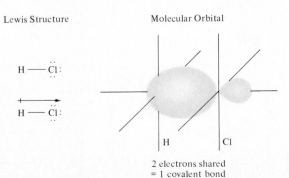

FIGURE 3.3 Hydrogen Chloride—a Polar Covalent Compound.

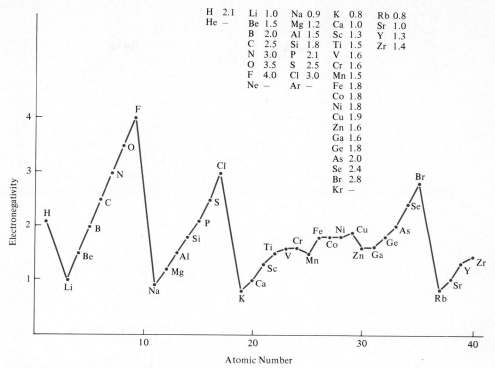

H 2.1	Li 1.0	Na 0.9	K 0.8	Rb 0.8
He –	Be 1.5	Mg 1.2	Ca 1.0	Sr 1.0
	B 2.0	Al 1.5	Sc 1.3	Y 1.3
	C 2.5	Si 1.8	Ti 1.5	Zr 1.4
	N 3.0	P 2.1	V 1.6	
	O 3.5	S 2.5	Cr 1.6	
	F 4.0	Cl 3.0	Mn 1.5	
	Ne –	Ar –	Fe 1.8	
			Co 1.8	
			Ni 1.8	
			Cu 1.9	
			Zn 1.6	
			Ga 1.6	
			Ge 1.8	
			As 2.0	
			Se 2.4	
			Br 2.8	
			Kr –	

FIGURE 3.4 Electronegativity—a Periodic Property.

left-hand corner of the table, attracts bonding electrons the least. Electronegativity is not defined for the noble gases because they typically are not reactive. Figure 3.4 gives the values of electronegativity for the first forty elements and shows the periodic nature of this property.

If two elements join in a covalent bond, the bond is considered polar if the electronegativity difference between the two bonded atoms is greater than 0.5 electronegativity units. If the difference is greater than 1.7, a covalent bond is not the principal type of bonding present. (The second major type of bonding, ionic bonding, will be discussed later in this chapter.)

■■■■■■ **EXAMPLE 3.2** ■■■■■■

Answer the following questions.
   a. Will the covalent bond between carbon and hydrogen be a polar covalent bond?
   b. Will the covalent bond between oxygen and hydrogen be a polar covalent bond?

Solution

To determine if a bond is polar, the difference in electronegativity between bonded atoms must be found. Figure 3.4 gives you the necessary electronegativity values. If the difference between the electronegativities of the two bonded atoms is greater than 0.5, the bond is considered polar.

a. Carbon's electronegatvity is 2.5, and hydrogen's electronegativity is 2.1. The difference is 0.4. Therefore, because the difference is less than 0.5, the carbon-hydrogen covalent bond *is not* considered polar.

b. Oxygen's electronegativity is 3.5, hydrogen's electronegativity is 2.1. The difference is 1.4. Therefore, because the difference is greater than 0.5, the oxygen-hydrogen covalent bond *is* considered polar.

Molecular Geometry and Polarity

Hydrogen chloride gas is composed of polar, covalently bonded molecules. Because there is just one bond in the molecule, the hydrogen chloride molecule itself is said to be a polar molecule. If more than one covalent bond is present in the same molecule, there is a second criterion, other than the presence of polar bonds, that must be considered before any molecule is judged to be a polar molecule. If each bond is polar but the overall symmetry of the molecule is such that the polarities "cancel" each other, then the molecule as a whole is nonpolar.

To better understand this idea, consider the difference between carbon dioxide (CO_2), and water (H_2O). In each case there is a central atom, carbon in the case of CO_2 and oxygen in the case of water. In each case the covalent bonds are polar because there is a difference in electronegativity greater than 0.5 units between bonded atoms. Yet in the case of carbon dioxide, the overall molecule is not polar, even though the bonds are. In the case of water, the molecule is polar. What accounts for this is the difference in geometry between the two molecules. You are not expected to be able to *predict* geometries so the correct shapes are shown in Figure 3.5.

Whether the molecule is polar or nonpolar has a large influence on the properties. This makes it important to know molecular polarity when working with drugs, solvents, or laboratory reagents. Particularly in the case of water, which is such a vital fluid for health, there are many ways in which physical and chemical interactions depend on its polar nature. Water is often referred to as the "universal solvent." Its superior ability to dissolve and interact with a wide variety of substances is based on the fact that the polar molecules are attracted to other polar substances and to charged species. A water molecule's polar regions also attract other water molecules. This particular attraction is a good deal stronger than other interactions between dipoles and goes by the name of **hydrogen bonding.** It can be illustrated by looking at the intermolecular attractions in an ice model, shown in Figure 3.6. Notice that the water molecules

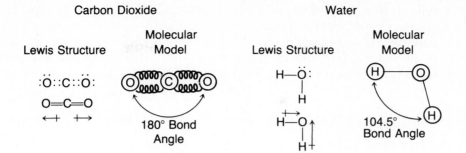

Carbon Dioxide   Water

| Lewis Structure | Molecular Model | Lewis Structure | Molecular Model |

180° Bond Angle   104.5° Bond Angle

The molecule is nonpolar.   The molecule is polar.

FIGURE 3.5 **Geometry Differences between Carbon Dioxide and Water. Even though all the covalent bonds in both molecules are polar, the overall molecule may be polar or nonpolar.**

are still internally bonded with covalent bonds but the molecules are held to each other with hydrogen bonds.

A hydrogen bond is weaker than a covalent bond but much stronger than most interactions between dipoles. Hydrogen bonding is the result of the highly

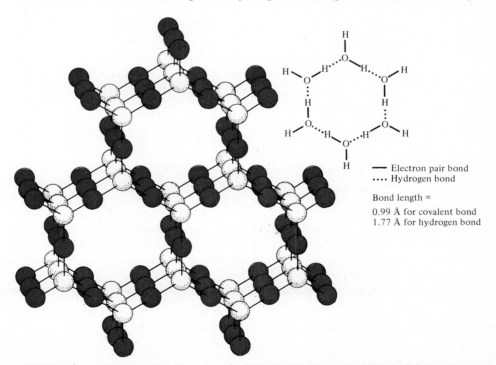

—— Electron pair bond
···· Hydrogen bond

Bond length =

0.99 Å for covalent bond
1.77 Å for hydrogen bond

FIGURE 3.6 **Hydrogen Bonding in Ice. The light spheres in the model represent oxygen, the dark spheres hydrogen.**

polar character of the bonds when hydrogen is covalently bonded to highly electronegative atoms such as oxygen, fluorine, or nitrogen. This type of bonding is very important in the interactions of water with biological compounds and also in understanding the unique properties of nitrogen-containing materials such as proteins, dioxyribonucleic acid (DNA), and ribonucleic acid (RNA).

EXAMPLE 3.3

For each of these molecules shown, decide if the bonds are polar and then if the molecule itself is polar. The correct geometry has been indicated.

    a. H—F̈: (hydrogen fluoride)

    b. :C̈l—Be—C̈l: (beryllium chloride)

Solution

To decide if the *bonds* are polar, you need to check the differences in electronegativity values for each set of bonded atoms. If the value is greater than 0.5, the bond is considered polar. To decide if the *molecule* is polar, you need to check the overall geometry, looking for cases in which the polarity will or will not cancel.

    a. The bond is polar because the difference in electronegativity between hydrogen and fluorine is $4.0 - 2.1 = 1.9$ units. Since there is just one bond, and therefore no possibility of symmetry, the molecule must be polar as well.

$$\overset{\longmapsto}{\text{H—F̈:}}$$

    b. The bonds are polar because the difference in electronegativity is $3.0 - 1.5 = 1.5$ units. The molecule is linear and so the polarities will cancel each other. The overall molecule is nonpolar.

$$\overset{\longleftarrow + \longmapsto}{\text{:C̈l—Be—C̈l:}}$$

Note that in this particular compound, the beryllium does not have an octet of electrons. It is a very small atom and is only surrounded by four electrons. It is an example of a stable compound in which the central atom does not obey the octet rule.

Coordinate Covalent Bonds

When a covalent bond forms, you have seen that the bonded atoms do not necessarily share the electrons of the bond equally. It is also true that they do not even have to provide equal numbers of electrons for creating the bond.

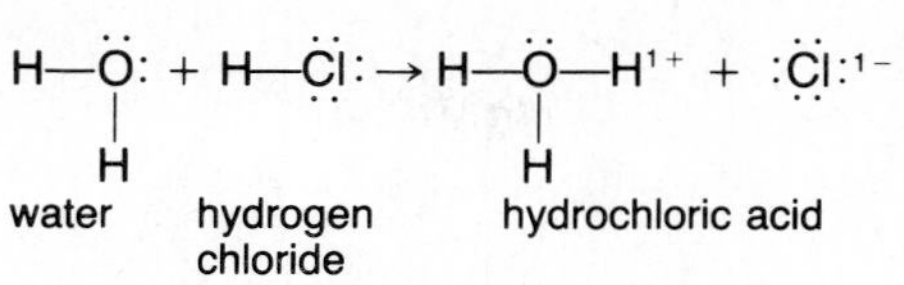

$$H—\overset{..}{O}: + H—\overset{..}{\underset{..}{C}l}: \rightarrow H—\overset{\underset{|}{H}}{O}—H^{1+} + :\overset{..}{\underset{..}{C}l}:^{1-}$$

water   hydrogen   hydrochloric acid
         chloride

FIGURE 3.7 **Coordinate Covalent Bond Formation.**

This possibility creates a type of bond called a **coordinate covalent bond.** In this bond, both the electrons that are shared come from one atom.

An example is provided by the interaction of hydrogen chloride gas and water. The polar water molecules attract the proton away from the polar hydrogen chloride molecule. The result is the aqueous form of hydrogen chloride, known as hydrochloric acid. The process is shown in Figure 3.7.

Once formed, a coordinate covalent bond resembles a typical covalent bond in every respect. The bond is still composed of a pair of shared electrons and the only difference lies in the fact that both electrons originated from one of the atoms.

In Figure 3.7, chlorine had been left with a full octet of electrons and a charge of -1. Because the proton has been transferred to the water, this group now has a charge, a $+1$ charge. Coordinate covalent bonds play an important role in protein tissue. An amino group ($R - NH_2$) on the protein molecule binds a proton by means of this reaction. Note that the product has a coordinate covalent bond.

$$R—\overset{\underset{|}{H}}{\underset{..}{N}}—H + H^+ \rightarrow R—\overset{\overset{H}{|} \; \text{coordinate covalent bond}}{\underset{\underset{H}{|}}{\oplus N}}—H$$

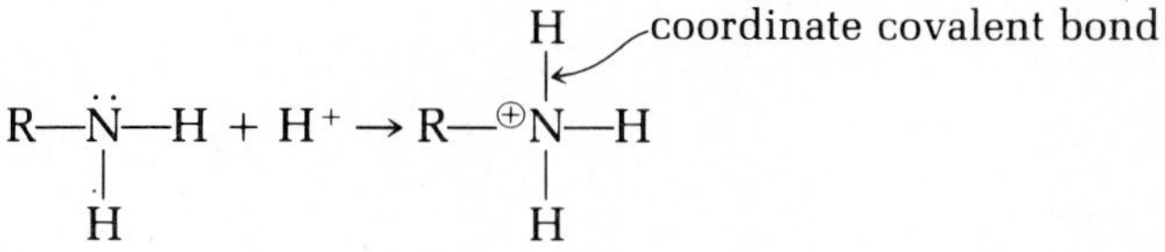

We will study proteins in more detail in Chapter 11. The next section in this chapter will further explain the formation of charged chemical species.

Ionic Bonding

When different atoms combine, the result is not always a molecule held together with chemical bonds of shared electrons. Electrons can actually be given away by one atom and accepted by another. This results in the formation of charged species called **ions,** which are then attracted to each other because opposite electrical charges attract. When the attraction between oppositely charged ions is the force holding the substance together, it is called **ionic bonding.** Ionically bonded substances are composed of at least one positive ion, called a **cation,** and one negative ion, called an **anion.** If the positive ion is not the hydrogen ion (H^+) and the negative ion is not the hydroxide ion (OH^-), then the combination of oppositely charged ions is called a **salt.**

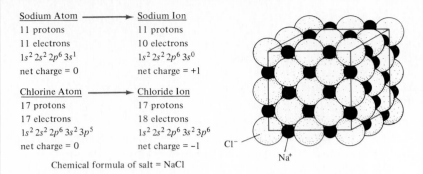

Sodium Atom ⟶ Sodium Ion

Sodium Atom	Sodium Ion
11 protons	11 protons
11 electrons	10 electrons
$1s^2 2s^2 2p^6 3s^1$	$1s^2 2s^2 2p^6 3s^0$
net charge = 0	net charge = +1

Chlorine Atom	Chloride Ion
17 protons	17 protons
17 electrons	18 electrons
$1s^2 2s^2 2p^6 3s^2 3p^5$	$1s^2 2s^2 2p^6 3s^2 3p^6$
net charge = 0	net charge = –1

Chemical formula of salt = NaCl

FIGURE 3.8   Sodium Chloride—an Ionically Bonded Salt.

Sodium chloride is an example of a salt. It is composed neither of atoms nor of molecules; it is made of ions. To see how this salt is bonded, we need to look at what has happened to the electrons of sodium and chlorine atoms in order to form ions. This information is summarized in Figure 3.8.

One electron from the sodium atom has been transferred to the chlorine atom. Since the nuclei are not affected, the sodium atom is left with eleven protons and only ten electrons, making it a cation with a charge of + 1. The numerical value + 1 may also be referred to as the **valence** of the sodium ion or its **oxidation number.** It represents the combining capacity of the sodium ion. Looking at the chlorine, you can see that the ion still has seventeen protons but now it has a total of eighteen electrons, giving it a net charge of − 1. This value of − 1 may also be called the chloride ion's valence, combining capacity, or oxidation number.

The exchange of the electron is advantageous to both the sodium and the chlorine, for each of the resulting ions has an octet of electrons in the outermost energy level. The sodium ion, with a total of ten electrons, has an electron distribution that is the same as that of neon. The chloride ion, with a total of eighteen electrons, has the same electron distribution as argon. In spite of the electronic resemblance of the sodium ion to neon, remember that they are not at all alike in any other properties. Neither is the chloride ion similar to argon in its properties. They are quite different because the ions carry charges and, of course, the nuclei are also different.

Once the ions have formed and are attracted because of their opposite charges, an array of ions forms into a crystal lattice. This particular regular association of ions in sodium chloride is cubic in shape (see Figure 3.8). Many other arrangements are possible for ionic solids. The ionic formula does not tell you what crystalline arrangement is used by the salt. It is just the simplest formula unit for the crystal, in this case one sodium ion and one chloride ion.

In order to dissolve a crystalline salt, the solvent must overcome the forces of attraction holding the ions in the crystalline lattice. Water can often do this because of its polarity, described earlier in this chapter. The water molecules will arrange themselves so that their partially positive regions are nearest a negatively charged chloride ion and their partially negative regions are near a positively charged sodium ion. If enough water molecules can arrange themselves in this way, they often can overcome the attraction of the other ions within the lattice and carry the trapped ions off into solution. A covalently bonded liquid such as benzene or hexane cannot do this as they are not themselves polar molecules.

Many cations and anions are vitally important in biological fluids. Sodium ion (Na^+) is the principal cation in blood while potassium ion (K^+) is the main cation in fluids inside body cells. Calcium ion (Ca^{2+}) is required for the formation of bone and teeth. Magnesium ion (Mg^{2+}) is needed for proper nerve conduction and enzyme function. Lithium ion (Li^+) is used in the treatment of manic-depressives and iron ion (Fe^{2+}) is needed for hemoglobin formation. The principal anion in blood is the chloride ion (Cl^-). Iodide ion (I^-) is required by the thyroid gland in order to function properly.

ELECTRONEGATIVITY AND BOND TYPE

We have now seen examples of covalent bonding to form molecular compounds and ionic bonding to form ionic compounds such as salts. The division of chemical substances into two types—molecular and ionic—is another example of a classification system. As with all systems, there needs to be a basis for predicting which type of bonding a compound will use.

One way of predicting which type of bonding a compound will use is to build up many examples of each type so that you begin to make the associations when you meet new substances. You have already begun to do that by studying the examples chosen so far. You have probably observed that salts are composed of a metallic cation with a nonmetallic anion but that molecular compounds are usually composed on nonmetals bonded to themselves or of nonmetals joined with other nonmetals. You could predict, based on your experience so far, that sulfur trioxide gas would be covalently bonded.

A more reliable way to predict bonding type is to make use of the property of electronegativity once again. You remember that, earlier in this chapter, the difference in electronegativity between bonded atoms helped to predict the polar nature of covalent bonds. The most polar bond will actually be the case in which the electron or electrons are not just pulled in one direction or another but are actually transferred. The division of chemical substances into molecular and ionic is not quite as simple as it first appears, but rather there is a gradual transition from one type to the other. Substances such as sodium chloride and diatomic chlorine gas represent the opposite ends in what is really a continuum of bonding types.

The correlation between differences in electronegativity of bonded atoms and bond type is shown in Table 3.3. Refer to Figure 3.4 to find the values of electronegativity. For any bonded pair, the greater the difference in electronegativity, the more polar the bond becomes until there is a point at which the bond is considered to be predominately ionic. This occurs if the difference in electronegativity is greater than approximately 1.7. The bond is then said to have more than a 50 percent ionic character. If the ionic character of the bond exceeds 50 percent, the compound will yield predominately ions, not molecules, when melted or when dissolved in a polar solvent such as water.

EXAMPLE 3.4

Will the substance KF be ionically bonded and exist as ions or will it be covalently bonded and exist as molecules?

Solution

First, check the electronegativity of each element. Then find the difference in electronegativity for the KF bond and look up that difference in Table 3.3. Use your common sense in considering the answer, because the compound mentioned is a combination of a metal and a nonmetal. You should be expecting ionic bonding for this compound.

The electronegativity of K is 0.8 and the electronegativity of F is 4.0, making the difference 3.2 units. From Table 3.3, the percent ionic character is 92 percent. The substance will be almost totally ionically bonded.

PROPERTIES AND BOND TYPE

One of the advantages of being able to predict the type of bonding in a compound is that predictions can then be made about typical properties of that compound. Most ionically bonded compounds have relatively high melting points, higher than either atomic or molecular solids. Ionic solids are relatively hard. They tend to shatter under pressure unless they actually melt first because of the pressure. In the solid state, they are not good conductors of either heat or electricity.

Molecular compounds as a class have lower melting and boiling points

Table 3.3. Electronegativity Differences and Percent Ionic Character

Difference in electronegativity	Percent ionic character
0.1	0.5
0.2	1
0.3	2
0.4	4
0.5	6
0.6	9
0.7	12
0.8	15
0.9	19
1.0	22
1.1	26
1.2	30
1.3	34
1.4	39
1.5	43
1.6	47
1.7	51
1.8	55
1.9	59
2.0	63
2.1	67
2.2	70
2.3	74
2.4	76
2.5	79
2.6	82
2.7	84
2.8	86
2.9	88
3.0	89
3.1	91
3.2	92

than ionically bonded compounds. Many molecular substances are soft solids, liquids, or gases at room temperature and pressure. They also are not good conductors of heat or electricity. Hydrogen bonding may alter the expected properties of covalently bonded compounds, as we have discussed earlier.

Some of the typical properties of ionic compounds and molecular compounds are summarized in Table 3.4.

Table 3.4. Properties of Typical Ionic and Covalent Compounds[a]

Ionic compounds (formed when metals join with nonmetals)	Covalent compounds (formed when nonmetals join with nonmetals)	
	If the atoms only form one or two covalent bonds per molecule	If the atoms have a tendency to form three or more covalent bonds per molecule
Ionic compounds	*Molecular compounds*	*Network covalent compounds*
Simplest unit: ions	Simplest unit: molecule	Simplest unit: atom or molecule
Solids at room temperature	Force within unit: covalent bonding	Forces between units: covalent bonding
Hard	Forces between units: polar attractions, hydrogen bonding	Hard, extremely high melting point
Brittle	Soft, low-melting-point solids	
Crystalline structure	Liquids	
High melting points	Gases	
Many translucent or transparent	Properties affected by hydrogen bonding	
Nonconductors in solid state	Di- and triatomic molecules of the first 17 elements are gases, unless there is hydrogen bonding	
Conductors if melted or dissolved		
More likely to dissolve in polar solvents (example, H_2O) than in nonpolar solvents (example, hexane)		
Forces between units—oppositely charged ions attract; ionic bonding		
Examples:	Examples:	Examples:
NaCl	Gases: CO, CO_2, NO_2, H_2S	Diamond (carbon)
$CaCl_2$	Liquids: H_2O, SCl_2	SiC (silicon carbide)
$CuSO_4$	Solids: S_8, P_4O_{10}	Si_3P_4
$MgCO_3$		

[a] If metals combine with other metals, the compounds retain many of the properties associated with metals. Metallic compounds will have high electrical and heat conductivity and will have typical metallic luster.

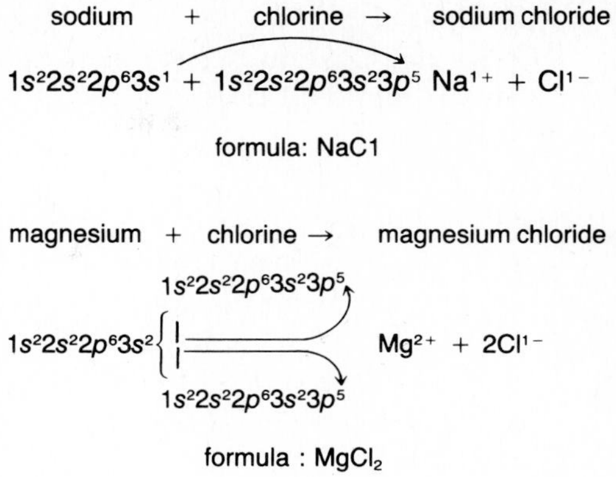

sodium   +   chlorine →   sodium chloride

$$1s^22s^22p^63s^1 + 1s^22s^22p^63s^23p^5 \quad Na^{1+} + Cl^{1-}$$

formula: NaC1

magnesium + chlorine →   magnesium chloride

$$1s^22s^22p^63s^23p^5$$

$$1s^22s^22p^63s^2 \left\{ \begin{array}{c} \\ \\ \end{array} \right. \quad\quad Mg^{2+} + 2Cl^{1-}$$

$$1s^22s^22p^63s^23p^5$$

FIGURE 3.9   **Transfer of Electrons to Form Ionic Compounds.**

formula : $MgCl_2$

PREDICTING FORMULAS

In the last chapter we saw ways in which the periodic table can be used to organize information about the elements. In this chapter we have been using the periodic property of electronegativity to predict bond types and properties. The periodic table can also be useful in helping to understand why sodium chloride joins in a one-to-one ratio of sodium to chlorine but magnesium would join in a one-to-two ratio with chlorine. Sodium, found in Group IA, has just one electron to give up in order to attain an octet of valence electrons. Magnesium, in Group IIA, has two electrons in excess of an octet. Since chlorine still needs only one electron to achieve its stable electronic configuration, two chlorines are needed in order to receive the two electrons available from magnesium. The simplest ratios for combination are reflected in the chemical formula in each case. A **chemical formula** will show the types of elements present and their simplest ratio by atoms when combined by either ionic or covalent bonding. This information is summarized in Figure 3.9 for sodium chloride (NaCl) and magnesium chloride ($MgCl_2$).

Notice in Figure 3.9 that the chemical formula shows the correct ratio of atoms that have joined. This ratio, if not just one-to-one, is indicated by the use of an appropriate subscript. A chemical formula in general will show the types of elements present and their simplest ratio when combined.

▰▰▰▰**EXAMPLE 3.5**▰▰▰▰

Predict the formulas for ionic compounds formed when each of these combinations of elements join.

a. Aluminum with chlorine.
b. Sodium with sulfur.

Solution

Check the location of each element in the periodic table in order to see how many electrons it needs to lose or gain to achieve an octet of electrons.

a. Aluminum is a Group IIIA element and will tend to lose three electrons in order to achieve a noble gas electronic configuration. Chlorine is a Group VIIA element and needs to gain one electron to have an octet. Therefore, three chlorines will be needed to accept the three electrons donated by Al. The formula is $AlCl_3$.

b. Sodium, a Group IA element, needs to lose one electron to achieve a noble gas electronic structure. Sulfur, a Group VIA element, with an electron configuration ending with $3s^2 3p^4$, needs to gain two electrons in order to fill its outermost energy level and have the electron configuration of argon. Therefore, two sodium atoms will be required to supply the two electrons required by each sulfur, and the formula is Na_2S.

INTRODUCTION TO CHEMICAL NOMENCLATURE

As we continue through our study of chemistry as it relates to the health professions, we will continually have the need to correctly identify substances, both by formula and by name. Many substances were originally given common names by early scientists, often before the true identities of the substances were known. Some of these names still persist and may be used today. Therefore, potassium nitrate may be referred to as saltpeter and copper sulfate as blue vitriol.

Common names are convenient. Often they describe something about the origin of the material such as wood alcohol, which comes from wood, or citric acid which is obtained from citrus fruits. For two reasons, however, common names leave much to be desired. First, they give no indication of the composition of the substance. The other disadvantage is that there are just too many compounds to be dealt with effectively by common names. It is more reasonable in the interest of good chemical communication to have the name reflect as much information as possible about the substance.

Correctly naming all substances is a continuing process and in this section, we will start by focusing on compounds of two types: those composed of two elements and those composed of three or four elements.

Naming Binary Compounds

Compounds composed of only two elements are called **binary compounds.** Examples of binary compounds are sodium chloride (NaCl), magnesium chloride ($MgCl_2$), aluminum sulfide (Al_2S_3), and sulfur dioxide (SO_2). Notice that in each case there are just two *elements*, although the total number of atoms per molecule may be more than two. These examples illustrate the general rules for systematically naming binary compounds. These rules are divided into two major categories, depending on the nature of the two elements present in the binary compound.

Category I: A metal is joined to a nonmetal (ionically bonded compounds).

1. The metallic element of the compound is named first; the nonmetallic is named second.
2. The first element is named just as it is in the uncombined state, for example, *sodium* chloride.
3. The ending of the element named last is modified to end in the suffix -*ide*. Thus, chlorine becomes chlor*ide*, and oxygen becomes ox*ide*.
4. If two different compounds exist for the same two elements, something which happens particularly if the metal is a transition metal with variable combining capacity, the same three preceding rules are followed but the following must be done as well. A Roman numeral is added to the name of the metal. The numeral is placed in parentheses and located just after the name of the first element. The Roman numeral indicates which valence state of the metal is used. Examples:

$SnCl_2$—tin(II) chloride[1]
$SnCl_4$—tin(IV) chloride

This distinction in the oxidation state of the metal is necessary *only* if there is some doubt as to which combining capacity the metal is using. Metals in Groups IA and IIA are never named with the use of Roman numerals because there is never any question as to their combining capacity in common compounds. Similarly, the elements silver, zinc, and aluminum never utilize Roman numerals, because their capacity to combine is always +1, +2, and +3 in common compounds.

Category II: Nonmetal joined to nonmetal (covalently bonded compounds).

1. If both elements in a binary compound are nonmetals, the least metallic element will be the element with the highest electronegativity value and will appear last in the compound.
2. Same as Step 2 in Category I.
3. Same as Step 3 in Category I.
4. The rule for Category I is actually correct for this type of compound also, but you will find an alternative method still in very common use here. A prefix is placed before the name of each element to denote the

---

[1] Another system has been used for naming two compounds of the same elements. This older system uses the ending -*ous* for the element in its lower valence and the ending -*ic* for the element in its higher valence. With this system, the two tin compounds are stannous chloride and stannic chloride, respectively. This system often uses the Latin root for the metal, as has been done here with the name *stannum*.

relative number of atoms of that element. If just one atom of the first element is present, that prefix is usually omitted. Examples:

CO—carbon *mon*oxide
CO_2—carbon *di*oxide
N_2O_5—*di*nitrogen *pent*oxide

━━━━━ **EXAMPLE 3.6** ━━━━━

Name each of these compounds.

    a. SO_3
    b. Na_2S
    c. $CuCl_2$

Solution

First, you should identify whether the binary compound is composed of a metal and a nonmetal or two nonmetals. This will tell you whether you should use the rules of Category I or Category II. The rules are different for these two types of binary compounds. Example a is composed of two non-metals, and examples b and c are metal–nonmetal combinations. Also observe if the metal is a transition element, in which case a Roman numeral will be required to indicate combining capacity.

    a. Sulfur trioxide (Category II, Rules 1–4).
    b. Sodium sulfide (Category I, Rules 1,2,3).
    c. Copper(II) chloride (Category I, Rules 1–4).

Naming Ternary Compounds

Compounds composed of three different elements are called **ternary compounds.** Their names often end in *-te*, the first two letters of the word *ternary*. Compounds with more than two components are more generally called **polyatomic compounds.** The names of many of these polyatomic compounds end in *-te*.

Ternary compounds are often formed by the combination of a metallic cation with a polyatomic anion, rather than a simple anion. A **polyatomic ion** is a group of atoms covalently bonded. The entire unit carries a charge and usually stays together as a unit in a chemical reaction. Some of the most common polyatomic ions are given in Table 3.5. A more complete list of simple and polyatomic ions is found in Appendix II.

Notice in this table that there are two common endings; *-ate* and *-ite*. In two related polyatomic ions such as the sulfate and the sulfite ions, observe

Table 3.5.   Common Polyatomic Ions

1 − charge		2 − charge	
Nitrate	NO_3^-	Sulfate	SO_4^{2-}
Nitrite	NO_2^-	Sulfite	SO_3^{2-}
Bicarbonate (or hydrogen carbonate)	HCO_3^-	Carbonate	CO_3^{2-}
		Dichromate	$Cr_2O_7^{2-}$
Chlorate	ClO_3^-	Chromate	CrO_4^{2-}
Chlorite	ClO_2^-		
Hypochlorite	ClO^-	3 − charge	
Permanganate	MnO_4^-	Phosphate	PO_4^{3-}
Acetate	$C_2H_3O_2^-$	Phosphite	PO_3^{3-}
Hydroxide	OH^-	Arsenate	AsO_4^{3-}
Cyanide	CN^-	Arsenite	AsO_3^{3-}
1 + charge			
Ammonium	NH_4^+		
Hydronium	H_3O^+		

that the ionic charge is the same in both cases (2 −) and that the *-ite* form has one fewer oxygen atom than the *-ate* form. Also notice that Table 3.5 shows only two common polyatomic ions with a positive charge. These are the ammonium ion (NH_4^+) and the hydronium ion (H_3O^+).

There are two common radical ions that follow typical binary ending rules: the hydroxide and cyanide ions. Since each of these is already composed of two elements, they will form ternary compounds when joined with a metallic ion, but their *-ide* endings give the impression that the resulting compounds are binary.

You will find it worthwhile to memorize Table 3.5, for with this information and the rules for naming compounds, you can write literally hundreds of correct names and formulas for compounds.

Naming ternary compounds is accomplished by following Rules 1, 2, and 4 for Category I binary compounds. The rules for naming ternary compounds are that the name of the first element remains unchanged. The Roman numeral system for metals with variable oxidation numbers is also used in naming ternary compounds. The anion is named with its appropriate polyatomic name, usually ending in *-ate* or *-ite*. The following example will illustrate the process.

EXAMPLE 3.7

Name each of these compounds.

    a. $Al_2(SO_4)_3$
    b. $CuCO_3$
    c. $Zn(ClO_2)_2$

Solution

Observe that they are all ternary compounds. Identify the metallic ion and decide if a Roman numeral is required. Find the correct name of the poly-atomic ion from Table 3.5.

a. Aluminum sulfate
b. Copper(II) carbonate
c. Zinc chlorite

In each of the preceding examples, note that the formula represented is electrically neutral. Aluminum sulfate must be composed of two aluminum (3 +) ions and three sulfate (2 −) units so that a 6 + charge is balance by a 6 − charge. Zinc chlorite must be composed of one unit of zinc ion (2 +) for every two chlorite units (1 −) in order to be electrically neutral.

There are many other rules for naming compounds, and we will continue to work on this skill throughout the text. For example, we will look at acids in Chapter 8 and the correct names for organic compounds as they come up in subsequent chapters.

Table 3.6 gives you the formulas, common names, and systematic names of important binary, ternary, and other common polyatomic compounds. Note that certain common names such as ammonia and water are so frequently used that you will find them even in the chemical literature. In general, the systematic name is preferred, but we will continue to use common names if they are in widespread use.

There is an old saying in medicine that helps the health professional distinguish between two particular substances with similar names. One of these substances is given quite safely internally as an x-ray contrast medium in diagnosing bowel cancer; the other similarly named substance is a poison. The saying goes: "barium sulfate, he ate; barium sulfide, he died." With the information presented in this chapter, can you write the correct chemical formulas for barium sulfate and barium sulfide?

Table 3.6.   Some Binary and Polyatomic Compounds with Health Applications

Name	Formula	Use
Aluminum oxide	Al_2O_3	Common name is alumina. Abrasive agent in toothpaste.
Ammonium carbonate	$(NH_4)_2CO_3$	Component of "smelling salts."
Barium sulfate	$BaSO_4$	Common name is barite. Insoluble in water, so administered orally in a slurry. Forms a radio-opaque image for x-ray studies of the gastrointestinal tract.
Calcium chloride	$CaCl_2$	Provides the correct salt content for "artificial tears."

Table 3.6. Some Binary and Polyatomic Compounds with Health Applications (*Cont.*)

Name	Formula	Use
Hydrogen peroxide	H_2O_2	Is a weak antiseptic. Releases oxygen when in contract with tissues. Stronger solutions are used as bleaching agents.
Iron(II) sulfate	$FeSO_4$	Also called ferrous sulfate. Used orally to treat iron-deficient anemia.
Mercury(I) chloride	Hg_2Cl_2	Also called mercurous chloride or calomel. Cathartic (stimulates evacuation of the bowels) in correct dosage but toxic in larger amounts.
Mercury(II) chloride	$HgCl_2$	A poison. Never taken internally. Also called mercuric chloride. Useful in dilute solution to disinfect equipment that cannot be boiled. Organic mercury compounds used as seed disinfectants in agriculture.
Potassium permanganate	$KMnO_4$	Used as a disinfectant in treatment of local wounds and infections. Also may be used to oxidize some types of impurities in drinking water treatment.
Silver nitrate	$AgNO_3$	Acts as a bactericide (an agent that destroys bacteria) in 1:1000 dilutions. Placed in infants' eyes to prevent conjunctivitis transmitted by gonorrheal infections from the mother.
Sodium hydrogen sulfite	$NaHSO_3$	Also called sodium bisulfite. Used as an antioxidant to stabilize various ophthalmic (eye) solutions. Sodium thiosulfate ($Na_2S_2O_3$) and sodium sulfite (Na_2SO_3) may also be used for their preservative properties.
Tin(II) fluoride	SnF_2	Also called stannous fluoride and the trade name "Fluoristan." Used, often in combination with sodium fluoride (NaF), as an additive in toothpaste to combat tooth decay.
Zinc oxide	ZnO	Used as a base for calamine lotions, often combined with a small amount of iron(III) oxide, (Fe_2O_3). Acts as a mild astringent (acts to contract body tissues and blood vessels).
Zinc sulfate	$ZnSO_4$	Used to treat certain skin conditions such as eczema.

SUMMARY

Compounds form through the processes of chemical bonding. The major driving force for chemical bonding is the increased stability that accompanies filled valence electron orbitals. This phenomenon is called the "octet rule."

If atoms share electrons in order to obey the octet rule, the bonding is referred to as convalent bonding. A covalent bond is generally formed between two nonmetals. It may either be nonpolar or polar, depending on the bonded atoms' relative ability to attract the shared electrons in the covalent bond. Even if the bonds are polar, the molecule may be nonpolar if the geometry of the molecule is symmetrical.

If some atoms donate and others accept electrons in order to obey the octet rule, the bonding between those atoms is referred to as ionic bonding. An ionic bond is generally formed by the attraction of a metallic cation for a nonmetallic anion. The anion may either be a simple ion or a polyatomic ion.

The property of electronegativity is used to help predict the type of bonding present in a compound. If the difference in electronegativity between the two bonded atoms is less than 0.5, the covalent bond is classified as nonpolar. If the difference is greater than 0.5 but less than 1.7, the bond is polar covalent. If the difference in electronegativity is more than 1.7, the bond is considered predominantely ionic. The type of bonding present is correlated with the predicted properties of the compound.

KEY TERMS

Check your understanding of this chapter. Can you explain what is meant by each of these terms?

anion	ion
binary compound	ionic bond
cation	Lewis structure
chemical bond	molecular orbital
chemical formula	molecule
compound	nonpolar covalent bond
coordinate covalent bond	oxidation number
covalent bond	polar covalent bond
covalent bond length	polyatomic compound
diatomic molecule	polyatomic ion
dipole	salt
double bond	structural formula
electron dot structure	ternary compound
electronegativity	triple bond
hydrogen bonding	valence

STUDY QUESTIONS

1. The noble gas family elements are all monoatomic but the halogen family elements are diatomic. Explain why this difference is reasonable, based on their respective electronic structures.

2. Draw a Lewis structure and a molecular orbital representation for the fluorine molecule, F_2. Is this a polar or nonpolar molecule?

3. Here is a list of the common diatomic molecules: H_2, N_2, O_2, F_2, Cl_2, Br_2, I_2. Answer the following questions.
   a. Which of these molecules contain double or triple bonds?
   b. Which of these molecules are in the gas state at room temperature?

4. Would you guess that it would take more energy to break up the bonds in diatomic oxygen molecules or the bonds in diatomic nitrogen molecules? Consider the Lewis structures and the bond lengths in supporting your guess.

5. Which of these molecules would have bonds that are more polar: HCl or HF? Explain.

6. Why is it that the presence of polar bonds does not always mean that the molecule itself is polar?

7. Methane, CH_4, has this Lewis structure:

$$\begin{array}{c} H \\ H : \overset{..}{\underset{..}{C}} : H \\ H \end{array}$$

   a. Does each element obey the octet rule?
   b. Does each element have a noble gas structure?

8. The geometry of the methane molecule (see Question 7) is tetrahedral.
   a. Are the bonds in methane polar or nonpolar?
   b. Is the molecule polar or nonpolar?

9. Why is water such a good solvent material for dissolving salts and polar covalent substances?

10. Which element is the most electronegative? Which the least? Where are these two elements located on the periodic table?

11. Why are there no values of electronegativity given for the noble gases?

12. Carbon dioxide (CO_2) is a nonpolar molecule, whereas sulfur dioxide (SO_2) has measurable polarity. What does this suggest about the shapes of these two molecules?

13. Water is a polar molecule. Which atom will have a greater share of the bonding electrons: oxygen or hydrogen?

14. What type of bond (ionic, polar covalent, or nonpolar covalent) do you predict will form between each of these pairs of elements?
    a. Na and Br   b. C and Cl
    c. I and I   d. N and O
    e. Zn and Cl   f. Fe and F

15. Would you expect the following compounds to contain ionic, polar covalent, or nonpolar covalent bonds?
    a. K_2O b. H_2O
    c. CO   d. $CaCl_2$
    e. NaH   f. NH_3

16. Based on your answers to Question 15, predict the state (solid, liquid, or gas) of each of the compounds in Question 15 at room temperature and pressure.

17. Sodium and potassium are both in Group IA. Each element has a tendency to lose one electron in order to form a stable chemical bond.
    a. Which of these elements has the greater electronegativity?
    b. Can you offer any possible reason for this difference in electronegativity?

18. Based on the electron structure of calcium, what is its most usual valence? What would be the formula of its chloride? Its sulfide? Its nitride?

19. Based on the electron structure of aluminum, what is its most usual valence? What would be the formula of its chloride? Its sulfide? Its nitride?

20. Explain the electron transfer that takes place in the formation of potassium sulfide.

21. Explain the electron sharing that takes place in the case of the ammonium ion, NH_4^+. Are there any coordinate covalent bonds within this ion?

22. Write the correct name for each of these compounds.
    a. K_2O b. $CaCl_2$
    c. CO_2 d. MgS
    e. $FeCl_3$ f. NaBr
    g. BaI_2

23. Write the correct formula for each of these compounds.
    a. Zinc oxide   b. Barium sulfide
    c. Sulfur trioxide   d. Water
    e. Sodium phosphide   f. diphosphorus pentoxide
    g. Tin(II) fluoride   h. Mercury(II) chloride

24. Write the correct name for each of these compounds.
    a. Na_2CO_3 b. $Zn_3(PO_4)_2$
    c. $Zn(OH)_2$ d. $Mg(CN)_2$
    e. $(NH_4)_2SO_4$ f. $Cu(HCO_3)_2$
    g. $Fe(NO_3)_3$ h. $Ca(C_2H_3O_2)_2$

25. Write the correct formula for each of these compounds.
    a. Barium permanganate   **b.** Aluminum carbonate
    c. Silver chlorate   **d.** Sodium dichromate
    e. lithium sulfite   **f.** Tin(II) hydroxide
    g. Lead(II) sulfate   **h.** Ammonium phosphate

4

Quantifying Chemistry

What do you observe as you look at this industrial scene? Your eyes may be drawn by the clouds of gases and particles coming out of the stacks of this steel mill. You cannot see, however, the emissions coming from the trucks, cars, and motorcycles shown in the photograph. Both types of emissions should be of concern to you as a future health professional and also as a citizen.

All of these emissions are the result of chemical changes. All of the starting materials and the end products can be represented by chemical formulas. The process can be quantified and conveniently expressed using chemical equations. The change may be the combustion of gasoline in an automobile engine or it may be the extraction of iron from iron ore. You may be called upon in your future profession to measure the results of such chemical changes and then to evaluate their effects on living systems. Particularly those of you interested in public health or occupational health services will have the need to understand these examples of chemical change.

Quantifying and representing chemical changes are skills that apply equally to materials and processes in your car, in a steel mill, or in your body; the ground rules are the same. In this chapter we will look at chemical formulas in more detail and then see how the formulas can be arranged in a meaningful format to give even more quantitative information. We will get a start at understanding chemical changes, a process that continues in Chapter Five and, in fact, throughout all the rest of your study.

INTRODUCTION

Now that you have studied the submicroscopic details of chemical bonding and how to correctly name some of the pure chemical substances, it is time to move from the theoretical world of atoms and molecules to the practical laboratory world of chemical reactions and measurement. This chapter will help to give you the necessary practical tools to describe quantitative relationships for chemical substances.

We will start by looking once again at chemical formulas. This time we will study the quantitative information they reveal to us, beyond the simplest ratio of atoms present. We will use the formulas to find the weight of the formula unit represented. A new unit called the mole will be introduced and its usefulness discussed.

Formulas are combined into chemical equations, a chemist's shorthand to represent chemical reactions. Chemical reactions are taking place constantly in the living organism. Each hour we live, food molecules are chemically broken up into smaller molecular fragments that we can assimilate. Molecules we need for fat and protein tissue are synthesized from our body's pool of raw materials. Drug molecules are detoxified in the liver and in other tissues. Oxygen from the air is bound chemically so that it can be transported to all body parts.

In this chapter we will learn how to describe some types of chemical reactions by writing both word equations and balanced chemical equations. These symbolic statements can then be used to extend our quantitative understanding of chemical reactions.

CHEMICAL FORMULAS REVISITED

We have been writing many chemical formulas based on our knowledge of the symbols for the elements, position on the periodic chart, and common combinations of these elements. Remember that a chemical formula is a representation of the simplest relative number of atoms of one element to those of another element in the compound. The symbol for the element is followed by an appropriate **subscript** to indicate how many atoms of each are present in this representation. As you have observed, the number 1 is never indicated in a chemical formula. Thus water, which can be shown experimentally to have twice as many hydrogen atoms as oxygen atoms, is represented by the formula H_2O. Similarly, carbon dioxide, as its name indicates, is CO_2.

Subscripts in chemical formulas apply only to the element preceding them unless there is a parentheses in the formula. Therefore in the formula $AlBr_3$, there is a ratio of one aluminum atom for every three bromine atoms. In the substance $Ca(OH)_2$, the formula shows one calcium atom, *two* oxygen atoms, and *two* hydrogen atoms, because of the parentheses. The subscript after the parentheses indicates that each quantity within the parentheses is doubled.

FINDING FORMULA WEIGHTS

We have seen that each element has its own atomic weight. To find the **formula weight** of a pure substance, we add together the atomic weights of the component elements, taking into account the relative number of atoms present. The formula weight is just the mass of the pure substance expressed in atomic mass units(u), since that is the same unit used for atomic weights. The following example will illustrate this simple procedure.

EXAMPLE 4.1

Find the formula weight of $(NH_4)_2S$ (ammonium sulfide).

Solution
Examine the formula carefully, especially noting the parentheses. The formula shows that in each formula unit, there will be two nitrogen atoms, eight hydrogen atoms, and one sulfur atom. These are the factors that will

need to be used, together with the atomic weights, to find the formula weights.

$$2 \times 14.0 = 28.0 \text{ u (mass of nitrogen)}$$
$$8 \times 1.01 = 8.1 \text{ u (mass of hydrogen)}$$
$$1 \times 32.1 = \underline{32.1 \text{ u}} \text{ (mass of sulfur)}$$
$$68.2 \text{ u (formula weight)}$$

No specific directions were given in Example 4.1 about the number of significant figures to be included in the answer. You really could have decided to report the formula weight to the nearest whole number of atomic mass units or to carry everything out to a few more decimal places. If there are no limits specified, then keeping values such as those shown in the example is a reasonable choice.

We know from earlier chapters that chlorine gas exists as molecules, elemental sodium as atoms, and sodium chloride as ions. These substances can, in turn, be more specifically said to have a molecular weight, an atomic weight, and an ionic formula weight. These terms more explicitly represent the type of particle(s) present. All of these weights are more generally called *formula weights*. The advantage of using this term is that you do not have to decide exactly what type of bonding is present. For example, what about the bonding in the ammonium sulfate shown in Example 4.1? You may recall from the last chapter (or check in Table 3.4) that the ammonium ion is a positively charged polyatomic ion that is bonded covalently within the ion but then bonded ionically as a unit to the negatively charged sulfide ion. We found the formula weight correctly without worrying about whether the substance existed as molecules or ions, so we will continue with this practice.

Gram Formula Weight and the Mole

The formula weight, if expressed in atomic mass units, is very helpful in comparing the relative masses of pure substances, but rather hopelessly impractical for use in the real world. One atomic mass unit is so small (on the order of 10^{-27} kg) that it is not likely to be utilized in most laboratories. Therefore a very closely related value, called the **gram formula weight,** is commonly used. The gram formula weight is just the formula weight expressed in grams rather than in atomic mass units; this same weight in grams is also known as a **mole.** Therefore the gram formula weight of ammonium sulfide is 68.2 g and the weight of 1 mol is 68.2 g. The formula weight and the gram formula weight are the same *numerically* but they have different units and represent far different amounts of matter. (We will return to this point in the next section of this chapter.) The term *gram formula weight* may be abbreviated as gfw, and the correct SI abbreviation for the mole is mol.

▆▆▆▆▆▆ **EXAMPLE 4.2** ▆▆▆▆▆▆

Find the gram formula weight of sucrose (common table sugar). Its chemical formula is $C_{12}H_{22}O_{11}$. Report your answer to the nearest whole number of grams.

Solution

Use the periodic table or the list of elements to find the atomic weights for each element. Observe the subscripts in the formula. Find the formula weight in atomic mass units and then just change the unit to grams. Round the value to the nearest whole number.

$$12 \times 12.0 = 144 \text{ u (mass of carbon)}$$
$$22 \times 1.01 = 22 \text{ u (mass of hydrogen)}$$
$$11 \times 16.0 = \underline{176 \text{ u}} \text{ (mass of oxygen)}$$
$$342 \text{ u (formula weight)}$$
$$342 \text{ g (gram formula weight)}$$

The gram formula weight might also be expressed using scientific notation as 3.42×10^2 g.

▆▆▆▆▆▆ **EXAMPLE 4.3** ▆▆▆▆▆▆

How many moles of sodium chloride are present in 68.2 g of sodium chloride?

Solution

The fundamental relationship here is that 1 mol is the same as 1 gfw of a substance. By adding the atomic weights and changing to grams, you can find the gram formula weight and use it to find the number of moles present.

Answer

$$68.2 \text{ g NaCl} \times \frac{1 \text{ mol NaCl}}{58.5 \text{ g NaCl}} = 1.17 \text{ mol NaCl}$$

Notice in this problem that since the original problem was given with three significant digits, the answer is also reported with three significant digits. Beware of your calculator, which has tendency to give you many more numbers than really have any meaning for a chemistry problem. You must be smarter than it is and remember to round off your answers to match the number of significant figures in the problem as given.

Table 4.1. The Meaning of the Mole Concept

Substance	Formula	Weight of 1 mol (g)	Meaning of Avogadro's number for this substance
Sodium	Na	23.0	6.022×10^{23} atoms/mol
Oxygen gas	O_2	32.0	6.022×10^{23} molecules/mol
Methane	CH_4	16.0	6.022×10^{23} molecules/mol
Sodium chloride	NaCl	58.5	6.022×10^{23} formula units/mol
Calcium hydroxide	$Ca(OH)_2$	74.1	6.022×10^{23} formula units/mol

Avogadro's Number

The weight of one formula unit of sugar was calculated to be 342 u. The weight of one formula unit of sodium chloride was 58.5 u. When the same numbers are followed by the unit *grams*, these values can no longer represent just one formula unit but rather a much larger collection of formula units. Because of the extremely small size and mass of atoms, ions, and molecules, this number of particles present in 1 gfw (1 mol) of any substance can be expected to be very large, and it is. Experimentation puts this number at 6.022×10^{23}, a number that is very large, indeed! The value is called **Avogadro's number**, after the Italian scientist Amedeo Avogadro (1776–1856), who first clarified the concept through experimental work with gases. Table 4.1 shows how the principle of Avogadro's number can be applied to different types of substances—atomic, molecular, and ionic. As you study this table, consider that the mole may be viewed as a special kind of measure that categorizes matter not according to weight or volume, but by number of particles. It might be likened to an egg carton that contains twelve eggs, regardless of their size. It might also contain twelve other objects, such as the limes my neighbor delivers from his productive tree! The difference between this egg carton and Avogadro's is that there are 6.022×10^{23} spaces, not twelve spaces in Avogadro's carton.

As Table 4.1 shows you, there is a constant number of formula units in a mole of any substance. Just as one dozen of any objects would be twelve objects, 1 mol always tells you there are 6.022×10^{23} formula units present. One of the most useful things about this fact is that you can use the relationship in solving problems. This process is shown in Example 4.4.

■■■■■■■ **EXAMPLE 4.4** ■■■■■■■

How many atoms are there in 1.00 g of pure copper metal?

Solution
To find the number of particles, you always must find the number of moles. Only if you know the number of moles can you know the number of particles. You will need to use the atomic weight of copper and Avogadro's

number. You will be expecting a very large number, knowing the size of Avogadro's number.

$$1.00 \text{ g } \cancel{Cu} \times \frac{1 \cancel{\text{ mol Cu}}}{63.5 \text{ g } \cancel{Cu}} \times \frac{6.02 \times 10^{23} \text{ atoms Cu}}{1 \cancel{\text{ mol Cu}}} = 9.48 \times 10^{21} \text{ atoms Cu}$$

This is over 9 billion trillion atoms! From calculations such as these you can begin to get a feel for the extremely small size and mass of individual atoms.

EXAMPLE 4.5

What is the mass of a single water molecule?

Solution
One mole, or 1 gfw is the mass that contains Avogadro's number of molecules for this molecular substance. Use this fact to find the mass of one molcule. You should be expecting a very small mass for a single water molecule.

$$\frac{18.0 \text{ g } H_2O}{1 \cancel{\text{ mol }H_2O}} \times \frac{1 \cancel{\text{ mol }H_2O}}{6.02 \times 10^{23} \text{ molecules } H_2O} = \frac{2.99 \times 10^{-23} \text{ g } H_2O}{\text{molecule } H_2O}$$

EXAMPLE 4.6

Serotonin ($C_{10}H_{12}N_2O$, molecular weight 176 g/mol) is a brain chemical involved in regulation of sleep-wake patterns in humans. How many molecules are available in 1.0 microgram of serotonin for attachment to receptor sites in the brain?

Solution

$$1.0 \cancel{\mu g} \times \frac{1 \cancel{g}}{10^6 \cancel{\mu g}} \times \frac{1 \cancel{\text{ mol}}}{176 \cancel{g}} \times \frac{6.02 \times 10^{23} \text{ molecules}}{\cancel{\text{mol}}} = 3.4 \times 10^{15} \text{ molecules}$$

WHY WRITE EQUATIONS?

The most interesting aspects of chemistry involve not just the static properties of individual pure substances but the chemical changes that occur when these substances interact. Symbolic statements called **chemical equations** can be used to compactly describe the changes that occur. An equation is a shorthand representation of a chemical reaction but a chemical equation does far more then just condense qualitative information about chemical change. It also represents a large amount of quantitative information because formulas are present in the equation for all substances involved in the reaction. The correct mole ratios are also shown. This allows you to calculate, for example, the mass of a substance that will be required to produce a certain amount of a desired product. As you might imagine given the complexity of matter, there are many different kinds of chemical equations. We will study some of these and explain how to write and interpret them.

Word Equations

To communicate about any type of chemical reaction, we can begin with a word description of what is happening. A minimum description would include identification of the substances you start with and identification of the resulting products. A **word equation**, then, is just an explanation of a chemical reaction. Some of the words you would use in a sentence to describe the reaction are replaced by representative symbols.

Consider the reaction that occurs when a simple hydrocarbon burns. (A **hydrocarbon** is a compound containing only hydrogen and carbon.) Methane is such a substance. It is a gas at normal temperatures and pressures. When it burns in air, methane combines with oxygen gas in the air and produces carbon dioxide gas and water droplets. All of this could be expressed in this word equation.

methane gas + oxygen gas → carbon dioxide gas + water droplets

The materials on the left-hand side of this equation are called **reactants** and the materials on the right-hand side are referred to as the **products**. The plus sign on the left side of the equation may be interpreted to mean "reacts with," whereas the same sign on the right means "and." The arrow means "forming" or "to yield."

Formula Equations

Although word equations are clear and simple, they lack sufficient detail to be very useful for quantitative purposes. It is much more useful to replace the words with the appropriate chemical formulas and to put in some additional representative symbols to convey more information. As a start on this process, the same reaction of methane with oxygen gas can be represented this way by a **formula equation**. (Note: this equation is not yet complete.)

$$CH_4(g) + O_2(g) \rightarrow CO_2(g) + H_2O(l)$$

The new information that has been added, other than the correct formulas for each of the substances, are the states of the materials, indicated by placing the correct letter (g for gas, l for liquid, or s for solid) in the parentheses.

There is, however, something still incomplete about this representation of the reaction. We have not been very careful with the quantities of substances shown in the reaction. Right now, we show that there are four hydrogen atoms on the reactant side of the reaction and only two atoms of hydrogen on the product side of the reaction. Similarly, we seem to have created an extra oxygen atom in the products that does not exist in the reactants. This cannot be, for it violates the **law of conservation of mass**, which says that matter can neither be created nor destroyed in a chemical reaction. The matter in a chemical reaction can be rearranged but the same number of atoms of each kind that go into a chemical reaction must come out of it, just rearranged into different chemical combinations. Only in a nuclear reaction, such as those discussed in Chapter 19, can some matter be converted into energy.

To solve this apparent dilemma, the equation must be **balanced**. A chemical equation is said to be balanced if the same number of atoms of each type appear on each side of the equation. You always balance an equation *after* you are sure that you have written all the formulas correctly. Once the formulas are fixed, the only changes that can be made are to alter the numbers in front of the respective formulas; these numbers are called **coefficients**. By altering these numbers, you are changing the relative number of formula units that react and form, but not the correct ratio of atoms within each substance. For the combustion of methane, the correct coefficients are

$$1 \ CH_4(g) + 2 \ O_2 \ (g) \rightarrow 1 \ CO_2(g) + 2 \ H_2O(l)$$

Check this equation by counting atoms on each side of the equation. There is one carbon atom, four hydrogen atoms, and four oxygen atoms on each side. Even though they are arranged differently on each side of the equation, all atoms are accounted for. If the coefficient is equal to one, it is usually omitted from the balanced equation, just as a subscript of 1 is omitted from a formula.

You can check the law of conservation of mass another way in order to determine that the mass of the reactants is equal to the mass of the products.

$$CH_4 + 2O_2 \quad \rightarrow CO_2 \ + 2H_2O$$

$$\underbrace{16.0 \ g + 2(32.0 \ g)}_{80.0 \ g} \rightarrow \underbrace{44.0 \ g + 2(18.0 \ g)}_{80.0 \ g}$$

Since the total mass of the reactants is the same as the total mass of the products, the law of conservation of mass has been obeyed in this chemical reaction. Be sure to notice that methane reacts in a ratio of 1 mol to *2 mol* of oxygen gas but it is *not* true that 1 g of methane reacts with 2 g of oxygen. Coefficients give the ratio of formula units but not the ratio of mass. You must use the mole weights to obtain the mass ratio.

Rules for Balancing Equations

It is helpful, particularly for more complex chemical reactions, to have a set of rules to follow for writing and balancing chemical equations. As you learn to write equations, try following these steps in order.

1. Write the word equation, including all the reactants and products. Include any additional information such as the states of the substances, if desired.
2. Write the correct formulas for all substances, and translate any additional information into symbols if you have included it in your word equation.
3. Check to see if the equation is balanced by counting atoms of each element on each side of the equation.
4. Balance the equation by changing the *coefficients* before the formulas, never by changing the formula. Fractional coefficients are possible. (After all the elements have been balanced, you can multiply the entire equation by a suitable whole number if it is desired to remove fractions.)
5. Check again to be sure that the same number of atoms of each element are represented on both sides of the equation.

The following example will illustrate the use of these steps. Keep in mind that to be comfortable with writing and balancing equations, the most helpful thing is to just keep practicing this skill.

EXAMPLE 4.7

When hydrogen gas burns with oxygen gas, water droplets are observed forming. Write a balanced chemical equation for this reaction.

Solution
Start by translating the facts into a word equation.

hydrogen gas + oxygen gas → water droplets

Then, following the steps given earlier, you will need to write the correct formulas for each substance. Remember that both hydrogen gas and oxygen gas exist in diatomic molecules.

$$H_2(g) + O_2(g) \rightarrow H_2O(l)$$

The final step is to balance the equation. Remember at the end of balancing, you should check once again to see if the same number of each type of atom occurs on each side of the equation.

$$2H_2(g) + O_2(g) \rightarrow 2H_2O(l)$$

━━━━━━━ **EXAMPLE 4.8** ━━━━━━━

Balance this equation and verify that the law of conservation of mass is followed.

sodium metal + chlorine gas → sodium chloride salt

Solution
Since the word equation is given, start by writing the correct formulas for each substance. Then find the necessary gram formula weights to verify that the total mass on the left-hand side of the equation is equal to the total mass on the right-hand side.

$$2Na(s) + Cl_2(g) \rightarrow 2NaCl(s)$$

$$\underbrace{2(23.0)g + (71.0)g} \rightarrow \underbrace{2(23.0 + 35.5)g}$$

$$117.0\,g \qquad \rightarrow \qquad 117.0\,g$$

Additional Symbols Used in Equations

There are many ways that the symbolism in an equation can be expanded to provide even more information. As we have seen, the states of the reactants and products may be included. Whether or not a substance is in solution is often important. A common way to indicate that a solution has been made up using water is to show *aq* in parentheses. This stands for an **aqueous** solution.

The energy changes associated with a reaction may also be of interest. If the reaction required heat, a triangular sign called **delta** (Δ) is placed over the arrow in a chemical equation.

Not all equations are written in molecular form; it is often useful to write them in ionic form. We will look at some of these options in the next chapter and in subsequent chapters as the need arises.

Quantitative Interpretation of Equations

Earlier we looked at the chemical equation for the combustion of methane. Now we will examine a related system to understand quantitative aspects of equations. As with the reaction with methane, this reaction also involves combination with oxygen gas. If iron combines with oxygen gas, the product is most likely to be iron (III) oxide. This is a process you surely have observed, for it is the main reaction occurring when an iron object rusts. (Water plays a role in the reaction, too, but we will simplify the system by not including the water here.)

The word equation for this reaction is

iron solid + oxygen gas → iron(III) oxide solid

The balanced chemical equation is

$$4\ Fe(s) + 3O_2(g) \rightarrow 2Fe_2O_3(s)$$

Once the equation is balanced, it can be used for quantitative interpretation. The simplest way to use this equation is in terms of the number of atoms and molecules represented. (The states of the reactants and products will be omitted as we continue. They are not essential to the following discussion.)

$$4\ Fe \quad + 3\ O_2 \quad \rightarrow 2\ Fe_2O_3$$

4 atoms of iron + 3 molecules of oxygen → 2 formula units of iron(III) oxide

Because of the extremely small size and mass of atoms, molecules, and formula units, we know that it is not practical to convert the preceding statement directly into the absolute number of grams. Instead, we can use the mole relationship and restate the interpretation of the equation in this way.

$$4\ Fe \quad + 3\ O_2 \quad \rightarrow 2\ Fe_2O_3$$

4 mol + 3 mol → 2 mol

Since a mole is the same as a gram formula weight, this can also be stated in terms of the mass of the substances reacting and forming, as shown in this relationship.

$$4(55.8\ g) + 3(32.0\ g) \rightarrow 2(159.6\ g)$$
$$223.2\ g + 96.0\ g \quad \rightarrow \quad 319.2\ g$$

$$319.2\ g \quad \rightarrow \quad 319.2\ g$$

As we expect, the law of conservation of mass is obeyed here, too. The sum of the masses of the reactants is equal to the total of the masses of the products. One important thing to note is that the total number of moles on one side of the equation does not *have* to equal the total number of moles on the other side. There are 7 mol of reactants on the left-hand side of the equation and only 2 mol of products on the right-hand side in this example. The actual number of moles will depend on the relative complexities of the reactants and products. In some reactions, such as this one, the product contains many more atoms per formula unit, so the number of moles of products is smaller than the number of moles of reactants. It could easily be the other way around if the particular reaction under consideration involved material breaking down into simpler forms from a greater complexity. In some cases, there may coincidentally be the same number of moles on both sides. That is not commonly to be expected.

As an example of the usefulness of the mole in finding quantitative relationships in a balanced chemical equation, consider this problem based on the same reaction of iron with oxygen.

EXAMPLE 4.9

How many grams of iron(III) oxide will be produced if 55 g of iron completely rusts?

Solution

First you will need to convert grams of iron to moles of iron. This is done so that you can use the mole ratio between iron and iron(III) oxide found in the balanced chemical equation. By consulting the balanced chemical equation given previously, you will see that 2 mol of iron(III) oxide will be produced for every 4 mol of iron metal that is consumed. Once you have found the number of moles of product, you can find the number of grams by using the number of grams per mole.

$$55 \text{ g Fe} \times \frac{1 \text{ mol Fe}}{55.8 \text{ g Fe}} \times \frac{2 \text{ mol Fe}_2\text{O}_3}{4 \text{ mol Fe}} \times \frac{160. \text{ g Fe}_2\text{O}_3}{1 \text{ mol Fe}_2\text{O}_3} = 79 \text{ g Fe}_2\text{O}_3$$

Each of the steps in Example 4.8 could be carried out separately. However, if there is no specific need for the answers to the intermediate steps, it is always better mathematically to perform all the steps sequentially and obtain just the final answer. There is no point in giving yourself and your calculator any additional opportunities for error by increasing the number of times you enter and read out numbers.

The reasoning required for Example 4.8 is quite typical of problems based on the quantitative aspects of chemical equations. Here is a generalized version of a problem-solving flow chart, emphasizing the use of the mole ratio from a balanced chemical equation. You will find that we can add to this plan in the chapters ahead, particularly when we study gases and solutions. After studying Figure 4.1, work through the remaining examples in this chapter.

EXAMPLE 4.10

If 26.8 g of sodium entirely reacts with chlorine gas, how many grams of sodium chloride will be produced?

Solution

You will need to change from grams of sodium to moles of sodium. Then you will need the mole ratio in the balanced chemical equation to relate the number of moles of sodium to the number of moles of sodium chloride.

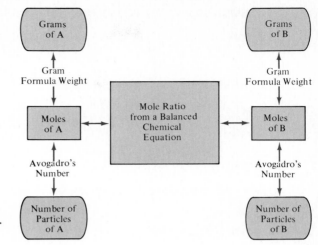

FIGURE 4.1 Problem-Solving Flow Chart.

Finally, you will need to use the gram formula weight of sodium chloride to obtain the number of grams of sodium chloride.

The balanced chemical equation is

$$2\ Na\ +\ Cl_2 \rightarrow 2\ NaCl$$

$$26.8\ \text{g Na} \times \frac{1\ \text{mol Na}}{23.0\ \text{g Na}} \times \frac{2\ \text{mol NaCl}}{2\ \text{mol Na}} \times \frac{58.5\ \text{g NaCl}}{1\ \text{mol NaCl}} = 68.2\ \text{g NaCl}$$

━━━━━━ **EXAMPLE 4.11** ━━━━━━

If 4.5×10^{24} atoms of iron react with oxygen gas to form iron(III) oxide, how many moles of iron(III) oxide will be formed?

Solution
Since the number of atoms of iron are given, you will first have to use Avogadro's number to find the number of moles of iron present. Then you can use the mole ratio in the balanced chemical equation to find the number of moles of the product, iron(III) oxide. You can stop there for the number of grams of product is not asked for. The balanced equation is found earlier in this chapter.

$$4.5 \times 10^{24}\ \text{atoms Fe} \times \frac{1\ \text{mol Fe}}{6.02 \times 10^{23}\ \text{atoms Fe}} \times \frac{2\ \text{mol Fe}_2\text{O}_3}{4\ \text{mol Fe}} = 3.7 \times 10^0\ \text{mol Fe}_2\text{O}_3$$

■━━━━━**EXAMPLE 4.12** ■━━━━━

An astronaut on a space flight eats a candy bar containing one avoirdupois ounce (28.35 g) of dextrose, $C_6H_{12}O_6$. How many grams of water will his body produce from the dextrose? The representative equation for the metabolism of dextrose is

$$C_6H_{12}O_6 \rightarrow 6\ CO_2\ +\ 6\ H_2O$$

Solution

$$28.35\ \text{g}\ C_6H_{12}O_6 \times \frac{1\ \text{mol}\ C_6H_{12}O_6}{180.2\ \text{g}\ C_6H_{12}O_6} \times \frac{6\ \text{mol}\ H_2O}{1\ \text{mol}\ C_6H_{12}O_6} \times \frac{18.02\ \text{g}\ H_2O}{1\ \text{mol}\ H_2O} = 17.01\ \text{g}\ H_2O$$

HINTS FOR PROBLEM SOLVING

Keep in mind these hints about solving problems such as those in this chapter. In each case it is necessary to read the problem very carefully. Determine what quantity you know (Is it grams? Moles? Number of particles?) and what quantity you need to know. Proceed step by step, starting with what is given, converting to moles, using the mole ratio in the balanced chemical equation to find the new quantity of moles, and finally convertng to the unit needed. All units should be fully labeled. A final check is then in order to be sure that all units cancel except the quantity asked for in the problem. At the same time, remember to handle the numerical part of the problem so that it agrees with the original data given in the number of significant figures. Finally, be sure that the answer makes sense.

As you do in writing equations, you gain facility in solving problems by practice. This is why many examples and problems are provided for you to try. Check the Study Guide that accompanies this text for additional practice. You may also find it helpful to refer to one of the many supplementary books, audiovisual programs, or computer-assisted instructional programs available to help chemistry students learn problem-solving skills. Your teacher will be able to provide you with some specific resources.

SUMMARY

This chapter has provided an introduction to the quantitative relationships found in chemical formulas and in balanced chemical equations. The chemical formula shows the simplest ratio of atoms present in the compound. The formula can be used to determine the gram formula weight of the compound. This value is also the weight of 1 mol of the compound and contains Avogadro's number of particles.

Chemical formulas are combined into formula equations. Although word equations can also be used to represent chemical change, formula equations are preferred because of the additional quantitative information available from them. Balanced formula equations give not only the correct chemical formulas but also the ratios of moles of reactants and products present. Other representative symbols can be used in chemical equations to show the state of matter, energy changes, or the presence of an aqueous solution.

Problem-solving skills have been practiced in this chapter. These skills are essential for all practical applications of chemistry. The relationships between mass, moles, and number of particles present are all key ideas for quantitative use of chemical formulas and equations. We have seen once again how cancellation of units can help in problem solving. It is also important to keep in mind a common sense estimate of the answer.

KEY TERMS

Check your understanding of this chapter. Can you explain what is meant by each of these terms?

aqueous	gram formula weight
Avogadro's number	hydrocarbon
balanced equation	law of conservation of mass
chemical equation	mole
coefficients	products
delta	reactants
formula equation	subscripts
formula weight	word equation

STUDY QUESTIONS

1. How many atoms of each type are represented by one molecule of each of the following substances?
   a. Na_3PO_4 b. $Ca(C_2H_3O_2)_2$
   c. H_2SO_4 d. $(NH_4)_3AsO_4$
   e. C_2H_6 f. $C_{12}H_{22}O_{11}$

2. Find the gram formula weight of each substance in Question 1. Express each value to the nearest whole number.

3. Find the mass of 1 mol of each of these substances. Express the mass to the nearest tenth of a gram.
   a. H_2O b. $CuSO_4$
   c. $Fe(NO_3)_3$ d. CH_4N_2O
   e. adrenaline, $C_9H_{13}NO_3$ f. dopamine (a brain amine), $C_8H_{11}NO_2$

4. Calculate the mass of a single carbon dioxide molecule.

5. Is there a larger number of atoms in a gram of sodium or in a gram of helium? Explain.

6. Are there more atoms present in 25 g of zinc or in 25 g of copper(II) sulfate? Explain.

7. How many molecules are there in 38 g of sulfur dioxide?

8. Find the mass in grams of the following:
   a. 0.75 mol of H_2O b. 2.4 mol of C_2H_5OH
   c. 0.001 mol of NaCl   d. 4.5 mol of CH_4
   e. 0.40 mol of $CaCl_2$ f. 5.525 mol of N_2

9. Find the number of moles in the following:
   a. 5.5 g of $Ca(OH)_2$ b. 825 g of $C_{12}H_{22}O_{11}$
   c. 0.44 g of SO_2 d. 2.25 g of $Ca(HCO_3)_2$
   e. 4.0×10^{14} g of NaOH   f. 7.7×10^{-3} g of H_2SO_4

10. Calculate the number of atoms in the following:
    a. 1.0 g of gold
    b. 1.0 mol of gold
    c. 1.0 gram atomic weight of gold
    d. 1.0 gram formula weight of gold

11. How many formula units are there in the following:
    a. 1.00 g of NaOH   b. 3 mol of NaOH
    c. 155 kg of NaOH   d. 5.6×10^2 mg of NaOH

12. Compare the number of molecules in a mole of carbon monoxide with the number of molecules in a mole of carbon dioxide. Then compare the number of atoms present in a mole of each of the same two substances.

13. What information does a formula equation give you that is not found in a word equation?

14. Why are the rules for balancing equations given in this chapter not applicable to such things as the radioactive decay of radium to produce radon and helium?

15. Convert the following word equations into balanced formula equations.
    a. aluminum solid + hydrogen chloride gas → aluminum chloride solid and hydrogen gas
    b. sodium solid + water → hydrogen gas and sodium hydroxide solution
    c. sulfur dioxide gas + oxygen gas → sulfur trioxide gas
    d. copper(II) carbonate solid → copper(II) oxide solid + carbon dioxide gas
    e. hydrogen phosphate + ammonium hydroxide → water + ammonium phosphate

f. sodium sulfite + hydrogen chloride → sodium chloride + water + sulfur dioxide

16. Show that the law of conservation of mass is observed for this equation.

$$Cu + 4 HNO_3 \rightarrow Cu(NO_3)_2 + 2 NO_2 + 2 H_2O$$

17. Balance each of these equations.

a. $PbO_2 \rightarrow PbO + O_2$
b. $Al + Br_2 \rightarrow AlBr_3$
c. $Cr_2O_3 + H_2 \rightarrow Cr + H_2O$
d. $KClO_3 \rightarrow KCl + O_2$
e. $C_4H_{10} + O_2 \rightarrow CO_2 + H_2O$

18. Convert each of the equations in Question 17 back into word equations. (In part e, the substance C_4H_{10} is named butane, another example of a hydrocarbon.)

19. When iron(III) chloride solution reacts with barium hydroxide solution, the products are solid iron(III) hydroxide and barium chloride in solution. Write a balanced chemical equation for this reaction.

20. Here is the balanced equation for the reaction of hydrogen gas with nitrogen gas to produce ammonia gas, NH_3.

$$N_2(g) + 3 H_2(g) \rightarrow 2 NH_3(g)$$

Use this equation to answer each of the following questions.

a. How many moles of nitrogen gas are required to react with 6 mol of hydrogen gas?
b. How many moles of ammonia gas will be produced if 8 mol of nitrogen exactly react with 24 mol of hydrogen gas?
c. How many grams of nitrogen will be required to produce 100. g of ammonia?
d. How many grams of hydrogen will react with 55 g of nitrogen?
e. How many grams of ammonia gas will be produced if 10. g of nitrogen react with sufficient hydrogen?

21. If 1.55 g of sodium entirely reacts with chlorine gas, how many grams of sodium chloride will be produced?

22. Calculate how many grams of oxygen are required to completely burn 2.00×10^2 g of pentene. The balanced equation is

$$2 C_5H_{10}(g) + 15 O_2(g) \rightarrow 10 CO_2(g) + 10 H_2O(l)$$

23. Silver metal can be formed by means of this reaction:

$$2 AgNO_3(aq) + Cu(s) \rightarrow Cu(NO_3)_2(aq) + 2 Ag(s)$$

How many grams of copper are required to produce 5.00 g of silver? Assume that you have more than enough silver nitrate solution to react with the copper.

24. "Gasohol" is a fuel containing ethyl alcohol, C_2H_5OH. When it burns with oxygen gas, the products are carbon dioxide and water. How many grams of oxygen gas will be required to completely burn 75 g of ethyl alcohol?

25. If you vary the concentration of the HNO_3 used in Question 16, the products will be changed to include nitrogen monoxide rather than nitrogen dioxide. (The rest of the products remain the same.) Write the balanced equation for this reaction and use it to predict how many grams copper will be required to produce 25 g of nitrogen monoxide.

5 Chemical Change

Chemical changes occur all around us every day. When we operate an automobile, strike a match, bake a cake, or paint a wall, we are changing things chemically.

In industrial laboratories, too, chemical changes occur in reactions carefully designed to yield products valuable in medicine or to our society. The technician in our photo is carrying out a hydrogenation reaction in which hydrogen gas (H_2) is made to add to an unsaturated molecule in the presence of a catalyst. This is one of the ways useful drugs are obtained. Note that in a chemical reaction such as hydrogenation, one or more substances disappear as new substances are formed.

More important to you are the chemical reactions that occur inside living organisms. These are given the special name *biochemical reactions.* In this chapter we shall introduce you to the principles that govern chemical change; in Chapter 16, we will examine closely many of the biochemical reactions occurring in the human.

This chapter is especially important because here you will learn to recognize the types of chemical reactions that occur in the laboratory and in the organism. Generally speaking, these include oxidation, reduction, hydrolysis, combination, and decomposition. (Photo courtesy of Pfizer, Inc.)

INTRODUCTION

All of the concepts, laws, calculations, and structural concerns of chemistry are preliminary to the primary concern of chemistry, the chemical reaction. We have seen many different examples of chemical change in the preceding chapters, and now we need to develop a better understanding of the types of reactions that can take place and the changes in energy that accompany them.

The health professional can apply knowledge of chemical processes in deciding the appropriate dose of a drug to administer. Drugs are chemicals, too, and are broken down in the body (biotransformed) by chemical reactions to inactive products. The health professional knows that drugs have a half-life in the body, that is, the time required for one-half of the original dose to be chemically changed and excreted from the body. This leads us to the knowledge that additional doses of the drug must be timed to maintain blood levels high enough to result in the desired drug action.

Many questions can be asked concerning chemical reactions. For example, why do reactions occur? Can the products of a reaction be predicted? How fast do reactions take place and can you influence the rate? These are some of the questions that we will start to answer in this chapter.

To do this, we will first classify chemical reactions. If we recognize the type of reaction present, the pattern associated with that category can be used to help predict the products of the reaction. A major type of chemical change in health-related chemistry is the redox reaction. We will learn several ways to recognize this type of reaction and see why it is important.

Energy changes in chemical reactions are central to answering the basic question of why reactions take place. The role of energy changes will be examined. Catalysts can be used to speed up chemical change. Catalysts play an essential role in many biologically important reactions.

Finally, in this chapter, we will look at one more factor that helps to answer our questions about chemical change. That is the role played by the tendency of chemical and physical processes to maximize randomness in a system.

TYPES OF CHEMICAL REACTIONS

We saw earlier that it was helpful to classify matter. So it is with chemical reactions. There are various systems for organizing the huge number of observed reactions into more manageable groupings. There are so many different chemical reactions that it would be impossible to remember each one. However, if you can recognize the patterns in chemical change, you can often determine the type of chemical product to expect. Without calling attention to the specific names, we have already seen typical examples of many different types of reactions. For example, in the last chapter we looked at the rusting of iron.

$$4 \text{ Fe(s)} + 3 \text{ O}_2\text{(g)} \rightarrow 2 \text{ Fe}_2\text{O}_3\text{(s)}$$

This illustrates a **composition** reaction which is always characterized by two or more elements or compounds combining to form one new compound. This type of reaction may also be called **combination** or **synthesis.** It can be easily recognized because there will be just *one* product, forming from two or more reactants.

Decomposition reactions, as you might expect from the name, are just the opposite. They may be recognized from their equations because *one* reactant breaks down to form two or more products. Here is an example of a decomposition reaction.

$$2 \text{ KClO}_3\text{(s)} \rightarrow 2 \text{ KCl(s)} + 3 \text{ O}_2\text{(g)}$$

Redox Reactions

Both of the preceding equations are also equations representing **redox reactions**. The term redox is a shortening of the terms *reduction* and *oxidation*. A redox reaction always involves changes in oxidation number (valence) for at least some of the atoms or ions in the products. This means that a redox reaction always involves gain and loss of electrons.

In Chapter 3, we saw that oxidation number was a means of expressing the combining capacity of that particular element. It is just the charge on any single ion. The oxidation number of any free element is always zero for it has not yet lost or gained electrons. The following sequence will help to explain changes in oxidation number.

$$Co^0 \xrightarrow[\text{2 electrons}]{\text{Loss of}} Co^{2+} \xrightarrow[\text{1 more electron}]{\text{Loss of}} Co^{3+}$$

elemental cobalt	cobalt (II) ion	cobalt (III) ion
oxidation number = 0	oxidation number = 2+	oxidation number = 3+

The oxidation number of any atom within a formula unit can be determined by inspection. Remember that if combined oxygen is present, it usually has an oxidation number of 2−. If combined hydrogen is present, it usually has an oxidation number of 1+. The oxidation numbers of other atoms can be found by noting the charge on the formula unit and using the laws of algebra. The examples in this section will help you to practice finding oxidation numbers.

The vocabulary of redox reactions can be illustrated by looking more carefully at the iron reaction.

$$4 \ Fe(s) \ + \quad 3 \ O_2(g) \ \rightarrow \quad 2 \ Fe_2O_3(s)$$

oxidation number of Fe = 0	oxidation number of O = 0	oxidation number of Fe = 3+ O = 2−
substance oxidized is the reducing agent	substance reduced is the oxidizing agent	

The iron has changed from an oxidation number of zero as a free element to its combined capacity of 3+. An increase in oxidation number, caused by the loss of electrons, defines the process of **oxidation.** Three negatively charged electrons are being transferred from each iron atom in this case, leaving it with a 3+ charge. Oxygen, on the other hand, is gaining electrons and each atom changes from an oxidation number of zero to 2−. The oxidation number is decreasing because the gain of an electron. This defines the process of **reduction.** Furthermore, the iron is the reactant that donates the electrons and so is responsible for oxygen undergoing reduction. This means the iron can be called the **reducing agent.** Oxygen gains the electrons from the iron. The oxygen gas is, therefore, called the **oxidizing agent,** because it makes possible the oxidation of the iron.

These relationships always holds true:

substance oxidized = reducing agent
substance reduced = oxidizing agent

These four terms can only apply to the *reactants* in a chemical change. You must examine the products in order to compare them with the reactants but the substance oxidized and related terms are used to describe the reactants. The two processes of oxidation and reduction must take place *together,* for when electrons are transferred, one substance must lose electrons so that the other can gain those electrons.

Some redox processes, such as the rusting reaction first given, can be recognized because there is at least one element on one side of the equation and only compounds on the other side. This must mean that a change in oxidation numbers has occurred, for all free elements have an oxidation number of zero; all compounds will be made up of atoms that are combined and so have an oxidation number different from zero.

Another common way to recognize redox reactions is also illustrated by the rusting process. In many cases, oxygen gas serves as an oxidizing agent. This fact was the basis for the original meaning of the term oxidation. Although we now apply the term to a much broader spectrum of reactions, oxygen gas is still an important and common oxidizing agent. Study this example to see how the basic definitions for redox reactions are applied when oxygen is not present.

EXAMPLE 5.1

Identify the oxidizing agent in this reaction.

$$2 \text{ Na(s)} + \text{Cl}_2(\text{g}) \rightarrow 2 \text{ NaCl(s)}$$

Solution
You will need to find which element is decreasing in oxidation number, for that will be the substance that is gaining electrons and defined to be the oxidizing agent. Start by observing that both reactants are elements and so have oxidation numbers of 0. The product sodium ion has a charge of $1+$ and the product chloride ion has a charge of $1-$.

Each atom in chlorine gas is reduced in oxidation number from 0 to $1-$ in changing to the chloride ion. Therefore chlorine gas, Cl_2, is reduced and is acting as the oxidizing agent.

There are some very important applications of the concept of redox reactions to health and disease. Consider, for example, the serious blood condition called *methemoglobinemia*. When blood is exposed to various drugs or other oxidizing agents such as nitrates or nitrites, the Fe^{2+} ion that normally binds O_2 in the hemoglobin molecule is oxidized to the Fe^{3+} ion, forming *methemoglobin*. Methemoglobin is incapable of transporting O_2.

Mercury in the $1+$ oxidation state can be safely taken internally in the form of Calomel (a cathartic). Mercury in the $2+$ oxidation state ($HgCl_2$, for example) is a dangerous poison. Chlorine in the elemental state is a dangerous-to-handle greenish yellow gas, used to kill bacteria in municipal water supplies. Chlorine in the $1-$ oxidation state is a harmless ion we all commonly take into our bodies every day. In body cells, the coenzyme nicotinamide adenine dinucleotide exists in its oxidized form (NAD+) and in its reduced form (NADH). This coenzyme functions in a number of biological oxidations and reductions. In Chapter 16 we will discuss some of these reactions, such as the process of oxidative deamination.

Organic Redox Reactions

All of the preceding definitions apply to carbon compounds as well. The only thing that may make this type of redox difficult to recognize is that you are unfamiliar with how to assign oxidation numbers to complex organic compounds. You still may have the important clue of elemental oxygen on one side of the equation but not on the other side. Previously, we looked at this reaction.

$$CH_4(g) + 2\ O_2(g) \rightarrow CO_2(g) + 2\ H_2O(l)$$

This is classified as a **combustion** reaction, which is characterized by rapid combination with oxygen, accompanied by the release of heat and light. A great deal of energy is released when any hydrocarbon burns, which is why hydrocarbons are so useful as fuels.

Look closely at this combustion reaction, for it is also a redox reaction. The reactant methane (CH_4) has combined with oxygen gas and there have been changes in oxidation number. These are summarized in the next equation.

$CH_4(g)$	$+$	$2\ O_2(g)$	$\rightarrow$	$CO_2(g)$	$+$	$2\ H_2O(l)$
oxidation number of		oxidation number of		oxidation number of		oxidation number of
H = 1+		O = 0		C = 4+		H = 1+
C = 4−				O = 2−		O = 2−
substance oxidized		substance reduced				
reducing agent		oxidizing agent				

Notice that the oxygen decreases in oxidation number from zero to $2-$, a process of reduction. Oxygen gas is both the substance reduced and the oxidizing agent. The carbon in the methane molecule increases in oxidation number from $4-$ to $4+$. The methane is the substance oxidized and serves as the reducing agent. Once again, the processes of oxidation and reduction are occurring together in this reaction.

Another way to recognize the process of oxidation in an organic compound is to see which substance has lost hydrogen and subsequently gained oxygen. That substance will be the one undergoing oxidation. Another possibility utilizing the broader concept of oxidation is that oxygen will not be involved at all. It is possible to lose or gain only hydrogen from an organic compound and still have a redox reaction. Table 5.1 summarizes some of the different ways by which you can recognize oxidation and reduction in chemical reactions.

EXAMPLE 5.2

"Benedict's test" is used to detect the presence of glucose in urine. The presence of glucose in the urine is associated with the disease diabetes mellitus. The equation for the test reaction is

$$2\ Cu^{2+}\ +\ 4\ OH^-\ +\ C_6H_{12}O_6\ \rightarrow\ Cu_2O(s)\ +\ 2\ H_2O(l)\ +\ C_6H_{12}O_7$$

Benedict's solution (major reactants only shown), is blue in color.   glucose   reddish brown solid forms   gluconic acid

Identify the oxidizing agent and the reducing agent in this reaction.

Solution
Use Table 5.1 to help you identify the characteristics of a redox reaction. The oxidizing agent will be the substance that decreases in oxidation number. The reducing agent will increase in oxidation number. For the organic substances involved, look for the loss/gain of oxygen or hydrogen to help identify the process.

Table 5.1. Recognizing Redox Reactions

Reduction must involve:	Oxidation must involve:
decrease in oxidation number	increase in oxidation number
gain of electrons	loss of electrons
Reduction may involve:	**Oxidation may involve:**
loss of oxygen	gain in oxygen
and/or	and/or
gain in hydrogen	loss in hydrogen

Copper ion has decreased from an oxidation number of $2+$ to an oxidation number of $1+$ in the compound Cu_2O. Therefore Cu^{2+} is itself reduced and must be the *oxidizing agent*.

Glucose has gained one oxygen atom per molecule in reacting to form gluconic acid. This means the glucose is itself oxidized and must be the *reducing agent*. (Glucose may be called a "reducing sugar" because of its chemical behavior in this type of reaction. In Chapter 12 we will consider the structure and reactions of sugars.)

Replacement Reactions

Another large category of reactions are the **replacement** reactions. (They may be called **displacement** reactions as well.) When at least one ion, atom, or group of atoms actually takes the place of another ion, atom, or group of atoms in a compound, a replacement reaction has occurred.

If just one replacement takes place, the reaction is called a **single replacement** reaction. Here is an example.

$$Mg(s) + 2\ HCl(aq) \rightarrow MgCl_2\ (aq) + H_2(g)$$

The magnesium has replaced the hydrogen ion from the aqueous solution of hydrogen chloride, forming magnesium chloride and releasing hydrogen gas. This single replacement reaction is also a redox reaction, with magnesium metal being oxidized and the hydrogen ion being reduced.

Nonmetals may undergo single replacement reactions too.

$$Cl_2(g) + 2\ NaBr(aq) \rightarrow 2\ NaCl(aq) + Br_2(l)$$

The nonmetal chlorine replaces the bromide ion in the compound. The free element bromine is produced and the new substance sodium chloride is formed. Once again, this is a redox reaction with the chlorine gas serving as the oxidizing agent.

If a single replacement reaction takes place, this means that the element acting as the oxidizing agent is able to successfully decrease its oxidation number as the reducing agent increases its oxidation number. The reaction would *not* take place if tried in the reverse direction. Therefore, if you try to combine bromine with sodium chloride solution, nothing will happen. Experiments show that Cl_2 is able to oxidize the bromide ion. Bromine, on the other hand, cannot oxidize the chloride ion.

Another common type of replacement reaction is a **double replacement.** This type of reaction is also referred to as a double displacement or a metathesis reaction. Instead of just one replacement, two have taken place. There is a "switching of partners" as positive ions from one reactant join with the negative ions of the other reactant. The remaining ions then form the second product. For example

$$\overline{AgNO_3}(aq) + \overline{NaCl}(aq) \rightarrow NaNO_3(aq) + AgCl(s)$$

As this reaction shows, a double replacement reaction is *not* a redox reaction, for the ions have all retained their usual charges, but just changed positions. In order for a double replacement reaction to take place, some type of product must form that removes ions from solution. In this case, the formation of a **precipitate,** an insoluble solid, has served that purpose.

In other cases of double replacement reactions, there may be a molecular substance formed, such as water. This happens, for example, if your reactants are an *acid* (containing hydrogen ion in solution) and a *base* (containing a hydroxide ion in solution). The water formed is only very slightly ionized and therefore serves to remove H^+ and OH^- from solution. Double replacement reactions involving H^+ and OH^- ions are known as a **neutralization** reactions. In any neutralization reaction, an acid and a base react to produce a salt and water.

$$HCl(aq) + NaOH(aq) \rightarrow NaCl(aq) + H_2O(l)$$

    acid   base   salt   water

Another circumstance in which ions are removed from a reaction mixture, and therefore are no longer free to react, is the case in which there is subsequent decomposition of an unstable product to form a gas. This is what happens in the following example.

$$Na_2CO_3(s) + 2\,HCl\,(aq) \rightarrow 2\,NaCl(aq) + H_2CO_3(aq)$$
$$\rightarrow 2\,NaCl(aq) + H_2O(l) + CO_2(g)$$

The last three reactions have shown us that double replacement reactions take place if at least one of the products is capable of removing ions from the reaction. The common reasons for a double replacement reaction taking place are as follows:

1. The formation of an insoluble solid.
2. The formation of a gas.
3. The formation of an only slightly ionized product.

EXAMPLE 5.3

Balance each equation. Classify each of these equations into as many different reaction types as possible.

   a. $Zn + H_2SO_4 \rightarrow ZnSO_4 + H_2$
   b. $C_3H_8 + O_2 \rightarrow CO_2 + H_2O$
   c. $FeCl_3 + NH_4OH \rightarrow Fe(OH)_3(s) + NH_4Cl$

Solution
The balancing may be done either before or after assigning the reaction types. The chemical formulas are correct and they are sufficient to determine reaction type.
   Examine each of these equations, looking for patterns of reaction type. Notice that both a and b have free elements on one side, but not the other, a strong clue that redox reactions have occurred. Check for replacements, either single or double. Since there are two reactants and products in each equation, you can eliminate either composition or decomposition as reaction types.
a. The equation is balanced. It is a single replacement reaction, the zinc replacing the hydrogen ion. It is also a redox reaction with the zinc being oxidized and the hydrogen ion in the sulfuric acid (H_2SO_4) being reduced.
b. To balance the equation, start with either the carbon atoms or the hydrogen atoms, for they each appear once on each side of the equation. The balanced equation is

$$C_3H_8 + 5\,O_2 \rightarrow 3\,CO_2 + 4\,H_2O$$

This is an organic redox reaction, with the oxygen serving as the oxidizing agent. It can also be called a combustion reaction.
c. The balanced equation is

$$FeCl_3 + 3\,NH_4OH \rightarrow Fe(OH)_3(s) + 3\,NH_4Cl$$

This is a double replacement reaction. The solid iron(III) hydroxide removes ions from solution and is the reason that the reaction takes place.

ENERGY CHANGES IN CHEMICAL REACTIONS

Now we are at the point that we can classify many types of chemical change and use balanced equations as the basis for calculations. All of the equations we have shown so far represent reactions that do actually take place, but be warned that it is perfectly possible to write beautiful equations and perform wonderful calculations that have no scientific meaning. This can happen unless we have a better way of predicting if a reaction will actually take place. One way to approach this problem is to look at the energy changes that accompany a chemical reaction.
   Energy is defined by chemists as the ability to do work. It can be thought

of as just the ability to cause change in a system. Energy exists in various forms. Common forms include the following:

Heat energy
Light energy
Sound energy
Electrical energy
Magnetic energy
Chemical energy
Food energy

Energy may be stored, in which case it is called **potential energy,** or it may be associated with motion of large or small particles, in which case it is called **kinetic energy.** There is a corresponding statement to the law of conservation of mass which is the **law of conservation of energy.** This law says that energy is neither created nor destroyed in an ordinary chemical reaction, but just transformed from one form to another. For example, our bodies are capable of taking the energy stored in a molecule of starch and converting it into the energy to hit a tennis ball, mend a broken bone, or synthesize a molecule of an important chemical such as a neurotransmitter. Remember that neither the law of conservation of energy nor the law of conservation of mass includes nuclear changes, but only chemical changes involving the electrons of an atom.

It has been observed that many, but not all, chemical reactions that do take place release energy. The energy released is most commonly in the form of heat. Heat-releasing reactions are called **exothermic** reactions. Physical processes that release heat are also termed exothermic. (The more general term is **exergonic,** meaning a change that yields some form of energy.) There appears to be a generalized drive in our universe for systems to strive toward maximum stability by reaching minimum energy content. One means through which this can be accomplished is a chemical reaction that releases heat energy. Many physical and chemical changes in which energy escapes from a system tend to occur spontaneously. Hot objects tend to give up heat to their surroundings. Explosive substances decompose with the evolution of great amounts of energy. Water flows downhill, losing potential energy. Moving objects eventually stop, losing their kinetic energy.

In the case of chemical reactions, those reactions in which the products represent a lower energy content than the reactants tend to occur spontaneously. If the reaction requires energy, forming products with greater energy content than the reactants, the reaction is termed **endergonic.** Given that the most common form of energy will be heat energy, the correct term is an **endothermic** reaction for those reactions requiring an increase in heat energy. Figure 5.1 shows schematic energy diagrams for typical exothermic and endothermic reactions.

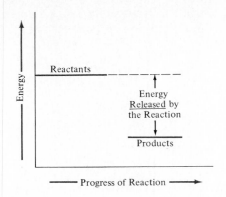

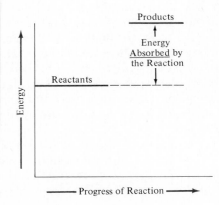

FIGURE 5.1 Energy Diagrams. (Top) In an *exothermic reaction,* heat energy is released as the products are formed. For the combustion of methane

$$CH_4(g) + O_2(g) \rightarrow CO_2(g) + 2\ H_2O(l)$$

890 kJ of heat energy are released per mole of methane burned at 298 K, 1.00 atm pressure. (Bottom) In an *endothermic* reaction, heat energy is absorbed as the products are formed. For the decomposition of limestone

$$CaCO_3(s) \rightarrow CaO(s) + CO_2(g)$$

178 kJ of heat energy are absorbed per mole of limestone (calcium carbonate) decomposed.

Notice from Figure 5.1 that we are using the SI unit of energy, the **joule** (J). This is the SI unit for all forms of energy. You may be more familiar with the **calorie** (cal), which is a metric unit of heat energy. One calorie is equivalent to 4.184 J; it is equally true that 1 J is equivalent to 0.2390 cal. A calorie is the amount of heat necessary to raise the temperature of 1 g of water by 1 degree Celsius. Heat content associated with the metabolism of foods are often quoted in the unit **Calorie,** which is a kilocalorie (kcal).

What is the difference between the two reactions in Figure 5.1? Why is the first one exothermic and the second one endothermic? This question is closely related to how the energy is stored in chemical bonds. Remember that energy can neither be created nor destroyed in a chemical reaction, only changed. In the case of the combustion of methane, the energy stored in the carbon–oxygen bonds of carbon dioxide and the hydrogen–oxygen bonds of water is *less* than the energy stored in the carbon–hydrogen bonds of methane and the oxygen–oxygen bonds of oxygen gas. The difference in stored energy is what is being released when the methane burns. A continuous output of energy results as long as the reaction has raw materials. It is an exothermic reaction.

In the case of limestone decomposing, the energy stored in the limestone is a good deal more than the energy in the calcium–oxygen bonds holding together the calcium oxide or the carbon–oxygen bonds holding together the carbon dioxide. Therefore, overall, the second reaction requires a continuous input of energy in order to proceed. It is an endothermic reaction.

Activation Energy

Still another factor must be considered in understanding why and how reactions take place. This factor is necessary in order to resolve an apparent conflict in information. For example, when hydrogen gas and oxygen gas unite, they form water and release 286 kJ (68.4 kcal) of energy per mole of water. However promising this is for predicting that the reaction should take place spontaneously, the fact is that if you bring hydrogen and oxygen gases together, nothing may happen at all without some external energy to initiate the reaction. This is because there is an **activation energy** that must be supplied. The activation energy is the minimum amount of energy that must be supplied in order for a particular chemical reaction to be initiated. It represents an energy barrier that must be climbed. In this case, the bonds within hydrogen and oxygen must be broken before they can rejoin to form water molecules. The activation energy is unrelated to the overall energy change in the reaction. The activation energy is an energy requirement that must be met to initiate either an exothermic or an endothermic reaction. Both the combustion of methane and the decomposition of limestone will have an energy barrier. (The activation energy does influence the rate of the reaction, as we will discuss shortly.) Figure 5.2 illustrates this energy relationship for an exothermic chemical reaction.

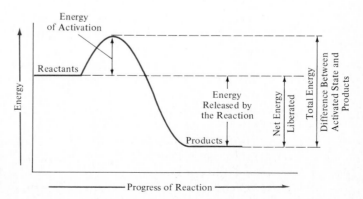

FIGURE 5.2 Activation Energy for an Exothermic Chemical Reaction.

Unless this activation energy is provided or available, no reaction will occur, no matter how great the potential heat energy that may be released. Wood just does not burn spontaneously even though it gives off a great deal of heat once it is ignited. It is often necessary to initiate a reaction even though the reaction is largely exothermic once started. Even the most powerful explosive may need a detonator to set it off. Explosives with small activation energies are obviously "touchy," whereas those with high activation energies are safer to use. Trinitrotoluene is (TNT) is an example of an explosive with high activation energy. It can be safely hammered, sawed, cut, and even burned without danger of explosion. On the other hand, nitroglycerin was too dangerous to use at all until Alfred Nobel (1833–1896) discovered a way to stabilize it in a form called dynamite. The profits from this discovery are the source of the funds for the Nobel prizes given each year. A nonexplosive formulation of nitroglycerin has been used medically for more than 100 years because it can produce rapid relaxation of the smooth muscle of blood vessels, while causing dilation of all large and small arteries of the heart. This makes nitroglycerin important for the treatment of angina pectoris, a heart disorder characterized by severe pain.

Catalysts

It is possible to influence the amount of activation energy that a reaction requires. This can be done by means of a catalyst. A **catalyst** is a substance that provides an alternative course for the reaction to proceed along, one in which the activation energy is lower. The effect this usually has is to increase the **rate of the reaction.** Catalysts are not themselves consumed in the reaction. They do not change the nature of the overall reaction nor do they change the net energy released or required overall for the reaction to proceed. A catalyst cannot make a nonspontaneous reaction suddenly spontaneous, but it can make a spontaneous reaction occur in a reasonable amount of time. A catalyzed pathway is also shown in Figure 5.3 for the same exothermic reaction shown in Figure 5.2.

For the hydrogen and oxygen reaction, a workable catalyst is finely divided platinum metal. The identity of the catalyst is often shown on or over the arrow in a chemical equation.

$$2 \ H_2(g) \ + \ O_2(g) \ \overset{Pt}{\longrightarrow} \ 2 \ H_2O(l)$$

Among the most important catalysts in biological systems are **enzymes.** These proteins act as catalysts and perform a number of important and specific functions in both plants and animals, including changing the rates of digestion and assimilation, respiration, oxidation and reduction, and numerous other chemical changes. All of these reactions could occur without the presence of catalysts but their rate would not be useful at normal body temperatures. It is no overstatement to say that we could not live if we did not have properly functioning enzymes in our bodies.

One of the first biocatalysts known was the enzyme urease. This material catalyzes the decomposition of the organic material urea, $(NH_2)_2CO$, into ammonia and carbon dioxide. You may have detected this reaction by its odor if you have ever had the occasion to smell accumulated pet or human urine. This is an example of a **hydrolysis** reaction, meaning reaction with water.

$$(NH_2)_2CO + H_2O \xrightarrow{\text{urease}} 2\ NH_3 + CO_2$$

In the presence of this enzyme, the reaction will take place about 100 trillion times faster than without the urease present! Remember that the overall energy change for the reactants to the products is not altered, but the activation energy is lowered in the presence of the enzyme, which is what changes the rate. The enzyme is specific for this reaction.

Another example of a biologically important enzyme is peroxidase, which is the enzyme that catalyzes the decomposition of peroxides. Using hydrogen peroxide, H_2O_2, as an example, water and oxygen gas are produced as described in this reaction.

$$H_2O_2(l) \xrightarrow{\text{peroxidase}} H_2O(l) + \tfrac{1}{2} O_2(g)$$

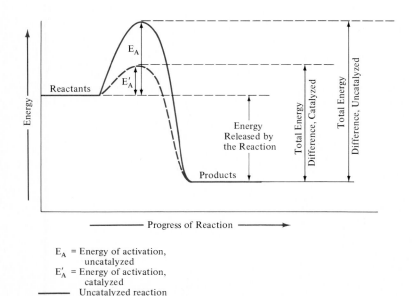

E_A = Energy of activation, uncatalyzed
E'_A = Energy of activation, catalyzed
——— Uncatalyzed reaction
– – – Catalyzed reaction

FIGURE 5.3 **The Effect of a Catalyst on Activation Energy.**

Once again, the enzyme is specific for this reaction and changes the rate dramatically, by a factor of 10 to the tenth power in this case. This is fortunate for the correct functioning of cells, because oxygen gas in its role as a good oxidizing agent is often reduced to hydrogen peroxide. (The oxygen in hydrogen peroxide has the unusual oxidation number of $1-$, not its usual combining capacity of $2-$.) If this resultant hydrogen peroxide were to remain in the cells, it would produce symptoms due to its toxicity to the cells. Instead, the H_2O_2 is quickly decomposed by the enzyme.

EXAMPLE 5.4

Use the given information to draw an energy diagram similar to Figure 5.2 and 5.3 to reflect the following information about the hydrogen peroxide decomposition reaction just discussed.

1. The activation energy for the uncatalyzed reaction is 18 kcal/mol of H_2O_2.
2. The activation energy for the catalyzed reaction is 0.5 kcal/mol of H_2O_2.
3. The overall reaction is exothermic and releases 23.5 kcal/mol of H_2O_2.

Solution

You will need to show the reactants on the left-hand side of the graph and the products at a lower energy on the right-hand side. Two pathways will be shown, the higher one being the uncatalyzed pathway and the lower one, the catalyzed pathway. Keep your diagrams roughly to scale. (See Figure 5.4)

We will see many examples of catalysts as we continue our study of chemistry and will return to the fascinating study of enzymes in Chapter 16. Not only are catalysts important biologically, they are crucial to industry as well. Most commercially important chemical reactions such as the formation

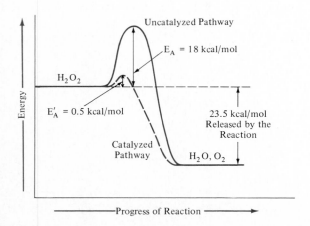

FIGURE 5.4 Energy Diagram for the Decomposition of Hydrogen Peroxide.

of high-molecular-weight plastics are carried out with the use of catalysts. The ways in which industrial catalysts work are not always clearly understood, but very detailed experimental evidence provides working knowledge of the best catalysts. Very often this information is carefully guarded by the company that has developed the process.

Entropy

We have seen that many spontaneous chemical reactions are governed by a drive for minimum energy. Products of chemical reactions strive to achieve the most stable final state. This is why so many exothermic reactions do actually take place. However, as is often the case, this cannot be the full story, for there are examples of changes that take place in which energy is absorbed overall or in which there is no energy change apparent. For example, what happens if a drop of ink is allowed to fall into a glass of water? The ink will swirl into the water and eventually spread out uniformly throughout the water. There is no heat associated with this process and yet it does happen spontaneously. Furthermore, you do not ever reasonably expect that the reverse process by which a dilute solution will separate itself into a small drop of ink and a large quantity of water will take place, unless you count running the film backwards!

We conclude that there is a second driving force associated with change, one that is evident in both physical and chemical processes. This is the natural tendency toward maximum randomness in a system, called **entropy**. This means there is a unidirectional aspect to many processes, driven by the favorable change in the system that comes about if the result is more highly disordered. Anyone who has a messy desk, garage, or tool box can relate to this natural tendency for disorder!

Many important processes in health-related chemistry are also driven by this same tendency for randomness in the system. Gases diffuse, osmosis and hemodialysis take place, and a drug becomes distributed throughout the body. All of these are cases in which delocalization means stability.

For chemical changes, those reactions in which the products represent a greater randomness *and* a lower energy content will occur spontaneously, provided that the activation energy is present. If *both* the energy of the products and their orderliness increases in relation to the reactants, then the reaction will not take place under those conditions. These statements and their meaning for chemical processes are summarized in Table 5.2.

There are, of course, cases in which the entropy changes are favorable but the energy change is not, or vice versa. In such cases, whether or not the reaction will occur depends on which factor is dominant. For example, if the energy change is favorable and not completely offset by an unfavorable entropy change, then the reaction will take place spontaneously. If, however, the unfavorable entropy change is just too large for the size of the favorable heat

Table 5.2. Energy and Entropy in Chemical Reactions

Changes	Meaning for chemical processes
Energy absorbed	The reaction is endothermic and is usually not spontaneous. Bond breaking reactions are endothermic.
Energy released	The reaction is exothermic and is often spontaneous.
	The formation of stable chemical bonds is exothermic.
Entropy increase	There is an increase in randomness in the system. It is becoming more disordered.
	An increase in randomness leads to greater chemical stability for the system.
	A reaction in which entropy increases tends to happen spontaneously.
Entropy decrease	The reaction is causing a decrease in randomness in the system, becoming more ordered.
	A decrease in entropy leads to less chemical stability.
	A reaction in which entropy decreases does not tend to happen spontaneously; energy must be supplied.

energy change, then the whole reaction will not take place. Keep in mind these potentially conflicting predictors of chemical change, energy and entropy, as we examine other reactions in the chapters to come.

SUMMARY

We have presented several major ideas about the nature of chemical change. First, we looked at ways to categorize chemical change into composition, decomposition, redox, single replacement, and double replacement reactions. These categories are not mutually exclusive because many reactions fall into more than one category.

The redox reaction is a particularly important class. We learned how to recognize the reactant serving as the oxidizing agent, which is also the substance that is itself undergoing reduction. We also saw that the reducing agent is the substance undergoing oxidation. Both inorganic and organic reactions were examined within this category.

Double replacement reactions are not redox reactions. They are ionic reactions and for a reaction in this category to be completed, something must happen to remove ions. This may be the formation of a precipitate, the formation of a gas, or the formation of an essentially molecular compound.

Energy changes are a key to understanding why chemical change takes place. Reactions take place when minimum energy corresponding to maximum stability can be achieved. Often this means the release of heat energy in the reaction. It may also mean the tendency toward maximum randomness.

Catalysts influence the activation energy for a chemical reaction. This will change the rate of a chemical change but not the outcome. Enzymes are important natural catalysts in biological systems.

KEY TERMS

Check your understanding of this chapter. Can you explain what is meant by each of these terms?

activation energy	hydrolysis
calorie	joule
Calorie	kinetic energy
catalyst	law of conservation of energy
combination	neutralization
combustion	oxidation
composition	oxidizing agent
decomposition	potential energy
displacement	precipitate
double replacement	reaction rate
endergonic	redox
endothermic	reducing agent
energy	reduction
entropy	replacement
enzyme	single replacement
exergonic	synthesis
exothermic	

STUDY QUESTIONS

1. Identify the oxidizing agent in each of these reactions.
   a. $2\ PbO\ +\ O_2 \rightarrow 2\ PbO_2$
   b. $Fe_2O_3\ +\ 3\ C \rightarrow 2\ Fe\ +\ 3\ CO$
   c. $3\ Mg\ +\ N_2 \rightarrow Mg_3N_2$
   d. $2\ NO\ +\ O_2 \rightarrow 2\ NO_2$

2. Classify each of these reactions.
   a. $Cl_2\ +\ 2\ NaI \rightarrow 2\ NaCl\ +\ I_2$
   b. $CaF_2\ +\ H_2SO_4 \rightarrow CaSO_4\ +\ 2\ HF$
   c. $FeC_2O_4 \rightarrow FeO\ +\ CO\ +\ CO_2$
   d. $2\ P\ +\ 3\ Cl_2 \rightarrow 2\ PCl_3$

3. Consult Question 17 in Chapter 4. Classify each of the reactions in that question.

4. Identify the oxidizing agent, the substance oxidized, the reducing agent, and the substance reduced in the reaction

$$C_3H_8 + 5 O_2 \rightarrow 3 CO_2 + 4 H_2O$$

5. This reaction takes place if zinc metal is placed in aqueous hydrogen chloride.

$$Zn + 2 HCl \rightarrow ZnCl_2 + H_2$$

What will happen if hydrogen gas is bubbled into a zinc chloride solution? Explain.

6. Here is the balanced equation for the reaction of hydrogen gas with nitrogen gas to produce ammonia gas.

$$N_2 + 3 H_2 \rightarrow 2 NH_3$$

(The combined hydrogen in ammonia has an oxidation number of $1+$, even though it stands second in the formula.)
   a. What is the oxidation number of nitrogen combined in ammonia?
   b. Identify the reducing agent in this reaction.

7. Balance and classify each of these reactions.
   a. $AgNO_3 + H_3PO_4 \rightarrow Ag_3 PO_4(s) + HNO_3$
   b. $AgNO_3 + Mg \rightarrow Mg(NO_3)_2 + Ag$

8. Ammonium nitrate is used as a fertilizer. It can be prepared by the reaction of ammonia gas (NH_3) with nitric acid (HNO_3, aqueous). Write a balanced chemical equation for this process and classify the type of reaction represented.

9. Marble, a decorative stone used in making statues and buildings, is composed principally of calcium carbonate. When marble is exposed to rainfall that is acidic because of the presence of sulfuric acid in polluted air, the acid rain reacts with the marble and forms carbon dioxide and water, leaving behind calcium sulfate. Write the chemical equation for this process which is responsible for the pitting and destruction of marble. Then classify the type of reaction represented.

10. Which of these changes is expected to be exothermic and which endothermic?
   Physical changes: *Chemical changes:*
   a. Water evaporating   **c.** Wood burning
   b. Water freezing   **d.** Iron rusting

11. Which of the same changes listed in Question 10 do you expect will involve an increase in entropy? Explain.

12. When sulfur dioxide combines with oxygen gas, sulfur trioxide gas forms; 23.5 kcal of heat energy will be released per mole of sulfur trioxide formed. Write a balanced equation for the process, classify the reaction, and discuss whether or not you expect this reaction to take place spontaneously.

13. When silver chloride solid decomposes, silver metal and chlorine gas are the products. When silver iodide solid decomposes, silver metal and iodine solid are the products. Which process would you expect has a greater entropy change? Explain.

14. When ice melts, the reaction is endothermic but it is still spontaneous. How do you account for this observation, given that most endothermic reactions are not spontaneous?

15. Is an endergonic reaction always an endothermic reaction? Explain your answer.

16. The chemical energy from foods is essential for many functions within the body, including muscular activity, maintaining body temperature, and building new structural materials. The average fuel value of fats is about 9 kcal/g. How many kilojoules per gram is this?

17. Cheddar cheese has a fuel value of approximately 20 kJ/g. How many kcal of food energy will you gain by eating 250 g of cheddar cheese?

18. If you require 3000 kcal of energy a day and plan to gain this food energy exclusively from fudge (not recommended!) that has a fuel value of 4.1 kcal/g, how many pounds of fudge would you have to eat per day?

19. Pepsin is a natural enzyme in gastric juice that accelerates the otherwise very slow conversion of proteins in our food to amino acids. This is a useful mechanism for our bodies. Sketch a diagram similar to that in Figure 5.3, showing the effect of pepsin on the activation energy.

20. Chymotrypsin is an enzyme that catalyzes the hydrolysis of proteins in the small intestine. What would be the problem if this enzyme were not functioning properly?

6 | Gases

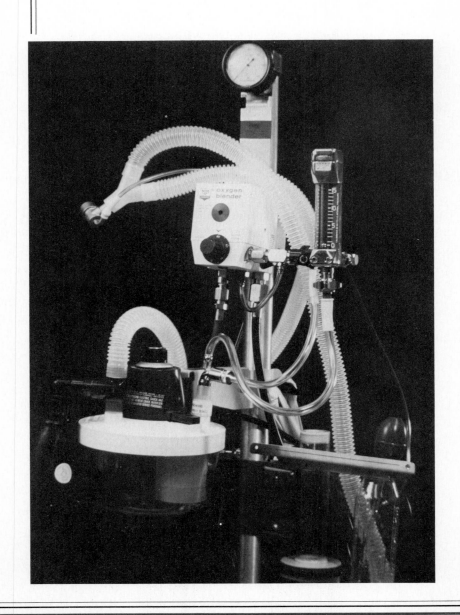

Our figure shows an apparatus used in hospitals to insure constant positive airway pressure (CPAP) in patients whose lungs are not inflating effectively. Newly born babies are frequently treated with this device.

A respiratory therapist has an expert knowledge of gases and how to use them in therapy and diagnosis. She or he knows that the volume of a gas is governed by the temperature and pressure under which the gas is held and that the pressure on a gas regulates how much of it will dissolve in a fluid to which the gas is exposed.

A knowledge of gases and the laws that govern their behavior is essential for the health professional because gases are administered to patients in the hospital and home setting. They are also utilized in athletic endeavors and as propellants in many commercial products. In this chapter we shall study the kinetic molecular theory of gases, and the laws that define behavior and clinical uses of gases. (Photo courtesy of Mercy Hospital, San Diego, Calif.)

INTRODUCTION

Gases may not be as obvious to us in our surroundings as solids and liquids. Our dependency on the properties of gases, however, continues until our last breath. We become aware of gases if we happen to be exposed to one with a characteristic odor, color, or health effect. The quality of the air we breathe is a subject of increasing concern. Our daily activities include transportation by automobiles that ride on air-filled tires. Our living spaces are lighted by various devices ranging from incandescent bulbs filled with argon to mercury vapor arc lamps, neon signs, and fluorescent lights, all of which use gases. Beverages are carbonated with pressurized carbon dioxide. Natural gas or bottled gas may heat our homes or cook our foods. Compressed gases propel sprays of deodorant, room fresheners, perfume, shaving cream, and a myriad of other products. This "invisible" state of matter is actually a very important part of our daily experience. Gases dissolved in fluids, such as oxygen in blood, play an essential role in the proper functioning of our bodies. Gases are indispensible for many types of therapy important to the health professional.

In this chapter we will look at some of the qualitative and quantitative laws that describe the behavior of gases. We will examine how the volume, temperature, pressure, and number of moles of a gas are related. We will learn how to solve problems based on these relationships. Mixtures of gases will also be studied. We will end this chapter by describing some of the applications of gases in health-related situations.

THE KINETIC MOLECULAR THEORY

The remarkable thing about gases is that they resemble each other in physical behavior much more closely than is the case for either solids or liquids. Gases show many more similarities than dissimilarities in their physical behavior. This is why so much of the early experimentation in chemistry was performed on gases, and through their use, explanations of many of our present laws of chemistry were developed.

The similarity in behavior can be rationalized by the assumptions of the **kinetic molecular theory** as it applies to gases. This theory says that gases are made up of very small particles that are always in motion within large areas of empty space. These moving particles possess considerable kinetic energy, the energy associated with motion. Their average kinetic energy is directly proportional to the Kelvin temperature of the gas. If the temperature is higher, the particles will move faster on the average and have a greater average kinetic energy. If the temperature drops, the particles will move slower on the average and have less average kinetic energy. In fact, if the particles lose all of their thermal energy, then the temperature would drop to **absolute zero,** a value never achieved but closely approached experimentally. This temperature is $-273.15°C$ and is defined to be zero on the Kelvin temperature scale.

The particles of a gas are assumed to collide continually with each other but not to lose energy in the process. When the particles collide with the walls of their container, the effect is what is called the pressure exerted by the gas. **Pressure** is defined as the force exerted per unit area. Mathematically, it is stated as

$$\text{pressure} = \frac{\text{force}}{\text{area}}$$

If there are more collisions on the wall of the container per unit time, then the pressure increases. If the size of the container increases for the same number of moles of gas at the same average kinetic energy, then the particles will hit the walls less frequently and the result will be a decreased pressure. If the size of the container were to decrease under the same conditions of constant temperature and number of moles of gas, then the pressure would be expected to increase as the frequency of the collisions with the walls increased. Gases can be easily compressed, compared with solids or liquids, just because there is so much empty space between the particles.

Because of their constant motion, the particles of a gas will quickly fill any container and escape from it, given the opportunity. If you have ever experienced a flat tire, you know the natural tendency of gases to move from higher pressure to lower pressure, leaving you the problem of reversing this tendency. Gases and liquids are both classified as **fluids,** sharing the characteristic that they will assume the shape of their container and will flow, not separate, when under pressure.

Boyle's Law

In the preceding discussion, four factors were used to describe the gas—temperature, pressure, volume, and number of moles of gas present. If any three of these conditions are specified, the fourth one can be calculated for any gas. You do not need to know chemically what the gas is for this process, just that it *is* a gas. It is often convenient to separate out just two factors at a time and hold the other two factors constant. This allows you to design an experiment to test the influence of one factor on the other.

This was the type of experiment performed originally by Robert Boyle (1627–1691). He was the first to describe the effect of pressure on the volume of a gas. This relationship is now known as **Boyle's law.** It states that the volume of a given sample of gas at a constant temperature varies inversely with its pressure. That is, the higher its pressure, the smaller its volume. Conversely, if a gas is allowed to expand in volume, its pressure will decrease. Notice that in order for this simple relationship to hold, you must not change the number of moles of gas or the temperature.

Boyle's law is illustrated by the proper functioning of many common objects, including the behavior of air in a bicycle pump. (See Figure 6.1.) In a bicycle pump, you push down on the piston in the barrel of the pump to compress the air in the barrel to a smaller volume. What happens when the volume is decreased? Boyle's law tells that the pressure increases and this higher air pressure forces air into the tire.

Filling a medical syringe also illustrates Boyle's law. When the plunger is drawn back from its starting position, the increase in volume inside the chamber of the syringe results in decreased pressure inside the chamber. The liquid, being pushed by atmospheric pressure, flows into the syringe. Since the principal content of the syringe is now noncompressible liquid, the liquid can be expelled by pushing on the plunger.

Another example of Boyle's law in the area of medicine is the use of mechanical ventilation for patients who have lost voluntary control of breathing. The principle of using pressure to change gas volume is also illustrated by the procedure of introducing air under pressure into the pneumothorax, bringing about the intentional collapse of a lung. This may be carried out as a treatment for pulmonary tuberculosis.

Mathematical Statement of Boyle's Law

Once we understand the inverse relationship between volume and pressure, it is convenient to express Boyle's law in mathematical form that can be used for calculation. The simplest statement of the inverse relationship between volume and pressure is this:

$$V \propto \frac{1}{P} \qquad T \text{ and } n \text{ (number of moles) both being constant}$$

Pumping air into a tire is an application of Boyle's law. A bicycle pump adds air to a tire without changing the volume of the tire itself very much. The pump allows you to compress the air in the cylinder of the pump and then forces the air into the tire.

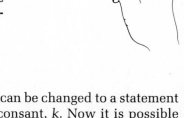

Giving an injection with a syringe puts Boyle's law into action. As the plunger is drawn out of the syringe, the pressure drops inside the syringe because of the increase in volume inside the chamber of the syringe. This results in liquid coming into the syringe. To reverse the process, the plunger is pushed back into the chamber, ejecting the liquid.

FIGURE 6.1   Applications of Boyle's Law.

This is a statement of inverse proportionality and can be changed to a statement of equality by the insertion of a proportionality consant, k. Now it is possible to state Boyle's law as follows:

$$V = k \cdot \frac{1}{P} \qquad T \text{ and } n \text{ being constant}$$

Using the rules of algebra, you can multiple both sides of this last equation by P and obtain still another mathematical statement of Boyle's law.

$$P \cdot V = k \qquad T \text{ and } n \text{ being constant}$$

It may be easiest for you to see the inverse nature of this relationship in this last form of the equation. In order for the constant product to be maintained, the volume must go down if the pressure goes up. Study the values in Figure 6.2 to understand this relationship. As long as the temperature and the number of moles of gas remain constant, the product of the pressure and volume will equal the same numerical constant, k.

If the same sample of a gas is to be examined at two different pressures or volumes, assuming that the temperature and the number of moles of gas are constant, then Boyle's law can be used to make this mathematical statement. If

$$P_1 \cdot V_1 = k \qquad \text{and} \qquad P_2 \cdot V_2 = k$$

then

$$P_1 \cdot V_1 = P_2 \cdot V_2$$

The only requirement about the use of units here is that you should use the same units for comparing two different conditions of volume and pressure. Gas volumes are typically measured in liters or in related subunits, such as milliliters or microliters. The cubic centimeter or the cubic meter are also used.

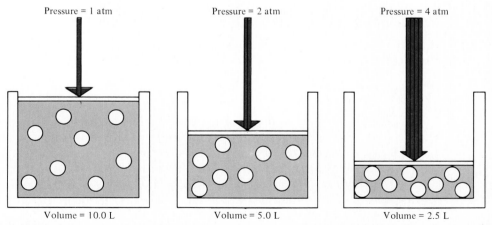

FIGURE 6.2   **Pressure and Volume Relationships. A fixed sample of a dry gas is kept at constant temperature. The pressure is varied by means of a piston. The change in volume with the change in pressure is observed.**

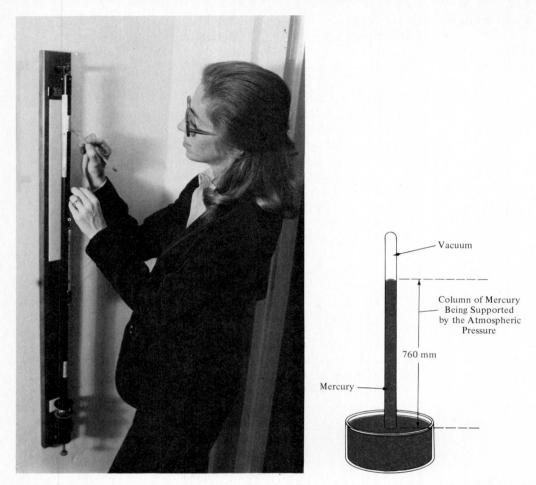

FIGURE 6.3   The Mercury Barometer. Notice that the "normal" atmospheric pressure can support a column of mercury 760 mm high. This is called a pressure of 760 mm Hg or 760 torr.

Gas pressures are measured in atmospheres, millimeters of mercury (mm Hg), torr, pascals and other related units. There are actually quite large differences in the sizes of these units, although they are all in use for specific purposes. You are probably most familiar with an **atmosphere.** This is the unit used to describe the pressure exerted by a column of air above the earth's surface from sea level to the top of the atmosphere. Pressure is also commonly measured in height units, because the earth's column of air can support a column of mercury to a height of 760 mm in a barometer. Figure 6.3 shows this relationship.

Table 6.1.   Units of Pressure for Gases

1.00 atm	= 14.7 lb per square inch (psi)
	= 760 mm Hg
	= 760 torr
	= 29.9 in. of mercury
	= 1.01×10^5 Pa
1.00 bar	= 1.00×10^5 Pa
	= 100 kPa

Table 6.1 shows how the different pressure units are related to 1 atm of pressure. Note the extremely small size of the SI unit, the **pascal (Pa).** There are approximately 10,000 Pa per atmosphere of pressure. The **bar** is the unit recently designated as the international unit of standard pressure. Notice that it represents almost the same pressure as the atmosphere. Most pressure measurements in clinical settings are made in millimeters of mercury, which is the same as the **torr.** Very often the unit is assumed in clinical work and you will not see any unit reported. For example, a blood pressure of 120/80 really means that the maximum pressure of the blood in the arteries is 120 mm Hg greater than atmospheric and the minimum pressure when the heart is relaxed between beats is 80 mm Hg greater than atmospheric, even though the units are not specifically stated.

EXAMPLE 6.1

If an inflated balloon with a volume of 2.5 L at a pressure of 760 mm Hg is taken to a higher elevation where the pressure is only 650 mm Hg, what will be the new observed volume of the balloon? Assume that the temperature does not change and that the balloon is perfectly elastic.

Solution
First make sure that Boyle's law applies. The temperature is constant and it is assumed that the balloon is not leaking so the number of moles of confined gas must be constant. Since the pressure of the atmosphere is decreasing, the volume of the balloon is expected to increase according to Boyle's law. Use that common sense check for your answer, no matter which method you choose to solve the problem.

Method I

$$2.5 \text{ L} \times \frac{760 \text{ mm Hg}}{650 \text{ mm Hg}} = 2.9 \text{ L}$$

Notice that the pressure factor is a fraction more than 1. If you wish to increase the 2.5-L value, as you need to do in this problem, you must put the larger value of pressure in the top of the fraction used.

Method II
Use the algebraic statement that

$$P_1 \cdot V_1 = P_2 \cdot V_2$$

Substituting the values

$$(760 \text{ mm Hg})(2.5 \text{ L}) = (650 \text{ mm Hg})(V_2)$$

Solving algebraically for V_2

$$\frac{(760 \text{ mm Hg})(2.5 \text{ L})}{(650 \text{ mm Hg})} = 2.9 \text{ L}$$

Notice that this final statement is exactly what we achieved by using Method I. The final volume is more than the initial volume, as we predicted.

We recommend Method I because it does not require memorizing a formula. It only depends on knowing the principle that volume and pressure are inversely related. Method I also lends itself to more complex problems in which a number of conversion factors might be used. You may want to try both methods on any given problem until you become more familiar with the gas laws.

Charles' Law

Gases also respond to changes in temperature. When air is heated during the day, it expands. As a result of the expansion, the density of the air decreases and it tends to rise, producing convection currents in the air. On a smaller scale, a gas-filled balloon placed in the sun will warm and the gas inside will expand. Because the rubber balloon is elastic and can readily stretch, the volume of the balloon increases, restoring the pressure inside the balloon to its original value.

The law relating the volume of a gas to its temperature was discovered nearly 100 years after Boyle's law. It is called Charles' law after the French scientist Jacques Charles (1746–1823). **Charles' law** states that the volume of a gas varies directly with the Kelvin temperature, assuming that the pressure and the number of moles of confined gas remain constant. This means that if the temperature increases, so does the volume, and if the temperature decreases, so does the volume. The effect of the change in temperature is to change the average kinetic energy of the gas particles.

Charles' law can also be expressed in mathematical forms. This expression shows the direct variation of Charles' law.

$$V \propto T \quad P \text{ and } n \text{ being constant}$$

We can also rewrite this same proportionality as we did with Boyle's law, changing it into an equality statement with the insertion of a constant.

$$V = k' \cdot T \qquad P \text{ and } n \text{ being constant}$$

This time we have called the constant k' (read k prime) to distinguish it from the constant used with Boyle's law. You may symbolize this constant with any letter you choose, just as long as you realize it is a constant and a different constant from that used in Boyle's law.

It is also true that for the same sample of a gas examined at different volumes and Kelvin temperatures,

$$\frac{V}{T} = k' \qquad \text{and that} \qquad \frac{V_1}{T_1} = \frac{V_2}{T_2}$$

Notice, in particular, how the form of the direct proportion expressed in Charles' law differs from the form of the inverse proportion of Boyle's law. In this case it is the *ratio* of the volume to the Kelvin temperature that is constant, not the *product* of two factors. Study Figure 6.4 to understand this type of proportionality more fully.

EXAMPLE 6.2

A balloon full of air has a volume of 2.5 L at 25°C. It is placed in the sun and the air warms it to 65°C. Given that the balloon can expand to 2.7 L before it will burst, will this balloon survive its day in the sun?

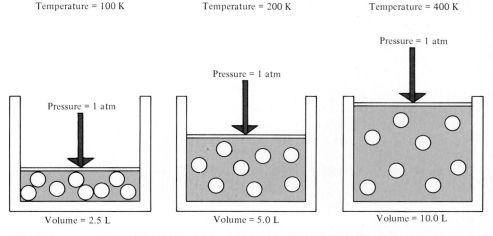

Temperature = 100 K   Temperature = 200 K   Temperature = 400 K

Pressure = 1 atm

Pressure = 1 atm

Pressure = 1 atm

Volume = 2.5 L   Volume = 5.0 L   Volume = 10.0 L

FIGURE 6.4 Temperature and Volume Relationships. A fixed sample of a dry gas is kept at a constant pressure of 1 atm by means of a piston. The change in volume with the change in temperature is observed.

Solution

First make sure that Charles' law applies. It is assumed that as the temperature rises, the volume of the balloon will increase, keeping the pressure inside the balloon constant. It is assumed that the number of moles of confined gas remains constant. If the new volume at the higher temperature is calculated to be greater than 2.7 L, the balloon will break. Remember that the volume is directly proportional to the *Kelvin* temperature, not the Celsius temperature, so you will first need to change the two temperature readings to Kelvin.

Method I

$$25°C + 273 = 298 \text{ K}$$
$$65°C + 273 = 338 \text{ K}$$
$$(2.5 \text{ L})\frac{(338 \text{ K})}{(298 \text{ K})} = 2.8 \text{ L}$$

The new volume is predicted to be 2.8 L, larger than the maximum volume of the balloon. It will break.

Note in this method that the temperature fraction is set up with the larger value in the top of the fraction because you need to increase the volume of air in the balloon as the temperature increases.

Method II

$$\frac{V_1}{T_1} = \frac{V_2}{T_2} \quad \text{so} \quad \frac{(2.5 \text{ L})}{(298 \text{ K})} = \frac{(V_2)}{(338 \text{ K})}$$

Solving for V_2,

$$V_2 = \frac{(2.5 \text{ L})(338 \text{ K})}{(298 \text{ K})} = 2.8 \text{ L}$$

Naturally, the same answer and the same result for the balloon apply. Again notice that Method I is more direct and you do not need to worry about remembering a specific algebraic formula, only the general statement of Charles' law that tells you the volume must be increasing if the temperature is increasing.

Avogadro's Law

Avogadro's law is a logical outcome of the discussion in Chapter 4 dealing with Avogadro's number. If 1 mol of any substance contains the same number of particles, then 1 mol of any gaseous substance contains the same number of particles as well. **Avogadro's law** states that the volume of the gas is directly proportional to the number of moles of gas present, given constant pressure and temperature. If there is an increase in the number of moles of gas present, there will also be an increase in the volume.

Mathematically, Avogadro's law can be stated in these equivalent expressions.

$$V \propto n \qquad P \text{ and } T \text{ being constant}$$

$$\frac{V}{n} = k'' \qquad k'' \text{ is a constant, different from previous constants}$$

$$\frac{V_1}{n_1} = \frac{V_2}{n_2}$$

A useful number to remember is called the **molar volume.** It is the volume that 1 mol of any gas will occupy under standard conditions of 1.00 atm of pressure and a temperature of 273 K. Numerically, the molar volume is equal to 22.4 L. **Standard conditions** of temperature and pressure (abbreviated STP) may be defined for different applications. For the gas laws, the values stated here are those commonly used in the United States.

Gay-Lussac's Law

The last law in this series of four governing the physical behavior of gases is again a direct relationship. In the event that a gas were heated in a rigid container that could not expand, such as in an aerosol can, the effect of the temperature change would be to increase the pressure of the gas rather than its volume. If the increase in pressure exceeds the strength of the container, the result might well be an explosion or blowout. This is why you are warned on the label never to put an "empty" aerosol can into a fire. (See Figure 6.5.) The directly proportional increase in pressure with an increase in (Kelvin) temperature is called **Gay-Lussac's law.** Joseph Louis Gay-Lussac (1778–1850) was a French scientist who like Charles, worked with hot air balloons to learn about the relationship of volume change to temperature.

$$P \propto T \qquad V \text{ and } n \text{ being constant}$$

$$\frac{P}{T} = k''' \qquad k''' \text{ is a constant, different from previous constants}$$

$$\frac{P_1}{T_1} = \frac{P_2}{T_2}$$

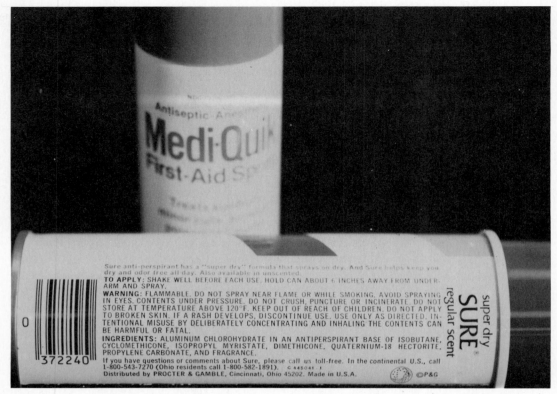

FIGURE 6.5   **Practical Applications of Gay-Lussac's Law. Aerosol products are required to be labeled concerning the hazards of heating or puncturing any container with pressurized gas inside.**

Gay-Lussac's law has direct meaning for an emphysema patient who keeps a cylinder of oxygen gas in the home for emergency use. The cylinder must be stored in a cool place because external heat could raise the internal pressure of the oxygen gas to a dangerously high level.

The gas laws attributed to Boyle, Charles, Avogadro, and Gay-Lussac can be used together. The combination of gas laws can be carried out by deriving a new formula to express the direct and inverse relationships involved. More simply, the combination can be accomplished by applying a sequence of conversion factors that describe the changes taking place. The next example illustrates one possible combination of gas laws.

■ EXAMPLE 6.3 ■

A gaseous sample is known to contain 0.33 mol of pure oxygen and has a volume of 3.57 L at a temperature of 298 K. What will be the volume of the gas if the number of moles of gas is doubled but the temperature drops to 280 K?

Solution
First, identify the gas laws that apply to the question. Since the number of moles is doubling, the volume will also increase by the same factor since Avogadro's law says volume and number of moles of gas are directly proportional. The pressure is not changing, so you will not use Boyle's law. The volume is changing, so you will not use Gay-Lussac's law in which the volume is assumed constant. Because the Kelvin temperature is decreasing, according to Charles' law, the volume will also decrease. Overall the increase due to the number of moles is probably going to be much larger than the decrease due to the temperature drop.

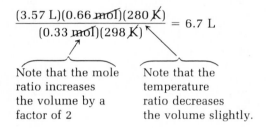

$$\frac{(3.57\ \text{L})(0.66\ \text{mol})(280\ \text{K})}{(0.33\ \text{mol})(298\ \text{K})} = 6.7\ \text{L}$$

Note that the mole ratio increases the volume by a factor of 2

Note that the temperature ratio decreases the volume slightly.

When more than one factor is involved, multiplying by sequential conversion factors is far easier and less prone to error than doing the problem by formulas in separate steps.

The Ideal Gas Law

We have now seen the four important measurements used to describe a gas: its temperature, pressure, volume, and number of moles. If the gas is following all of the relationships expressed in Boyle's, Charles', Avogadro's, and Gay-Lussac's laws, then it is said to be an *ideal gas*. Such a gas would have no attractions between particles and each collision between particles would leave each gas molecule with the same amount of energy that it had before collision. An ideal gas is also assumed to remain in the gaseous state at temperatures all the way down to absolute zero. This behavior is not very typical of gases encountered in the real world. Rather, real gases will sooner or later condense into liquids, which, upon further cooling, will eventually change into solids.

Even though the perfect ideal gas does not exist, all gases behave almost ideally as long as the temperature and pressure conditions are not too extreme. If the gas is placed under very high pressure or low temperatures, or both, the particles in the gas will come closer together and the attractions between particles become important. Conversely, gases often exhibit near-ideal behavior under low-pressure and high-temperature conditions, for then the particles of a gas are relatively far apart. Despite the fact that real gases deviate somewhat from ideal behavior, it is still very useful to combine all of the relationships developed so far into an **ideal gas law.** All of these relationships can be combined simply by placing all of the mathematical statements together in one summarizing statement. Therefore, if

$$V \propto \frac{1}{P} \quad \text{at constant } T, n \quad \text{(Boyle's law), and}$$

$$V \propto T \quad \text{at constant } P, n \quad \text{(Charles' law) and}$$

$$V \propto n \quad \text{at constant } T, P \quad \text{(Avogadro's law), and}$$

$$P \propto T \quad \text{at constant } V, n \quad \text{(Gay-Lussac's law), then}$$

$$V \propto \frac{nT}{P}$$

As before, this statement of proportionality can be changed to an equality statement by the use of an appropriate constant, usually called R rather than k in this case. The relationship becomes

$$V = \frac{RnT}{P}$$

or, cross multiplying, the usual form of the ideal gas law is

$$PV = nRT$$

The constant R is called the ideal gas law constant and has a numerical value that depends on the units used to measure the variables in the equation. For example, if the pressure is measured in atmospheres and the volume of the gas in liters, the value of R is

$$R = 0.0821 \frac{\text{liter} \cdot \text{atmospheres}}{\text{mole} \cdot \text{Kelvin}}$$

If the pressure is measured in millimeters of mercury or torr, the value of R is

$$R = 62.4 \frac{\text{liter} \cdot \text{millimeters of mercury}}{\text{mole} \cdot \text{Kelvin}}$$

The usefulness of the ideal gas law is that you can find any one of the variables if you know the rest of the information about the conditions of the gas.

▬▬▬▬▬▬**EXAMPLE 6.4** ▬▬▬▬▬▬▬▬▬▬

Chlorine gas is often stored in a container under pressure for use as a disinfectant for drinking water. What is the pressure (in atmospheres) of 3.25 mol of chlorine gas if it is confined in a 10.4 L container at 28.0°C?

Solution

Three of the four variables that describe a gas are given, therefore, the fourth can be calculated from the ideal gas law. Since the pressure is asked for in atmospheres, the appropriate value for R is 0.0821 L · atm/mole · K. The temperature must be changed to Kelvin, as is true for all the gas laws. The pressure of the gas is expected to be considerably more than 1 atm because the molar volume of a gas at standard conditions is 22.4 L. In this problem we have more than a mole of gas confined in a relatively small container at a temperature greater than standard, so the pressure should be greater than 1 atm.

$$PV = nRT$$

$$(P)(10.4\,\cancel{L}) = (3.25\,\cancel{mol}) \frac{(0.0821\,\cancel{L} \cdot atm)}{\cancel{mol} \cdot \cancel{K}} (301\,\cancel{K})$$

$$P = 7.72\ atm$$

Dalton's Law of Partial Pressures

Many of the gases important to health studies are not single substances but rather mixtures of gases. If we can once again assume that the gases act nearly ideally, then all interactions between particles of different types of gases can be ignored. In such cases, the gases behave independently, even if mixed together. John Dalton, whose ideas we discussed in Chapter 2, noted this independence for the pressure of gases mixed together. Dalton found that the total pressure of a gas mixture is just the sum of the individual pressures of each component gas. The individual pressures are called the **partial pressures.** This statement is called **Dalton's law** and can be stated symbolically in this way:

$$P_T = p_1 + p_2 + p_3 + \ldots$$

P_T is the total pressure of a gas mixture, and p_1, p_2, p_3, . . . , are the partial pressures of components 1, 2, 3, and so on.

The air we breathe is a mixture. The atmospheric pressure is the sum of the partial pressures of the gaseous components present in air. Table 6.2 shows the composition of air by percentage volume, as well as by partial pressure.

The atmosphere also contains smaller amounts of neon, helium, methane, krypton, hydrogen, dinitrogen oxide, xenon, ozone, sufur dioxide, nitrogen

Table 6.2. The Composition of Air[a]

Component	Volume (%)	Partial pressure (mm Hg)
Nitrogen	78.084	593.4
Oxygen	20.948	159.2
Argon	0.934	7.1
Carbon dioxide	0.031	0.2
Trace gases	0.003	0.02
	100.000	759.9(2)

[a] Dry air at 1.00 atm pressure.

dioxide, ammonia, and carbon monoxide. Notice that the sum of the percentages of the gases given in Table 6.2 is 99.997 percent and so the rest of the gases can usually be ignored for all but the closest work. Water vapor is also a significant but variable component of the atmosphere; it is not included in Table 6.2.

The partial pressures of the gases in the atmosphere are not the same as the partial pressures of the gases once they enter the body. From Table 6.2 you can see that the partial pressure of oxygen in the atmosphere is 159.2 mm Hg. When you inhale, you deliver oxygen to the air sacs (the alveoli) in your

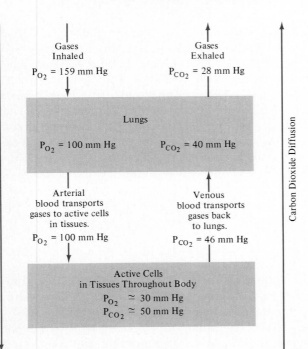

FIGURE 6.6 Transport of Oxygen and Carbon Dioxide in the Body.

lungs. At this point, the partial pressure of oxygen is about 100 mm Hg. The arterial blood now picks up oxygen and delivers it, both in solution and in combination with hemoglobin, to all the body tissues that must utilize oxygen. When the oxygenated blood reaches the tissues, oxygen is released because the partial pressure of oxygen is only about 30 mm Hg in the tissues.

As a result of the processes of metabolism, carbon dioxide gas is produced. From a partial pressure of approximately 50 mm Hg in active cells throughout the body, the carbon dioxide in solution and in chemical combination with hemoglobin travels through the venous blood back to the lungs. The partial pressure of the carbon dioxide in the venous blood is about 46 mm Hg and falls to 40 mm Hg in the lungs. Because the external atmosphere contains carbon dioxide only to a partial pressure of 0.2 mm Hg, we exhale the carbon dioxide. The partial pressure of exhaled carbon dioxide is about 28 mm Hg.

Both oxygen and carbon dioxide show that gases will flow from a region of higher partial pressure to a region of lower partial pressure. **Diffusion** is the name given to this process in which gas molecules spontaneously spread throughout another substance. Figure 6.6 summarizes the diffusion of the two gases in the body.

Henry's Law

The solubility of a gas in a solution is greatly affected by pressure. This fact is clear anytime you open a soft drink or a champagne bottle. By exposing the contents to atmospheric pressure, the carbon dioxide gas, which has been dissolved under higher pressure, comes bubbling out of solution. A similar effect with changing pressure is not observed for solids or liquids dissolved in solution. Their solubility is not so sensitive to pressure changes.

William Henry (1775–1836) was an English chemist who made this generalization quantitative by observing that the solubility of a gas is directly proportional to the partial pressure of the gas above the solution. In symbols, this is represented as

$$c_g \propto p_g$$

where c_g = solubility of the gas in the given solution
p_g = partial pressure of the gas above the solution.

Using a constant of proportionality as we have done before, the statement becomes the following:

$$c_g = k \cdot p_g$$

The units of the constant k will depend on those chosen for concentration and pressure. It is specific for each gas.

EXAMPLE 6.5

The solubility of nitrogen gas in water at 25°C and a partial pressure of 583.4 mm Hg, its normal partial pressure in the atmosphere, is 5.3×10^{-3} mol/L. What will happen to the solubility of nitrogen gas if the partial pressure of the nitrogen is tripled?

Solution
Henry's law shows a direct proportion between the partial pressure of the gas above the solution and its solubility in that solution. Therefore, as the partial pressure is increased, so is the solubility. The solubility should triple.

$$\frac{5.3 \times 10^{-3} \text{ mol}}{\text{liter}} \times \frac{3(593.4 \text{ mm Hg})}{(593.4 \text{ mm Hg})} = \frac{1.6 \times 10^{-2} \text{ mol}}{\text{liter}}$$

In practical terms for human health, **Henry's law** becomes important if a person experiences a change in pressure from normal atmospheric pressure. Astronauts or deep-sea divers are examples of types of persons who must cope with the realities of gas laws. Divers use compressed air but it may be quite different in composition from the air we breathe normally. The compressed air may contain as little as 2 percent oxygen, but at the increased pressures the diver experiences this is enough to maintain a near "normal" oxygen partial pressure of 159 mm Hg so that the correct pressure of oxygen exists in the body. Remember that the solubility of a gas in fluids increases with pressure. A lower percentage of oxygen in the compressed air is able to produce the same result as breathing a higher percentage of oxygen at 1 atm of pressure.

The rest of the compressed air is usually helium rather than nitrogen. Helium has a much lower solubility in blood than does nitrogen. Use of normal air under high pressures results in too much nitrogen being forced into the body's biological fluids. Oxygen is consumed by various processes in the body, but the inert nitrogen remains and can only be eliminated by the respiratory system. If the diver then is suddenly brought back to atmospheric pressure, the nitrogen will come out of solution, as Henry's law predicts. Gas bubbles form and are trapped in the bloodstream. This causes the "bends" or decompression sickness. Slow decompression, often in a high-pressure chamber as described in the next section, is usually the only treatment.

CLINICAL USES OF GASES FOR THERAPY

Various techniques are often used as a means to increase the partial pressure of oxygen available to a person experiencing difficulty in breathing. The oxygen is provided by means of a nasal tube, tent, incubator, or enclosed chamber that may actually assist in the mechanics of the contraction and relaxation of

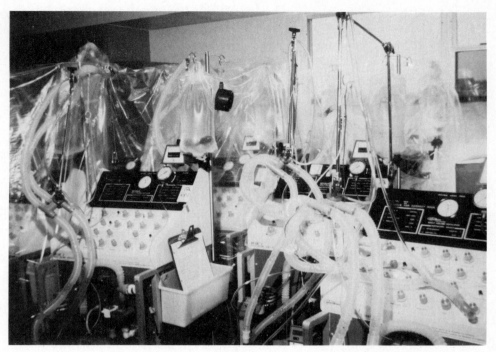

FIGURE 6.7 Respiratory Therapy. **This equipment is used by a respiratory therapist to ventilate patients who have lost voluntary control of breathing. For example, these ventilators are often moved to the post-operative area for use. The respiratory therapist must have an intimate knowledge of the gas laws to make use of these machines.**

the diaphram. An inhalation therapist is trained to administer gases and operate respirators. Figure 6.7 shows one type of apparatus in use in hospitals today.

Hyperbaric (pressures greater than normal) oxygen therapy refers to the administration of 100 percent oxygen at up to three times sea level atmospheric pressure. It must be carried out in a heavy walled chamber in which it is possible to closely control the pressure. This procedure is also called hyperbaric oxygenation (HBO). The use of HBO has been shown to be effective in situations where it is necessary to infuse high concentrations of oxygen. By increasing the partial pressure of the oxygen administered, the concentration of oxygen in body fluids is increased, an application of Henry's law. HBO is a primary treatment for acute carbon monoxide poisoning, decompression sickness in divers and aviators, gangrene, smoke inhalation, cyanide poisoning, and exceptional blood loss anemia.

HBO can be used for the treatment of air embolism, a gas bubble formed and trapped in the bloodstream. At the increased pressures in a hyperbaric chamber, the air embolism becomes smaller and has a better chance of dissolving. This means that the embolism, rather than acting like an obstruction or a clot, can move more easily through major blood vessels. If the bubble then diffuses into smaller blood vessels, this makes reabsorption of the gas even more likely. If there is no way for additional air to enter the bloodstream, the embolism is effectively eliminated and the pressure can be returned to normal.

There are a number of other applications that have been tested in research settings that may also be valuable in specific cases. One interesting use is in cancer therapy. Highly oxygenated cells are destroyed more readily by radiation. A promising use of this fact is to combine HBO with radiation therapy. The same end result can also be achieved without HBO by administering molecules that protect normal oxygen-rich cells or by the use of chemical radiation sensitizers that increase the interaction of radiation with the target cells.

Oxygen used in HBO is a drug and, as with any drug, there are safe and effective ways to use it. Even oxygen can be toxic when administered improperly, particularly if the patient is also taking certain medications that will lower the threshold of seizures. Prolonged breathing of oxygen can result in coughing, sore throat, and chest pains. If the partial pressure of oxygen is too high over a period of time, the carbon dioxide formed will not be removed effectively from the body, leading to carbon dioxide poisoning.

CASE HISTORY—ARTIFICIAL BLOOD

"Artificial blood" is a synthetic preparation able to carry oxygen to the tissues and then surrender the vital oxygen to those tissues, carrying away the carbon dioxide. This is the function that is normally carried out by hemoglobin in the blood, and its ability to reversibly bind oxygen has been difficult to duplicate in a synthetic material. Recently an experimental material called Fluosol-DA has been tested. It is a mixture of organic compounds called **perfluorocarbons** which contain ring structures of carbon with the element fluorine attached. The main component of the mixture is called perfluorodecalin, and its structure is shown here.

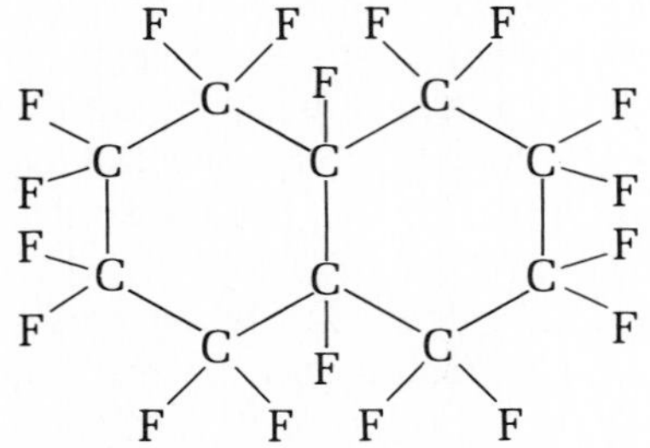

Perfluorocarbons can transport and release oxygen. Their ability to carry oxygen can be enhanced by increasing the partial pressure of oxygen, such as in the hyperbaric chamber described previously. There are also small amounts of other fluorine organic compounds, inorganic salts, and other ingredients forming this mixture. It is an emulsion, not a true solution. In the emulsion, the rather insoluble perfluorocarbons are dispersed evenly as fine particles that can be carried in the blood without causing clotting.

Animal studies with Fluosol date back to 1966, but the first human uses were reported in 1979. For example, a patient suffering from a massive bleeding duodenal ulcer needed surgery but refused to undergo this procedure for religious reasons because of the need for blood transfusions. After receiving emergency approval from the Food and Drug Administration, Fluosol-DA was administered, starting 8 hours before surgery. The total dose given was 20 mL of the artificial blood for each kilogram of body weight; a total of 1500 mL was administered. Following this infusion, the surgery was performed successfully and, at the time of discharge, the patient had a normal hemoglobin reading and no adverse secondary effects were noted.

You should realize that the term *artificial blood* is really quite an exaggeration. These new materials fulfill only *one* of the crucial functions of blood and cannot assume any of its other complex functions. Still, most blood transfusions are given for the express purpose of replacing lost blood volume and its vital hemoglobin, so in that sense, Fluosol-DA can be very effective as an emergency alternative to blood in the case of shortages or if there is difficulty in matching blood types. The Fluosol compounds can also be used as the priming fluid for heart-lung machines and as an acceptable alternative for those who need to receive blood but have restrictive religious principles that prevent the use of whole blood.

SUMMARY

This chapter has dealt with the laws explaining the behavior of gases. It also has shown some practical examples of how those laws are related to health. The kinetic-molecular theory is a set of generalized assumptions that helps us to rationalize many of the physical laws of gases. Gas particles are assumed to be constantly in motion, creating pressure as they hit the container walls. Boyle's law shows us there is an indirect relationship between volume and pressure of a gas. Charles' law relates volume to the Kelvin temperature; this is a direct relationship. Avogadro's law is also a direct relationship, this time between the number of moles and the volume of a gas. There is a direct relationship between the Kelvin temperature of a gas and its pressure, a statement of Gay-Lussac's law.

The ideal gas law summarizes the four laws into one statement.

$$PV = nRT$$

This relationship allows you to calculate any one of the variables that describe a gas, as long as you know the values of the other three variables and the gas law constant, R.

Dalton's law of partial pressures tells us that the summation of individual pressures will equal the total pressure of a mixture of gases. Henry's law directly relates the partial pressure of a gas above a solution to the solubility of the gas in the solution. In practical terms for human health, Henry's law governs the transport and concentrations of gases in the body.

Gases have many clinical applications, including respiratory therapy and hyperbaric oxygen therapy. Carrying oxygen in the body by means of artificial blood is a current subject of research.

KEY TERMS

Check your understanding of this chapter. Can you explain what is meant by each of these terms?

absolute zero	hyperbaric
atmosphere	ideal gas law
Avogadro's law	kinetic molecular theory
bar	molar volume
Boyle's law	partial pressure
Charles' law	pascal
Dalton's law	perfluorocarbons
diffusion	pressure
fluid	standard conditions
Gay-Lussac's law	torr
Henry's law	

STUDY QUESTIONS

1. Use the kinetic molecular theory to explain why gases are easily compressed.

2. You have probably seen a beam of strong light going through a room containing dust or smoke. Explain how this observation could be used to illustrate the kinetic molecular theory.

3. What is the theoretical meaning of the temperature "absolute zero"?

4. Explain using the kinetic molecular theory why there is pressure on the walls of a container holding pressurized gas.

5. For each of these cases, tell what gas law or laws are represented by the situation.
   a. A balloon, first filled with hot air, is allowed to cool. It shrinks in volume.
   b. Too much air is pumped into a bicycle tire and it blows out.

c. An aerosol can is thrown into a bonfire and it explodes, sending dangerous bits of metal in all directions.

d. A partially inflated rubber raft is driven from sea level over the mountains on the way to a nearby river for use; the raft is observed to be fully inflated at the top of the mountain pass.

e. A fully inflated rubber raft is placed into cold river water; the raft no longer appears to be fully inflated.

f. A diver comes up too quickly from a deep sea dive and develops the "bends."

g. A climber carries a bottle of champagne to the top of a 14,000-ft peak but when she opens the bottle, all of the contents gush out of the bottle and are lost.

6. If the volume of a confined sample of gas is to expand, assuming that the temperature remains constant, will the pressure increase or decrease? Explain your answer.

7. At what pressure, if any, will the volume of a real gas at a constant temperature of 25°C become zero? Explain your answer.

8. How is it possible for a person to drink liquid through a straw? Draw a diagram and use it to help explain the process.

9. Use the conversion factors listed in Table 6.1 to answer each of these questions.
   a. Express a pressure of 2.45 atm in millimeters of mercury.
   b. Is a pressure of 1.00×10^6 Pa greater or less than a pressure of 780 mm Hg?
   c. If a gas cylinder is designed to withstand a maximum pressure of 55 psi, will it be able to hold a gas at a pressure of 100 kPa?
   d. If you are told that your blood pressure is 125/82, how would you report this same blood pressure in atmospheres?

10. What is the actual range of fluid pressure, expressed in millimeters of mercury, if a blood pressure is reported as 130/82?

11. If the pressure on a confined volume of 25 L of a gas is increased from 760 to 1500 mm Hg, what will happen to the volume of the gas?

12. If the Kelvin temperature and the pressure of a 500 ml sample of gas are both doubled, what effect will this have on the volume of the gas?

13. When a confined gas sample with a volume of 5.0 L at 1.0 atm pressure is subject to 15 atm of pressure, what happens to the volume?

14. What new volume is occupied by 500. mL of a gas if it is heated at constant pressure so that the temperature changes from 70°C to 170°C?

15. Eighteen liters of a gas at 25°C are cooled until the volume is only 9 L.

What is the new temperature of the gas, assuming that the pressure remains constant?

16. If a balloon full of air has a volume of 2 L at 25°C, what will happen to its volume if it is heated in the sun to 80°C?

17. On a warm day in San Diego, the temperature was reported as 93°F. On the same day in Tijuana, Mexico, the temperature was given as 38°C. Which city was warmer on that day?

18. Which of these measurements represents the highest temperature: 50 K, 50°F, or 50°C?

19. If a gas exhibits a pressure of 500 mm Hg and occupies a volume of 50 mL at 20°C, what volume will it have at 50°C and a pressure of 700 mm Hg?

20. Express "normal body temperature" of 98.6°F in degrees Celsius and in Kelvin.

21. Your patient has a temperature of 34°C. Assuming the patient is a human, should you be concerned by this temperature?

22. Assume that oxygen gas is being stored under high pressure. Under the right conditions, some oxygen gas will combine to form ozone, as shown in this equation.

$$3\ O_2(g) \rightarrow 2\ O_3(g)$$

oxygen   ozone

If this reaction takes place in the stored oxygen gas, how will the total pressure in the stored container be changed?

23. What is the total pressure in a container if it holds nitrogen gas at a pressure of 27 mm Hg, oxygen gas at a pressure of 300 mm Hg, and hydrogen gas at a pressure of 65 mm Hg?

24. A dentist may use an anesthetic mixture composed of 20 percent by volume dinitrogen oxide and 80 percent by volume oxygen gas. If the total pressure of the gas mixture is 2.5 atm, what are the partial pressures of each gas?

25. Which gas is the major component of the atmosphere? Will it be present in our exhaled breath to the same percentage? Explain.

26. What types of situations can lead to placing a patient in a hyperbaric chamber? What are the advantages of such a chamber?

27. Is the partial pressure of oxygen gas higher in the lungs or in active cells of body tissues? Use the value in Figure 6.6 to help explain.

28. Why is the partial pressure of carbon dioxide higher in our lungs than in the atmosphere? Explain.

29. Why and how does the composition of compressed air used by divers differ from the composition of air at sea level?

30. **a.** Why is it not safe for divers to use tanks of normal air under high pressure?
    b. Why do divers perfer to use a mixture of helium-oxygen rather than nitrogen-oxygen?

31. Helium has a concentration in blood of 3.7×10^{-4} mol/L at 25°C and a partial pressure of He equal to 1 atm. Would the concentration of the helium increase or decrease if the partial pressure of helium were to decrease? Explain.

32. Calculate the solubility of nitrogen gas at 25°C and a partial pressure of 760 mm Hg. Use the solubility data given in Example 6.5.

33. What is "artificial blood"? Can it replace all the normal functions of whole blood?

34. Find out what is meant by the "Heimlich maneuver." What principles of gas law behavior are illustrated by this procedure?

7

Chemistry and Mathematics of Solutions

INTRODUCTION

No topic in this book touches our lives more than the topic of solutions. First of all, we are walking solutions ourselves. We adults are about 50 to 60 percent water by weight, and dissolved in that water are many, diverse chemicals, ranging from trace elements such as copper and zinc to large molecules such as glycogen and insulin. Blood is a solution of many chemicals, as is urine. Even lacrimal fluid (tears) is a solution consisting mostly of salt in water.

Second, the health professional works with solutions almost constantly. Medicines are administered as syrups (sugar in water solution), elixirs (alcohol solution), hypodermic injections, eye, ear, and nose drops, and retention enemas. A solution of a drug in water may be applied to the skin for germicidal, analgesic (pain-relieving), or cooling effects. The nurse is often called upon to administer a solution of a drug by intravenous methods (Figure 7.1) or sometimes to make a dilution of an existing solution. The professional may be in the position to judge the quality or potency of a solution that has stood on the shelf for a long period.

Whenever the chemist tests for the presence of an enzyme in the blood, or when the lay person uses one of the do-it-yourself test kits that detects HCG (a hormone of pregnancy), solutions of specific chemicals are being used.

141

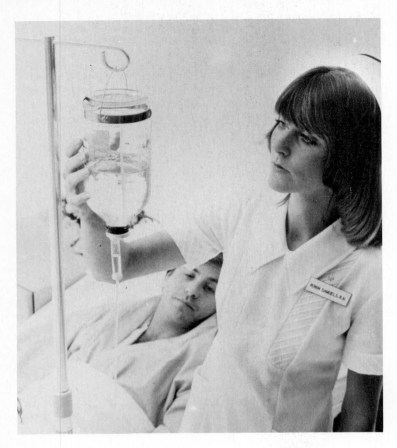

FIGURE 7.1   Drugs, such as dextrose, can be administered as a 5 percent solution in water. (Photo courtesy Travenol Corporation.)

Anesthetics administered by inhalation are often solutions of one gas in another.

In this chapter we shall explain what is meant by a solution and how solutions differ from suspensions and colloids. We shall examine solvents commonly used in biomedical work, types of solutions, and the units used to express strength of solutions. We shall spend much time studying the mathematics of solutions, especially the mathematics involved in making dilutions (an area that demands knowing what you are doing). Our discussions will focus on the beginning level and will center on practice problems, with answers worked out in the text. We urge you to work all the practice problems given at the end of the chapter—especially those that deal with the mathematics of solutions.

In Appendix III, you will find a glossary of solution terminology in which we describe thirty-three different kinds of solutions the health professional can be expected to encounter.

DEFINITION OF A SOLUTION

In its broadest sense, a **solution** is the dispersal of one substance within another, to yield a homogeneous, single-phase preparation. The substance that is dispersed is dispersed as atoms, ions, or molecules.

According to this broad definition, alloys such as brass are solid solutions of one metal dissolved in another, and air is a solution of various gases dispersed in nitrogen and in each other. Whatever the example, in a true solution the species that dissolves becomes invisible in the solution. We can not tell it is there by looking at it (except, of course, if it has color).

In a practical sense, almost all of the solutions you will encounter will be dispersals of a chemical (drug, germicide, electrolyte) into a liquid. And the liquid most often encountered is water.

The chemical that dissolves is called **solute.** The fluid that holds the solute in solution is the **solvent.** Usually the solvent is present in the greater amount. Solutes that ionize in solution are often termed **electrolytes.**

EXAMPLE 7.1

In Figure 7.1, identify the solute and the solvent.

Solution
Dextrose is the solute and distilled water is the solvent.

In a true solution, the dissolved substance is invisible (unless it has color), does not settle out, and cannot be filtered out. One method to recover a dissolved substance from a liquid solution is to evaporate the solvent; if the solute is nonvolatile it will remain as a residue. This is how sea salt is obtained.

Although water is the most common solvent in biomedical work, we also sometimes encounter ethyl alcohol, glycerin, various oils, and organic solvents such as ethyl acetate, acetone, and chloroform. The last three solvents mentioned are not used internally but may be encountered in products for external use.

In Syrup USP[1] the solute is sucrose and the solvent is water. Peanut oil and cottonseed oil, solvents that are official in the USP, are used to dissolve compounds such as estradiol (an estrogen) and menadione (an antihemorrhaging factor).

[1] USP stands for *United States Pharmacopoeia*, a book that lists official descriptions, properties, and tests for widely used drugs and preparations. See your library.

POLAR SOLVENTS

Water and alcohol are polar solvents. A polar solvent consists of molecules that are **dipoles**—that is, having positively and negatively charged ends (poles). Alcohol and water are dipoles because the oxygen they contain is more electronegative than the hydrogen atom attached to it (Figure 7.2). The more electronegative oxygen attracts the bonding pair of electrons more closely to itself. This unequal sharing results in the charge separation shown in Figure 7.2. Other examples of polar solvents are glycerin, ethylene glycol, chloroform, and rubbing alcohol. Examples of low or zero polarity solvents are ether, carbon tetrachloride, and hydrocarbons such as butane and benzene. Table 7.1 shows the polarity of some common solvents in terms of debye units. The larger the debye value, the more polar the solvent.

There is an axiom in chemistry: like dissolves like. This means that polar solvents tend to dissolve polar substances and nonpolar solvents tend to dissolve nonpolar substances. Water, a polar solvent, dissolves ionic compounds such as NaCl, KNO_3, morphine sulfate, sodium phenobarbital, and sodium benzoate. Water also dissolves polar covalent compounds such as sucrose, vitamin C, the B complex vitamins, acetic acid, and phenol.

Table 7.1   Polarity of Common Solvent Molecules in the Gas Phase. (The greater the debye value the more polar the solvent.)

Solvent	Debyes
Dimethylsulfoxide (DMSO)[a]	3.9
Acetone	2.88
Ethylene glycol	2.28
l-chloro-1,1-difluoroethane	2.14
Water	1.85
Ethyl acetate	1.78
Acetic acid	1.74
Methyl alcohol	1.70
Ethyl alcohol	1.69
Isopropyl alcohol	1.66
Dichloromethane	1.60
Ethyl ether	1.15
Chloroform	1.01
Toluene	0.36
Benzene	0
Carbon tetrachloride	0

[a] Measured in liquid state.

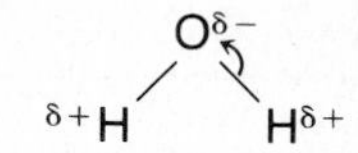

$$O^{\delta-}$$
$$\delta+H \qquad H^{\delta+}$$

charge separation in water

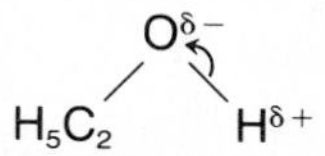

FIGURE 7.2 Water and ethyl alcohol are dipolar sub-stances. The more electronegative oxygen atom pulls the bonding pair of electrons closer to itself, creating a positive end and a negative end to the molecule.

$$O^{\delta-}$$
$$H_5C_2 \qquad H^{\delta+}$$

charge separation in ethyl alcohol

A low-polarity solvent such as cottonseed oil dissolves nonionic, low-polarity compounds such as vitamins A, D, E, and K, petrolatum, triglycerides, DDT, and other fats and oils.

▰▰▰▰▰ EXAMPLE 7.2 ▰▰▰▰▰

The oft-mentioned blood–brain barrier consists in part of a sheath of fat surrounding capillaries in the brain and spinal cord. Thus, for solutes in the blood to get to the brain, they must be soluble in the fatty sheath. Barbiturates can easily penetrate the blood–brain barrier but penicillin cannot. What does this tell us about the polarity of barbiturates and of penicillin?

Solution
Barbiturates have relatively low polarity and penicillin has relatively high polarity.

Ethyl alcohol is a unique, versatile, and very important solvent for industrial and medicinal purposes. Ethyl alcohol will be discussed more fully in Chapter 10.

MISCIBILITY

In Chemistry the word **miscibility** is used to indicate the ability of two liquids to mix with each other or to dissolve each other. Water and acetone are miscible, as are water and ethyl alcohol. Water and chloroform are **immiscible**—that is, if shaken together and allowed to stand, they will separate into two layers.

In general, polar solvents are miscible with other polar solvents but low-polarity solvents do not mix with polar solvents. Table 7.2 lists miscible and immiscible pairs.

For the health professional, it is important to know which liquids are miscible and which are not in order to avoid possible physical incompatibilities. For example, attempting to add an aqueous (water) solution to cottonseed oil

Table 7.2. Miscibility or Immiscibility of Some Solvent Combinations[a]

	Water	Ethyl alcohol	Acetone	Chloroform	Isopropyl alcohol	CCl₄	Ethyl ether
Water	—	M	M	I	M	I	I
Ethyl alcohol	M	—	M	M	M	M	M
Acetone	M	M	—	M	M	M	M
Chloroform	I	M	M	—	M	M	M
Isopropyl alcohol	M	M	M	M	—	M	M
CCl₄	I	M	M	M	M	—	M
Ethyl ether	I	M	M	M	M	M	—
Ethylene glycol	M	M	M	I	M	I	I

Key to abbreviations:
[a] M = miscible
I = immiscible

would result in the separation of two phases or layers, with the less dense cottonseed oil floating on the water. If this were a medication for a patient and he took a dose from the top, he would get only the oil and any drug it might contain.

■■■■■■ **EXAMPLE 7.3** ■■■■■■

Seeking a special solvent combination, a chemist mixes acetone with carbon tetrachloride. Are these two liquids miscible?

Solution
In Table 7.2, we find the line marked acetone, follow along to the column marked CCl₄, and read "M" for miscible.

HOW WATER ACTS AS A SOLVENT

Potassium chloride (KCl) is important in medicine because it is used in the therapy of potassium-deficient states. Potassium ion, K^+, is the principal intracellular cation. Its presence in the correct concentration is essential for proper nerve conduction.

When solid KCl is placed into water, the following events take place. The crystals of KCl (in a latticelike arrangement) are attacked by the polar water molecules, giving potassium ions and chloride ions. Energy is required to disintegrate the crystalline lattice, and this energy comes in part from the

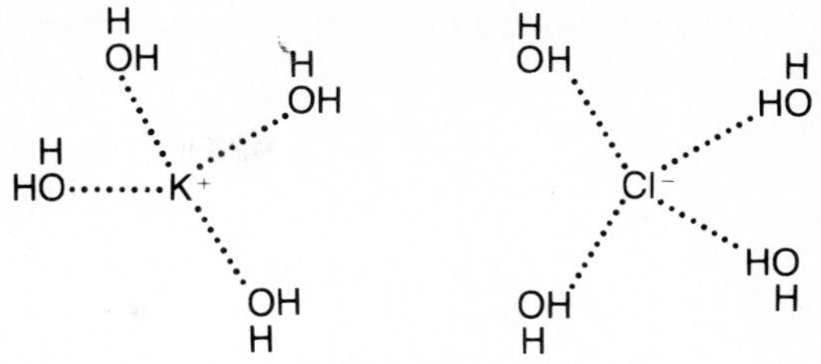

FIGURE 7.3 Solvation of K^+ and Cl^- ions aids in the solubilization of KCl. The dotted lines represent weak but significant ion-dipole attractive forces. (*Left*) A K^+ ion, freed from its crystal, and solvated by water molecules. (*Right*) A chloride ion, freed from its crystal, and solvated by water molecules.

energy gained when the freed K^+ and Cl^- ions are surrounded by, and bound to, polar water molecules. This process is termed *solvation* (Figure 7.3). Solvation represents ion–dipole bond formation between the ions and the water molecules. Ion–dipole bonds such as those depicted in Figure 7.3 are individually weak (a few kilocalories per mole) but collectively quite significant since there are so many of them.

In summary, water acts as a solvent by virtue of its ability to solvate solute ions, thus breaking down the crystal lattice. Water cannot dissolve compounds like paraffin because no ions exist in solid paraffin and no solvation can occur.

There are many rules for predicting the solubility of compounds in water. Table 7.3 is a brief listing of rules for some of the more important classes of salts.

■ EXAMPLE 7.4 ■

Gold compounds such as gold chloride ($AuCl_3$) have been used in the treatment of arthritis; ferrous sulfate is given to patients with iron-deficiency anemia; and barium sulfate is used internally as a radio-opaque contrast

Table 7.3. Water Solubility of Classes of Salts

Ion involved	Name	Solubility rule
$C_2H_3O_2^-$	Acetate	Generally, acetate salts are soluble.
NO_3^-	Nitrate	Generally, nitrate salts are soluble.
Cl^-	Chloride	Generally, chlorides are soluble; exceptions are AgCl, $PbCl_2$, and Hg_2Cl_2.
SO_4^{2-}	Sulfate	Most sulfates are soluble or slightly soluble; exceptions are $PbSO_4$, $BaSO_4$, and $SrSO_4$.
CO_3^{2-}	Carbonate	Most carbonates are insoluble; exceptions are Na_2CO_3, K_2CO_3, and $(NH_4)_2CO_3$.
S^{2-}	Sulfide	Most sulfides are insoluble; exceptions are $(NH_4)_2S$ and the alkali metal and alkaline earth sulfides.
PO_4^{3-}	Phosphate	Generally, phosphates are insoluble; exceptions are Na_3PO_4, K_3PO_4 and $(NH_4)_3PO_4$.
Na^+, K^+, Li^+, NH_4^+		Ammonium salts and all salts of the alkali metals are soluble.

medium in x-ray work. Predict the solubility of each of these compounds in water.

Solution

Predictions based on Table 7.3 turn out to be correct. All are soluble except $BaSO_4$, which fortunately is *not* soluble and is *not* absorbed from gastrointestinal (GI) tract.

Solutions of gases in liquids are commonplace. Rather unexpectedly, less gas dissolves in a liquid when the liquid is hot than when it is cold. This explains the likeliness of fish kills to occur in the hot summer months when lakes and streams can hold less dissolved oxygen. It also explains why open containers of carbonated beverages quickly go flat when allowed to stand at room temperature.

Our blood contains dissolved quantities of all the gases in the air we breathe. This means our blood will normally contain dissolved quantities of nitrogen, oxygen, carbon dioxide, the inert gases, and other gases found in clean air. It also means, for the smoker and those near the smoker, that dissolved amounts of carbon monoxide will be in the blood. The user of an aerosol hair spray is getting a dose of the spray by absorption from the air into the bloodstream (via the lungs).

EXAMPLE 7.5

Normally, we breathe sixteen times a minute, but in certain disease states or in panic, we may hyperventilate. Explain the body's extra gain of oxygen and extra loss of carbon dioxide during hyperventilation.

Solution

Air is about 21 percent by volume oxygen. Very rapid breathing acts to provide fresh supplies of air, with its 21 percent of oxygen, to the alveoli, thus increasing availability of O_2 for absorption into the bloodstream. Conversely, rapid breathing acts to remove exhaled CO_2 from the alveoli, facilitating further removal of CO_2, which has been dissolved in the blood.

SATURATED AND UNSATURATED SOLUTIONS

Health professionals sometimes work with solutions that are **saturated,** that is, holding in solution all of the solute that they can at the given temperature. An example is 5 percent boric acid solution. This solution at room temperature can hold no more solute. Attempts to dissolve additional boric acid in it will be futile.

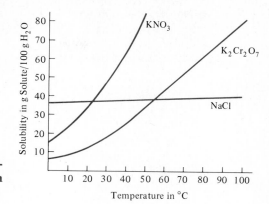

FIGURE 7.4 **The Influence of Temperature on Solubility of Three Common Salts.**

On the other hand, there are many situations in scientific work in which a solution is prepared that could hold more solute than is actually present. Consider 1 percent silver nitrate solution (instilled into newborn babies' eyes as a prophylaxis against gonorrheal infection). A 1 percent $AgNO_3$ solution is unsaturated: one could add more solid $AgNO_3$ and it would dissolve. In fact, the maximum solubility of $AgNO_3$ is 122 g per 100 mL of cold water. The solubility of dextrose (glucose) is a remarkable 1.1 g in 1.0 mL of water at 25°C. Thus, the solution referred to in Figure 7.1 is unsaturated.

Temperature is the critical factor in determining how much of a solute will dissolve in a given quantity of solvent. Usually, but not always, an increase in the temperature results in an increase in solubility of a solute, but the effect is variable (Figure 7.4). Calcium butyrate is one of those rare salts that is actually more soluble in cold than in hot water. The solubility of calcium sulfate dihydrate $(CaSO_4 \cdot 2H_2O)$ increases with rising temperature to a maximum at 40°C and then decreases.

EXAMPLE 7.6

According to Figure 7.4, the solubility of NaCl in water at 20°C is 36 g in 100 g water.

   a. How many grams of NaCl will dissolve in 1 g of water at 20°C?
   b. Would a solution of 3.1 g of NaCl in 10 ml of water at 20°C be saturated or unsaturated?

Solution

   a. $\dfrac{36 \text{ g NaCl}}{100 \text{ g } H_2O} \times 1.0 \text{ g } H_2O = 0.36 \text{ g NaCl}$

   b. $\dfrac{3.1 \text{ g NaCl}}{10 \text{ mL } H_2O} \times 100 \text{ mL } H_2O = 31 \text{ g NaCl}$

This quantity, the amount of NaCl dissolved in 100 mL of water, is less than the 36 g maximum that 100 mL of water can dissolve. Hence the solution is unsaturated.

Under certain unique conditions, some solutions can temporarily hold more solute than the solubility rules ordinarily permit. Such solutions are said to be *supersaturated*. The condition is a very unstable one, for if a seed crystal is added or if the inside of the container is scratched or even agitated, immediate precipitation of the excess solute occurs, leaving an ordinary saturated solution.

The concept of a saturated solution is important to the lay person who is predisposed to kidney stones (renal calculi). Occasionally, the concentration of calcium oxalate, calcium phosphate, or other salt dissolved in the urine may reach the saturation point and a solid mass may precipitate out of solution. These masses can be large enough to block the tubes carrying urine, and the resulting great pain may require surgical intervention. Interestingly, in medieval times, traveling "stone removers" would perform such operations in public to attract new patients (who often died from postoperative infection due to lack of sterile technique!).

DISTINCTION BETWEEN SOLUTIONS, COLLOIDS, AND SUSPENSIONS

We learned earlier that in a true solution, the substance that is dissolved is invisible. We cannot see that it is there (unless it has color). This is not the case, however, with **colloids** and **suspensions.** In the latter two cases, we are dealing with "solute" particles large enough to be physically noticeable. Thus, particle size becomes the determining factor in distinguishing between solutions, colloids, and suspensions.

In true solutions, solute particle size is approximately 1 nanometer (nm) (10 Angstroms [Å]) or less. This corresponds to the size of a molecule or ion. In colloidal dispersions, particle size is approximately 1 to 100 nm (10 to 1000 Å). If particle size is greater than 100 nm we are dealing with a suspension, and the particles are large enough to be visible to the eye, with or without a microscope. Just as a point of review, 1 nm = 1×10^{-9} m. One Å = 1×10^{-8} cm. Thus, there are 10 Å in each nanometer.

We see many examples of colloids in daily life: fog, smoke, dust in the air, volcanic ash, aerosol sprays, lead aerosols from auto exhausts, and foods such as whipped cream, mayonnaise (Figure 7.5), and gelatins. Colloidal sulfur, though no longer official in the USP, is a stable aqueous colloidal preparation of elemental sulfur. The large surface area of sulfur in the colloidal state helps impart high activity as a fungicide and in the treatment of various dermatological (skin) conditions such as acne.

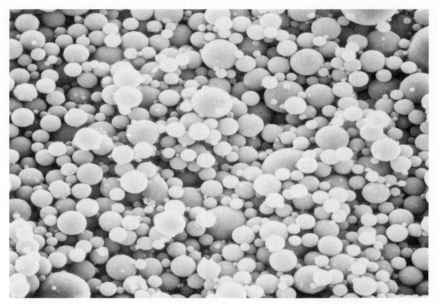

FIGURE 7.5 **Scanning Electron Micrograph of Colloidal-Sized Globules in Mayonnaise.** (Courtesy Kraft, Inc., Glenview, IL.)

Since the size of colloidal particles is very small (10–1000 Å), the particles do not settle out on standing and cannot be removed by filtration. Colloidal particles do not pass through a dialysis membrane. To help visualize a colloidal dispersion, we can apply the *Tyndall effect*, in which a light beam is scattered by reflection from the surface of the colloidal particles (Figure 7.6). True solutions do not give the Tyndall effect, but suspensions and colloids do.

Biochemically, we encounter colloidal systems in the blood where plasma proteins such as the globulins and albumin are colloidal in size. The proteins of milk serve as protective colloids to calcium phosphate aggregates suspended in the milk. The colloidal nature of bile salts and bile protein helps keep sparingly soluble cholesterol in suspension in the blood. Fat, after ingestion and absorption, is transported in the blood and lymph systems as colloidal-sized structures called micelles (of 0.4- to 10-nm diameters). There are many other types of colloids in our body; they play a fundamental role in the structure and function of body protoplasm.

The extremely small size of colloidal particles translates into a tremendous surface area. Physiologically, this is very important because chemical and physical events take place at surfaces. An example is the very rapid uptake of oxygen in red blood cells, facilitated by the great surface area of the proteins present.

In suspensions, particle size is large enough so that settling can occur and the particles can be removed by filtration. Although suspensions can be irksome

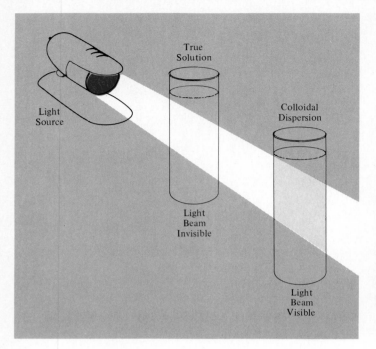

FIGURE 7.6 The Tyndall Effect. Presence of colloidal particles is proven by passing a light beam through a colloidal dispersion. Viewed from the side, the beam becomes visible owing to reflection from the colloidal particles. In passing through a true solution, the light beam is not scattered and remains invisible.

in some aspects of life (muddy water), they can be put to good use in medical therapy. For example, a sterile aqueous suspension of cortisone acetate is available for intramuscular injection and a sterile aqueous ophthalmic suspension of hydrocortisone is used to treat inflammations of the eye. Sterile epinephrine oil suspension USP uses a vegetable oil as the suspending medium. Antacid preparations of the magnesium oxide–aluminum oxide type are suspensions that must be shaken before being dispensed.

EXAMPLE 7.7

Sterile aqueous suspensions of drugs may be safely injected intramuscularly but must *never* be injected intravenously. Why not?

Solution
Suspensions are not solutions. Suspensions contain particulate matter that can lodge in internal organs, causing damage or irritation *if the suspension is injected directly into the bloodstream*. Drug addicts who mainline suspensions of heroin cut with talc and cornstarch suffer from talc and cornstarch emboli (clots) in their lungs and eyes.

Table 7.4. A Summary of the Properties of Solutions, Colloids and Suspensions

Property	True solution	Colloid	Suspension
Particle size	1 nm or less	1–100 nm	>100 nm
Tyndall effect	Not observed	Observed	Observed
Removable by filtration?	No	No	Yes
Particles will settle out	No	No	Yes
Particles will pass through dialysing membrane	Yes	No	No
Osmotic effect*	High	Weak	None

* Discussed later in this chapter.

Certain chemicals can be added to suspensions to slow the process of settling out. These chemicals, termed *suspending agents*, are often gums or mucilages that swell and thicken in the presence of water, thus increasing the viscosity of the solvent. Suspended particles do not settle out as quickly in a thick (highly viscous) solvent.

The characteristics of solutions, colloids, and suspensions are summarized in Table 7.4.

UNITS FOR EXPRESSING STRENGTHS OF SOLUTIONS

One of the most important things we can learn about solutions is how to express their strengths and how to use knowledge of the strengths of solutions in calculations of doses or dilutions.

The most important unit for strengths of solutions used in biomedical work is the percent weight to volume (% w/v) (and its variations, milligram percent (mg %) and "parts per"). Also encountered are percent volume to volume (% v/v), milliequivalents per liter, molarity, and osmolarity. We will not discuss percent weight to weight (% w/w) here.

Percent Weight to Volume

Percent weight to volume (% w/v) is generally used to express the strength of dilute solutions (10% or less) used in the health field. Solutions of drugs for internal and external use, intravenous and intramuscular injectables, and reagent solutions for testing are usually labeled in percent weight to volume units. For higher strength solutions of the 20, 30, and 40% levels, percent weight to weight is used. Percent volume to volume is used when one is working with solutions of liquids in liquids, for example 70% ethyl alcohol in water.

Percent weight to volume indicates the number of grams of solute dissolved in each 100 mL of solution. Thus a 1% w/v solution of any solute will contain 1 g of solute in each 100 mL of solution; a 5% w/v solution will contain 5 g in each 100 mL; and a 0.90% w/v solution will contain 0.90 g in each 100 mL. Figure 7.7 shows a label of 3% hydrogen peroxide solution. Even if the label did not indicate the units of strength used, one would take this to be percent weight to volume since it is a typical *dilute* drug solution. Checking in the latest USP, we find that, indeed, 3 percent H_2O_2 officially contains 3 g of H_2O_2 in each 100 mL of solution.

The actual preparation of a percent weight to volume solution deserves comment. To prepare 100 mL of a 3% aqueous w/v solution, we would weigh 3 g of the solute, dissolve it in some water, and then with mixing add enough water to make exactly 100 mL. Note that one does not *add* 100 mL of water, but enough to make 100 mL final volume, since the dissolved solute will occupy some volume in the final solution. Note also that in our example the final 100 mL of solution will not necessarily weigh 100 g.

Equation 7.1 can be used to solve all percent weight to volume problems in this book.

$$\frac{\% \text{ w/v}}{100} = \frac{\text{g solute}}{\text{mL of solution}} \tag{7.1}$$

Equation 7.1 tells us that for dilute solutions

$$\frac{\text{parts per hundred (\%)}}{100 \text{ parts}} = \frac{\text{grams of solute}}{\text{the volume of solution}}$$

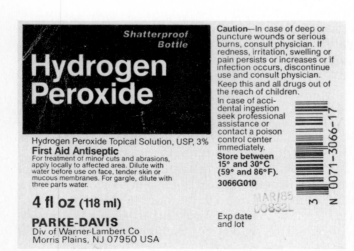

FIGURE 7.7 **The strength of all** *dilute* **solutions, such as 3 percent** H_2O_2**, is given in percent weight per volume units.**

Sometimes, in applying Equation 7.1, we will know the percent weight to volume and the volume of solution, and will calculate the grams of solute. Other times, we will know the grams and volume and will calculate the percentage. Here are some worked out examples.

EXAMPLE 7.8

What is the proper method of preparing 100 mL of a 1.75% w/v aqueous solution?

Solution
Since this is a percent weight to volume problem, use Equation 7.1. Check to see which of the three factors we know and which is unknown. We know the final volume (100 mL). We know the percentage (1.75). Only the grams of solute remains to be calculated. Insert 1.75 percent and 100 mL into Equation 7.1 to get

$$\frac{\% \text{ w/v}}{100} = \frac{\text{g of solute}}{\text{mL of solution}}$$

$$\frac{1.75\%}{100} = \frac{x}{100 \text{ mL}}$$

$$175 = 100x$$

$$1.75 \text{ g} = x$$

Take 1.75 g of solute, dissolve it in some water, and then add, with mixing, sufficient water to make a final volume of 100 mL.

EXAMPLE 7.9

What will be the percent weight to volume strength of a solution prepared by dissolving 37.5 g of a chemical in enough alcohol to make 500 mL?

Solution
Since this is a percent weight to volume problem, use Equation 7.1. Of the three possible variables, we know grams of solute (37.5) and the milliliters of solution (500), and we must calculate the percent weight to volume. Thus,

$$\frac{x\%}{100} = \frac{37.5 \text{ g}}{500 \text{ mL}}$$

$$x = \frac{3750}{500}$$

$$x = 7.5\%$$

■■■■■■ **EXAMPLE 7.10** ■■■■■■

A solution is found to contain 100 mg of solute in a total of 200 mL of solution. What is its strength in percent weight to volume?

Solution
This problem is similar to 7.9, for we know the grams of solute and the final volume and must calculate the percent weight to volume. However, in checking Equation 7.1, we see that the number of *grams* of solute is required, not milligrams. Always remember to use gram and milliliter units when applying Equation 7.1. Thus, we must first convert 100 mg to grams and then solve for percent weight to volume.

$$\frac{\% \text{ w/v}}{100} = \frac{0.100 \text{ g}}{200 \text{ mL}}$$

$$\% \text{ w/v} = 0.05$$

Dilution of a higher strength stock solution to a lower strength dispensing solution is a common requirement.

■■■■■■ **EXAMPLE 7.11** ■■■■■■

A stock solution of 5% w/v silver nitrate solution is available but a nurse requires 30 mL of a 1.0% w/v solution. How much of the stock solution and how much water should be used to obtain the desired preparation?

Solution
As we increase the volume of a solution by adding pure solvent, we decrease its strength proportionately. Doubling the volume halves the strength, and so on. Here we are diminishing the strength fivefold (5 → 1%), so we must dilute by a factor of 5. Thus take 6 mL of the stock solution and dilute to 30 mL with water. This solution to the problem is summarized in the following general equation:

$$(\text{conc}_{stk}) (\text{vol}_{stk}) = (\text{conc}_{dil}) (\text{vol}_{dil})$$

where stk means stock solution and dil means diluted solution. When you know three of the four factors in the equation, it is easy to solve for the fourth. In our problem, the set-up would be

$$(5\%) (\text{vol}_{stk}) = (1\%) (30 \text{ mL})$$

$$(\text{vol}_{stk}) = \frac{(1\%) (30 \text{ mL})}{(5\%)} = 6 \text{ mL}$$

Occasionally the nurse is called upon to measure a *very small volume* of an injectable or other solution. This may occur in connection with an unusually small dose of a drug or in a dilution problem. Here is an example. Meperidine hydrochloride injection USP (Demerol) is available in various strengths, one of which contains 100 mg/2 mL (that is 5% w/v). If a child patient is to receive a dose of 30 mg, only 0.60 mL of the meperidine injection should be withdrawn from the vial. Such a small quantity is readily and accurately measured in a hypodermic syringe. Illustrations of two types of syringes are given in Figure 7.8. The use of a hypodermic syringe to measure very small volumes of liquids, whether injectables or other types, reduces the percentage of error that occurs when graduated cylinders are used for measuring very small volumes.

A variation of the percent weight to volume unit that is often used in clinical laboratory reporting is the *milligram percent (mg %)* unit. The "percent" in this case means milligrams of solute per 100 mL of blood or other liquid. For example, normal blood levels of cholesterol are reported as in the range 120 to 220 mg %. Fasting blood glucose levels fall in the range of 70 to 110 mg %. You may encounter the milligram percent method applied to blood, plasma, serum, cerebrospinal fluid, and urine. You may encounter 100 mL expressed in its equivalent unit, a deciliter (dL). A concentration of 400 μg/dL is equal to 4 mg/L.

Yet another method of expressing percent weight to volume is used in health professions work. It is the *"parts per" method*. You will see labels that state "one part per 750 parts," or simply, 1:750 w/v. It means just what you would expect, 1 g in a total 750 mL. Potassium permanganate, used for wound irrigation, is typically used in a 1:1000 w/v strength (stated as "1 to 1000"). You should interpret this to mean 1 g $KMnO_4$ in a total 1000 mL of solution. Of course, this is equal to 0.1 percent. If you cannot make that connection mentally, resort to Equation 7.1:

$$\frac{\% \text{ w/v}}{100} = \frac{\text{g solute}}{\text{mL solution}} \qquad \frac{\% \text{ w/v}}{100} = \frac{1 \text{ g}}{1000 \text{ mL}} \qquad \frac{100}{1000} = 0.1\%$$

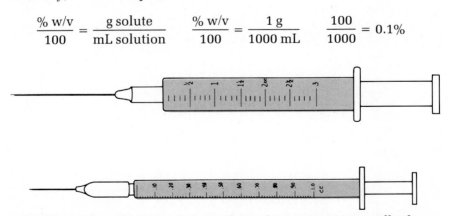

FIGURE 7.8 **Hypodermic syringes can be used to measure very small volumes of injectables or other solutions. (Top) 3-cc capacity hypodermic syringe graduated in tenths of 1 cc. (Bottom) 1-cc capacity hypodermic syringe (Tuberculin type) graduated in hundredths of 1 cc.**

Parts per million (ppm) and parts per billion (ppb) are similar units in wide use to express the concentrations of air and water pollutants. A 2-ppm solution would contain 2 g of solute in a million (10^6) mL of solution (or, 2 mg of solute in 1 L of solution). A 1-ppm solution is equivalent to 0.0001%, or 1×10^{-4}%.

■■■■■ **EXAMPLE 7.12** ■■■■■

Which is the stronger solution, a 1:2500 w/v or a 1:5000 w/v? Express each in their equivalent percent weight to volume units.

Solution

"Parts per" simply is another way to express percent weight to volume. In parts per, the first number is the grams of solute and the second is the volume of solution. For a 1:2500, 1 represents grams and 2500 represents volume; now solve for percent weight to volume. Hence, a 1:2500 solution, at 0.04 percent, is twice as concentrated as a 1:5000 at 0.02 percent.

Percent Volume to Volume

Percent volume to volume (% v/v) is used to indicate strengths of liquid–liquid mixtures such as alcohol in water, and gas–gas mixtures such as carbon dioxide in oxygen. A 1% v/v solution contains 1 mL of solute in each 100 mL of solution. In this discussion we will limit our percent volume to volume examples to dilutions, for these are most typically encountered.

■■■■■ **EXAMPLE 7.13** ■■■■■

Ethyl alcohol is usually purchased as a 95% v/v solution. However, health professionals know that 70% alcohol in water is best for germicidal purposes since the 30% of water helps to denature bacterial protein better. How should one prepare 1 L of 70% alcohol from stock 95% alcohol?

Solution

We can use the same approach we used earlier in the $AgNO_3$ dilution problem. Since strength is inversely proportional to dilution,

$$(conc_{stk}) (vol_{stk}) = (conc_{dil}) (vol_{dil})$$

$$(95\%) (vol_{stk}) = (70\%) (1000 \text{ mL})$$

$$(vol_{stk}) = \frac{(70\%) (1000 \text{ mL})}{95\%} = 740 \text{ mL (2 significant figures)}$$

Measure 740 mL of 95 percent alcohol, and, with mixing, add *enough water to make* exactly 1 L. Note that it is wrong to add 260 mL of water, because water and alcohol contract a little when mixed.

Molarity

For expressing strengths of medicinal solutions, **molarity** is not used as often as percent weight to volume. Yet we should learn about molarity units because they are used in hospitals in the United States and in countries around the world. In fact, there exists an international system of biomedical units that is based on the mole, millimole (mmol), and micromole (μmol). One can obtain more information on this *Systeme International d'Unites* by writing SI for the Health Professions, World Health Organization, Office of Publications, Geneva, Switzerland.

Before proceeding, be sure you know what the prefixes *milli* and *micro* mean, and how many micromoles there are in 1 mmol.

A *one-molar solution of any substance contains one gram-molecular-weight (1 mol) of solute in each 1000 mL of solution.* The symbol M stands for molarity; hence a solution containing 0.25 mol/L may be referred to as 0.25 M. To prepare a 1-*M* solution of glucose ($C_6H_{12}O_6$, molecular weight 180), dissolve 180 g of glucose in sufficient water to make a total 1000 mL of solution.

To work problems involving molar concentrations, we can use Equation 7.2:

$$\text{Molarity} \times \text{volume} \times \frac{\text{mass}}{\text{mole}} = \text{mass}$$

or

$$\frac{\text{mole}}{\text{liter}} \times \text{liter} \times \frac{\text{g}}{\text{mole}} = \text{g} \tag{7.2}$$

EXAMPLE 7.14

In the treatment of iron-deficiency anemias, ferrous sulfate can be used. An aqueous solution is found to be 0.10-M in iron(II) sulfate. How many grams of $FeSO_4$ will be supplied by 150 mL of this solution?

Solution
Since this is a molarity problem, check Equation 7.2 to see which factors are known and which is not. We know the mole/liter to be 0.1; we know the volume, but must be careful to express it as 0.150 L. We know we are dealing with $FeSO_4$ but must calculate the mass of $FeSO_4$ as 152 g per mole. Then, from Equation 7.2,

$$\frac{\text{mole}}{\text{liter}} \times \text{liter} \times \frac{\text{gram}}{\text{mole}} = \text{gram}$$

or

$$\frac{0.10 \, \cancel{\text{mole}}}{\cancel{\text{liter}}} \times 0.150 \, \cancel{\text{L}} \times \frac{152 \, \text{g}}{\cancel{\text{mole}}} = 2.3 \, \text{g}$$

■■■■■■■■■**EXAMPLE 7.15**■■■■■■

Recent research has implicated abnormally high Ca(II) ion levels as a factor in manic-depressive states in humans. If the normal plasma level of Ca(II) is 2.1 to 2.6 mmol/L and a patient is found to have a 20 mg/100 mL level of Ca(II), is his level significantly outside the normal range?

Solution
One mole of Ca^{2+} is 40 g; 20 mg is 0.020 g; 100 mL is 0.1 L; then

$$\frac{mole}{liter} \times liter \times \frac{gram}{mole} = grams$$

or

$$\frac{mole}{liter} \times 0.1 \; liter \times \frac{40 \; g}{mole} = 0.020 \; g$$

We solve the equation for moles per liter and find the answer to be 0.005 mol/L. This is the same as 5 mmol/L and is about twice the normal plasma level for Ca(II).

■■■■■■■■■■■■■■■■■■■■

Equivalent Weight

The final unit we shall consider for measuring the concentrations of solutions is the **equivalent weight.** You will encounter this unit as you read clinical laboratory reports on blood, serum, and urine electrolytes and as you work with injectable solutions.

In biomedical work, 1 equivalent weight (Eq) is 1 mol of an ionized substance divided by its valence:

$$1 \; Eq = \frac{ionic \; weight \; (g)}{charge \; on \; the \; ion} \tag{7.3}$$

Since blood and urine electrolyte concentrations are very small, it is much more convenient to use the milliequivalent weight (mEq). One milliequivalent is 1 mmol of an ionized substance divided by its valence:

$$1 \; mEq = \frac{ionic \; weight \; (mg)}{charge \; on \; the \; ion} \tag{7.4}$$

Note that whenever the valence on the ion is 1, 1 mol = equivalent weight. But when the valence is 2, 1 mol = 2 Eq. And when the valence is 3, 1 mol = 3 Eq.

We need a unit of concentration like the equivalent weight that takes into consideration the charge on a solute so that we can compare the electrical equivalence of solutes. In blood and in cells, *1 Eq of anions is equal to 1 Eq of cations,* but 1 *mol* of anions is not necessarily equal to 1 *mol* of cations. For example, if a patient lost, say, 500 mEq of urinary Na^+ and the physician desires to replace all of that with Ca^{2+}, he bases his calculation on equivalent weights, not moles. Here are the calculations, using Equation 7.4:

$$1 \text{ mEq Na}^+ = \frac{\text{ionic weight (mg)}}{\text{valence}} = \frac{23 \text{ mg}}{1} = 23 \text{ mg}$$

$$1 \text{ mEq Ca}^{2+} = \frac{\text{ionic weight (mg)}}{\text{valence}} = \frac{40 \text{ mg}}{2} = 20 \text{ mg}$$

Thus 500 mEq of Na^+ is 500 mEq $\times$ 23 mg/mEq or 11,500 mg. In the body, this is electrically equivalent to 500 mEq Ca^{2+} $\times$ 20 mg/mEq or 10,000 mg calcium. We see that less calcium is required because the calcium cation has a 2+ charge whereas the sodium cation has only a 1+ charge.

EXAMPLE 7.16

What are the milliequivalent weights of Mg^{2+} ion, K^+ ion, and Al^{3+} ion?

Solution
By application of Equation 7.4, we find that 1 mEq of Mg^{2+} is 12.2 mg; 1 mEq of K^+ is 39 mg; 1 mEq of Al^{3+} is 9 mg. These different masses all contribute 1 mEq of cations to a solution.

In body fluids there must be an electrical balance between cations and anions, or the fluid would carry an electrical charge. In Chapter 15 we will learn that the chief ions of the body are Na^+, K^+, Cl^-, and HCO_3^-. Other inorganic ions for which clinicians routinely report values are Ca^{2+}, Li^+, SO_4^{2-}, Pb^{2+}, Mg^{2+}, and Fe^{2+}. We will also learn to express normal reference values of these ions in the blood. For example, our bodies normally maintain a blood chloride level of 100–106 mEq/L.

Summary of Units Used in Expressing Strength of Solutions

% w/v   Used for dilute solutions (10% or less); used in drug dosage forms and in dilute stock solutions.

% v/v   Used for solutions of liquids in liquids and gases in gases.

% w/w   Used for concentrated solutions (>10%); applies mainly to the more concentrated stock solutions (not discussed in this chapter).

mg %   A special way of expressing percent weight to volume; used to express concentrations of solutes in the blood or plasma.

parts per   A special way of expressing percent weight to volume; examples are parts per hundred, parts per thousand, parts per million, parts per billion.

mmol (μmol, mol)   Molarity units are applied to solutes in the blood, urine, and other body fluids.

equivalent weight   Takes into account any electrical charge on the solute particle; milliequivalent weight is more common; used for expressing electrical equivalence of ions in the blood and urine.

OSMOSIS

When we consider biological solutions, we must consider **osmosis,** a phenomenon of very great importance to body physiology. In fact, osmosis is basic to life because it helps regulate the distribution of water and chemicals to all body cells and fluids. Osmosis is a special example of the general process of **diffusion,** that is, the tendency toward the equalizing of concentrations of solutes in all parts of an organism or in all parts of a biological system. Diffusion occurs as solute particles migrate from areas of high concentration to areas of lower concentration.

To illustrate osmosis, one needs a **semipermeable membrane.** Examples of semipermeable membranes are cell walls, the lining of the GI tract, and certain man-made films such as modified cellulose film, parchment, and collodion. Some authorities reserve the term *selectively permeable* for membranes that are living; they use semipermeable for membranes that are not alive.

A membrane is semipermeable when it allows passage of certain molecules (e.g., water) but excludes other substances (e.g., large proteins). For our discussion of osmosis, a membrane is semipermeable if it *allows passage of water but of no other cell constituents.* The passage of water can be in both directions. Such a membrane is often termed an *osmotic membrane.*

The permeability of membranes is not absolutely fixed, but can vary under certain circumstances, and can vary from one type of membrane to another.

As an example of an osmotic membrane in action, picture the tiny cell membranes in the root hairs of a tree (Figure 7.9). Outside of the membrane there is water in the earth around the tree. Inside the cells and passages of the tree is the sap, in which salts, sugars, and other molecules are dissolved in water. In effect, we have two compartments, with water molecules in each, but a higher concentration of solutes in one compartment (the sap) than the other (the earth). Under such circumstances, nature has decreed that solvent water molecules will pass from the earth through the membrane into the root hair cell, *in an attempt to dilute the more concentrated solution.* That is, nature is attempting to bring the two solutions to the same concentration. This passage

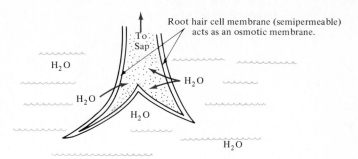

FIGURE 7.9 Osmosis in a Root Hair Cell. **Water flows into the more concentrated solution in an attempt to dilute it, thereby increasing the pressure inside the root system.**

of water across a semipermeable membrane from an area of lower solute concentration to an area of higher solute concentration is termed *osmosis* (from the Greek word for push).

So great is the tendency for water to cross the membranes in the root hairs and to dilute the sap that a pressure arises that is sufficient to push sap high up into the tree. This pressure is termed the *osmotic pressure*.

Note that the key to osmosis is the semipermeable membrane that allows for passage of solvent but restricts passage of solute particles. When such a membrane separates water solutions of different concentrations, osmosis can occur. Note also that nature is constantly striving to reduce orderliness. That is why the more concentrated solution (more ordered) becomes a more dilute solution (less ordered). Physical chemists use the word *entropy* to indicate disorderliness or randomness. Thus, they say, the entropy of the universe is constantly increasing.

Next, consider two human red blood cells (RBCs or erythrocytes), one placed into a high salt concentration solution (Figure 7.10) and the other into pure distilled water (Figure 7.11). The cell membrane of the RBC is the

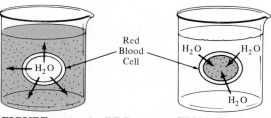

FIGURE 7.10 An RBC Placed in Highly Concentrated NaCl Solution. **Water is leaving the cell. The RBC shrinks in volume.**

FIGURE 7.11 An RBC Placed in Pure Water. **Net flow of water molecules is into the RBC, which swells to breaking point.**

semipermeable membrane, which allows passage of water but effectively restricts passage of solute particles already inside the cell. In the situation depicted in Figure 7.10, water molecules will leave the RBC in an attempt to dilute the more concentrated salt solution outside the cell; the RBC will shrink in volume and assume a notched appearance, a process called *crenation*. In the situation depicted in Figure 7.11, there will be a net flow of water molecules into the RBC because there is more salt inside the cell than outside it. It is possible in this situation for enough pressure to build up inside the cell for the RBC to burst, a destructive process termed **hemolysis** (*hemo* = blood, *lysis* = splitting).

The salt solution depicted in Figure 7.10 is said to be physiologically **hypertonic** (*hyper* = more than the normal) compared with the red cell, that is, it shows a greater osmotic effect than the solution inside the cell. The plain water in Figure 7.11 is **hypotonic** (*hypo* = less than normal) for it shows less osmotic effect than the solution inside the RBC. Biochemists have learned that if one prepares a 0.9% w/v solution of NaCl in water, or a 5% w/v solution of dextrose (glucose) in water, the solution will be physiologically nearly **isotonic** (*iso* = the same) with red cells, and RBCs suspended in this solution will neither shrink nor swell. Another way of saying this is that all of the solute particles dissolved inside a normal RBC exert the same osmotic pressure as that exerted by the sodium chloride in a 0.9% NaCl solution. Lacrimal fluid is isotonic with blood (although the eye tolerates 0.6 to 1.8 percent sodium chloride or equivalent without damage).

■ EXAMPLE 7.17 ■

Sodium chloride injection, 5 percent dextrose in distilled water (D5W), lactated Ringer's solution, and water for injection all have one property in common. What is it? (*Hint:* They are all suitable for intravenous injection.)

Solution
Though differing in solutes, they are all isotonic, or nearly isotonic, with blood and therefore can be injected intravenously.

Is NaCl the only solute that produces an osmotic effect in water? By no means! Any dissolved species—cation, anion, undissociated molecule, protein or atom—will produce the effect. In fact, for osmotic purposes, it is not as important *what* the particle is as *how many* particles are present. In solutions, an effect such as osmosis, which depends mainly on how many particles of anything are present, is termed a *colligative* effect. Other colligative effects in solutions are the lowering of the vapor pressure, the depression of the melting point, and the elevation of the boiling point. Some common colligative effects are salt sprinkled on ice melting the ice and salt added to water raising the boiling point of the water.

Blood circulating in our cardiovascular system contains dissolved gases, electrolytes, organic molecules, and proteins. All of these dissolved particles contribute to the osmotic effects of the blood; that is, they attract and hold water, thereby maintaining the total volume of blood. Loss of any of these dissolved particles can mean loss of osmotic effect. For example, loss of blood proteins as in burns, liver disease, or kidney disease can cause water to shift out of the vascular system and into the tissues. The albumins are the most abundant of the blood proteins, and they make a very important contribution to blood's osmotic activity.

EXAMPLE 7.18

The congestive heart failure patient has *too much* body fluid because his heart is not working effectively to send blood to the kidneys. There may be fluid accumulation in his lungs and in his extremities (often an abnormal accumulation around the ankles, called *edema*). Salt intake is severely restricted in patients with congestive heart failure. In simple terms, explain why salt must be avoided.

Solution

We know from our discussion of osmosis that salt attracts and holds water. That is, water will cross membranes into cells and compartments in an attempt to dilute the NaCl solutions. This will make the body retain even more water, placing a greater strain on the heart to remove it. A higher blood pressure is usually a consequence. Remember this: Na^+ plays a key role in determining total volume of body water.

Osmolarity

The unit for expressing osmotic strength is the *osmole (Osm)*, defined as the molecular weight of a substance divided by the number of particles (ions, atoms, or molecules) the substance liberates in solution. (Recall that it is not *what* particles but *how many* that affects colligative properties.) Glucose and fructose, which do not ionize, provide only one particle in solution. Hence a 1-*M* glucose solution is the *same* as a 1-Osm glucose solution. KCl provides two ions in solution (K^+ and Cl^-) and $CaCl_2$ provides three (Ca^{2+}, Cl^-, Cl^-). Hence a 1-*M* KCl solution is a 2-Osm solution and a 1-*M* $CaCl_2$ is 3-Osm. This can be summed up in the statement

osmolarity = molarity × number of particles formed per formula unit

One one-thousandth of an osmole, the *milliosmole (mOsm)*, is the unit used to express the strength of solutions that are injected into the human body in fluid and electrolyte therapy. Here are some examples:

Solution	Milliosmole per liter
5% dextrose in water	250
5% dextrose and 0.45% NaCl in water	405[a]
lactated Ringer's with 5% dextrose in H_2O	525
10% dextrose and 0.9% NaCl in water	815
(for comparison, human blood has an osmolarity of approximately 300 mOsm/L)	

[a] Adding 0.45% w/v NaCl (that's 4.5 g/L) to 5% dextrose in water adds 0.077 mol of NaCl, or 0.154 Osm (NaCl ionizes to give two particles in solution). Multiply the 0.154 Osm by 1000 to get milliosmoles. This added to the 250 mOsm from the 5% dextrose gives approximately 405 mOsm/L.

DIALYSIS

Dialysis, a special example of diffusion, is a process in which a semipermeable membrane permits passage of water, dissolved ions, and certain smaller organic molecules (according to their concentration gradient), but blocks passage of colloids, high molecular weight proteins, and polymers. The majority of membranes in our body are *dialyzing membranes* (cell and capillary walls, placental barrier) as are man-made films such as cellulose (viscose).

Dialysis occurs across cell walls because these membranes are practically impermeable to intracellular protein (colloidal in nature) and organic anions but permeable to water, Na^+, Cl^-, and especially K^+.

Research indicates that semipermeable membranes in the body have ion channels through which Ca^{2+}, Na^+, or K^+ migrate, subject to the channel opening in response to a drug or a hormone. Nifedipine and verapamil are two drugs used in the treatment of tachycardia; they work by blocking the transmembrane flux (dialysis) of extracellular Ca^{2+} into cardiac and vascular smooth muscle. (In heart tissue, Ca^{2+} flowing through channels into the cell's interior triggers heart muscle contraction.)

Man-made semipermeable membranes such as cellulose have "pores" that vary in size; one of the commercial products has pores large enough to permit molecules no larger than about 12,000 molecular weight to pass.

Osmosis and dialysis are similar processes since both involve a semipermeable membrane. However, they are distinctly different because in osmosis only solvent can pass through the membrane. In dialysis, many electrolytes can pass in addition to water.

Hemodialysis, a process for removing toxic wastes from the blood, is described in Chapter 15 on body fluids.

In your study of dialysis, keep in mind that in organs such as the kidney, electrolytes can be moved through membranes *against* their concentration and electrical gradients by a process called *active transport*. This process requires energy and is the explanation for the fact that the kidneys can excrete urine with a high concentration of, say, urea, but with a zero concentration of glucose. (The glucose is saved from excretion in the urine by reabsorption back into the general circulation.)

SEMIPERMEABLE MEMBRANES IN DRUG DELIVERY SYSTEMS

Scientists have developed a means of administering drugs through the skin (transdermally) using a patch worn on any convenient flat skin surface area. The key to the system is a semipermeable membrane made of an ethylene-vinyl acetate copolymer (Figure 7.12), through which the drug diffuses from the area of high concentration (the drug reservoir) to the area of low concentration (subdermal tissues). The system is well suited to the administration of drugs such as the vasodilator nitroglycerin, the requirements for which can be prolonged.

SURFACE TENSION OF SOLUTIONS: SURFACTANTS

Have you seen the interesting trick of floating a needle on water? Well, it can be done, for water actually has a tension across its surface, a kind of "skin" or barrier to entry of outside objects. This barrier is termed water's *surface tension*.

To be sure, water's surface tension is not high, but it is significant enough to permit insects to walk on and needles to float on water. It also makes a drop of water bead up into a sphere when the drop is placed onto a water-repellent surface. The water molecules seem to be pulling themselves together through forces that attract water molecules to each other (dipole–dipole bonds). Those on the surface are attracted inward without any opposing outside force.

The surface tension of water can sometimes be a problem to the health professional. If we attempt to use water to wash away grease or oils, to remove and destroy bacteria, to clean contact lenses, or to disperse sulfur particles for use on the skin, or in any situation where we want the water to penetrate, we

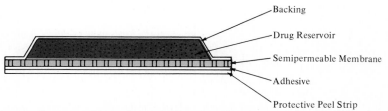

Backing

Drug Reservoir

Semipermeable Membrane

Adhesive

Protective Peel Strip

FIGURE 7.12 **Transdermal Therapeutic System.** **Five milligrams of nitroglycerin is absorbed from the skin patch over 24 hours.**

may find it does not do this. In other words, water is sometimes a poor "wetting agent."

Science has overcome water's sometimes poor penetrating power by creating synthetic wetting agents, chemicals that will reduce the surface tension of water. **Surfactants,** or *surface active agents,* are synthetic compounds that, when dissolved in water, interfere with the forces that attract water molecules to each other and thus reduce surface tension. Water molecules are thus free to act individually. Vastly improved penetrating power and wettability result.

You should know that a surfactant exists naturally in our bodies. A complex mixture of lipids and proteins, this endogenous surface active agent helps reduce surface tension in the water film that lines the tiny air spaces in the lungs (the alveoli). This helps prevent the airsacs from collapsing. Hyaline membrane disease in newborn infants is due to a deficiency of this natural surfactant.

Three classes of man-made surfactants exist:

1. Negatively charged (anionic) compounds such as ordinary soaps (medicinal soft soap, green soap), salts of bile acids, alkyl sulfates, alkyl sulfonates (e.g., Colace[R]), and detergents of the alcohol-sulfate and alcohol-phosphate ester types (e.g., sodium lauryl sulfate, USP).
2. Positively charged (cationic) compounds such as benzalkonium chloride solution USP (Zephiran Chloride), methylbenzethonium chloride USP (Diaparene, Bactine), and cetyl pyridinium chloride USP (Ceepryn). Chemically, these compounds are all quaternary ammonium salts (see Chapter 10).
3. Nonionic compounds such as polyoxyl 40 stearate USP (Myrj 52), polysorbate 80 USP (Tween 80), and methyl cellulose.

Table 7.5. Medicinal Uses of Various Surfactants

Surfactant type	*Uses*
Anionic	Soaps and detergents are used as cleansing agents; they help stabilize oil-in-water emulsions and aid in grease removal. Dioctyl sodium sulfosuccinate is a wetting agent that is used as a fecal softener.
Cationic	Has potent germicidal properties. Kills bacteria by altering their protective protein coat. Used to sterilize surgical instruments and other inanimate objects. Used for presurgical antisepsis and for washing dirty wounds. Used in mouthwashes. Not used internally.
Nonionic	Helps form oil-in-water and water-in-oil emulsions (see hydrophilic ointment USP). Used in the preparation of suppositories that will dissolve in water. Used in wetting solutions for soft contact lenses. Used to disperse oil-soluble vitamins in water.

Some of the medicinal uses of surfactants are listed in Table 7.5. Of special interest are the cationic type of surfactant such as the widely used Zephiran and Ceepryn. These compounds exhibit germicidal as well as detergent properties because of their ability to damage the protein in the surface membranes of bacteria. With its protective coat damaged, a bacterial cell undergoes lysis, and its cell contents seep out. Cationic surfactants do not kill spores or some fungi or viruses. Their germicidal activity is reduced by contact with food residues, blood proteins, or other body tissues. Cationic surfactants are incompatible with anionic surfactants such as soap.

In summary, we have seen that surface active agents can greatly increase the penetrating and wetting power of water and that this has application to health science, especially in the areas of emulsification, cleansing and germ killing. All of these uses are external, for surfactants become deactivated when administered internally.

PARENTERAL SOLUTIONS

A **parenteral solution** is one that is intended to be administered by injection—most often by intravenous injection (IV), but (in the broadest sense of the term) also by other routes such as intramuscular (IM) and subcutaneous. Intravenous solutions are used to provide fluid, nutrients, and electrolytes when oral and other routes of administration are not feasible. What, when, and how much to give must be determined in each individual case.

Patients who are candidates for parenteral solution therapy are the preoperative, the postoperative, the debilitated, the diarrheic, and those with persistent vomiting, renal (kidney) failure, and hormonal imbalance.

With regard to volume, 2000 to 3000 ml/day is a typical IV intake, providing IV intake is the only source of fluid and providing the kidneys are functioning normally. Caution must be exercised when administering parenteral solutions lest too much sodium be given. For example, isotonic saline should not be used to supply the entire daily fluid requirement because this will result in an excess of sodium. Dextrose in water (5 or 10 percent) provides no sodium but does supply water and calories.

Especially if IV administration is continued for several days, attention should be paid to the possible need for K, Mg, and Ca ions.

As a source of injectable amino acids, protein hydrolyzates (such as hydrolyzed casein from milk) can be used. These are termed hyperalimentation solutions.

Parenteral solutions containing protein, vitamins, and other nutrients usually must be sterilized by filtration since autoclaving would destroy heat-sensitive ingredients.

Table 7.6. Some Commonly Used Parenteral Solutions

Name of solution	Ingredients	Use in medicine
5% dextrose injection USP	Dextrose (glucose) and water for injection	Supplies fluid and calories (170 kcal/L)
5% dextrose and 0.45% sodium chloride injection USP	Dextrose, NaCl, and water for injection	Supplies fluid, electrolytes, and calories
Lactated Ringer's with 5% dextrose	Na^+, K^+, Ca^{2+}, Cl^-, lactate, and water	Supplies water, electrolytes, and calories at a pH of 6.3; often used by paramedics
Isolyte M (McGaw) with 5% dextrose	Dextrose, Na^+, K^+, Cl^-, acetate, HPO_4^{2-}	For daily water and electrolytes, sweating, lack of water intake, potassium deficiency due to chronic pyloric obstruction, ulcerative colitis, burns (in healing phase), prolonged infusion of K-free solutions, vomiting and diarrhea, mild metabolic acidosis

═══ **EXAMPLE 7.19** ═══

Whole blood and plasma should not be used as vehicles for IV feeding because they provide excess sodium. Give another general disadvantage to using whole blood or plasma. (*Hint:* The answer deals with disease transmission.)

Solution
In the use of whole blood or plasma, there is increased risk of infectious hepatitis transmitted by the blood of a carrier.

Table 7.6 lists commonly used, standard parenteral solutions, their composition, and their applications.

STORAGE OF SOLUTIONS: SHELF LIFE

The health professional who is in charge of a patient is also responsible for the medications that are to be given to that patient. When these "meds" are in the form of solutions, a special responsibility applies, for solutions generally are much more susceptible to deterioration with age than are solid preparations.

And medications do deteriorate. A spot check of paraldehyde (a nonbarbiturate sedative and hypnotic) in one large hospital revealed that only 11 percent of the samples collected from the wards met USP standards. One sample was almost 50 percent decomposed.

Four important factors can determine how long a medicinal solution retains its potency:

1. Temperature at which it is stored.
2. Exposure to sunlight or other radiant energy.
3. Presence of atmospheric oxygen.
4. Presence of moisture (in nonaqueous preparations).

Generally speaking, the higher the storage temperature, the shorter the shelf life. This is true because an increase in temperature generally means an increase in chemical reactivity. Solutions or suspensions of antibiotics in water are best stored at refrigerator temperatures, as are biologicals such as immune globulin, toxoids, and vaccines. Suppositories, designed to melt at body temperature (38°C), should be stored in the refrigerator, especially in hot weather.

Amber-colored glass containers are used to filter out solution-deteriorating light rays.

It is wise to keep solution containers tightly closed when not in use, as this reduces the chance of oxidation by atmospheric oxygen.

If the medicinal solution is nonaqueous (examples: external use only preparations using mineral oil, hexane, or ether as the solvent; peroxides in glycerin), care should be taken to exclude water, for the presence of water can result in possible **hydrolysis** of the active ingredient (*hydro* = water, *lysis* = splitting).

Many solutions supplied by a manufacturer or pharmacist will have a stated expiration date beyond which the preparation should not be used. Be sure to check for expiration dates and to adhere to them.

SUMMARY

The length of this chapter on solutions reflects how important solutions occurring naturally in the body and administered as drugs are to the health professional.

After we defined solutions, we examined types and behavior of solvents. The concept of solubility (solids in liquids, liquids in liquids, gases in liquids) was discussed. We next defined colloids and suspensions and contrasted their behavior with that of true solutions.

A major portion of this chapter was devoted to units for expressing concentrations of solutions. This is because you as a health professional often will be working with concentration units when you administer drugs or read clinical laboratory reports. We learned how to work with percent weight to volume, milligram percent, parts per, percent volume to volume, molarity, and equivalent weight. We learned that any ionic charge on a solute must be taken into consideration if we wish to compare electrical equivalence of solutions.

Under the general topic of diffusion, we discussed osmosis and dialysis. Here the key concept was the semipermeable membrane and a spin-off concept was tonicity (hyper-, hypo-, iso-). We learned how dialysis differs from osmosis and how to relate molarity to osmolarity.

We examined the idea of surface tension in water and how this surface tension can be reduced by the use of surfactants. We use surfactants to improve the wettability of water for the purpose of cleaning and emulsifying or for destroying bacteria.

We discussed parenteral solutions, a type of solution very often used in health-related work.

Finally, we identified factors that can affect the shelf life of drugs and chemicals in solution. Drugs are subject to deterioration if improperly stored or used.

KEY TERMS

Check your understanding of this chapter. Can you explain what is meant by each of these terms?

colloid	molarity
dialysis	osmosis
diffusion	parenteral solution
dipole	percent volume to volume
electrolyte	percent weight to volume
hemolysis	saturated solution
hydrolysis	semipermeable membrane
hypertonic	solute
hypotonic	solution
immiscible	solvent
isotonic	surfactant
miscibility	suspension

STUDY QUESTIONS

1. Define solution, solute, solvent, nonaqueous solution, otic solution, polar solvent, miscibility, and cationic surfactant.

2. Contrast true solutions, colloids, and suspensions from the standpoint of particle size, filterability, settling out, and Tyndall effect.

3. What is the difference between a spirit, an elixir, and a syrup?

4. What is the solute in each of the following solutions?
   a. Maple syrup
   b. Radiator antifreeze
   c. Whiskey
   d. Vinegar
   e. Honey
   f. Soda water
   g. Smelling salts (see Appendix I)
   h. Limewater
   i. Formalin
   j. Colorado River water

5. Which of the following solvents are properly classified as polar and which as nonpolar?
   a. Water
   b. Glycerol
   c. Gasoline
   d. Ethyl alcohol
   e. Benzene
   f. Toluene
   g. Isopropyl alcohol
   h. Carbon tetrachloride

6. Define semipermeable membrane, osmosis, hemolysis, hypertonic, and isotonic.

7. Assume that a passenger is in an aircraft flying at 10 km (30,000 ft) with the cabin pressurized to equal sea level pressure. If a sudden depressurization were to occur, what would happen to the oxygen and nitrogen gases dissolved in the passenger's blood?

8. On the basis of the information presented in this chapter, predict which of the following are water soluble and which are water insoluble: potassium acetate, ammonium sulfate, mercury(I)chloride (calomel), cobalt(II)-chloride, calcium carbonate, silver sulfide, zinc nitrate, morphine sulfate, morphine (free base), ephedrine sulfate, sodium phenobarbital, phenobarbital.

9. On the basis of "like dissolves like," which of the following compounds can we expect to be soluble in water?
   a. KCl
   b. Table sugar
   c. HCl
   d. Corn oil
   e. Helium gas
   f. CCl_4
   g. KCN
   h. Vitamin A

10. True or false: The majority of solutes become more soluble as the temperature of their solvent is increased.

11. Which of the following statements are descriptive of colloids? (a) They do not settle out on standing. (b) They are not removed by filtration. (c) They show the Tyndall effect. (d) They will not pass through a dialyzing membrane.

12. Highly purified water can still show the properties of a solution, as evidenced by a change in pH when the water is freshly boiled. What solute could be driven off during the process of boiling water?

13. Explain the correct method for preparing 1 L of a 0.9% w/v solution of NaCl in water.

14. How much water would be required to prepare a 0.25% w/v solution from 1.00 g of $NaHCO_3$?

15. What volume of 0.5% w/v solution can be made from 2.5 g of gentian violet?

16. A patient is required to soak his foot in 1 gal of saturated boric acid solution (5% w/v H_3BO_3 is saturated at room temperature). How many grams of boric acid will be required and exactly how should the solution be prepared?

17. If, instead of 5% boric acid solution, one prepared 5% w/v acetic acid solution, would the quantity of solute used be the same? In other words, for purposes of calculation, does it make any difference what the solute is in a 5% w/v solution?

18. Label information states that there is 0.50 mg of drug per 2.0 mL of solution.
    a. What is the equivalent percent weight to volume?
    b. Is this strength equivalent to 1:4000?

19. For each pair, indicate which has the larger numerical value:
    a. $\frac{1}{120}$ or $\frac{1}{180}$ b. $\frac{3}{2}$ or $\frac{3}{4}$
    c. 1:100 or 1:1000   d. 1:20 or 1:2
    e. $\frac{1}{4}$% or $\frac{1}{3}$%   f. 35 mL of 0.12 M or 50 mL of 0.080 M
    g. 0.25% or 1:500   h. 0.01% or 1:7500

20. Express 110 mg % as percent weight to volume.

21. Three percent weight to volume sodium bicarbonate solution is sometimes used in irrigation of the eyes. Which strength, 1:20 w/v or 35:1000 w/v, is closer to 3%?

22. Mathematically, prove that 1 g in 1 million mL and 1 mg in a liter are both 1 ppm.

23. A chemist takes 25 mL of a 2% w/v $KMnO_4$ solution and dilutes it to 5000 mL with distilled water. What is the percent weight to volume of the $KMnO_4$ in the diluted solution?

24. Mercury(II)chloride solutions are used to sterilize inanimate objects. (Mercury is a poison if taken internally.) State the correct procedure for preparing 1500 cc of 1:1000 w/v $HgCl_2$ solution if the only available source of $HgCl_2$ is 0.5-g tablets.

25. An asthmatic patient is to receive a 0.4-mg injection of epinephrine every 3 hours, but only a solution of epinephrine 1:1000 is available. How much of the 1:1000 solution should be used to supply the correct dose?

26. An injectable solution of digoxin (a powerful heart stimulant) is labeled 0.5 mg in 2.0 mL. How much of this solution should be injected to administer a dose of 0.020 mg?

27. How much water would one need to add to 10 mL of a 12% w/v benzalkonium chloride solution to obtain a 1:5000 w/v solution?

28. Five liters of 25% v/v isopropyl alcohol in water solution is to be prepared as a cooling sponge bath. State how to prepare this from 70% v/v isopropyl alcohol and distilled water.

29. How much water would be required to pepare a 40% solution from 400 mL of pure ethyl alcohol?

30. What is the correct procedure for preparing 250 cc of a 0.25-M solution of Glauber's salt (Na_2SO_4) in water?

31. A typical dose of iron(II) sulfate in the treatment of iron-deficiency condition is 0.30 g. How many milliliters of the solution described in Example 7.14 should be used to supply this dose?

32. In healthy humans, the normal Mg(II) iron serum level is considered to be 1.0 mmol/L. Express that as (a) milligrams per liter and (b) milliequivalents per liter.

33. True or false:
    a. For most ions, the equivalent weight is either once, twice, or three times the gram-ionic weight.
    b. In body fluids, 1 Eq of anions is equal to 1 Eq of cations.
    c. Inside a red blood cell, the (+) charges on all of the K^+, Na^+, and Mg^{2+} ions are just equal to the total (−) charge on all of the HCO_3^-, phosphate ions and other anions.

34. Consider two compartments, A and B, separated by an osmotic membrane. The NaCl solution in A is 0.9% w/v and the NaCl solution in B is 0.5% w/v. At the start, the volume of A = the volume of B. At equilibrium, will the water level in compartment A have risen or fallen?

35. Drinking seawater will not allay a sailor's thirst, but make it worse. Therefore, is seawater hypotonic, hypertonic, or isotonic with blood?

36. True or false:
    a. A red blood cell placed into 1.5% w/v NaCl solution will shrink in size.
    b. A red blood cell placed in 0.9% w/v NaCl solution will shrink.
    c. A 0.2% w/v NaCl solution is hypertonic with blood.
    d. Blood proteins called albumins attract water by osmotic effects.

37. An aqueous solution contains 0.50 g of NaCl in every 50 mL. Is the solution hypertonic, hypotonic, or isotonic?

38. Epsom salt (magnesium sulfate, $MgSO_4$) is used as a saline cathartic. In the GI tract, it produces a high concentration of osmotically active particles inside the lumen of the intestine. Explain how this results in catharsis (*Hint:* Intestinal walls can be thought of as semipermeable membranes.)

39. Explain the difference between an osmotic membrane and a dialyzing membrane.

40. Fill in the blank spaces. The first compound is the example. You may work backwards or forwards.

Substance	Mass	Number of millimoles	Number of milliosmoles
NaCl	58.5 mg	1	2
$MgCl_2$	9.5 mg	0.1	(a) _____
Glucose	(b) _____	1	(c) _____
$NaHCO_3$	(d) _____	1	(e) _____
NH_4Cl	(f) _____	(g) _____	2

41. We say that Zephiran, Bactine, soap, and so forth, are agents that are active at the surface of water. Explain what they do there.

42. Why are surfactants sometimes called "wetting agents"?

43. Which type of surfactant—anionic, cationic, or nonionic—has potent germicidal properties?

44. With a check, indicate the probable effect of the condition on the shelf life of a typical drug.

Condition	Prolong	Shorten
a. Higher temperatures		
b. Exposure to moisture		
c. Keeping container well closed		
d. Exposure to oxygen		
e. Storage under nitrogen		
f. Storage in amber bottles		
g. Storage in refrigerator		
h. Exposure to radiation		

ADVANCED STUDY QUESTIONS

(You may have to consult your library or subsequent chapters in this book for the answers to these questions.)

45. Our text mentions a gas mixture containing carbon dioxide and oxygen. Why would a small volume (5%) of CO_2 gas deliberately be added to oxygen gas for use in hospital patients?

46. What damage to the human system could result if plain water were injected IV instead of the isotonic water for injection? (Assume both are sterile and pyrogen-free.)

47. Prolonged vomiting, excessive perspiration, and diarrhea can account for much loss of body water and electrolytes. Potentially dangerous dehydration can result. To compensate for this type of dehydration, would the human body tend to (a) conserve or excrete sodium ion? (b) increase or decrease urine output?

48. What is the apparent relationship between the milliequivalent and the milliosmole?

49. The ancient Egyptians prepared the bodies of their dead pharaohs for burial by immersing them for an extended period in "natron" fluid. Explain the physiochemical principles involved.

8

Acids, Bases, and Salts

All of the items shown in this collection are examples of acids, bases, or salts commonly encountered in your life. The range of materials is really rather amazing—everything from pool acidifier and citrus fruits to table salt to window cleaner. These same three classes—acids, bases, and salts—represent biologically important compounds as well. The balance between acids and bases and the concentration of salts is crucial to the proper functioning of most bodily systems.

You will need to learn to recognize acids, bases, and salts by their formulas and by their properties. These skills will help you to understand the reactions and uses of these classes of compounds.

Some of the materials shown in this photograph are actually very dangerous if directly consumed and yet others are part of a normal, well-balanced diet. What makes the difference? This is another of the key ideas you will need to know as a health professional. Some acids, for example, are very strong and would be dangerous to directly consume. Others are much weaker and make up an important part of our diets. For example, vitamin C is an expected part of our daily diets but you would not consider drinking concentrated hydrochloric acid! The difference in strength of acids is a very important idea that helps to determine the potential use of the acid. The same may be said for the use of bases.

The interaction of acids and bases is another very important idea for the health professional. You will learn in this chapter how these classes of materials react. You will also see how acids and salts (or bases and salts) produce systems called buffers. These buffers are most important in regulating the function of most biological systems.

INTRODUCTION

We have already seen many examples of acids, bases, and salts in earlier chapters. These three categories are so important that it would be nearly impossible to conduct even a very limited study of chemistry without including them. Almost every biochemical reaction in our bodies is dependent on the level of acids, bases, and salts present. Significant changes in the levels of these substances can alter the proper functioning of cells and therefore affect our health. Many industrial manufacturing processes also require the use of acids, bases, and salts. Sulfuric acid, sodium hydroxide, and sodium carbonate are among the most used chemicals in industry. Acids are present in the food we eat and bases are in many common cleaning agents. Salts serve a variety of purposes such as water treatment and flavor enhancement.

This chapter groups these three categories together because they represent the major types of ionic compounds. If we find experimentally that ions are present in water solution, these ions are most likely to have been produced by an acid, base, or salt. Do not conclude from this statement, however, that all acids, bases, and salts have the same tendency to produce ions. We shall see very soon that there are significant differences among the members of each category in their ability to produce ions.

In this chapter, we will first look at experimental evidence for detecting the presence of ionic substances. We will study the general properties of each major category. Alternative definitions of acids and bases will be introduced. The reaction of an acid with a base to produce a salt and water will be studied. We will examine the theory behind experimental procedures to determine the concentration of acids or bases and we will learn how to predict the extent of an acid–base reaction.

The interactions of salts with water are part of the basis for controlling the level of acidity in the body. We will study these biologically important systems, called buffers, at the conclusion of this chapter.

ELECTROLYTES

If ions are present in water solution, the ions enable the solution to carry an electric current. Such a solution is referred to as an **electrolyte.** (The substance releasing the ions may also be called an electrolyte.) If this solution is an excellent conductor, even at low concentrations, it is called a *strong* electrolyte. If no ions are present, the solution is a *nonelectrolyte. Weak* electrolytes represent a middle ground in which some current is carried but the number of ions released is not as great as with a strong electrolyte at the same initial concentration. Figure 8.1 shows an experimental plan for determining if an electrolyte is present.

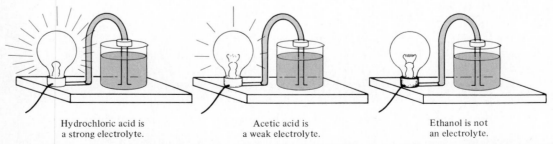

Hydrochloric acid is a strong electrolyte.

Acetic acid is a weak electrolyte.

Ethanol is not an electrolyte.

FIGURE 8.1 Finding Electrolytes Experimentally. **Each beaker holds the same volume of a 0.10M solution. Notice that the light bulb is brightly lit in the first case, dimly lit in the second case, and not lit in the third case, illustrating the differences in the ability of the substances to produce ions that carry electricity.**

GENERAL PROPERTIES OF ACIDS AND BASES

All acids and bases have general characteristics that can be used to help identify them. You are all familiar with the sour taste of lemons, oranges, grapefruit, and limes. These fruits all contain the organic acid called citric acid. The sour taste you associate with these fruits is a general property of all acids, not just this one. A concentration of about 5 percent acetic acid, another organic acid, is what makes vinegar sour. Sour milk contains lactic acid, and there are recognizable amounts of tartaric acid in grapes, malic acid in apples, and oxalic acid in rhubarb.

Very few foods contain strongly basic substances. Egg white is slightly basic and common antacids such as milk of magnesia are basic. Most pure bases have a somewhat bitter taste, quite different from the sour taste of acids, but this is not a property that you will necessarily find useful for identification. It is possible that you have felt the slippery or "soapy" feeling associated with bases present in cleaning materials in your home. This sensation is caused by the base, commonly sodium hydroxide, reacting with the fat in your skin tissue. (*Caution:* strong bases are caustic because of this reaction.) Acids do not have this same sensation of slipperiness and, in fact, will destroy the soapy feeling by reacting away the bases.

Because it is not safe to taste and feel potentially caustic and corrosive substances, it is better to rely on a simple chemical test to detect the presence of an acid or a base. This is practical because many colored substances of organic origin respond to the presence of acid or base by changing color. You may have observed that adding lemon to tea lightens the color of the tea. Another such organic substance is a dye called **litmus,** which is found in a certain variety of lichen.[1]

Litmus turns red in the presence of an acid and blue in the presence of a base. Litmus is often used in the form of litmus paper, which is made by soaking the paper in the dye and then drying and conveniently packaging the paper. The color change is fully reversible. Substances such as tea or litmus are called **indicators** because they indicate the acidity or alkalinity of a solution.

When acids and bases react chemically with one another, each one loses its characteristic properties. This reaction is called **neutralization,** and its products are water and a salt. This type of reaction was discussed in Chapter 5 as an example of a double replacement reaction.

These properties and other common properties of acids and bases are summarized in Table 8.1.

---

[1] You may have seen colorful patches of lichen on rocks, branches, or on the ground. These small plants actually consist of both a fungus and an alga growing together.

Table 8.1. General Properties of Acids and Bases

Property	*Acids*	*Bases*	*Example equations*
Taste	Sour, like lemons	Bitter	
Feel	Grating, but may burn if concentrated	"Soapy," but can also burn if concentrated	
Red litmus	Stays red	Turns blue	
Blue litmus	Turns red	Stays blue	
Reaction with each other	Neutralizes bases to form water and a salt	Neutralizes acids to form water and a salt	$HCl + NaOH \rightarrow H_2O + NaCl$ acid — base → water — salt
Reaction with metals	Active metals react to produce hydrogen gas and a salt	Certain metals (e.g., Zn, Al, Cr) will react to produce hydrogen and a ternary salt; most metals will not directly react	$2\,HCl + Mg \rightarrow H_2 + MgCl_2$ acid — active metal → hydrogen gas — salt $2\,NaOH + Zn \rightarrow H_2 + Na_2ZnO_2$ base — metal → hydrogen gas — salt
Reaction with certain salts	Carbonates and acid react to give carbon dioxide and water; sulfites to give sulfur dioxide and water	Salts will react with bases if an insoluble hydroxide results	$2\,HCl + Na_2CO_3 \rightarrow CO_2 + H_2O + 2\,NaCl$ acid — carbonate salt → carbon dioxide gas — water — salt $NaOH + AgNO_3 \rightarrow AgOH(s) + NaNO_3$ base — salt solution → ppt — salt

━━━━━**EXAMPLE 8.1**━━━━━

Predict the reaction of nitric acid, HNO_3, with

    a. Both colors of litmus paper
    b. Zinc metal
    c. Potassium hydroxide

Solution

Nitric acid can be expected to have typical acidic properties as shown in Table 8.1.

    a. Blue litmus should turn red and red litmus will remain red.
    b. Zinc will react with nitric acid to produce hydrogen gas and zinc nitrate solution.
    c. Potassium hydroxide will neutralize nitric acid, producing water and potassium nitrate in solution.

Acids and Bases Defined

One of the first theories dealing with the chemical nature of acids and bases was proposed by the Swedish scientist Svante Arrhenius (1859–1927). He suggested that an **acid** must produce hydrogen ions in water solution and that this ion is responsible for the typical acid taste, feel, and other physical and chemical properties. Today chemists consider that the hydrogen ions are not single free ions in solution but are so strongly surrounded by water molecules that a hydrated hydrogen ion called a **hydronium ion** forms.

$$H^+(aq) + H_2O(l) \rightarrow H_3O^+(aq)$$

    hydrogen   water   hydronium
      ion   molecule   ion

Remembering the Lewis structure for water, you can see that the hydrogen ion now shares one of the previously unshared electron pairs on the water molecule.

$$H^+ \quad + \quad H-\ddot{O}: \quad \rightarrow \quad \left[H-\ddot{O}-H \right]^+$$
$$\qquad\qquad\qquad\quad | \qquad\qquad\qquad |$$
$$\qquad\qquad\qquad H \qquad\qquad\quad H$$

    hydrogen   water   hydronium
      ion   molecule   ion

As a convenience, the hydrogen ions are often shown in the free, unhydrated form, but the presence of the hydronium ion is always understood in water solution. Therefore, in a dilute solution of hydrochloric acid, the formation of the strong electrolyte can be represented in either of these two ways.

$$HCl(aq) + H_2O(l) \rightarrow H_3O^+(aq) + Cl^-(aq)$$

$$HCl(aq) \rightarrow H^+(aq) \quad + Cl^-(aq)$$

For an acid that is a weak electrolyte, one that is not fully separated into ions in solution, the ionization may be represented in either of these two ways, shown here for acetic acid, the weak electrolyte of Figure 8.1.

$$HC_2H_3O_2(aq) + H_2O(l) \rightleftarrows H_3O^+(aq) + C_2H_3O_2^-(aq)$$

$$HC_2H_3O_2(aq) \rightleftarrows H^+(aq) \quad + C_2H_3O_2^-(aq)$$

The placement of the double arrows indicates that the reaction does not go completely to the right and that the acetic acid does not completely ionize. This is an example of an **equilibrium** reaction. Both the forward reaction and the reverse reaction are occurring at the same time. When equilibrium is established between these two reactions, there are more acetic acid molecules present than ions.

The double arrow does not give you a quantitative measure of how weak the acid really is but rather alerts you to the fact that it is not a strong acid. Table 8.2 gives you some of the common **strong and weak acids** that you may find important for your studies.

Do not confuse the terms *strong* and *weak* with the terms *concentrated* and *dilute*, which we studied in the last chapter. The terms strong and weak always refer to the degree of ionization of the electrolyte in water. The terms concentrated and dilute always give you an indication of how much solute is dissolved in the solution. Hydrochloric acid is a strong acid, no matter how concentrated or dilute its solution may be. Carbonic acid is a weak acid at any concentration.

Arrhenius also proposed that a **base** will produce hydroxide ions, OH^-, in solution. These ions are not usually shown in hydrated form as the connection between the hydroxide ion and water is not as strong as it was for the single hydrogen ion. Bases may also be strong or weak electrolytes, depending upon their ability to release **hydroxide ions.** A common **strong base** is sodium hydroxide and a common **weak base** is ammonium hydroxide. Their ionization can be represented in these equations.

$$NaOH(aq) \rightarrow Na^+(aq) \quad + OH^-(aq)$$

$$NH_4OH(aq) \rightleftarrows NH_4^+(aq) + OH^-(aq)$$

Representation of Neutralization

We have stated that one of the main chemical characteristics of acids and bases is their neutralization reaction with each other. The reaction of hydrochloric acid with sodium hydroxide can once again be used to illustrate this process and then the equation represented in ionic form as well.

Table 8.2. Common Strong and Weak Acids

Strong Acids:

Hydrochloric acid	$HCl \rightarrow H^+ + Cl^-$
Nitric acid	$HNO_3 \rightarrow H^+ + NO_3^-$
Sulfuric acid	$H_2SO_4 \rightarrow H^+ + HSO_4^-$

Weak Acids:

Acetic acid[a]	$CH_3COOH \rightleftarrows H^+ + CH_3COO^-$
Adipic acid	$(CH_2)_4(COOH)_2 \rightleftarrows H^+ + (CH_2)_4(COOH)COO^-$
Carbonic acid	$H_2CO_3 \rightleftarrows H^+ + HCO_3^-$
Citric acid	$HOC(CH_2CO_2H)_2COOH \rightleftarrows H^+ + HOC(CH_2CO_2H)_2COO^-$
Formic Acid	$HCOOH \rightleftarrows H^+ + HCOO^-$
Lactic acid	$CH_3CHOHCOOH \rightleftarrows H^+ + CH_3CHOHCOO^-$
Phosphoric acid	$H_3PO_4 \rightleftarrows H^+ + H_2PO_4^-$
Salicylic acid	$C_6H_4OHCOOH \rightleftarrows H^+ + C_6H_4OHCOO^-$
Sulfurous acid	$H_2SO_3 \rightleftarrows H^+ + HSO_3^-$
Glycine[b]	$CH_2NH_2COOH \rightleftarrows H^+ + CH_2NH_2COO^-$
Alanine[b]	$CH_3CHNH_2COOH \rightleftarrows H^+ + CH_3CHNH_2COO^-$

[a] Acetic acid has two carbon atoms, two oxygen atoms, and four hydrogen atoms per formula unit. An inorganic chemist tends to write this formula as $HC_2H_3O_2$. This form emphasizes the presence of one acidic hydrogen with the rest of the formula staying together as a charged radical ion, the acetate ion. The form of acetic acid given above emphasizes its structure. The —COOH group is typical of an organic acid. Notice the presence of this group in each of the organic acids above.

[b] Glycine and alanine are examples of amino acids. They are also organic acids but with an additional structural group, —NH_2. This type of acid will be discussed in detail in Chapter 10.

Molecular equation

$$HCl(aq) + NaOH(aq) \rightarrow H_2O(l) + NaCl(aq)$$

Ionic equation

$$H^+(aq) + Cl^-(aq) + Na^+(aq) + OH^-(aq) \rightarrow H_2O(l) + Na^+(aq) + Cl^-(aq)$$

Both of the reactants are substances that are highly ionized in water and, therefore, they are shown in their ionized form in the second equation. The product water is written in molecular form, even in the "ionic" equation, because water is largely molecular (non-ionized) and is therefore not an electrolyte by itself. Sodium chloride, on the other hand, is entirely ionized in solution and is written in separated ions in the second equation. The ions that have not participated in the reaction can now be eliminated from the equation, leaving the actual reacting ions in the net ionic equation.

Net ionic equation

$$H^+(aq) + OH^-(aq) \rightarrow H_2O(l)$$

The advantage of the net ionic equation is that it shows exactly the reason for the reaction. In this case, the hydrogen ion (really the hydronium ion, remember) reacts with the hydroxide ion to form the molecular substance, water. The formation of the water removes ions from the reaction and so completes the reaction. The neutralization of any strong acid with any strong base will give the same net ionic equation. As evidence of this fact, the heat of reaction for the neutralization of any strong acid with any strong base is exactly the same, 13.7 kcal per mole of water formed. This heat energy is released because of the new hydrogen–oxygen bond that forms to make the water molecule. Remember that in any neutralization, the reason that the reaction proceeds to completion is the formation of the predominately molecular water molecules.

SELF-IONIZATION OF WATER

Water is a most amazing molecular substance and one of its interesting properties is that it is capable of producing small amounts of both hydronium ion and hydroxide ion by the process of self-ionization. **Self-ionization** is the term used to describe the ability of a compound to react with itself to produce ions. In all of the collisions between water molecules, some of them are energetic enough to result in the formation of ions. Figure 8.2 shows this process schematically.

The reaction illustrated in Figure 8.2 does not take place to a large extent. In fact, the concentration of ions produced is very small (<0.1%) and the vast majority of water molecules stay in covalently bonded form. However, the concentration of ions produced is constant at a constant temperature, and no amount of purification or special treatment is able to remove these ions.

There is a very small but constant amount of electrical conductivity remaining when pure water is tested. The conductivity is too small to show up on the apparatus pictured in Figure 8.1. The equilibrium represented in Figure 8.2 is highly in favor of the nonionized molecular water. The position of the equilibrium is shifted to the left in the reaction as written.

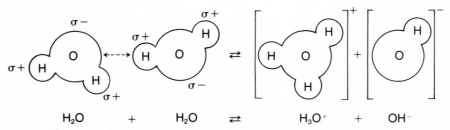

$$H_2O \quad + \quad H_2O \quad \rightleftharpoons \quad H_3O^+ \quad + \quad OH^-$$

FIGURE 8.2   The Self-Ionization of Water. **Water's polar molecules act on themselves to produce hydronium and hydroxide ions.**

$$H_2O + H_2O \leftrightharpoons H_3O^+ + OH^-$$

By comparing the residual conductivity of water with that of strong acids and bases of known strength, it can be found that in pure water, the concentrations of the hydronium and hydroxide ion, which must be equal to each other, are both 1×10^{-7} mol/L at 25°C.

The Meaning of K_w

The constant concentration of hydronium and hydroxide ions in water, each being 10^{-7} mol/L, can be used in calculating a constant called K_w. It is an equilibrium constant relating the extent of the forward reaction, in which water breaks into ions, with the extent of the reverse reaction, in which the ions reform to molecular water. In pure water there are very few ions and this is reflected in the very small value of K_w for pure water. K_w is mathematically defined in this manner.

$$\begin{aligned} K_w &= [H_3O^+] \cdot [OH^-] \\ &= (1 \times 10^{-7}) \cdot (1 \times 10^{-7}) \\ &= 1 \times 10^{-14} \end{aligned}$$

K_w is a significant value, for it holds for all water solutions at 25°C. This includes aqueous solutions of acids and bases in which the two ions, hydronium and hydroxide, will be present in unequal concentrations. It includes all aqueous body solutions such as blood, urine, bile, and tears. If something happens to upset the balance between the hydrogen ions and the hydroxide ions, then the resulting solution is no longer neutral. Remember that the *product* of the concentrations will still be constant and equal to 1×10^{-14} at 25°C.

▬▬▬▬ EXAMPLE 8.2 ▬▬▬▬

Find the hydronium ion and hydroxide ion concentration in 0.01 *M* sodium hydroxide solution.

Solution
Sodium hydroxide is a strong base, so the concentration of the hydroxide ion released will be assumed to be the concentration of the base originally dissolved, a reasonable assumption if 100 percent ionization takes place. The concentration of the hydronium ion will be calculated from the fact that in a water solution at 25°C the ion product for hydroxide and hydronium is constant and equal to 1×10^{-14}. $[OH^-] = 0.01$ *M*, which is 1×10^{-2} *M*.

Therefore, since

$$[H^+] \times [OH^-] = 1 \times 10^{-14}$$

then

$$[H^+] \times [1 \times 10^{-2}] = 1 \times 10^{-14}$$

and

$$[H^+] = \frac{1 \times 10^{-14}}{1 \times 10^{-2}}$$

$$[H^+] = 1 \times 10^{-12} M$$

Notice that the concentration of the hydroxide ion is far greater than the concentration of the hydronium ion in this case. This is reasonable in this basic solution. (Remember that M means molar concentration and stands for moles of solute present per liter of solution.)

If the solution is neutral, the fundamental relationship that must exist is that the concentration of the hydronium ion must equal that of the hydroxide ion. If one or the other is in excess, then the solution is either acidic or basic, depending on which ion is present in greater concentration. These statements are summarized in the following way.

A solution is **neutral** if $[H^+] = [OH^-]$

A solution is **acidic** if $[H^+] > [OH^-]$

A solution is **basic** if $[H^+] < [OH^-]$

Remember that if the solution conducts electricity and yet does not produce hydronium or hydroxide ions, it must be a salt solution. The behavior of salt solutions will be discussed later in this chapter.

THE pH SCALE

We could continue to express hydrogen ion concentrations in moles per liter, but a simpler scale is in common use. This is the **pH** scale, a numerical scale that easily expresses the wide range of values possible for the concentration of the hydrogen ion. This scale was devised by a Danish biochemist, S. Sorensen (1868–1939). The abbreviation *pH* comes from the original expression *pouvoir hydrogene* or hydrogen power. The power referred to is the exponential power used to express the hydrogen ion concentration.

To find the pH, the hydrogen ion concentration must first be known in units of molarity (moles per liter) and then expressed in scientific notation. The absolute value of the exponent on the power of 10 is then used to define

Table 8.3. Relationships in Aqueous Solutions[a]

	pH	$[H^+]$	K_w	$[OH^-]$	
↑ Acidity ↑	0	1×10^0	1×10^{-14}	1×10^{-14}	↓ Basicity ↓
	1	1×10^{-1}	1×10^{-14}	1×10^{-13}	
	2	1×10^{-2}	1×10^{-14}	1×10^{-12}	
	3	1×10^{-3}	1×10^{-14}	1×10^{-11}	
	4	1×10^{-4}	1×10^{-14}	1×10^{-10}	
	5	1×10^{-5}	1×10^{-14}	1×10^{-9}	
	6	1×10^{-6}	1×10^{-14}	1×10^{-8}	
Neutral	7	1×10^{-7}	1×10^{-14}	1×10^{-7}	Neutral
↑ Increasing ↑	8	1×10^{-8}	1×10^{-14}	1×10^{-6}	↓ Increasing ↓
	9	1×10^{-9}	1×10^{-14}	1×10^{-5}	
	10	1×10^{-10}	1×10^{-14}	1×10^{-4}	
	11	1×10^{-11}	1×10^{-14}	1×10^{-3}	
	12	1×10^{-12}	1×10^{-14}	1×10^{-2}	
	13	1×10^{-13}	1×10^{-14}	1×10^{-1}	
	14	1×10^{-14}	1×10^{-14}	1×10^0	

[a] Notice that the product of the $[H^+]$ and $[OH^-]$ ion concentrations is always constant at 1×10^{-14}.

the pH. For example, if the hydrogen ion concentration is 1×10^{-7} M, as it is in pure water at 25°C, then the pH is equal to 7. If the hydrogen ion concentration is 1×10^{-4} M, as it is in some soft drinks, then the pH is 4 and the solutions are acidic. The definition that we are using here is as follows:

$$[H^+] = 1 \times 10^{-pH}$$

This means that the pH of a solution is the negative power to which the number 10 must be raised in order to correctly express the hydrogen ion concentration. As we saw in the two examples above, the exponential power of the hydrogen ion concentration, expressed with the opposite numerical sign, gave the value on the pH scale.

Study the relationships between hydrogen ion concentration, pH, and hydroxide ion concentration in Table 8.3. Table 8.4 gives the pH of some common substances so you can begin to have a common sense feeling for the values of the pH measurement scale.

As you become familiar with the use of the pH measurement scale, keep in mind the relationships shown in Table 8.3. You have observed that pH values below 7 represent **acidic solutions,** a pH of 7 represents a **neutral solution,** and that a pH above 7 represents a **basic solution.** Notice that the pH values expressed here are positive numbers, even though the concentrations they represent have negative exponents. Remember, too, that every pH unit represents an acidity that is another power of 10 greater than the next higher

Table 8.4.   Values of pH for Selected Solutions

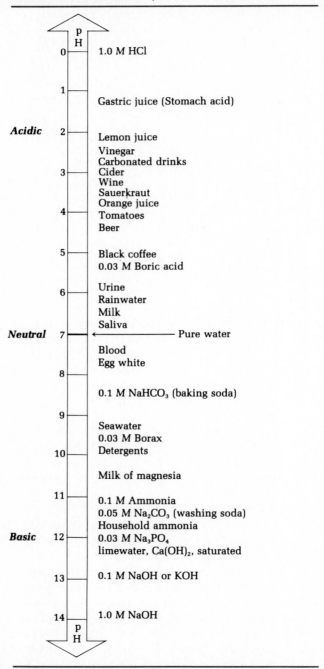

pH unit. This means that a solution with a pH of 3 has ten times the concentration of hydrogen ions as does a solution with pH of 4. You already know that in an aqueous solution at 25°C, the product of the hydrogen ion and hydroxide ion concentrations are constant and equal to 1×10^{-14}.

EXAMPLE 8.3

A solution of milk of magnesia is found to have a hydrogen ion concentration of 1×10^{-10} M. Find the pH of this solution. Is the solution acidic, basic, or neutral?

Solution
The hydrogen ion concentration is given and so the power of 10 can be used to find the pH, remembering to reverse its sign to give a positive value for the pH. Once the numerical value of the pH is known, the material can be classified by comparing the value with the neutral value of 7.

pH = 10

The milk of magnesia solution is basic, a useful feature for a product designed to neutralize "excess stomach acid."

EXAMPLE 8.4

Find the pH and the [OH$^-$] of a solution of lime juice in which the hydrogen ion concentration is equal to 1×10^{-2} M.

Solution
Since the hydrogen ion concentration is given, the pH can be found from the power of 10. Remember that the product of the hydrogen ion and hydroxide ion concentration must equal 1×10^{-14}.

If [H$^+$] = 1×10^{-2}, then the pH = 2. The product of the hydrogen ion concentration and the hydroxide ion concentration is equal to 1×10^{-14} only if the [OH$^-$] is equal to 1×10^{-12}.

Supplementary Study

If you have studied the concept of logarithms, you know that it is also possible to rewrite the preceding exponential expression in related forms. If

$$[H^+] = 1 \times 10^{-pH}$$

then

$$pH = -\log [H^+]$$

In this form, you can use the given hydrogen ion concentration to find the pH or you can use the given pH to find the hydrogen ion concentration. If you know how to use a table of logarithms, consult that table. Chances are that your calculator can be helpful here and will allow you to find logarithms and retrieve exponential values from logarithms. Ask your instructor for help with your calculator and also consult the instruction manual that accompanied your calculator. Here are examples of each process.

If lemon juice has a hydrogen ion concentration of 4.0×10^{-3} M, what is its pH?

$$pH = -\log (4.0 \times 10^{-3})$$
$$= 2.4$$

Note how acidic this is compared with pure water. Check Table 8.3 to see how this value fits between a pH of 2 and 3, given that the hydrogen ion concentration of 4×10^{-3} is between 1×10^{-2} and 1×10^{-3}. Also find lemon juice on Table 8.4.

If saliva has a pH of 6.4, find the hydrogen ion concentration.

$$pH = -\log [H^+]$$
$$6.4 = -\log [H^+]$$
$$antilog (-6.4) = 4.0 \times 10^{-7} M$$

Note that this concentration is larger than the concentration of hydrogen ion in pure water. This means you have an acidic material. Again, check the location of this value in Tables 8.3 and 8.4.

EXAMPLE 8.5

Which of these solutions is more acidic—a solution with pH 6.0 or a solution containing a hydrogen ion concentration of 1.5×10^{-6} M?

Solution
You can compare only if you know the same type of information about each solution. You have the choice, therefore, of converting the pH for the first solution to an equivalent hydrogen ion concentration or find the pH corresponding to the second solution's given hydrogen ion concentration and then comparing pH values.

Method 1: A solution with a pH of 6.0 has an hydrogen ion concentration of 1.0×10^{-6} M. This is less than the hydrogen ion concentration of the second solution, so the second solution is more acidic.

Method 2: If the hydrogen ion concentration of the second solution is 1.5×10^{-6} M, then the pH of the solution must be between 5 and 6. (You may solve for the exact pH, which is 5.82, or you may find an approximate value from Table 8.3.) Since this pH is lower than the pH of the first solution, the conclusion once again is that the second solution must be more acidic.

Laboratory Measurement of pH Values

Experimental determination of the exact pH value is often an important measurement. The health professional may need to know the pH of a patient's blood or urine in order to follow the progress of recovery from disease. There are many commercially available instruments that perform this task reliably. These instruments are known as pH meters and are designed to measure the difference in electrical potential between two electrodes, this measurement being proportional to the hydrogen ion concentration when specifically designed electrodes are used. A typical digital pH meter is shown in Figure 8.3.

For many purposes, an exact reading of the pH is not as important as the information you might obtain more quickly by the use of chemical indicators. You may only want to know *if* the solution being tested is acidic or basic, and not the specific pH. Acid–base indicators typically exist in two or more forms with different colors. These substances are, therefore, sensitive to the concentration of hydrogen or hydroxide ion in solution and are chosen for their ability to noticeably change color as the pH changes. We have aleady discussed the use of litmus indicator to tell if a solution is acidic or basic, but there are many other possible acid–base indicators that change colors in different pH ranges. Table 8.5 gives you some of the options commonly available.

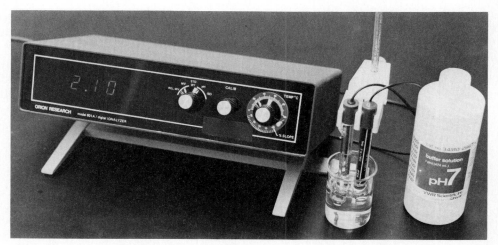

FIGURE 8.3   This digital pH meter is able to accurately read the pH of the solution being tested. As with all instruments, it must be properly calibrated before use, the function of the buffer solution shown.

Table 8.5.   Color Changes and pH Ranges for Acid–Base Indicators

pH	0	1	2	3	4	5	6	7	8	9	10	11	12	13	14

Litmus ← Red ⟶ T.R.[a] (6.0 – 8.0) Blue →

Phenolphthalein ← Colorless ⟶ T.R. (8.2 – 10.0) pink →

Methyl orange ← Red ⟶ T.R. (3.2 – 4.4) Yellow →

Bromocresol green ← Yellow ⟶ T.R. (3.8 – 5.4) Blue →

Bromothymol blue ← Yellow ⟶ T.R. (6.0 – 7.6) Blue →

Thymolphthalein ← Colorless ⟶ T.R. (9.4 – 10.6) Blue →

[a] T.R. = transition range.

Chemical acid–base indicators may be used in combination to help determine the pH of a solution more closely but without the need for the use of a pH meter. For example, if a certain solution turns litmus red, we know only that the pH of that solution is below 6. If a separate portion of this same solution is tested with methyl orange and produces a yellow color, this tells us that the pH of that solution must be greater than 4.4. Together these two indicators tell us that the solution's pH must be between 4.4 and 6. Paper can be saturated with combinations of indicators and then used as "universal" indicators. Color charts are printed on the dispenser of such indicator papers to help interpret the color changes.

Clinical urinary pH values are widely reported for hospitalized patients and you should be familiar with the following points. Normal kidneys are capable of producing urine that can vary from a pH of 4.5 to slightly higher than 8.0. Freshly voided urine from patients on a normal diet is acidic and has a pH of about 6.0. Contrast this pH level with that of normal blood, which is between 7.34 and 7.42.

More acidic urine may be excreted by patients on high protein diets, those taking urinary acidifiers such as NH_4Cl, or those suffering from uncontrolled diabetes mellitus. Alkaline urine may be seen in people consuming diets high in vegetables or milk or those taking heavy doses of antacids like $NaHCO_3$. Certain antibiotics (neomycin, kanamycin, streptomycin) are most effective in the treatment of urinary tract infections when they are excreted in an alkaline urine.

For routine analysis, urinary pH may be measured with indicator paper strips and a color chart. Some of the commercial products sold for this purpose are Combistix and Multistix. One simply dips the paper strip into the urine (or touches the wet diaper) and then reads the pH by comparison with the color chart supplied by the manufacturer.

Titrations

In addition to determining the pH of a solution, it is sometimes necessary to determine its total acidity or alkalinity. This is a measure of its ability to neutralize a given acid or base. The pH does not necessarily reflect this capacity for neutralization, especially in the cases of weak acid or a weak base. As an example, if the pH of a 0.1 M solution of acetic acid is taken, it is found to be approximately 3. On the other hand, the pH of a 0.1 M solution of hydrochloric acid is equal to 1. This difference is because the strong acid HCl is totally ionized in solution, releasing all of its hydrogen ions; the weak acid $HC_2H_3O_2$ is not fully ionized. Despite this difference in pH, a given volume of each acid will neutralize the *same* amount of base.

The determination of the total acid or base in a solution can be accomplished by the procedure called **titration.** In an acid–base titration, a measured volume of an acid or a base of known concentration is exactly reacted with a measured volume of a base or acid of unknown concentration. These values can then be used to calculate total acid or base present. Titrations are possible if the end point of the reaction can be easily and clearly detected by means of a color change in a chemical indicator, change in electrical conductivity, or some other method. Acid–base titrations are the most common titrations, although other types are also possible, such as oxidation–reduction titrations. Figure 8.4 shows equipment necessary for a typical titration in the laboratory. Titrations that are done repeatedly in a clinical laboratory may also be carried out in automated fashion by the use of specifically designed equipment.

The goal in any titration is to locate the **equivalence point** of the titration. This is the point at which exactly the same number of moles of hydrogen ion have been added to equal and therefore neutralize the same number of moles of hydroxide ion. The use of a chemical indicator actually allows you to find the **end point,** literally the end of the titration. The indicator cannot change color, however, until the equivalence point has been passed, and you now have a very small amount of excess acid or base. It is important when titrating, therefore, to go very slowly as the end point is approached so that you can stop the titration just as the drop causing the excess of acid or base is added. When you try this in the laboratory, you will see that your skill in locating the end point will grow as you practice the technique.

After completing a titration, we can calculate the unknown value for which we performed the titration. Here is an example of that process.

▰▰▰▰▰**EXAMPLE 8.6**▰▰▰▰▰

A titration was carried out on a 25.00-mL sample of nitric acid of unknown concentration. The titration required 35.75 mL of 0.1025 M sodium hydroxide.

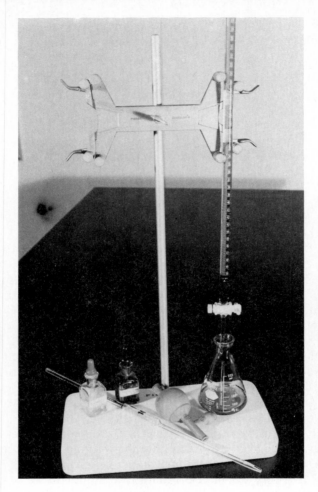

FIGURE 8.4 Assembled here are the necessary equipment and materials to carry out an acid-base titration. The volumetric pipet and bulb have been used to deliver a measured volume of acid to the flask. Distilled water and a few drops of indicator have then been added. Now base of known concentration can be added from the buret until the characteristic color change for that indicator is observed.

   a. Write the equation for the titration reaction.
   b. Calculate the concentration of the acid.

Solution
This is an acid-base neutralization reaction so the product will be water and a salt, in this case sodium nitrate.

   To solve the quantitative part of the question, remember that 1 mol of hydrogen ion reacts with 1 mol of hydroxide ion. Because molarity is a ratio of moles to liters, the product of volume and molarity will yield the number of moles of each ion.

   a. $HNO_3 + NaOH \rightarrow H_2O + NaNO_3$

b. At the equivalence point of the titration, the number of moles of hydrogen ion will equal the number of moles of hydroxide ion.

$$\text{moles of hydrogen ion} = \text{moles of hydroxide ion}$$

$$(0.02500 \text{ L}) \frac{(X \text{ mol HNO}_3)}{\text{L}} = (0.03575 \text{ L}) \frac{(0.1025 \text{ mol NaOH})}{\text{L}}$$

$$X = \frac{(0.03575)(0.1025)}{0.02500}$$

$$X = \frac{0.1466 \text{ mol HNO}_3}{\text{L}}$$

$$X = 0.1466 \text{ M HNO}_3$$

Remember to check this value with your common sense. It is reasonable that the acid is *more* concentrated than the base because it took more volume of base (35.75 mL) to neutralize a smaller volume of the acid (25.00 mL).

If the acids and bases used contain only *one* hydrogen or hydroxide ion per formula unit, then this generalized equation for the titration calculation can be used.

(volume of acid, L) (concentration of acid, mol/L) =
(volume of base, L) (concentration of base, mol/L)

This relationship will hold for strong or weak acids and bases, as long as they contain only *one* hydrogen ion per formula unit and only *one* hydroxide ion per formula unit. In the circumstance that an acid or a base contains more than one of its characteristic ion, it is still true that a number of moles of hydrogen ion must equal the number of moles of hydroxide ion at the equivalence point. A suitable ratio must be used in the calculation in order to maintain the fundamental idea that one hydrogen ion reacts with one hydroxide ion.

BRONSTED-LOWRY ACIDS AND BASES

In 1923, the Danish chemist Johannes Bronsted (1897–1947) and the English chemist Thomas Lowry (1874–1936) simultaneously proposed a broader definition of acids and bases. In Bronsted-Lowry theory, a **Bronsted-Lowry acid** is defined as a **proton donor** and a **Bronsted-Lowry base** is defined as a **proton acceptor**. A proton, of course, is the same as a hydrogen ion, for if a hydrogen atom loses one electron to become an ion, all that is left is a single proton. Note that this set of definitions does not depend on aqueous solution and also does not depend on the presence of any hydroxide ions.

In general, a Bronsted-Lowry acid-base reaction is represented in this manner:

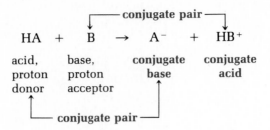

The reaction is not limited to neutral starting materials. A Bronsted-Lowry acid may be positive, negative, or neutral in charge as long as it can donate a proton. A Bronsted–Lowry base is usually neutral or negatively charged to make accepting a proton possible.

One of the advantages of the broader Bronsted-Lowry definitions of acids and bases is that they greatly extend the concept of an acid-base reaction. For example, the reaction of hydrogen chloride gas dissolving in and reacting with water can be considered as a B-L acid-base reaction.

The HCl in the gas phase is a very good proton donor and for this reason, is a Bronsted-Lowry acid. Water is not considered a classic base because it does not release significant amounts of hydroxide ions in solution. Water does act as a B-L base in this reaction by accepting the proton donated by HCl(g).

As you look at several Bronsted-Lowry reactions, you will find that water is capable of either donating or accepting a proton, depending on the relative proton donating or accepting abilities of other reactants present. Water is said the be **amphoteric** because of its ability to act as either a B-L acid or a B-L base, as the circumstances require.

EXAMPLE 8.7

Which of the substances shown in the following equations are Bronsted-Lowry acids? Which are Bronsted-Lowry bases? Identify the conjugate pairs in each reaction.

a. $H_2CO_3(aq) + H_2O(l) \rightleftharpoons H_3O^+(aq) + HCO_3^-(aq)$

b. $NH_3(g) + H_2O(l) \rightleftharpoons NH_4^+(aq) + OH^-(aq)$

Solution

Bronsted-Lowry acids donate a proton and B-L bases accept the proton. The conjugate pairs can be identified after the individual reactants and products are identified as acids or bases.

a. In the first reaction, H_2CO_3 is a B-L acid. Its conjugate base is HCO_3^-. Water is acting as a B-L base by accepting the proton donated by the carbonic acid. Its conjugate acid is the hydronium ion, H_3O^+. This exchange can be summarized this way.

$$
\underbrace{H_2CO_3 + H_2O}_{} \rightleftharpoons H_3O^+ + HCO_3^-
$$

$$
\text{acid} \qquad \text{base} \qquad \text{acid} \qquad \text{base}
$$

(conjugate pair: H_2CO_3 / H_3O^+; conjugate pair: H_2O / HCO_3^-)

b. In the second reaction, water is acting as a B-L acid by donating a proton to the ammonia. Ammonia is acting as a B-L base in this reaction by accepting the proton donated by the water. This exchange can be summarized this way.

$$
NH_3 + H_2O \rightleftharpoons NH_4^+ + OH^-
$$

$$
\text{base} \qquad \text{acid} \qquad \text{acid} \qquad \text{base}
$$

(conjugate pair: NH_3 / NH_4^+; conjugate pair: H_2O / OH^-)

Predicting the Position of Equilibrium

Bronsted-Lowry acid-base theory can be used to show us how a reaction either proceeds to completion or establishes an equilibrium. Why is it, for example, that when hydrogen chloride gas and water react to form hydronium ion and chloride ion, the reaction goes nearly to completion? On the other hand, when ammonia gas reacts with water, the reactants do not fully combine to form 100 percent products; rather an equilibrium is established in which there are actually more reactants than products when the process is complete. These two reactions follow.

$$
HCl(g) + H_2O(l) \rightleftharpoons H_3O^+(aq) + Cl^-(aq)
$$

$$
\text{acid} \qquad \text{base} \qquad \text{acid} \qquad \text{base}
$$

(conjugate pair: HCl / H_3O^+; conjugate pair: H_2O / Cl^-)

$$\text{NH}_3(\text{aq}) + \text{H}_2\text{O}(\text{l}) \rightleftharpoons \text{NH}_4{}^+(\text{aq}) + \text{OH}^-(\text{aq})$$

base   acid   acid   base

(conjugate pair / conjugate pair)

The difference between these two reactions is that in the first case, the HCl interacts strongly with water, producing a high proportion of a fairly weak conjugate base, the chloride ion. In the second case, the water acts as the B-L acid and interacts fairly weakly with the ammonia. Water's conjugate base is the relatively strong hydroxide ion, and this species is able to attract protons so strongly that the reverse reaction is actually favored over the forward reaction.

Note that water, capable of being amphoteric, acts as a base in the first reaction but as an acid in the second reaction. Study these two reactions to understand how water's behavior in a Bronsted-Lowry acid-base reaction depends on the relative proton accepting or donating ability of the other substance present.

In order to help predict how substances will react, Bronsted-Lowry acids

Table 8.6. Relative Strengths of Bronsted-Lowry Acids and Bases

	Acid	Formula for acid	Conjugate base	
	Perchloric acid	$HClO_4$	$ClO_4{}^-$	
	Sulfuric acid	H_2SO_4	$HSO_4{}^-$	
	Hydroiodic acid	HI	I^-	
	Hydrobromic acid	HBr	Br^-	
	Hydrochloric acid	HCl	Cl^-	
	Nitric acid	HNO_3	$NO_3{}^-$	
	Hydronium ion	H_3O^+	H_2O	
	Bisulfate ion	$HSO_4{}^-$	$SO_4{}^{2-}$	
	Phosphoric acid	H_3PO_4	$H_2PO_4{}^-$	
	Hydrofluoric acid	HF	F^-	
	Acetic acid	$HC_2H_3O_2$	$C_2H_3O_2{}^-$	
	Carbonic acid	H_2CO_3	$HCO_3{}^-$	
	Hydrosulfuric acid	H_2S	HS^-	
	Dihydrogen phosphate ion	$H_2PO_4{}^-$	$HPO_4{}^{2-}$	
	Ammonium ion	$NH_4{}^+$	NH_3	
	Hydrocyanic acid	HCN	CN^-	
	Bicarbonate ion	$HCO_3{}^-$	$CO_3{}^{2-}$	
	Water	H_2O	OH^-	

Left margin label: Weak acid → Moderate acid → Strong acid →

Right margin label: Weak base ← Moderate base ← Strong base ←

and bases can be rated on a scale of chemical strength. This rating includes the conjugate acids and bases, because the stronger the acid, the weaker its conjugate base will be. By the same reasoning, the stronger a B-L base is, the weaker its conjugate acid will be. Table 8.6 lists the relative strengths of Bronsted-Lowry acids and bases.

We are now in a position to more fully understand the reaction of HCl with water. Both HCl and H_3O^+ are rated as strong acids. Chloride ion is rated as a very weak base for it has very little tendency to accept a proton. Water is a weak-to-moderate base. The reaction of HCl with H_2O is therefore seen to go essentially to completion because a strong acid and a moderately strong base are reacting. In the reverse direction, a strong acid and a weak base would be reacting; this competition is not sufficient to establish much of a reverse reaction.

The combination of ammonia and water is an example of how water can act as a Bronsted-Lowry acid rather than as a base. Experimental evidence shows that the solution is alkaline, indicating that hydroxide ions are formed. Notice that water has given up a proton to ammonia to form the ammonium ion. Thus water acts as an acid and ammonia as the base. In the reverse reaction, NH_4^+ is the conjugate acid and OH^- is the conjugate base. Because OH^- is a stronger base than NH_3, the reaction progresses further in the reverse direction. Consequently, ammonia solutions are only partially ionized and the solution is considered to be a weak classic base.

INTERACTIONS OF SALTS WITH WATER

Anions and cations may change the pH of an aqueous solution. This may seem surprising to you, for salts are often thought of as "neutral" substances. In fact, when a salt is placed in water solution, you may find that the solution remains neutral, becomes acidic, or becomes basic. We have already mentioned that NH_4Cl is used to acidify urine. These reactions can be understood by applying Bronsted-Lowry theory. You should have the opportunity to actually try these different reactions in the laboratory.

EXAMPLE 8.8

Why does an acidic solution result when sodium hydrogen sulfate, $NaHSO_4$, is placed in water?

Solution

There must be some interaction between sodium hydrogen sulfate and water. The sodium ion or the hydrogen sulfate (bisulfate) ion must be responsible. You will need to evaluate each of these substances for their ability to interact with water in a Bronsted-Lowry acid-base reaction.

According to Table 8.6, the hydrogen sulfate ion is capable of acting as a Bronsted-Lowry acid, releasing some of its protons to water. The reaction is

$$\overset{\text{acid}}{HSO_4^-} + \overset{\text{base}}{H_2O} \rightleftharpoons \overset{\text{acid}}{H_3O^+} + \overset{\text{base}}{SO_4^{2-}}$$

conjugate pair

conjugate pair

In this process, notice that some hydronium ions are released into solution, lowering the pH of the solution from the neutral value of 7. The reaction is shown as an equilibrium reaction. It will not go to completion as the hydrogen sulfate ion is only a moderately strong B-L acid according to Table 8.6.

The sodium ion is not capable of changing the pH of the water for Na^+ cannot gain or donate a proton. If the ion cannot upset the balance of hydronium and hydroxide ion in water, it cannot change the pH of the system.

The Bronsted-Lowry acid-base reaction shown in Example 8.7 is an example of hydrolysis. **Hydrolysis** of a salt takes place when the salt interacts with water such that solution with an excess of hydronium or hydroxide ions results.

EXAMPLE 8.9

Do you expect a solution of sodium cyanide to be acidic, basic, or neutral? Explain your reasoning.

Solution

To decide, you will need to see if there is any potential interaction between either the sodium ion or the cyanide ion and water. Table 8.6 will be helpful in finding the relative strengths of B-L acids and bases.

Sodium ion will not interact with water to change the balance of hydronium and hydroxide ion, but the cyanide ion will interact. The cyanide ion is the conjugate base of a weak acid, HCN. The interaction will be

conjugate pair

$$\overset{\text{base}}{CN^-(aq)} + \overset{\text{acid}}{H_2O} \rightleftharpoons \overset{\text{acid}}{HCN} + \overset{\text{base}}{OH^-(aq)}$$

conjugate pair

Therefore, the solution now has an excess of hydroxide ions and will be basic. This particular hydrolysis process was the basis for a simple and fast test for the possible presence of cyanide ion in contaminated Tylenol tablets.

BUFFER SYSTEMS TO CONTROL pH

We have now seen that three large classes of compounds—acids, bases, and salts—are all capable of affecting the pH of aqueous solutions. Maintaining the proper level of acidity is very critical in most biological systems, and so the control of pH is also important. **Buffers** are solutions formed from combinations of solutes that serve to hold the pH of a solution constant or nearly constant in the face of threatened change from the addition of either acid or base. Buffers are composed of a weak acid and its conjugate base or a weak base and its conjugate acid. These solutions can be understood by once again applying the principles of Bronsted-Lowry acid-base theory.

One of these natural buffer systems is the carbonic acid/bicarbonate buffer. This is the system partially responsible for maintaining a nearly constant pH of 7.3-7.4 in the blood. The solutes present are carbonic acid, H_2CO_3, and the bicarbonate ion, HCO_3^-. Find the position of this conjugate pair in Table 8.6. The equilibrium reaction involving this pair of solutes is

$$\underbrace{H_2CO_3 + H_2O \rightleftharpoons HCO_3^- + H_3O^+}_{}$$

conjugate pair

acid   base   base   acid

conjugate pair

This equilibrium can be threatened by the addition of either hydronium ions or hydroxide ions. If this happens, we say that a *stress* is put on the equilibrium system. One of the major strengths of an equilibrium system is that it can respond to that stress and restore the equilibrium, as long as the system is not stressed too severely. This ability to restore the equilibrium in response to stress is known as **Le Chatelier's principle**. In this system, the restoration of equilibrium means that the pH is maintained at approximately the same level.

If extra hydrogen ions enter the blood, they can react with the bicarbonate ion and produce more carbonic acid. However, the carbonic acid produced does not build up in the blood because the acid is unstable and decomposes into carbon dioxide gas and water.

$$H_2CO_3 \rightarrow H_2O + CO_2(g)$$

The gaseous carbon dioxide can then be eliminated through respiration, reestablishing the original ratio of hydrogen carbonate ion (bicarbonate ion) to carbonic acid.

On the other hand, if additional base is added to the carbonate buffer system, the hydroxide ion will react with the acid and produce more hydrogen carbonate ion until the excess hydroxide ions are neutralized. This reaction can be represented this way.

$$\text{H}_2\text{CO}_3 \;+\; \text{OH}^- \;\rightleftharpoons\; \text{HCO}_3^- \;+\; \text{H}_2\text{O}$$

acid   base   base   acid

The blood is called upon to use this buffer system any time that the blood pH falls above or below the usual level of 7.3-7.4. The carbonate buffer system is normally able to react to each type of stress and to reestablish equilibrium. It is possible to exceed the capacity of the buffer to respond if enough acid or base is added.

If the pH drops below 7.3, the condition is called **acidosis**. This may be caused by respiratory system problems that make the level of carbon dioxide in the blood temporarily rise and therefore the concentration of carbonic acid increase and the pH drop. Acidosis of the blood may also be caused by kidney failure or other metabolic conditions such as diabetes mellitus.

If the pH of the blood rises to above 7.4, the condition is called **alkalosis**. This also may be caused by certain respiratory problems or may be the result of metabolic dysfunctions such as kidney disease or even severe dehydration. In Chapter 15 we will again apply our knowledge of acids and bases as we study in more detail the conditions of acidosis and alkalosis.

SUMMARY

The three major categories of ionic substances—acids, bases, and salts—were the topics of this chapter. Experimentally, many members of this class have the ability to conduct electricity in water solutions. Acids may be recognized by their characteristic properties, including turning blue litmus to red. Bases turn the same indicator blue. Acids and bases react with each other in neutralization reactions to produce salt and water. Titration is a laboratory procedure for determining acid or base concentration. We have practiced the calculation of concentration from titration data.

Acids may be defined in the classic manner of Arrhenius as substances capable of furnishing hydrogen ion in solution. More generally, Bronsted–Lowry acids are proton donors. Bases may be considered as providing hydroxide ions in solution or as proton acceptors. The broader definitions of Bronsted-Lowry theory do not limit us to consideration of aqueous solutions and are more useful in explaining a wider variety of reactions.

The self-ionization of water into hydronium and hydroxide ions is expressed in the value K_w. This equilibrium constant of 1×10^{-14} is the product of the hydronium ion and hydroxide ion concentrations in any aqueous solution at 25°C. It can be used to calculate hydrogen or hydroxide ion concentration in any acidic, basic, or neutral aqueous solution if either ion concentration is known.

The pH scale is a convenient means of expressing the hydrogen ion concentration. A pH below 7 is acidic, 7 is neutral, and above 7 is basic. Each pH unit difference represents another power of 10 factor in the hydrogen ion concentration.

Bronsted-Lowry theory can be used to predict the position of equilibrium in an acid-base reaction. It also can be used to help understand the interactions of salts with water and the ways in which buffer systems help to control pH in our bodies.

KEY TERMS

Check your understanding of this chapter. Can you explain what is meant by each of these terms?

acid	hydrolysis
acidic solution	hydronium ion
acidosis	hydroxide ion
alkalosis	indicator
amphoteric	Le Chatelier's principle
base	litmus paper
basic solution	neutral solution
Bronsted-Lowry acid	neutralization
Bronsted-Lowry base	pH
buffer	proton acceptor
conjugate acid	proton donor
conjugate base	self-ionization
conjugate pair	strong acid
electrolyte	strong base
end point	titration
equilibrium	weak acid
equivalence point	weak base

STUDY QUESTIONS

1. Identify each of the following as a classic acid, a classic base, or neither.
   a. HNO_3 d. H_2SO_4
   b. CH_4 e. C_2H_5OH
   c. $Ca(OH)_2$

2. An unknown solution causes blue litmus to turn red, reacts with magnesium metal to produce hydrogen gas, and releases carbon dioxide gas when combined with potassium carbonate. Is the unknown solution an acid, a base, or neutral?

3. Zinc can react with hydrochloric acid to produce zinc chloride and hydrogen gas. Zinc can also react with sodium hydroxide to produce sodium zincate (Na_2ZnO_2) and hydrogen gas. Write a chemical equation for each reaction.

4. What do you know about an aqueous solution if it
    a. Causes red litmus to stay red and blue litmus to turn red.
    b. Causes red litmus to stay red and blue litmus to stay blue.
    c. Causes red litmus to turn blue and blue litmus to stay blue.

5. Classify each of these solutions as a strong electrolyte, a weak electrolyte, or a nonelectrolyte.
    a. Pure water
    b. Pure ethanol
    c. 0.1 M nitric acid
    d. 0.1 M lactic acid
    e. 0.1 M sodium hydroxide
    f. 0.1 M sodium chloride

6. Draw the Lewis structure for
    a. Formic acid
    b. Lactic acid

7. a. Define a classic acid and give an example.
    b. Define a Bronsted-Lowry acid and give an example.
    c. Give an example of Bronsted-Lowry acid that is not a classic acid.

8. May a given acid be described as both weak and dilute? Explain.

9. What is the difference between the terms *acidic* and *acetic*?

10. Write a molecular equation, a total ionic equation and a net ionic equation for the neutralization reaction of nitric acid with potassium hydroxide.

11. Magnesium carbonate is one of the ingredients in Tums. It reacts with the hydrochloric acid in your stomach. Explain, using an equation, why Tums can neutralize "excess stomach acid."

12. Sourdough bread is given in characteristic taste by the presence of lactic acid in sour milk. Write a chemical equation for the reaction of lactic acid with baking soda, $NaHCO_3$.

13. a. Find the hydronium ion and hydroxide ion concentration in 0.001 M hydrochloric acid.
    b. Why can you not answer the same type of question with this information for a 0.001 M acetic acid solution?

14. Classify each of these solutions as acidic, basic, or neutral. Assume all solutions are at 25°C.
    a. $[H^+] = 1.0 \times 10^{-3} M$
    b. $[H^+] = 1.0 \times 10^{-10} M$
    c. $[H^+] = 1.0 \times 10^{-7} M$
    d. $[OH^-] = 1.0 \times 10^{-3} M$
    e. $[OH^-] = 1.0 \times 10^{-7} M$

15. Arrange these solutions in order of increasing acidity: milk of magnesia, stomach acid, lemon juice, blood.

16. Arrange these solutions in order of increasing basicity: egg white, household ammonia, coffee, orange juice.

17. Find the hydrogen ion concentration and the hydroxide ion concentration of sample of orange juice with a pH of 4.

18. Which solution is more acidic—carrot juice with a pH of 5.0 or lime juice with a pH of 2.0?

19. Which solution is more acidic—a solution with pH of 5.0 or a solution containing a hydrogen ion concentration of $2.5 \times 10^{-5} M$? Explain your reasoning.

20. A solution gives a yellow color in methyl orange and is colorless in phenolphthalein. What does this tell you about the pH of the solution?

21. If a solution has a hydrogen ion concentration of $1.0 \times 10^{-4} M$, what color would you expect the solution to turn
    a. In bromocresol green?
    b. In bromothymol blue?
    c. In phenolphthalein?

22. Barbituric acid is the compound used as the starting material for several types of sleeping pills. If a 0.010 M solution has a pH of 3, is this a strong acid or a weak acid? Explain.

23. A 0.0250 M NaOH is used to titrate HCl of unknown concentration. If 25.6 mL of NaOH were needed to exactly react with 10.0 mL of HCl, what is the molarity of the HCl?

24. If 0.252 g of sodium carbonate (gram moleclar weight = 106) exactly reacts with 40.0 mL of HCl, what is the molarity of the acid?

25. Vinegar is acidic because of the presence of acetic acid. If 20.0 mL of 0.500 M NaOH are used to titrate a 10.0 mL sample of vinegar, find the percentage of acetic acid in the vinegar by weight.

26. "Household ammonia" is prepared by dissolving ammonia gas in water. In aqueous solution, an equilibrium is established releasing ammonium ions and hydroxide ions; 0.515 M HCl can be used to titrate household

ammonia. If 25.0 mL of ammonia exactly reacts with 32.4 mL of acid, what percentage of ammonia is the sample?

27. In each reaction, identify the Bronsted-Lowry acid, base, conjugate acid, and conjugate base.
    a. $CO_3^{2-}(aq) + H_2O(l) \rightleftarrows HCO_3^- + OH^-$
    b. $NH_3(liquid) \rightleftarrows NH_4^+ + NH_2^-$
    (Note: Reaction b does not occur in aqueous ammonia, only in liquid ammonia.)

28. Explain why each reaction does not go to completion but rather establishes an equilibrium.

    a. $H_2PO_4^-(aq) + H_2O(l) \rightleftarrows H_3PO_4(aq) + OH^-(aq)$
    b. $H_2S(aq) + Cl^-(aq) \rightleftarrows HS^-(aq) + HCl(aq)$

29. Predict whether each of these salt solutions will be acidic, basic, or neutral.
    a. Sodium chloride
    b. Sodium acetate
    c. Ammonium nitrate
    d. Potassium sulfide

30. List all ionic species present in a solution of sodium hydrogen carbonate, $NaHCO_3$. Which species on your list are present only in relatively low concentrations?

31. What is the purpose of a buffer solution?

32. Another important buffer solution in the body, in addition to the carbonate buffer, is the $HPO_4^{2-}/H_2PO_4^-$ system. How would this system respond to the addition of hydrogen ion? To the addition of hydroxide ion? Use equations as well as the information in Table 8.6 to explain your answers.

33. If a patient is diagnosed as being in a condition of alkalosis, the patient may be directed, as an emergency measure, to breathe into and out of a paper bag for a short period of time. What would be the purpose of this therapy?

9

A Brief Introduction
to Organic Chemistry

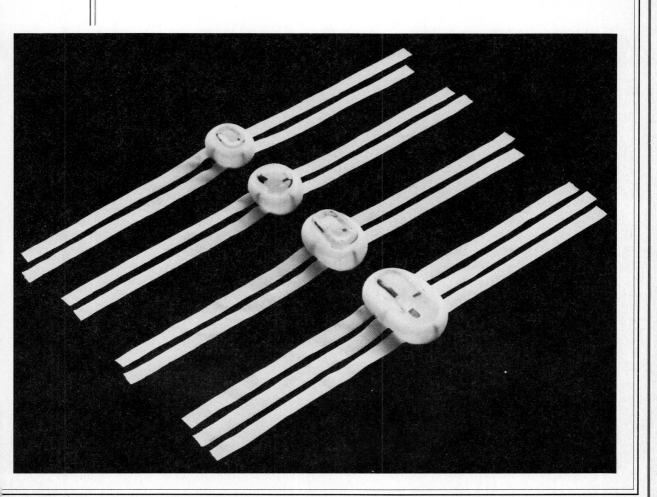

Organic chemistry, as a discipline, is only about 150 years old; it traces its development back to the Europe of the 1800s. Nevertheless, immense progress has been made in this area of chemistry, not the least of which is the synthesis of new, novel compounds—substances never before known to man.

From organic synthesis we obtain polymers such as those shown in our photograph. Each device is a Kaufman urinary incontinence prosthesis composed of a silicone gel polymer covered with a polyetherurethane foam. The device is implanted with the ties around the urethra.

Today we depend on organic chemistry to give us new substances with new applications: medical prostheses, IV tubing, dispensable syringes, and plastic containers of all sorts. Indeed, in the hospital and clinical setting, one sees evidence of organic chemical progress at every hand.

You will need a foundation in organic chemistry in order to understand the biochemical principles that follow in the rest of this book. Our goal in this and subsequent chapters is to prepare you for that understanding. (Photo courtesy of John S. Tiffany, American Heyer-Schulte Corporation.).

INTRODUCTION

It is likely that in your study of chemistry up to now, you have concentrated on inorganic chemicals—minerals, acids, bases, salts, and so on. With this chapter, however, you will begin the study of organic chemicals, and your appreciation of the world of chemistry will be immensely expanded. The fact that more than 3 million organic compounds are known to science tells us a lot about how extensively organic chemistry is involved in all aspects of our existence.

A knowledge of organic chemistry is prerequisite to the study of many disciplines in the health professions: biochemistry, nutrition, drugs from nature, synthetic drugs, problems of the environment, and enzymology. Without a knowledge of organic chemistry, our study of DNA, genes, and replication would be impossible.

Although we discuss organic compounds separately from inorganic compounds, thus emphasizing differences, we should be aware that both organic and inorganic compounds are found in living organisms, that both types can be used for various drug effects, and that both types contribute in important ways to human biochemistry.

THE NATURE OF ORGANIC COMPOUNDS

Organic compounds are carbon-containing molecules with accompanying hydrogen atoms, and often with other atoms such as oxygen and nitrogen. Organic compounds are characterized by

- Low melting points (usually well below 300°C).
- Volatility.
- Presence of covalent bonds.
- Ability to absorb radiant energy (UV, IR, and other).
- The potential for great complexity in structure.
- Often, significant pharmacological[1] action in the human and animal.

Many organic compounds are found in living organisms (man, animals, bacteria). Other organic compounds, like those in petroleum, are found in the earth. Of those that are naturally occurring, most all have been or could be obtained in the chemical laboratory by synthetic techniques. For example, most of the vitamin C on the market is synthetic in origin and is identical to naturally occurring vitamin C. Synthetic organic chemistry, in fact, has become extremely advanced. It is one of the most exciting achievements of modern science. Synthesis gives us the old chemicals and it gives us new drugs, new pesticides, new fibers, new paints, new lubricants and new polymers (and sometimes new problems!).

Although structures of organic compounds will not mean much to you at this point, we have listed a few of the more simple organic molecules in Figure 9.1. They are contrasted with a few inorganic compounds.

In this chapter we shall take up the class of organic compounds known as hydrocarbons—compounds that consist of covalently bound hydrogen and carbon. The principles of structure and naming that we learn here are very important and will apply to many of the compounds discussed in the following chapters. After our introduction to hydrocarbons, we shall begin the study of derivatives of hydrocarbons: alcohols, amines, carboxylic acids, esters, aldehydes, and ketones. Ultimately, our goal is to use our knowledge of organic chemistry in three special applications chapters (15, 16, and 17), which deal with body fluids, biotransformation, and body chemistry and diet.

THE TETRACOVALENT CARBON ATOM

The element that forms the backbone of the hydrocarbons is carbon. We can consider carbon to have four electrons in its valence shell (it lies in Group IVA of the periodic chart); thus each carbon atom needs four additional

[1] **Pharmacology** is the study of the actions of drugs in animals.

Organic Compounds

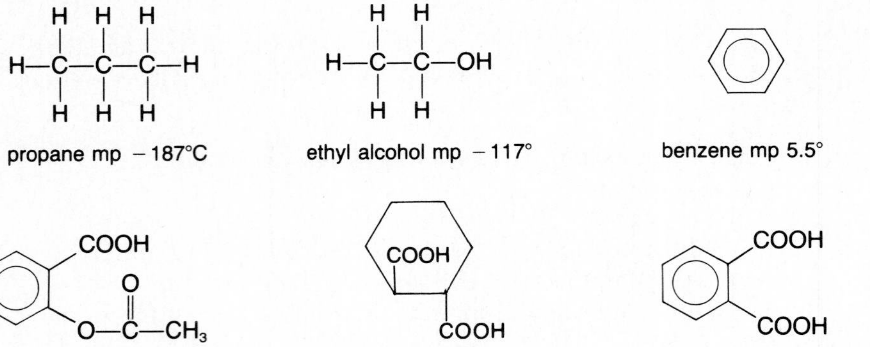

propane mp −187°C

ethyl alcohol mp −117°

benzene mp 5.5°

aspirin mp 135°

a cyclohexane dicarboxylic acid
mp 180°

phthalic acid mp 210°

Inorganic Compounds

$CaCl_2$	Na_2CO_3	UO_2
calcium chloride mp 772°	sodium carbonate mp 851°	uranium dioxide mp 2500°
H_3PO_4	Au	BI_3
orthophosphoric acid mp 42°	gold mp 1063°	boron triiodide mp 50°

FIGURE 9.1 A Few Organic Compounds Contrasted with Some Inorganic Compounds (mp = Melting Point).

electrons to achieve the stability of an octet. Four hydrogen atoms, each with one electron, can combine with carbon to give the stable octet

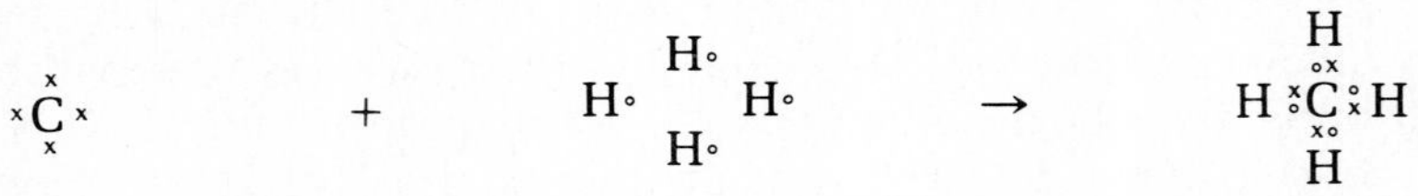

each C has 4 electrons with which to begin

four H atoms can supply the four additional electrons required for the octet

stability achieved: each C atom has an octet and each H has a duet of electrons

Carbon can also combine with other carbon atoms, sharing electrons to achieve the octet, and in the process forming chains:

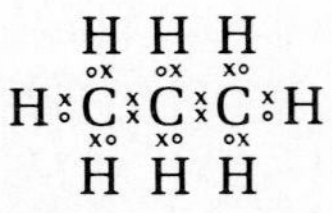

a chain of 3 C atoms

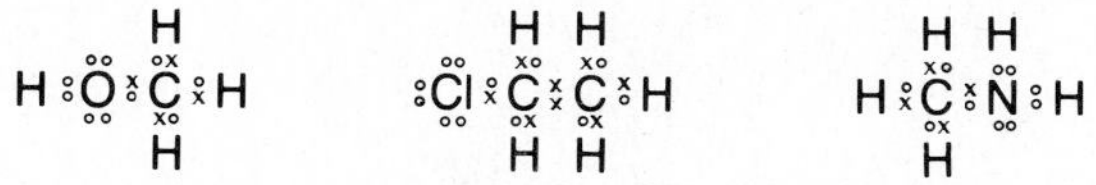

FIGURE 9.2 By bonding with four other atoms, carbon achieves an octet of electrons. Each carbon atom donates four electrons (denoted by an "x") to the covalent bonds. Count the number of "x"s in the molecules above to prove to yourself that each carbon contributes four.

In all of the carbon compounds we shall study in this book, carbon will show this tetravalence (valence of four). Examples are shown in Figure 9.2.

The bonds that carbon forms are covalent bonds, that is, shared electron pair bonds. This is because carbon lies in the middle of the periodic chart and has little tendency to either give up its electrons or take electrons away from other atoms. Hence carbon tends to *share* electrons in organic compounds. It is tetracovalent.

When we draw structures of organic compounds, we will not always use "dots" to represent the shared pair bonds. Rather, we usually use dashes (—) to represent a shared pair. Figure 9.3 illustrates the use of dashes for electron pairs and provides some additional examples of carbon's valence of four. Remember, four bonds equal eight electrons.

THE TETRAHEDRAL CARBON ATOM

The *shape* of a molecule can be an important factor if we are considering drug action. Drugs can act by attaching themselves to special cells in tissues (called the receptor site) and the shape or three-dimensional geometry of a molecule is critical for this attachment.

In most of the bonds it forms, carbon is said to be **tetrahedral**. Let us see what this means. We have learned that carbon is tetravalent; it forms four bonds with other atoms or with itself. These four bonds are not directed willy-nilly into space, but are arranged so *as to get out of each other's way as much as possible*. (Remember, electrons are negatively charged and repel each other.)

acetaldehyde   methyl ether   formic acid   allyl chloride

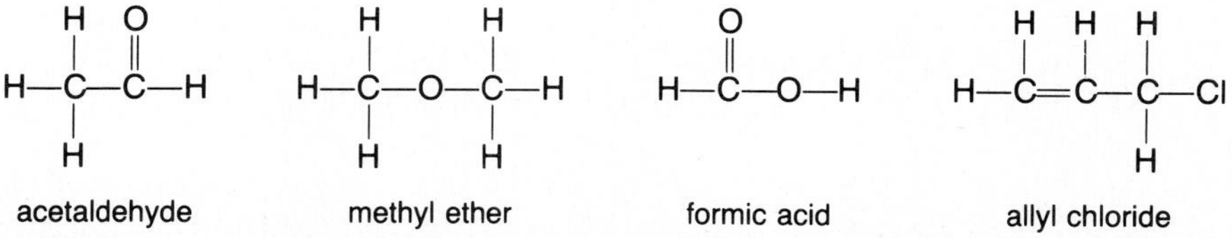

FIGURE 9.3 The four dashes around each carbon signify the octet. Where two dashes exist between atoms, a double bond (four electrons) is indicated. The total number of electrons around each carbon atom is eight.

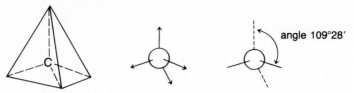

**FIGURE 9.4 In bonding, the carbon atom is tetrahedral.
The four covalent bonds it forms are directed to the four
corners of a regular tetrahedron.**

The optimal way to do this is to be directed to the four corners of a regular
tetrahedron. Try this. Pick up four pencils or four straws, all of the same
length. The four pencils will represent carbon's valence of four. Now, with all
four pencils held in one hand, arrange them so that the distance between them
is maximum. If you have done your arranging carefully, you will find that the
four pencils or straws are pointing exactly to the four corners of a regular
tetrahedron (Figure 9.4).

When tetrahedral carbon atoms bond, the bond angles between atoms are
approximately 109.5°. We can show this by drawing large circles to represent
carbon atoms, as in the two drawings on the right in Figure 9.5.

All through our study of hydrocarbons we must keep in mind the tetrahedral
shape of carbon atoms. In other compounds, carbon uses other types of
geometry; we shall examine these later in this chapter. When drawing hydrocarbon molecules, we must remember that even though we draw the structure
on a straight line (as in I), in reality the carbon chain is angular (as in II).

$$CH_3CH_2CH_2CH_3$$

I   II

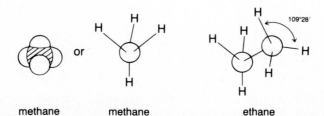

methane   methane   ethane

**FIGURE   9.5   Tetrahedral
Geometry of Methane and
Ethane (Large Circles Rep-
resent Carbon Atoms).**

THE ALKANES

We can now write structures for the simple carbon–hydrogen compounds with much greater sophistication and understanding. Methane, CH_4, can be written in four progressively more sophisticated ways:

$$CH_4 \qquad H:\overset{\displaystyle H}{\underset{\displaystyle H}{\overset{..}{\underset{..}{C}}}}:H \qquad H-\overset{\displaystyle H}{\underset{\displaystyle H}{C}}-H$$

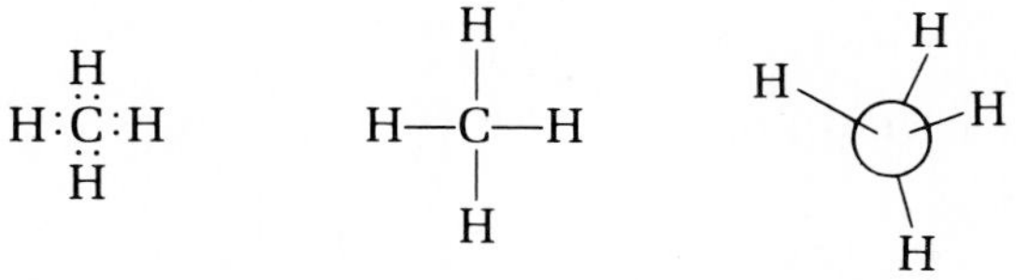

Actually, once we are thoroughly familiar with the idea of the tetrahedral carbon atom, we can use simple, shorthand expressions such as CH_4 and understand that the tetrahedral structure is implied.

Ethane, C_2H_6, is the next most complex hydrocarbon. It, like methane, is formed from tetrahedral carbon atoms. There are seven bonding pairs of electrons in ethane and a total of fourteen valence electrons.

═══ **EXAMPLE 9.1** ═══

Write the structure of ethane in four different ways: line formula, electron dot structure, dash-pair, and three-dimensional structure.

Solution

$$C_2H_6 \qquad H:\overset{\displaystyle H}{\underset{\displaystyle H}{\overset{..}{\underset{..}{C}}}}:\overset{\displaystyle H}{\underset{\displaystyle H}{\overset{..}{\underset{..}{C}}}}:H \qquad H-\overset{\displaystyle H}{\underset{\displaystyle H}{C}}-\overset{\displaystyle H}{\underset{\displaystyle H}{C}}-H$$

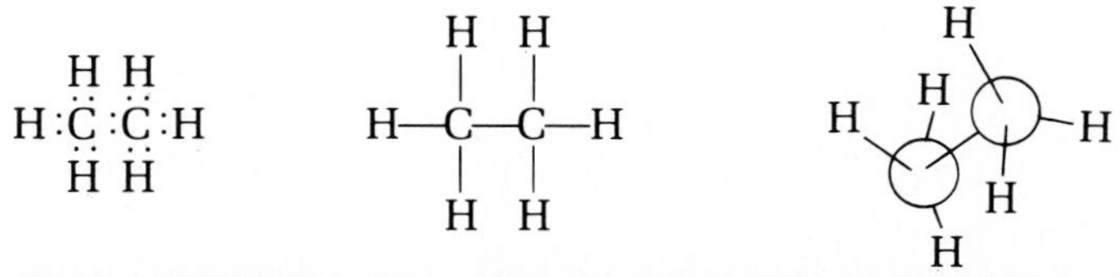

Propane, C_3H_8, is a hydrocarbon molecule in which three carbon atoms are attached to each other to form a chain:

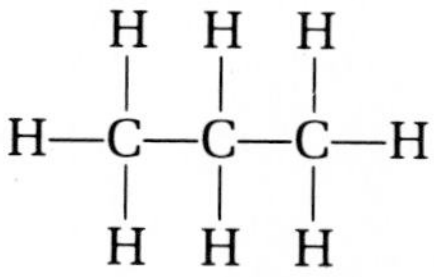

$$H-\overset{\displaystyle H}{\underset{\displaystyle H}{C}}-\overset{\displaystyle H}{\underset{\displaystyle H}{C}}-\overset{\displaystyle H}{\underset{\displaystyle H}{C}}-H$$

propane

Although two-dimensionally it may appear that all three carbon atoms in propane lie on a staight line, this is not the case. Remember, carbon atoms are tetrahedral; bond angles are 109°28′. Therefore, a more realistic way of representing propane is

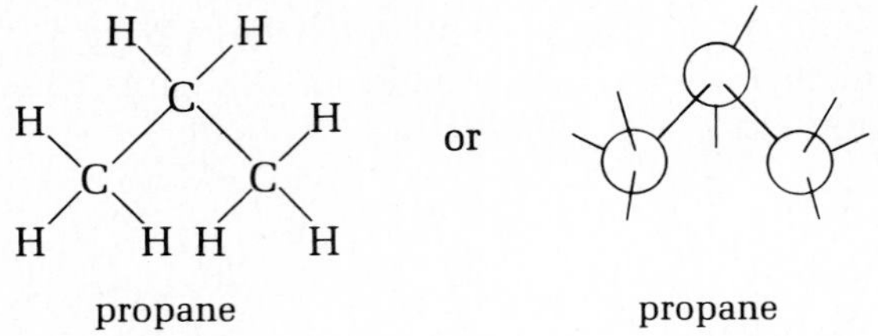

propane   propane

In both of these structures we see the tetrahedral nature of carbon. And we see that propane is an angular molecule, considering the chain of carbon atoms.

In butane, C_4H_{10}, the chain is four carbon atoms long, and again is angular. However, butane offers a new aspect of **alkane** geometry, for the chain in butane is free to be wide open, to fold back on itself, or to form hundreds of intermediate spatial arrangements (Figure 9.6). For us to understand how the three structures in Figure 9.6 can all represent the same molecule, butane, we need to recognize that a CH_3— group (called a *methyl group*) is free to assume many positions in space about its attachment to another carbon:

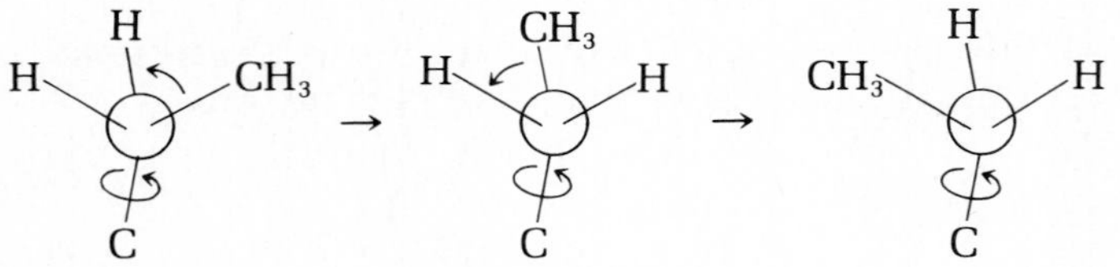

Thus we say there is *free rotation* about carbon–carbon single bonds. A C_2H_5— group (called an *ethyl group*), is also free to rotate about its attachment to another carbon:

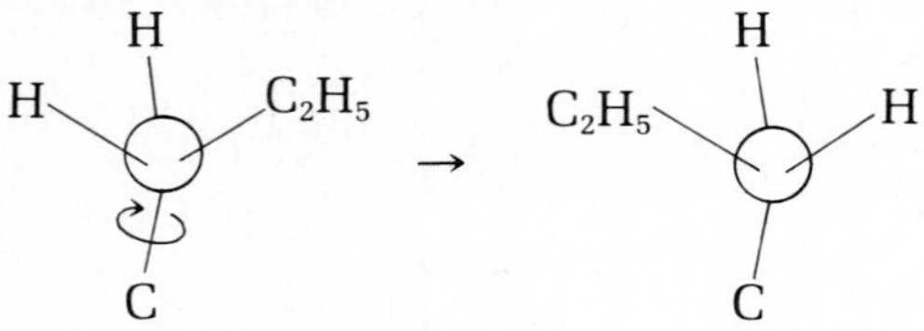

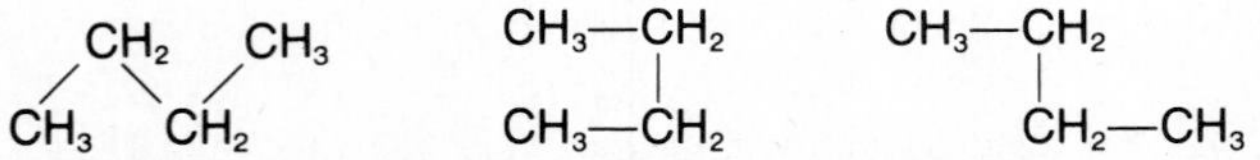

FIGURE 9.6   **Three Ways of Depicting the Same Molecule, Butane.**

EXAMPLE 9.2

How many different alkanes can be found among the following structures (We're using a special shorthand in which each dot represents a carbon atom; hydrogens are not shown).

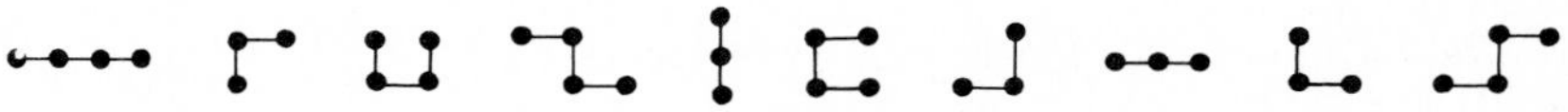

Solution

Only two alkanes, propane and butane, can be found among all of these structures.

Pentane, C_5H_{12}, can be expressed in a line formula as $CH_3CH_2CH_2CH_2CH_3$. We understand this to mean

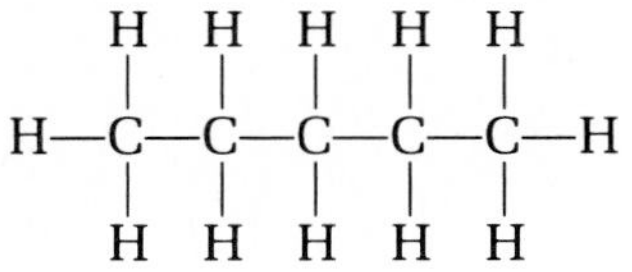

$$
\begin{array}{ccccc}
H & H & H & H & H \\
| & | & | & | & | \\
H-C-&C-&C-&C-&C-H \\
| & | & | & | & | \\
H & H & H & H & H
\end{array}
$$

However, our knowledge of free rotation about C—C bonds permits us to understand many other conceivable representations for the same substance, pentane:

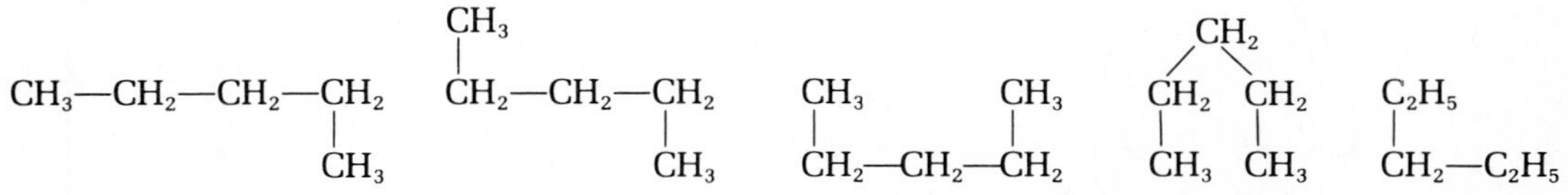

Table 9.1 summarizes the first five hydrocarbons we have studied and introduces the five additional that constitute the first ten. It is critical that you memorize all ten names as you will be using them repeatedly. Be sure you can correctly associate the prefix with the correct number of carbon atoms: *meth* for one, *eth* for two, *prop* for three, and so on.

Examine Table 9.1 and note that the structural difference between successive alkanes is constant: —CH_2—. A series of compounds in which each member differs from the preceding by a constant structure is termed a *homologous series*. Furthermore, the table tells us that we can ascribe a general formula to the alkanes: C_nH_{2n+2} (where n = number of carbon atoms in the alkane).

Table 9.1. The First Ten Unbranched Hydrocarbons

Structure	Formula	Name	Boiling Point (°C)	Phase at Room Temperature
CH_4	CH_4	Methane	-164	Gas
CH_3—CH_3	C_2H_6	Ethane	-88.6	Gas
CH_3—CH_2—CH_3	C_3H_8	Propane	-42	Gas
CH_3—$(CH_2)_2$—CH_3	C_4H_{10}	Butane	-0.5	Gas
CH_3—$(CH_2)_3$—CH_3	C_5H_{12}	Pentane	36	Liquid
CH_3—$(CH_2)_4$—CH_3	C_6H_{14}	Hexane	69	Liquid
CH_3—$(CH_2)_5$—CH_3	C_7H_{16}	Heptane	98	Liquid
CH_3—$(CH_2)_6$—CH_3	C_8H_{18}	Octane	126	Liquid
CH_3—$(CH_2)_7$—CH_3	C_9H_{20}	Nonane	151	Liquid
CH_3—$(CH_2)_8$—CH_3	$C_{10}H_{22}$	Decane	174	Liquid

Alkane Chain Branching

Branching in a chain of carbon atoms is inconceivable until we get to butane. For butane, two arrangements of a carbon skeleton are possible:

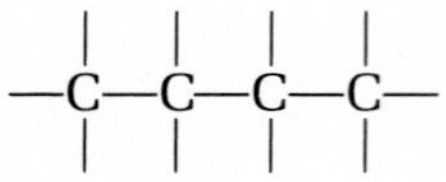

straight chain   branched chain

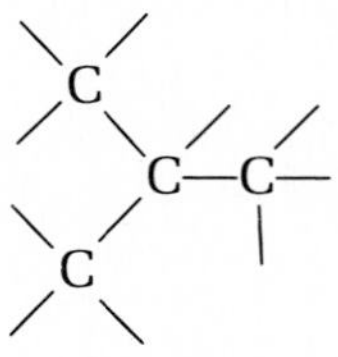

Branched-chain butane can be represented in many ways, but in fact there is only one branched chain butane (Figure 9.7).

Branched-chain butane is named isobutane because it is an **isomer** of straight chain butane. Isomers are compounds that have the same molecular formula but different arrangements of atoms; they are also called structural isomers. We shall encounter isomers of this and other types throughout our study of organic chemistry. The possibility of isomerism helps explain the great number of organic compounds known to science. To test your under-

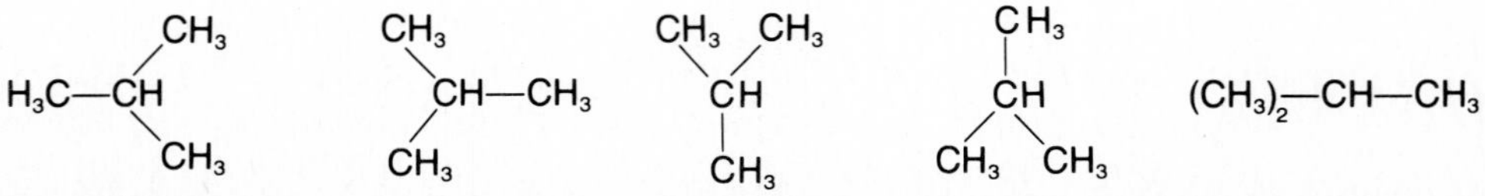

FIGURE 9.7 Five Ways of Drawing the Same Substance: Branched Chain Butane.

standing of structural isomerism, draw structures for all of the isomeric pentanes, C_5H_{10}. You should get three.

███████ **EXAMPLE 9.3** ███████

Examine all of the following structures and identify any isomeric pairs:

$(CH_3)_2CHCH_2CH_3$ $CH_3(CH_2)_3CH_3$ $CH_3(CH_2)_2CH(CH_3)_2$

       A   B   C

$(CH_3)_3CH$ $CH_3—CH(CH_3)_2$

    D   E

Solution

Compound A (C_5H_{12}) and compound B (C_5H_{12}) are isomers. They have the same number and kind of atoms, but the carbon chains are arranged differently. Compound C has six carbon atoms and D and E are both isobutane.

With all alkanes the prefix *iso* designates isomers in which a single methyl group is located on the next-to-last carbon atom of a straight chain, that is,

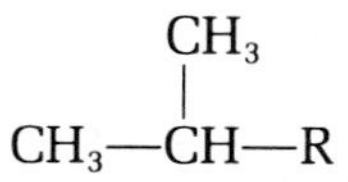

$$CH_3—\overset{\overset{\textstyle CH_3}{|}}{CH}—R$$

(R = remainder of chain.)

The prefix *normal*, abbreviated n, designates the unbranched or straight-chain isomer. If no prefix is specified, the normal is to be inferred.

You may be wondering if it is important that we make a point of distinguishing between isomers. Is a branched-chain form really different from a straight-chain form? The answer is emphatically *Yes*, they are different compounds, with different physical properties (like boiling point), and ofttimes with different drug effects in the human. The isomers we are considering here have different dimensions and different geometry, and this can greatly influence their ability to attach themselves to a receptor site and produce a drug action. Medicinal chemists know of many examples of slight, even subtle, changes in three-dimensional structure that result in major pharmacological differences.

Kinds of Carbon and Hydrogen Atoms in Alkanes

Because they can behave differently in chemical reactions, the carbon and hydrogen atoms in alkanes have been given special designations.

Primary carbons (1°) are those to which is directly attached only one other carbon group. *Primary hydrogen atoms* (1°) are those that are located on 1° carbon atoms.

Secondary carbons (2°) are those to which are directly attached two (and only two) carbon groups. *Secondary hydrogen atoms* (2°) are those located on 2° carbon atoms.

Tertiary carbons (3°) are those to which are directly attached three other carbon groups. *Tertiary hydrogens* (3°) are those located on 3° carbons. In the following examples, hydrogen atoms are designated as 1°, 2°, or 3°.

$$CH_3^{1°}$$
$$\diagdown$$
$$CH^{3°}\!\!-\!CH_2^{2°}\!\!-\!CH_2^{2°}\!\!-\!CH_3^{1°}$$
$$\diagup$$
$$CH_3^{1°}$$

$$CH_3$$
$$|$$
$$CH_3\!-\!CH$$
$$|$$
$$CH_3$$

nine 1°
and one 3°

$$CH_3^{1°}\!\!-\!CH_3^{1°}$$

$$CH_3^{1°}$$
$$|$$
$$CH_2^{2°}$$
$$|$$
$$CH_3^{1°}$$

$$CH_2$$
$$CH_2 \quad CH_2$$
$$CH_2$$

all 2°

Later on, we will add to our list of special designations for hydrogen atoms by identifying vinylic, allylic, benzylic, and aromatic hydrogen atoms.

Naming Alkanes and Their Derivatives

Table 9.1 provided us with the names of the first ten normal alkanes. We need additional names to identify complex branched chain alkanes and those with substituents such as halogen atoms or —OH groups.

When we remove one H atom from CH_4, we get a methyl group (CH_3—). Removal of any one of the six H atoms from ethane gives an ethyl group (C_2H_5—). In general, removal of one H atom from an alkane gives an alkyl group, "R." For naming, the rule is drop the *ane* from the alkane and add the suffix *yl*. Follow the name of the alkyl group by the name of the substituent.

■■■■ EXAMPLE 9.4 ■■■■

Provide a name for CH_3Cl, for C_2H_5I, and for RX.

Solution
CH_3Cl is methyl chloride; C_2H_5I is ethyl iodide; and RX is any alkyl halide. Remember, *R* means any alkyl group.

Propane, with two different kinds of H atoms (1° and 2°) can give rise to two different kinds of alkyl groups:

$$CH_3CH_2CH_2— \quad \text{and} \quad \overset{CH_3}{\underset{CH_3}{CH—}}$$

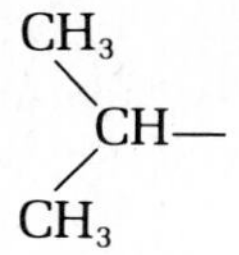

normal propyl group   isopropyl group

Note that the same n-propyl group results no matter which one of the six 1° H atoms is removed. Likewise, there is only one isopropyl group possible. Since both propyl groups have the $C_3H_7—$ formula, we can distinguish beween them by designating one n-$C_3H_7—$ and the other iso-$C_3H_7—$.

In n-butane, two kinds of hydrogen exist, and two kinds of butyl groups can arise:

$$CH_3CH_2CH_2CH_2— \quad \text{and} \quad CH_3\overset{|}{C}HCH_2CH_3$$
n-butyl group   secondary (sec-)butyl group

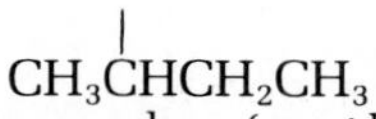

Isobutane, minus one hydrogen atom, gives rise to two more $C_4H_9—$ groups:

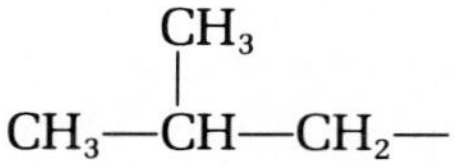

$$CH_3—\overset{CH_3}{\overset{|}{CH}}—CH_2— \quad \text{and} \quad CH_3—\overset{CH_3}{\underset{CH_3}{\overset{|}{\underset{|}{C}}}}—$$

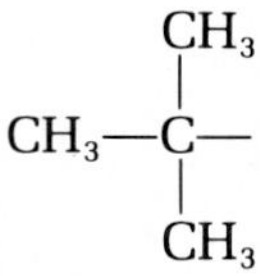

isobutyl group   tertiary-butyl group (t-butyl)

EXAMPLE 9.5

Alkane compounds with —OH groups are called alcohols. Provide names and structures for the four possible C_4H_9OH alcohols.

Solution

$CH_3CH_2CH_2CH_2OH$ $CH_3—\overset{|}{\underset{OH}{CH}}—C_2H_5$ $(CH_3)_2CH—CH_2OH$

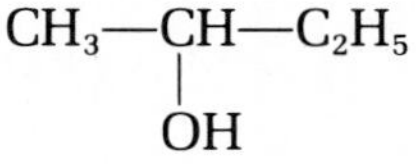

n-butyl alcohol   sec-butyl alcohol   isobutyl alcohol

$(CH_3)_3C—OH$

t-butyl alcohol

You should make a special effort to learn structures and names of all of the $C_3H_7—$ and $C_4H_9—$ alkyl groups because of their great importance. Take special care with the isobutyl group: removal of any one of the nine 1° hydrogen atoms from isobutane gives the same isobutyl group. This shows that the nine 1° hydrogens in isobutane are equivalent.

EXAMPLE 9.6

Which of the following structures correctly represents isobutyl bromide?

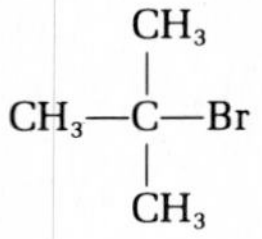

$$CH_3-\underset{\underset{CH_3}{|}}{\overset{\overset{CH_3}{|}}{C}}-Br \qquad CH_3(CH_2)_3Br \qquad CH_3CHBrCH_2CH_3 \qquad BrCH_2CH(CH_3)_2 \qquad CH_3CH_2CHBrCH_3$$

A   B   C   D   E

Solution
D. (*Note:* if the line formulas are confusing you, write out the completely expanded structural formula showing four bonds on each carbon.)

For pentane, hexane, and other alkanes, the number of possible alkyl groups become too great for us reasonably to assign individual names. We can, however, employ the *n-* designation in naming compounds substituted on the terminal carbon. Thus $CH_3(CH_2)_6CH_2Br$ is named *n*-octyl bromide.

We organic chemists use a special adjective when we refer to alkanes or compounds with hydrocarbon chains. We call them **aliphatic compounds,** to distinguish them from **aromatic** compounds (to be discussed later in this chapter).

The IUPAC System of Naming Alkanes and Their Derivatives
So far, all of the names we have learned are common names. A much more versatile system of organic nomenclature exists called the **IUPAC system** (for International Union of Pure and Applied Chemistry). Used throughout the world, this standard system of naming has become the common scientific language of organic chemists.

Here are the rules for naming any alkane or substituted alkane by the IUPAC system:

1. Examine the structure and identify the longest continuous chain of carbon atoms, no matter how it is written. Use Table 9.1 to find the name of the alkane that corresponds to that length. For example, if the chain is six carbons long, use the name "hexane." Once this parent name is selected, do not change it.
Example:

$$\underset{\text{CH}_3}{\overset{\text{CH}_3\qquad\qquad\quad\text{CH}_3}{\text{CH}-(\text{CH}_2)_3-\text{CH}}}\quad\underset{\text{C}_2\text{H}_5}{}$$

The longest C chain has eight carbons; name as an octane.

2. Locate and identify each and every group that is a substituent on the alkane you selected in Step 1. A substituent is any part of the molecule not included in the longest carbon chain.
Example:

methyl group substituents

$$\underset{\text{CH}_3}{\overset{\text{CH}_3}{\text{CH}}}-(\text{CH}_2)_3-\underset{\text{C}_2\text{H}_5}{\overset{\text{CH}_3}{\text{CH}}}$$

3. Number the longest carbon chain starting at one of the ends. This is done so that you can specify the location of the substituents.
Example:

$$\underset{\underset{1}{\text{CH}_3}}{\overset{(\text{CH}_3)}{\underset{2}{\text{CH}}}}-\overset{3}{\text{CH}_2}-\overset{4}{\text{CH}_2}-\overset{5}{\text{CH}_2}-\underset{\overset{7}{\text{CH}_2}\overset{8}{\text{CH}_3}}{\overset{(\text{CH}_3)}{\overset{6}{\text{CH}}}}$$

In our example, the substituents appear on carbons 2 and 6. That adds up to 8. **Always begin numbering from the end of the parent chain that gives you the smaller total of numbers for substituents.** If, in our example, we had begun numbering from the C_2H_5 end, the substituents would have appeared on carbons number 3 and 7. That adds up to 10, and that would be the wrong way to number because 10 is greater than 8.

4. Now write the name. Begin by indicating the location and name of each substituent. Use hyphens to make it all one word. Where the same substituent appears twice or more in the name, *condense* by using *di* or *tri* in the prefix for that substituent; use commas between successive numbers.
Example: 2-methyl-6-methyloctane condenses to 2, 6-dimethyloctane. (Note that the final IUPAC name is all one word.)

Here is another example of the use of the versatile IUPAC system. Consider the compound:

$$\underset{\underset{\text{Cl}}{\overset{\text{Cl}}{\text{Cl}-\text{C}-\text{CH}_2-\text{C}-\text{CH}_3}}}{\overset{\text{CH}_2\text{CH}_2\text{CH}_3}{}}$$

1. The longest continuous chain is six carbons long; thus hexane is the parent name.
2. There are five substituents: three chloro groups and two methyl groups.
3. Before condensing, the name is: 1-chloro-1-chloro-1-chloro-3-methyl-3-methylhexane. Note that numbering from the chloro end of the carbon chain gives the smaller total of numerical prefixes.
4. The final, condensed name is: 1,1,1-trichloro-3,3-dimethylhexane. For our purposes, it makes no difference if we list the chloros or the methyls first. (Most American chemists, however, follow the rule of listing the groups in alphabetical order.)

The following are some of the fine points to remember about IUPAC naming:

- There must be a number for every substituent. (Note in the previous example there are five numbers for five substituents.)
- Use commas between numbers.
- Use hyphens between numbers and names of substituents.
- When condensing, always insert *di* or *tri* or *tetra* for emphasis.
- Do not use the common system prefixes *n*, *sec*, or *ter* when naming by the IUPAC system

Here is a list of substituent groups most often encountered in naming substituted alkanes:

—F	fluoro	—NO_2	nitro
—Cl	chloro	—OH	hydroxy
—Br	bromo	—NH_2	amino
—I	iodo	—OCH_3	methoxy
		—OCH_2CH_3	ethoxy

Cycloalkanes

Not only can carbon atoms join to each other to form chains, the ends of the carbon chains can join to form rings of carbon atoms. Hydrocarbons of this type are called **cycloalkanes**

Two sizes of cycloalkane rings, the C_5 and the C_6, are especially stable and especially important in organic chemistry. These two are termed cyclopentane and cyclohexane (Figure 9.8).

The special stability of the five-membered and six-membered rings derives from the fact that the carbon-to-carbon bond angles in these two (puckered, nonplanar) cyclics are very close to the normal 109°28′ bond angles in the open-chain alkanes. By contrast, the carbon-to-carbon bond angles in cyclopropane and cyclobutane are, respectively, 60° and 90°. As a result, cyclopropane and cyclobutane are considered "strained" molecules, that is, their less

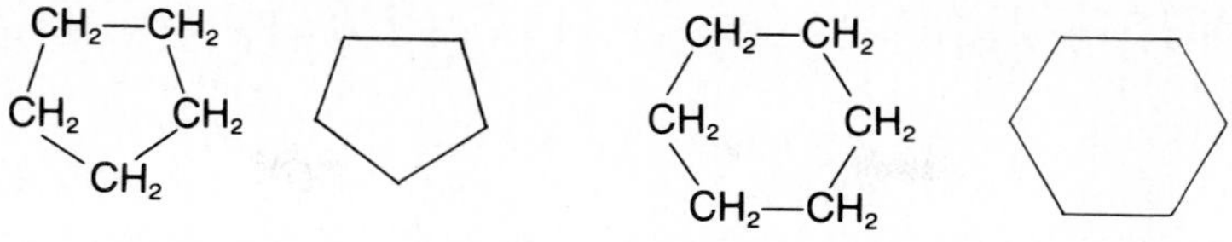

Two Representations of Cyclopentane   Two Representations of Cyclohexane

FIGURE 9.8   Two Important Cycloalkanes.

efficient sharing of electron pairs gives them a tendency to break their rings and revert to the open chain form.

Figure 9.8 shows the special shorthand used to represent cyclic structures. The pentagon is commonly used to represent cyclopentane: each of the five corners of the pentagon represents a —CH_2— group (methylene group). Hence a hexagon represents the six —CH_2— groups in cyclohexane. Naming is exactly like that for open-chain alkanes except that the prefix *cyclo* is used.

EXAMPLE 9.7

Name each of the following cyclic molecules:

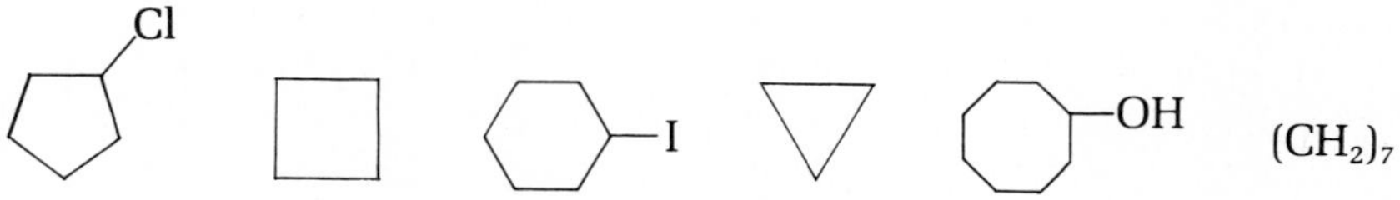

A   B   C   D   E   F

Solution

    a. cyclopentyl chloride.
    b. cyclobutane.
    c. cyclohexyl iodide.
    d. cyclopropane.
    e. cyclooctyl alcohol.
    f. cycloheptane.

Biological Importance of Alkanes

Methane, CH_4, occurs commonly in the human intestine as the result of bacterial action (other possible intestinal gases are CO_2, NH_3, H_2S, and H_2). Gases in the lumen of the bowel stretch it, reflexively stimulating peristalsis—a desirable occurrence. On the other hand, in a postoperative adynamic ileum (a loss of intestinal motility following surgery), the presence of a large stationary gas mass can be very painful. Methane, of course, is the gas with which we heat many of our homes.

Only a few alkanes find application as drugs *per se*. Cyclopropane, C_3H_6, was introduced in 1934 as a general inhalation anesthetic for surgical procedures. Because of its great potency, versatility, and low toxicity, it was widely used. Its flammability, however, has made it a second choice to the new anesthetics that are nonflammable and nonexplosive, and similarly potent.

Halogenated alkanes are useful anesthetics. A fluoroalkane, 1,1,1,2-tetrafluoroethane, has been synthesized as a general inhalation anesthetic, as has 1,1,1,-trifluoro-2-chloro-2-bromoethane. The latter, halothane USP (Fluothane), is used as a 2.0 to 2.5 percent solution in oxygen. These anesthetics are chemically similar to a class of refrigerants called Freons. Dichloro-difluoromethane is Freon-12. Teflon, another fluorocarbon, is chemically unreactive *in vitro* and *in vivo* (in the test tube and in the body).

Chloroform, $CHCl_3$, was introduced as an obstetric anesthetic in 1847. Though potent, it is also toxic, as are the chloroalkanes generally. Ethyl chloride, C_2H_5Cl, is a gas (boiling point 12°C), used to produce short-term anesthesia for brief, minor operations. After it is sprayed on unbroken skin, its rapid evaporation freezes (and anesthetizes) the tissue. Its use in sports to treat sprains, and other injuries, is to be condemned, as it can cause skin damage.

Mineral oil USP and *white petrolatum USP* are complex mixtures of alkanes; both are obtained from petroleum or crude oils. The carbon content of the alkanes in mineral oil ranges from C_{18} to C_{24}; for petrolatum it ranges from C_{25} to C_{30}. Mineral oil is chemically inert in the body. If taken by mouth it passes through the GI tract essentially unchanged and is excreted chemically intact. In the process it lubricates the intestine and softens the stool, and thus has had wide application as a laxative. If you are convinced by the ads that you have to take a laxative, and you select mineral oil, be aware that loss of fat-soluble vitamins A, D, E, and K can occur if mineral oil is consumed while these vitamins are in the intestine. They can dissolve in the oil and be excreted with it. If mineral oil is taken at bedtime, potential loss of fat-soluble vitamins is minimized.

Petrolatum is a mixture of semisolid hydrocarbons obtained from petroleum. It is sold in the yellow form and the more highly purified white form. The latter is used in household topical dressings, as a lubricant, as a base for ointments, and in the treatment of burns.

Paraffin is a purified mixture of solid C_{24} to C_{30} hydrocarbons obtained from petroleum. Its melting point range is 47 to 65°C. Chemically very similar to mineral oil and petrolatum, paraffin finds use as a stiffening agent in ointments, lipstick, and cosmetics.

Simple Chemical Reactions of Alkanes

Alkanes burn in the presence of oxygen to yield carbon dioxide, water, and heat. This reaction, termed combustion, is of profound importance to mankind because it is the source of much of the energy society requires for transportation, warmth, and industrial development. When we speak of the combustion of fossil fuels—oil, gas, and coal—we are, in effect, speaking of the combustion of hydrocarbons. Use of oil in America peaked at 18.4 million barrels a day (6.72 billion barrels a year) in 1978, and is expected to hold steady at about 17.2 million barrels a day through 1990.

Natural gas is mostly methane. The chemical equation for its combustion is

$$CH_4 + 2\ O_2 \rightarrow CO_2 + 2\ H_2O$$

Methane is the gas used to heat many homes. The gasoline we burn in our automobiles consists of alkanes in the C_5 to C_{10} range, plus other hydrocarbons. Kerosene (used as jet fuel) contains alkanes in the C_{11} to C_{12} range. Octane, a constituent of gasoline, burns according to the equation

$$C_8H_{18} + 12\tfrac{1}{2}\ O_2 \rightarrow 8\ CO_2 + 9\ H_2O$$

We can double every factor in the equation if we wish to eliminate half moles.

If insufficient oxygen is present in the combustion of hydrocarbons, carbon monoxide will form in place of carbon dioxide. Carbon monoxide is very dangerous if inhaled because it combines strongly with hemoglobin of the blood to form *carbonmonoxyhemoglobin* (it binds so strongly that it can displace oxygen from hemoglobin). The result is that the oxygen-carrying capacity of the blood is reduced and oxygen transport to the brain and other tissues is impaired. The consequences can be fatal. Combustion of tobacco and marijuana in cigarettes leads to significant levels of CO, which is inhaled with the smoke.

A second chemical reaction common to alkanes is *halogenation*. Let's use chlorine and *n*-butane in our example:

Word equation

$$\text{chlorine gas + butane gas} \xrightarrow{\text{heat}} \text{chlorobutane + HCl}$$

Chemical equation

$$Cl_2(g) + C_4H_{10}(g) \xrightarrow{\Delta} C_4H_9Cl + HCl$$

From our earlier discussion we know that *n*-butane offers two kinds of hydrogens for substitution: 1° and 2°. Hence two monochlorobutanes are possible:

$$\underset{\displaystyle\begin{array}{c}|\\ Cl\end{array}}{CH_2}\!\!-\!\!CH_2\!\!-\!\!CH_2\!\!-\!\!CH_3 \qquad \text{and} \qquad CH_3\!\!-\!\!\underset{\displaystyle\begin{array}{c}|\\ Cl\end{array}}{CH}\!\!-\!\!CH_2\!\!-\!\!CH_3$$

1-chlorobutane   2-chlorobutane

We see that 1-chlorobutane is the product no matter which of the six 1° hydrogens in *n*-butane is substituted and that 2-chlorobutane is the product no matter which of the four 2° hydrogens is substituted. In the actual case, chlorination of *n*-butane leads to a mixture of the two C_4H_9Cl isomers.

EXAMPLE 9.8

It is possible to substitute two chlorine atoms into *n*-butane to obtain dichlorobutane. How many different (isomeric) dichloro-*n*-butanes of the formula $C_4H_8Cl_2$ can theoretically exist?

Solution
Six: 1,1-dichlorobutane; 1,2-dichlorobutane; 1,3-dichlorobutane; 1,4-dichlorobutane; 2,2-dichlorobutane; and 2,3-dichlorobutane.

With regard to combustion and halogenation, cycloalkanes behave very much like open-chain alkanes.

STEREOISOMERISM IN ORGANIC COMPOUNDS: OPTICAL ACTIVITY

Examine your two hands, your right and your left, and then examine the following two structures for 2-chlorobutane (the large circles represent C No. 2).

I   II

What do your two hands and structures I and II have in common? The answer is: *a mirror-image relationship*. Look at your right hand in a mirror and you will see your left hand. Look at structure II in a mirror and you will see structure I! What's more, your right hand will not fit into a glove made for your left hand. That's because your hands, being arranged the way they are, three-dimensionally, are *not superimposable*. If you arrange your hands so that the backs and palms superimpose, then the thumbs and little fingers are opposite. And if you arrange the thumbs to coincide with the little fingers, then the backs and palms are opposite.

Nor are structures I and II superimposable. Try as you will, even with free rotation about C—C bonds, you will not get structure I to fit into a "glove" made for structure II.

But, you say, isn't there only one 2-chlorobutane? Yes, if we consider C_4H_9Cl isomers in the manner we learned earlier. No, if we consider the ways in which four different groups can be arranged in space around a tetrahedral C atom. (Now you know why it was so important to learn about carbon's tetrahedral geometry.)

There are two **stereoisomeric** forms of 2-chlorobutane (*stereo* = space); they have two different arrangements of groups on carbon atom No. 2—which is called the *asymmetric carbon atom* because it confers asymmetry on the whole molecule. To see the two possible arrangements of groups, rotate structures I and II until the H atoms are in the rear, out of sight. You then have III and IV:

CH$_3$ —— Cl   Cl —— CH$_3$

C$_2$H$_5$ C$_2$H$_5$

III   IV

Now, starting at the Cl and working clockwise, identify the sequences of groups. The two different sequences are $Cl \rightarrow C_2H_5 \rightarrow CH_3$ and $Cl \rightarrow CH_3 \rightarrow C_2H_5$. As long as we do not break any bonds, we clearly have two different three-dimensional structures. Their functional groups are identical. Their physical properties are identical (with one exception). Their formulas ($CH_3CHClCH_2CH_3$) are identical. But one is "right handed" and the other is "left handed"! They may behave identically in ordinary chemical reactions, but when it comes to fitting gloves (i.e., occupying a receptor site in a biological tissue), there can be all the difference in the world. For example, the drugs *quinine* and *quinidine* are mirror-image isomers, but the first is an effective antimalarial drug whereas the second is used in the treatment of cardiac arhythmias.

Stereoisomers that have the nonsuperimposable mirror-image relationship are termed **enantiomers**. The word **chirality** has been coined to denote their "handedness," (i.e., they are either right handed or left handed). Chirality is pronounced *kĭ-ral'-ity*. It is akin to the older term asymmetry.

To summarize: two space arrangements are possible for four different groups bonded to a tetrahedral carbon atom. The two stereoisomeric arrangements, termed enantiomers, are identical in all respects except when it comes to interacting with other stereoisomeric compounds or biologic tissues. Molecular chirality is the key to understanding enantiomerism. In general, a pair of enantiomers can be depicted as shown in Figure 9.9.

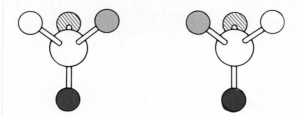

FIGURE 9.9 General Depiction of a Pair of Enantiomers. **Notice the mirror-image relationship.**

The rule for identifying uncomplicated chiral C atoms is the *four-different-group* rule. In the molecule you are examining, look for a carbon atom that bears four different atoms or groups. Use a large circle for the chiral carbon atom and draw four bonds as in the preceding structures I and II. Try to place all carbon groups above and below the circle; place hydrogen and other atoms to the right and left.

██████ **EXAMPLE 9.9** ██████

From the following, select the one compound having a chiral carbon atom. Draw the enantiomeric pair in three-dimensional structure.

$CH_3(CH_2)_6CH_3$ cyclohexane $CH_3CHOHCOOH$ CH_3I
CHI_3 $(CH_3)_2CH—CH(CH_3)_2$

Solution
$CH_3CHOHCOOH$ (lactic acid) can exist as a pair of nonsuperimposable mirror-image forms:

$$H \diagdown \overset{COOH}{\underset{CH_3}{\bigcirc}} \diagup OH \qquad HO \diagdown \overset{COOH}{\underset{CH_3}{\bigcirc}} \diagup H$$

Note: If you are having trouble visualizing enantiomers, try the old marshmallow-and-four-different-color-toothpick trick. (The marshmallows take the place of the circles in the drawing above.)

████████████████████████████████████

Although much more can be said about stereoisomerism, for our purposes only one aspect remains to be discussed. It is true that enantiomers behave identically in ordinary chemical reactions, and they do have identical melting points, boiling points, refractive indexes, and so on; however, there is one striking physical difference between them. To understand what it is, we must first understand what *plane-polarized light* is.

FIGURE 9.10   (*a*) **Before passing through Polaroid, light rays vibrate in all planes perpendicular to the axis of transmission. (*b*) After passing through Polaroid, light rays all vibrate in one plane; they have been plane-polarized.**

a   *b*

Light rays that pass through Polaroid sheets (as in sunglasses) become plane-polarized. This means that after passing through Polaroid, the light rays are all vibrating in the same plane (Figure 9.10).

The *polarimeter*, a more elaborate apparatus that accomplishes the same thing as Polaroid sheets, is used in research and in serious laboratory analysis.

About 150 years ago, two Frenchmen discovered that when plane-polarized light rays were passed through solutions of isolated enantiomers, the rays were rotated, to the left or to the right, depending on which enantiomer was present. Enantiomers that rotate plane-polarized light to the right are called **dextrorotatory** (and are given the symbol $+$). Enantiomers that rotate plane-polarized light to the left are called **levorotatory** (given the symbol $-$). Any stereoisomer capable of rotating plane-polarized light in either manner is said to be **optically active** (Figure 9.11).

Here is an important point to learn: a dextrorotatory enantiomer will rotate plane-polarized light to the right to the *exact same extent* that its levorotatory enantiomer will rotate it to the left, and vice versa. Obviously then, in a 50–50 mixture of enantiomers (called a *racemic mixture*), no optical activity can be demonstrated; the molecules are canceling out each other's effects. Conclusion: whenever there are just as many dextro isomers as levo isomers, the preparation will appear to be optically inactive.

EXAMPLE 9.10

What is the relationship between the principle of dextrorotatory isomerism and the well-known amphetamine *Dexedrine*?

Solution

Dexedrine is the *dextrorotatory* isomer of amphetamine $C_6H_5CH_2CH(NH_2)CH_3$. The enantiomeric pair is

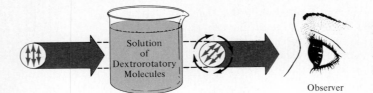

FIGURE   9.11   Plane-polarized light passed through dextrorotatory enantiomer molecule is rotated to the right. Optical activity is thus demonstrated.

Later on in this book, in the discussion of carbohydrates and amino acids, we will make important application of our knowledge of enantiomerism and optical activity. We will learn, for example, that the sugar in our blood (glucose) and the amino acids in our protein, are specific, noninterchangeable enantiomers. Their special three-dimensional geometry is the clue to their biological activity.

ALKENES

Up to now we have examined organic molecules that contain single carbon-to-carbon bonds, that is, a shared electron pair bond. When, however, two carbon atoms share *four* electrons, we say that a *double bond* exists. The organic compound is termed an **alkene** (also called an olefin). We also say that the molecule is *unsaturated*, either because it does not contain as many hydrogen atoms as the corresponding alkane (which is saturated with hydrogen), or because it can undergo addition reactions (see later). In Figure 9.12 you will see some examples of alkenes and some of the ways in which they are drawn.

Let's take a close look at the simplest alkene, ethene (widely known by its common name, ethylene).

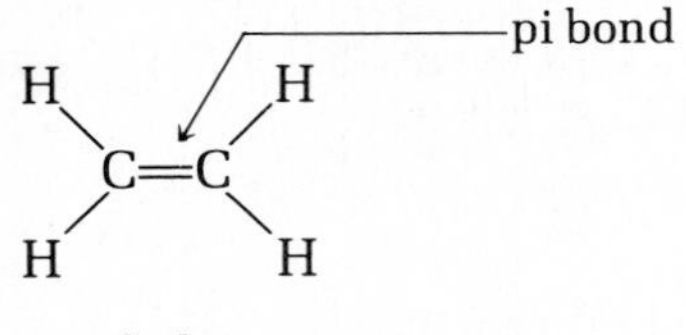

ethylene

One of the bonds between the carbon atoms forms in a special way and is given the name *pi bond*. The pi bond is the second bond between the carbon

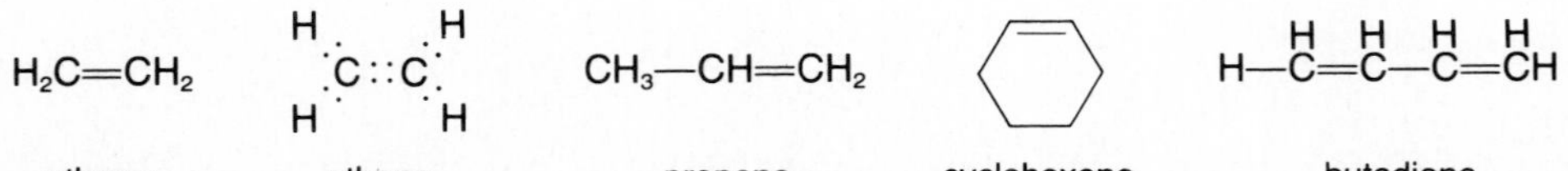

$H_2C=CH_2$ $CH_3-CH=CH_2$

ethene   ethene   propene   cyclohexene   butadiene

FIGURE 9.12   Some Examples of Alkenes.

atoms and its formation has a two fold effect: (1) there is no longer free rotation between the carbon atoms; they are locked into a rigid arrangement; and (2) the C═C portion of the molecule is planar (i.e., flat). Planarity in alkenes has major consequences, for it gives rise to the possibility of *geometric isomerism* (Figure 9.13). In a molecule of 2-butene, each carbon atom of the double bond is substituted by two different atoms or groups. Two isomers of 2-butene are therefore possible, one with like groups on the *same* side of the double bond (the *cis* isomer) and one with like groups on *opposite* sides (the *trans* isomer). (The Latin *cis* means "on this side," and the *trans* means "across.") Geometric isomerism could not occur without the fixed, nonrotatable character of a double bond.

Geometric isomers differ from each other in physical properties and usually in pharmacological properties. For example, consider *all-trans*-retinoic acid (Figure 9.13). This compound is widely used as tretinoin USP for the treatment of acne vulgaris where it causes skin peeling with resultant toughening.

Other examples of the importance of *cis-trans* isomerism in biochemistry are *cis*-aconitic acid and fumaric acid in carbohydrate metabolism; polyunsaturated fatty acids in fat metabolism.

Wherever we find carbon—carbon double bonds in molecules, we find areas of high electron density. The pi bond, the source of these elctrons, establishes the chemical behavior of the alkene molecule. For example, in the polymerization of ethylene to polyethylene (Figure 9.14), the pi electrons play the deciding chemical role. An organic group that determines the chemical functioning of a molecule is termed a **functional group**. The alkene double bond is a functional group, as are the —OH, —NH$_2$, —COOH, aldehyde, and ketone groups. Hydrogen atoms that are bound to olefinic (double-bond)

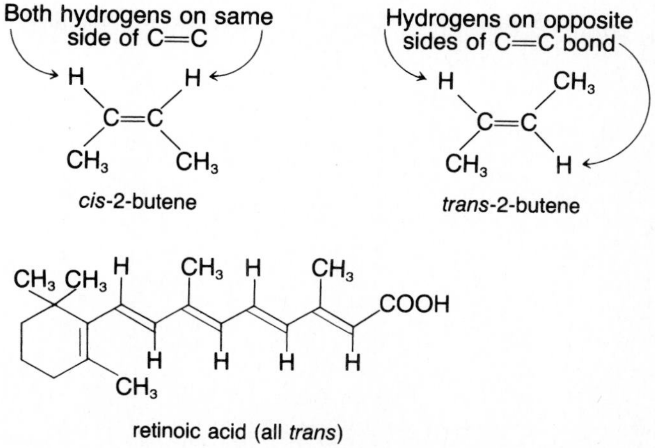

FIGURE 9.13 Geometric isomerism gives rise to isomers of the *cis*- and *trans*- type.

$$H_2C{=}CH_2 + H_2C{=}CH_2 + etc. \xrightarrow{\text{Polymerization}} {-}CH_2{-}CH_2{-}CH_2{-}CH_2{-}etc.$$

monomer   monomer   polymer

FIGURE 9.14   The Polymerization of Ethylene.

carbons are termed *vinylic* hydrogens. Hydrogens on carbons located imme-diately adjacent to olefinic carbons are termed *allylic* hydrogens.

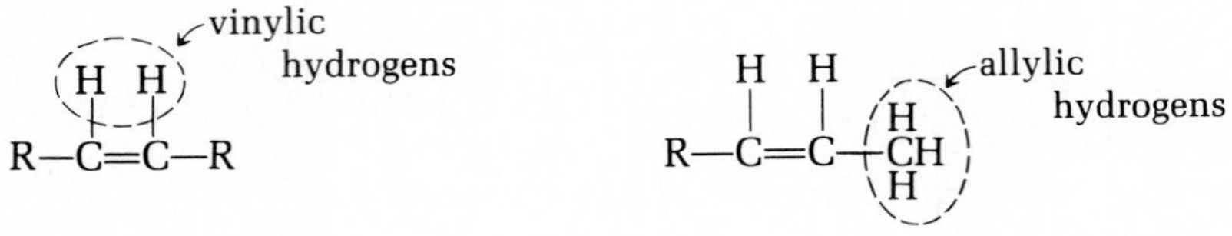

Common and IUPAC Names of Alkenes

Table 9.2 gives structures, common names, and IUPAC system names for some common alkenes.

The rules for assigning an IUPAC name to an alkene are as follows:

1. Identify the longest, continuous carbon chain *that includes the C=C.* Assign the corresponding alkane name.
2. Drop the *ane* suffix and add the *ene* suffix.
3. Start numbering carbon atoms from the end of the chain *nearest* to the C=C bond. The double bond is located by writing the number of the carbon where the double bond *begins*.
4. Name and number every substituent on the carbon chain, following the rules given for naming and numbering substituents on alkanes.

▬▬▬▬▬ **EXAMPLE 9.11** ▬▬▬▬▬

Ascribe IUPAC names to

  a. $(CH_3)_2C{=}CHCH_3$
  b. $CH_3CHCl{-}CH{=}CH{-}CHClCH_3$
  c. 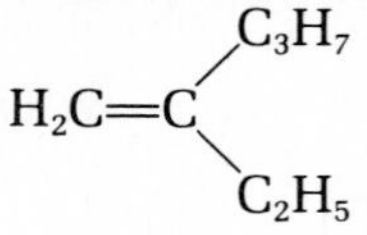 $H_2C{=}C\begin{array}{l} \diagup C_3H_7 \\ \diagdown C_2H_5 \end{array}$

Solution

  a. 2-methyl-2-butene
  b. 2, 5-dichloro-3-hexene
  c. 2-ethyl-1-pentene

Table 9.2. Structures and Names of Alkenes

Alkene	Common name	IUPAC name
$CH_2{=}CH_2$	Ethylene	Ethene
$CH_3{-}CH{=}CH_2$	Propylene	Propene
$C_2H_5{-}CH{=}CH_2$	Ethylethylene	1-Butene
$CH_3{-}CH{=}CH{-}CH_3$	Sym-dimethylethylene	2-Butene
$(CH_3)_2C{=}CH_2$	Isobutylene	2-Methyl-1-propene
$CH_3CH_2CH{=}CHCH_3$	—	2-Pentene

Simple Chemical Reactions of Alkenes

Addition reactions of alkenes are just what the name implies. Compounds such as the halogen acids, molecular hydrogen, or water literally add to the $C{=}C$ bond, removing the unsaturation. The pi electrons in the alkene are essential to addition reactions. Here are some examples:

$$CH_3{-}CH{=}CH{-}CH_3 + HCl \rightarrow CH_3{-}CHCl{-}CH_2{-}CH_3$$

$$CH_3{-}CH{=}CH_2 + H_2 \xrightarrow{\text{catalyst}} \text{propane}$$

$$CH_2{=}CH_2 + Br_2 \rightarrow CH_2Br{-}CH_2Br$$

In an **oxidation** reaction, oxygen attacks the $C{=}C$ double bond yielding oxygenated products, or actually rupturing the carbon chain at the point of unsaturation. Some examples of oxidation are

$$CH_3CH{=}CHCH_3 + \text{dilute } KMnO_4 \text{ solution} \rightarrow CH_3{-}\overset{\displaystyle HO}{\underset{\displaystyle |}{C}}H{-}\overset{\displaystyle OH}{\underset{\displaystyle |}{C}}H{-}CH_3$$

$$CH_3{-}CH{=}CH{-}CH_3 + O_3 \text{ (ozone)} \rightarrow 2\ CH_3COOH$$

Polymerization of olefins is of immense importance to society, because by polymerization we obtain new fabrics, cords, containers, wrapping material, insulators, industrial materials, medical prostheses, contact lenses, and even a form of artificial blood. By itself, one olefin molecule is termed a *monomer* (*mono* = one). When many monomers react with each other, the large molecule that forms is termed a *polymer*:

$$\text{many } H_2C{=}\underset{\displaystyle \underset{\displaystyle Cl}{|}}{C}H \xrightarrow{\text{catalyst}} \text{etc.}{-}CH_2{-}\underset{\displaystyle \underset{\displaystyle Cl}{|}}{C}H{-}CH_2{-}\underset{\displaystyle \underset{\displaystyle Cl}{|}}{C}H{-}CH_2{-}\underset{\displaystyle \underset{\displaystyle Cl}{|}}{C}H{-}\text{etc.}$$

vinyl chloride monomer   polyvinyl chloride (PVC)

The *etc.* in the equation means that the structure is repeated over and over again until we come to the end of the chain. Very large molecules can be made in polymerization reactions; the length of the chain influences the physical properties of the polymer.

EXAMPLE 9.12

Here are the structures of four famous monomers. Draw the structure of each corresponding polymer:

a. CH_2=CH—CN
b. CH_2=CH—CH_3
c. CH_2=CH—$COOC_2H_5$
d. CF_2=CF_2

Solution

a.

$$etc.—CH_2—\overset{\overset{\displaystyle CN}{|}}{CH}—CH_2—\overset{\overset{\displaystyle CN}{|}}{CH}—etc.$$

polyacrylonitrile
(used as the fabric Orlon)

b.

$$etc.—CH_2—\overset{\overset{\displaystyle CH_3}{|}}{CH}—CH_2—\overset{\overset{\displaystyle CH_3}{|}}{CH}—etc.$$

polypropylene
(used in sutures and catheters)

c.

$$etc.—CH_2—\overset{\overset{\displaystyle COOC_2H_5}{|}}{CH}—CH_2—\overset{\overset{\displaystyle COOC_2H_5}{|}}{CH}—etc.$$

acrylate polymer
(used in dental restorations)

d.

$$etc.—\overset{\overset{\displaystyle F}{|}}{\underset{\underset{\displaystyle F}{|}}{C}}—\overset{\overset{\displaystyle F}{|}}{\underset{\underset{\displaystyle F}{|}}{C}}—\overset{\overset{\displaystyle F}{|}}{\underset{\underset{\displaystyle F}{|}}{C}}—\overset{\overset{\displaystyle F}{|}}{\underset{\underset{\displaystyle F}{|}}{C}}—etc.$$

Teflon
(used in artery prostheses,
hemodialysis)

Man-made polymers have been put to use as prostheses (artificial substitutes for body parts used for medical or cosmetic reasons) and as valves, tubes, and bags for use in medical treatment. Figure 9.15 shows a bag used in the temporary external drainage of fluid in hydrocephalus patients.

ALKYNES

There is a prescription drug on the market called Eutonyl (generic name: pargyline hydrochloride), given in 10- or 15-mg doses for the treatment of high blood pressure.

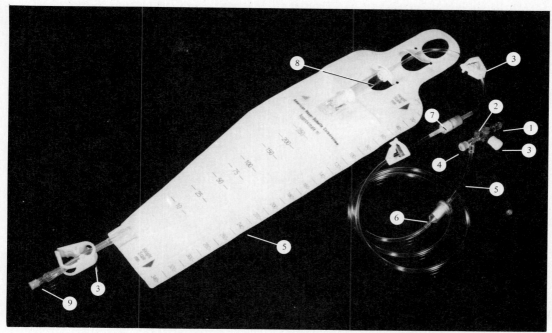

FIGURE 9.15 **The Use of Seven Different Synthetic Polymers (five of them polyalkenes) in an External Drainage Bag Used to Collect Cerebrospinal Fluid in Cases of Hydrocephalus. Key: 1-polycarbonate, 2-polyethylene, 3-polypropylene, 4-natural rubber, 5-poly (vinyl chloride), 6-acrylonitrile/butadiene/styrene terpolymer, 7-Nylon 6,6, 8-stainless steel, 9-poly (methyl methacrylate).** (Photo courtesy of John S. Tiffany, American Heyer-Schulte Corp., Goleta CA 93017.)

$$C_6H_5-CH_2-N\begin{matrix} CH_3 \\ \\ CH_2-C\equiv C-H \end{matrix}$$

Eutonyl[2]

In Eutonyl you will note the presence of a carbon—carbon *triple bond*. In such bonds, six electrons are shared by the bonding carbons. Triple bonds confer on a molecule an area rich in electrons. This can affect binding of the molecule to a receptor site.

Compounds that contain carbon—carbon triple bonds are called **alkynes**.

The simplest alkyne is ethyne (or, more commonly, acetylene), $H-C\equiv C-H$. Ethyne is a linear molecule: all four atomic nuclei lie on the same axis (i.e., in a straight line).

[2] C_6H_5 is the phenyl group; see next section.

As with alkenes, alkynes are considered to be unsaturated compounds, but more so, because each triple bond can add two moles of hydrogen:

$$CH_3\!-\!C\!\equiv\!C\!=\!H + 2\,H_2 \xrightarrow{\;catalyst\;} CH_3\!-\!CH_2\!-\!CH_3$$

propyne (C_3H_4) propane (C_3H_8)

Although triple bonds of carbon are not common in biologically important molecules, they do occur. An important example is the oral contraceptive (progestin) norethindrone:

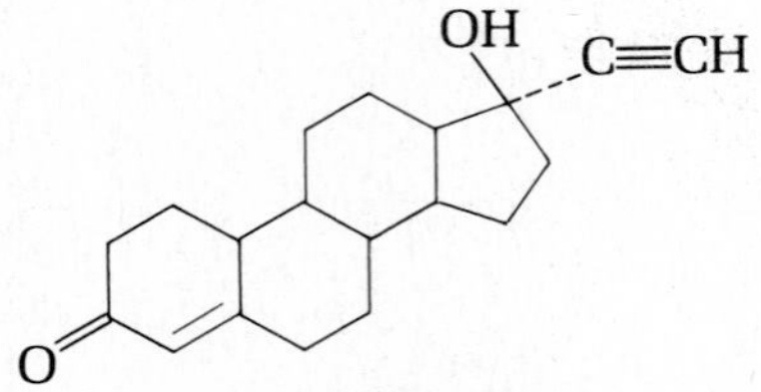

norethindrone

AROMATIC HYDROCARBONS

We twentieth-century students of chemistry are a sophisticated lot. We casually accept organic structures that baffled some of the great minds of a hundred years ago. Consider, for example, the structural formula for *benzene*, C_6H_6. In 1865, the great German organic chemist August Kekulé struggled to write a reasonable structure for six carbons and six hydrogens, as occur in benzene. Without looking ahead, pit your mind against Kekulé's and come up with a structure for benzene. Remember, each of the six carbons requires four bonds. Also, the molecule must be flat and have unusual stability toward chemical change.

Kekulé arranged the six carbons in a ring, with one hydrogen attached to each carbon. To satisfy carbon's valence of four, he placed alternate double and single bonds in the ring (I). Note the features of the resulting benzene molecule.

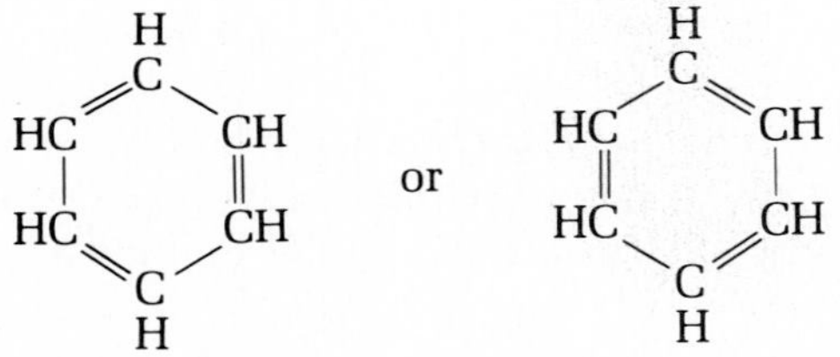

I

- Flat (or planar) structure.
- Six pi electrons.
- A six-membered ring.
- Unusual chemical stability.

Today we recognize that *any* organic molecule that shows these special characteristics belongs to the unique class called **aromatic** hydrocarbons. Rings other than six-membered can show aromaticity, too (five-membered rings with six pi electrons are common).

Perhaps the most important feature of an aromatic compound is its *aromatic sextet*—the six pi electrons that associate themselves with the six carbon atoms in the ring. These pi electrons (the same as we saw with alkenes) are not static; they are free to move around the ring. In fact, each electron is attracted to all six of the nuclei of the carbon atoms. We recognize this fact in the modern way of writing Kekulé's structure for benzene:

modern depiction of benzene

Here the hexagon represents the C_6H_6 skeleton and the circle represents the six pi electrons arranged in a kind of cloud on both sides of the flat ring. Because each pi electron is attracted to all six carbon nuclei, exceptional stability for the benzene ring results. Thus, it is difficult to disrupt the aromatic character of benzene; the ring is stable toward many biological and laboratory reagents that attack simple alkenes. Aromatic rings, for example, do not undergo addition reactions.

Organic chemists use Kekulé's formula for benzene as well as the modern symbol, interchangeably. We know what is meant, whatever structure for benzene we write.

EXAMPLE 9.13

---

a. How many different monochloro derivatives of benzene can exist?
b. How many different dichloro derivatives of benzene can exist? (*Hint:* The ring is flat and a perfect hexagon.)

Solution

a. Only one monochlorobenzene can exist since all six hydrogens in benzene are equivalent.
b. Three different dichlorobenzenes can exist: 1, 2-dichloro, 1,3-dichloro, and 1, 4-dichloro:

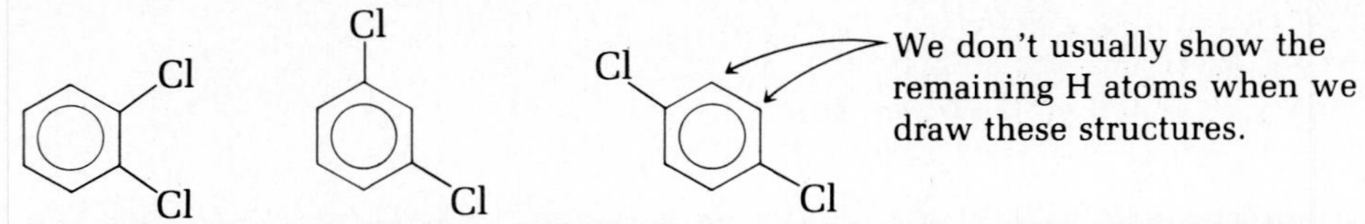

We don't usually show the remaining H atoms when we draw these structures.

A large proportion of the organic compounds we shall encounter exhibit both aliphatic and aromatic features. You can see examples of this dual nature in Figure 9.16, which gives structures and names of some important aromatic hydrocarbons. *Your should memorize the names of the first six structures in Figure 9.16.*

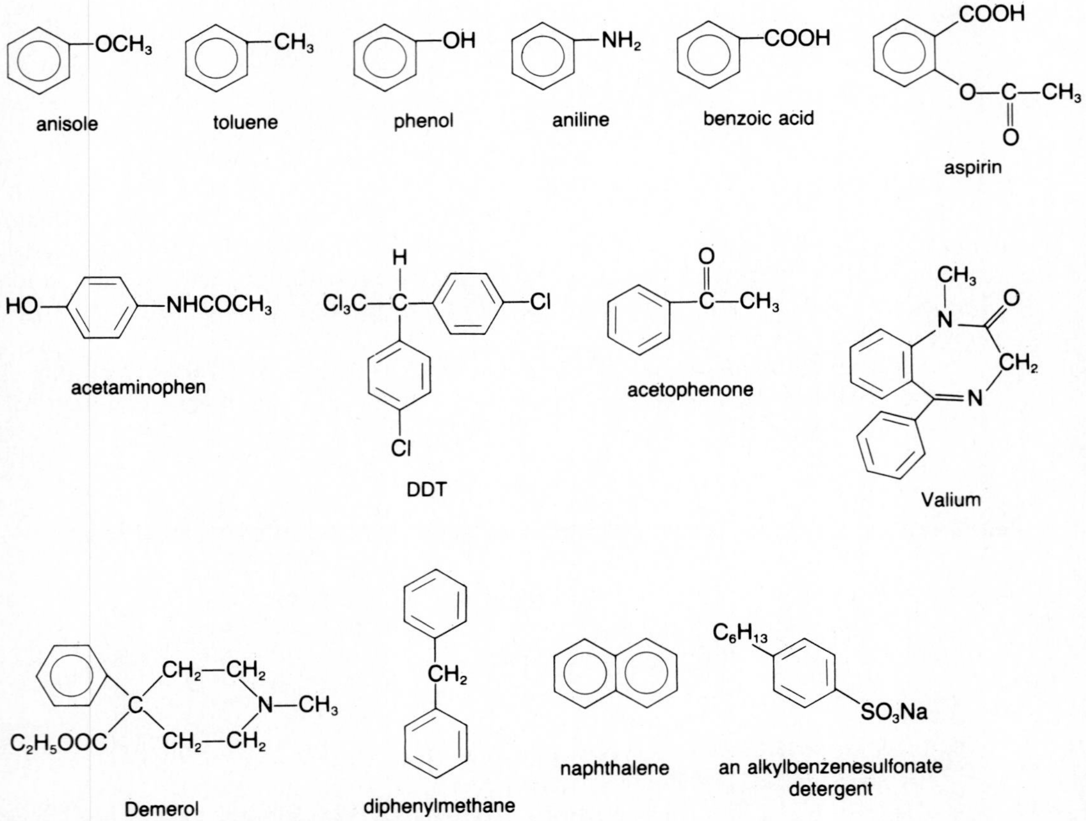

anisole   toluene   phenol   aniline   benzoic acid

aspirin

acetaminophen   DDT   acetophenone   Valium

Demerol   diphenylmethane   naphthalene   an alkylbenzenesulfonate detergent

FIGURE 9.16   Some Important Aromatic Hydrocarbons.

Simple Nomenclature of Benzene Derivatives

It is important that you become familiar with one simple system of naming derivatives of benzene, the *ortho-meta-para system*. These three prefixes designate positions on a benzene ring relative to a position *already substituted*. The position already substituted is the reference position:

The system is easily understood by considering the following five examples (refer to Figure 9.16 for parent names).

| ortho-nitro-phenol | meta-chloro-aniline | para-nitro-toluene | ortho-hydroxy-benzoic acid (salicylic acid) | para-phenylene-diamine |

Using the abbreviations *o* for ortho, *m* for meta, and *p* for para, the above names become: *o*-nitrophenol, *m*-chloroaniline, *p*-nitrotoluene, and so on. The name "*p*-phenylenediamine" is a special application of the common system; a nitro derivative of it, though toxic, is a commonly used hair dye.

An alternative to the ortho-meta-para system is the use of numbers to locate substituents. The ortho position corresponds to no. 2, the meta to No. 3, and the para to No. 4. This alternative system is more versatile than the *o*, *m*, *p* system. Examples are

| 2-nitrophenol | 2,4-dinitro-phenol | 3-bromo-5-chloro-nitrobenzene | 1,2,4-tribromo-benzene |

To use this numbering system, select a reference group and assign No. 1 to the carbon atom holding it. Reference groups are usually —OH, —NH_2, —CH_3, or —COOH. We start numbering the benzene ring at the reference group and proceed around the ring. We always number so as to obtain the smallest total of numerical prefixes. The number "1" is included in the name only when the substituents are all the same and the root name is benzene.

The word *phenyl* is commonly used in organic naming (pronounce it *feníll*). It means the C_6H_5— group, derived from benzene, as in phenyl chloride, C_6H_5Cl (synonymous with chlorobenzene). Hydrogen atoms located on carbons immediately adjacent to benzene rings are termed *benzylic hydrogens*. An example is toluene, which has three benzylic hydrogens.

Aromatic compounds such as benzene, toluene, and naphthalene burn in the presence of oxygen just as alkanes do:

$$C_6H_6 + 7\tfrac{1}{2} O_2 \rightarrow 6\ CO_2 + 3\ H_2O$$

A portion of the gasoline hydrocarbons that we burn in our automobile engines is aromatic. As mentioned previously, benzene is unusually stable and undergoes substitution with greater difficulty than alkanes do. Thus a catalyst such as $FeCl_3$ is required for chlorination of benzene:

$$C_6H_6 + Cl_2 \xrightarrow{\ FeCl_3\ } C_6H_5Cl + HCl$$

And what is more, once a Cl is bound to an aromatic ring, it is exceptionally difficult to remove, a fact that explains DDT's (figure 9.16) resistance to biodegradation. Because DDT tends to persist in our environment and because it has been found to be harmful to animals as well as to induce resistance in insects, its use has been banned in America.

Heterocyclic Aromatic Compounds

In chemistry, the prefix *homo* means "the same." An example is "homologous series." The prefix *hetero* means different from the ordinary," and an example is **heterocyclic** *aromatic compound*, that is, one that contains an atom other than the ordinary carbon in the ring. Most often, these hetero atoms are nitrogen and oxygen. Some famous examples are

pyridine   furan   pyrrole   pyran

pyrimidine   imidazole   purine   indole

You should know something about heterocyclics because they appear in some very important naturally occurring molecules, including Coenzyme A, ATP, atropine, morphine, azasteroids, amino acids, and DNA.

CARCINOGENS

Chemical compounds that are capable of causing cancer are termed **carcinogens**. Some of these (figure 9.17) are polycyclic aromatic hydrocarbons, that is, they have from two to five fused benzene rings. Tobacco and marijuana smoke contain polycyclic aromatic carcinogens. Many other types of carcinogens are recognized, including vinyl chloride, benzidine, propyleneimine, N-nitroso-piperidine, and nitrogen mustard:

$CH_2=CHCl$

vinyl chloride

H_2N-⟨◯⟩$-$⟨◯⟩$-NH_2$

benzidine

CH_3-N $\begin{array}{c} CH_2CH_2Cl \\ \\ CH_2CH_2Cl \end{array}$

nitrogen mustard

propyleneimine

N-nitrosopiperidine

SYNTHESIS

Organic chemists can synthesize compounds that occur in nature. They can also synthesize compounds that are totally unlike anything that occurs in nature. In a synthesis, the chemist performs the laboratory combination of the right chemicals, reagents, and conditions to create new arrangements of atoms.

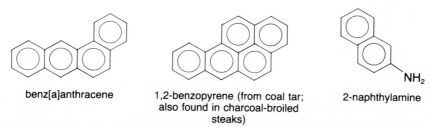

benz[a]anthracene

1,2-benzopyrene (from coal tar; also found in charcoal-broiled steaks)

2-naphthylamine

FIGURE 9.17 **Some of the well-known carcinogens are of the polycyclic aromatic hydrocarbon type.**

Although a grounding in synthetic organic chemistry is well beyond the scope of this book, we shall present three examples to help the reader gain appreciation of the importance of the synthetic approach (half of the prescription drugs in use today are synthetic).

Example 1. The synthesis of 2, 3-dimethyl-2-butene:

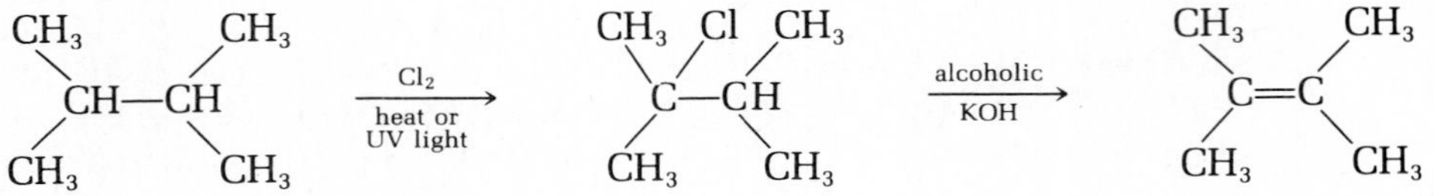

Example 2. The synthesis of p-chloroethylbenzene:

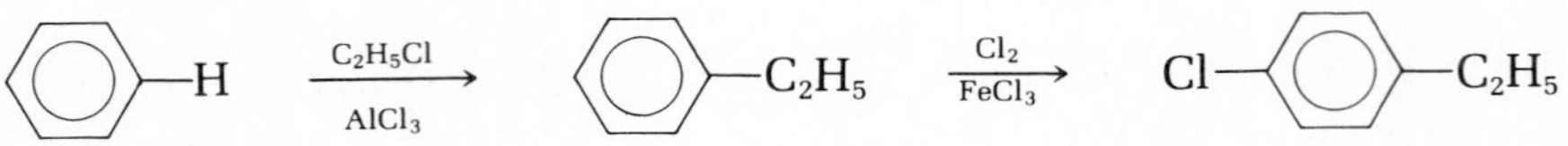

Example 3. The synthesis of aniline from benzene:

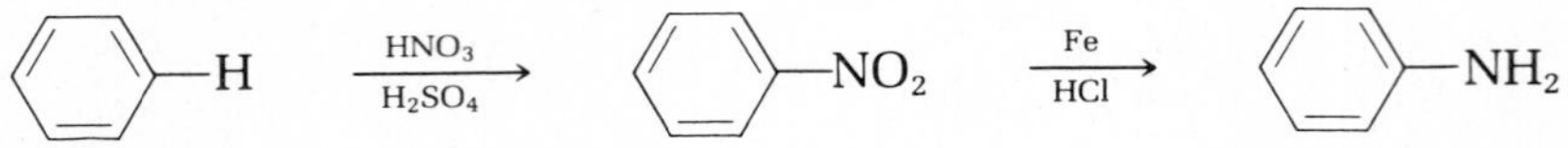

Great progress has been made in the development and application of organic synthesis. From it, we can obtain in the laboratory compounds that occur naturally in our body, in plants, and in microorganisms. We can also obtain totally new organic structures, that is, compounds that never existed before on the earth. As health professionals, we appreciate the application of organic synthesis in the preparation of drugs, vitamins, natural products, polymers, and laboratory tools and reagents.

SUMMARY

You have now been introduced to the subject of organic chemistry, and you have come to recognize the central role carbon and hydrogen play in organic compounds.

In this chapter you learned that in alkanes, carbon is tetracovalent and tetrahedral. The tetracovalent structure of carbon explains the great number and variety of hydrocarbon compounds because through it structural isomerism becomes possible. Carbon chains can be branched or unbranched; substituents can be located on a variety of positions on the chain.

The tetrahedral geometry of carbon explains the possibility of enantiomerism (optical isomerism) in alkanes having four different atoms or groups attached to the same carbon atom. We learned that the concept of enantiomerism is of great importance in certain drugs and naturally occurring compounds.

In this chapter we learned to name alkanes, alkenes, and alkynes and their derivatives by the common and IUPAC systems. This helped us to recognize alkyl groups and kinds of hydrogen atoms. We also learned the medicinal use of certain hydrocarbons and some of their derivatives. We examined a few simple chemical reactions of alkanes.

Unsaturation, in the form of carbon-to-carbon double and triple bonds, was a major topic in this chapter. Multiple bonds lock molecules into rigid shapes and this, we learned, is the explanation for *cis/trans* isomerism. Multiple bonds also account for the chemical behavior of alkenes and alkynes: addition reactions, oxidations, polymerizations. We studied synthetic polymers and some of their medical applications.

You began your study of aromatic compounds in Chapter 9. You will continue it for the remainder of this book, for aromaticity will occur repeatedly in the form of important biomolecules in steroids, foods, vitamins, coenzymes, proteins, and nucleic acids. You learned how to recognize and name benzene derivatives and heterocyclic compounds.

In this chapter, you were introduced to the art and science of organic synthesis by which the chemist can duplicate molecules in nature or create new molecules never before seen on this earth.

KEY TERMS

Check your understanding of this chapter. Can you explain what is meant by each of these terms?

aliphatic	functional group
alkane	heterocyclic
alkene	isomer
alkyne	IUPAC
aromatic	levorotatory
carcinogen	optically active
chirality	pharmacology
cycloalkanes	polymerization
dextrorotatory	stereoisomerism
enantiomers	tetrahedral C atom

STUDY QUESTIONS

1. List five characteristics of organic compounds.

2. Define the word *hydrocarbon* and give one example each of three different types of hydrocarbons (based on different chemical structures).

3. What does it mean to say that the carbon atom in methane is *tetrahedral*?

4. Another name for isopentyl chloride is 1-chloro-3-methylbutane. Draw (a) the electron dot structure for this compound, and (b) the line formula for this compound.

5. How many 1°, how many 2°, and how many 3° hydrogen atoms exist in n-pentane? In isobutane? In isopropyl bromide. In *t*-butyl bromide?

6. True or false: The formulas $(CH_3)_2CHCH(CH_3)_2$, $CH_3(CH_2)_3C_2H_5$, $(CH_3)_3C-C_2H_5$, and $(CH_3)_2CH(CH_2)_2CH_3$ all represent the same substance. (*Hint*: Completely draw out each structure.)

7. If C_nH_{2n+2} represents the general formula for an alkane, what does C_nH_{2n} represent?

8. There are three isomeric C_5H_{12} alkanes. Draw each structure and name each by the IUPAC system and by the common system.

9. Do the structures $(CH_3)_2CH-CH_2CH_2CH_3$ and $CH_3CH_2CH_2CH(CH_3)CH_3$ represent a pair of isomers or the same compound written twice?

10. Name each of the following compounds by the IUPAC system. Remember, only one name is correct.
    a. $CH_3CHICH_2CH_2CH_2CH_2I$
    b. Cl_3C-CH_3
    c.
$$CH_3 \qquad\quad CH(CH_3)_2$$
$$\diagdown\qquad\diagup$$
$$CH-CH$$
$$\diagup\qquad\diagdown$$
$$CH_3 \qquad\quad CH(CH_3)_2$$
    d. $(CH_3)_3CH$
    e.
$$CH_3 \qquad\qquad C_2H_5$$
$$\diagdown\qquad\quad\diagup$$
$$CH-CH-CH$$
$$\diagup\qquad|\qquad\diagdown$$
$$C_2H_5 \quad CH_3 \quad C_2H_5$$
    f.
$$\bigcirc\!\!-NO_2$$
    g. $F_2CH-CCl_2F$
    h. isobutylene: $(CH_3)_2C{=}CH_2$
    i. $(CH_3)_3C-CH{=}CH-C_2H_5$
    j. $(CH_3)_2CH-O-OH_3$

11. What is the source of mineral oil and what does mineral oil consist of?

12. Write and balance the equation to represent the combustion of decane to CO_2 and H_2O.

13. How many different (that is, isomeric) dibromopropanes can theoretically exist? (Exclude stereoisomers.)

14. Chemically, what is petrolatum and why is it important?

15. Optically active compounds are those that are able to rotate plane-polarized light. In which of the following compounds could we expect to find optical activity: *t*-butyl bromide; 1,1-dichloro-1,2-difluoroethane; *sec*-butyl alcohol; methylcyclohexane; 1-chloro-2-methylbutane; Demerol? (See Figure 9.16.)

16. **a.** Using the technique described in Example 9.10, draw the pair of glyceraldehyde enantiomers. Glyceraldehyde is CH_2OH—$CHOH$—C=O.
H

b. Another convention exists for depicting enantiomers in two dimensions. (See Chapter 12.) With it, glyceraldehyde enantiomers are drawn as

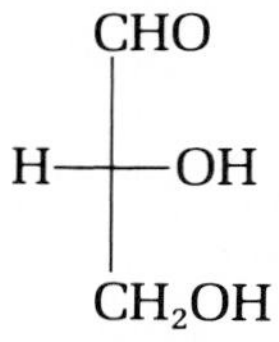

$$CHO$$
$$H\!\!-\!\!\!\!\!|\!\!-\!\!OH \qquad and \qquad HO\!\!-\!\!\!\!\!|\!\!-\!\!H$$
$$CH_2OH \qquad\qquad\qquad CH_2OH$$

D-(+)-glyceraldehyde   L-(−)-glyceraldehyde

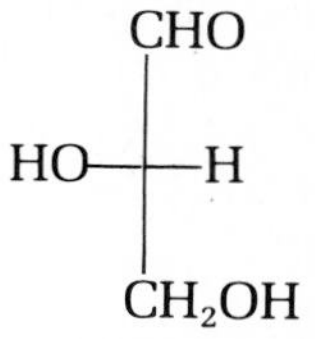

To which enantiomers in part a of this question do the above drawings correspond?

17. Which of the following compounds are properly termed "unsaturated": C_8H_{18}, C_2H_2, cyclopentene, isobutane, oleic acid, C_6H_6?

18. By drawing the appropriate structures, show that *cis/trans* isomerism is possible in 3-hexene. (Which is the *cis* and which is the *trans*?)

19. Write balanced equations to show
    a. The addition of HCl to ethene.
    b. The addition of H_2 to *cis*-2-butene.
    c. The polymerization of vinyl chloride monomer to PVC.

20. If the IUPAC names for C_2H_2 and C_3H_4 are, respectively, ethyne and propyne, what are the IUPAC names for the two isomeric C_4H_6 alkynes?

21. List three features common to all aromatic compounds.

22. Using the rules for aromaticity given in this chapter, classify each of the following as aromatic or nonaromatic?

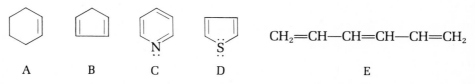

$$CH_2\!\!=\!\!CH\!\!-\!\!CH\!\!=\!\!CH\!\!-\!\!CH\!\!=\!\!CH_2$$

A   B   C   D   E

23. Examine Figure 9.16 and identify compounds that exhibit aliphatic features as well as aromatic.

24. Draw complete structures for the following:
    a. *o*-Chlorophenol   **b.** *m*-Iodobenzoic acid
    c. 2,4,6-Trinitrophenol   **d.** *p*-Nitroaniline
    e. 3,5-Dimethylaniline   **f.** Aspirin

25. Draw out the complete structural formula for the following:
    a. Furan
    b. Pyran
    c. Ethylene
    d. Propylene
    e. *Trans*-2-pentene
    f. 2-Hexyne
    g. C_6H_6
    h. Isobutylene

26. Create a hypothetical organic compound that possesses all of the types of hydrogen atoms we have discussed in this chapter: 1°, 2°, 3°, vinylic, allylic, and aromatic.

10

Medicinal Chemistry According to Functional Group

Sooner or later in your studies you will come across the term *receptor site*. It refers to specialized cells in some part of the body to which a drug or chemical attaches ("binds") because of a special fit. Once the fit is made, the drug action follows.

What governs whether a chemical will fit one receptor site, but not another? To a large degree, this is governed by the functional groups found on the organic molecule. Functional groups are special arrangements of atoms in a molecule that give the molecule its peculiar chemical and physical properties plus its three-dimensional shape.

In our photograph, a Merck & Co. scientist is examining a model he constructed of a receptor site that combines with drugs used in the treatment of cancer. The model represents 1300 atoms and was derived using x-ray diffraction.

The construction of three-dimensional molecular models of this type aids researchers in designing new, more effective drugs that will reach specific receptor targets. (Photo courtesy Merck and Co.)

INTRODUCTION

If humans did not have feet they could not wear shoes. If we did not have hands we could not wear gloves, and if we did not have ears we could not wear hearing aids. Very much of what we do is dictated by how we are constructed.

That's the way it is, too, with organic molecules. But the "appendages" on organic molecules have a special name: *functional groups*. We define functional groups as chemical arrangements in a molecule that can dictate how the molecule reacts chemically and physically with other molecules, with cells in tissues, or with itself.

Functional groups confer special properties on organic molecules. I once saw a picture of a calf born with two heads. Now that calf could do special things like eat twice as fast as any other calf. Likewise, some organic molecules have two carboxyl groups (—COOH). They can consume twice as much alkali as a monocarboxylic acid.

It is our goal in this chapter to examine seven functional groups: *alcohols, ethers, carboxylic acids, esters, aldehydes, ketones,* and *amines*—and to learn what special properties and chemical behavior these groups confer on organic molecules. Once gained, our knowledge of the chemistry of these functional groups will permit us to understand such interesting phenomena as replication in nucleic acids, solubility of morphine for intramuscular injection, the

Table 10.1. The Seven Functional Groups Discussed in This Chapter

Name of group	General formula	Example
Alcohol	R—OH	C_2H_5OH (ethyl alcohol)
Ether	R—O—R	C_2H_5—O—C_2H_5 (ethyl ether)
Carboxylic acid	RCOOH	C_6H_5COOH (benzoic acid)
Ester	RCOOR'	$CH_3COOC_2H_5$ (ethyl acetate)
Aldehyde	$R-\overset{\overset{\displaystyle O}{\|\|}}{C}-H$	CH_3CHO (acetaldehyde)
Ketone	$R-\overset{\overset{\displaystyle O}{\|\|}}{C}-R$	CH_3COCH_3 (acetone)
Amine	R—NH$_2$	CH_3NH_2 (methylamine)

synthesis of nylon, the antiseptic qualities of phenol, the cleansing effects of soaps, the effectiveness of Antabuse in alcoholism, the use of nitroglycerin in heart disease, and what happens to ethyl alcohol after we drink it.

Table 10.1 lists the seven functional groups we shall study in this chapter, with examples of each. The very important functional group, amides, is discussed in Chapter 11 in connection with peptides and proteins.

ALCOHOLS

Every **alcohol** has an OH group (a hydroxyl group) attached to an aliphatic carbon atom. The general formula is R-OH, where R = any alkyl group. Some very important alcohols are listed in Table 10.2. You should memorize their names and structures.

Alcohols must be clearly distinguished from **phenols.** In phenols, the OH group is bonded directly to an aromatic carbon (Figure 10.1) and this confers quite different chemical, physical, and pharmacological properties on the molecule. Phenols are much more acidic than alcohols. As seen in Figure 10.1,

Table 10.2. Some Important Alcohols

Common name	Synonym	Chemical structure	Boiling point	Comments
Methyl alcohol	Wood alcohol	CH_3OH	64.5	Poisonous in humans
Ethyl alcohol	Grain alcohol, ethanol	C_2H_5OH	78.3	The alcohol of alcoholic beverages
Isopropyl alcohol	Rubbing alcohol	$CH_3CHOHCH_3$	82.5	Miscible with H_2O
Benzyl alcohol	Phenyl carbinol	$C_6H_5CH_2OH$	205	Used as a preservative in injectables

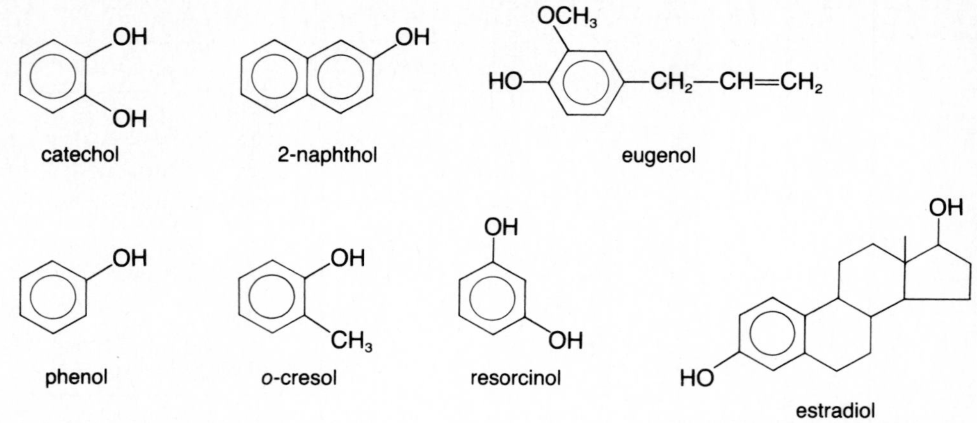

FIGURE 10.1   Some Important Phenols. **Phenols differ from alcohols by having the OH group bonded directly to an aromatic carbon.**

the female sex hormone estradiol has both an alcoholic and a phenolic OH group. Phenols are good disinfectants. Phenol itself is used for this purpose, but it is too damaging to the skin to be used on humans. Lysol owes its disinfecting properties to its active ingredients o-phenylphenol and 2-benzyl-4-chlorophenol. Another name for phenol (C_6H_5OH), is carbolic acid. Creosote, used as a wood preservative, is rich in the three isomeric cresols.

Hydrogen bonding, a phenomenon exhibited by alcohols, plays a major role in determining the physical properties of organic molecules. We learned in Chapter 7 that ethyl alcohol is a polar substance; its electronegative oxygen atom attracts the bonding pair of electrons more strongly than hydrogen does, creating a partial separation of charge.

$$\overset{\delta^+}{\text{H}}\!-\!\overset{\delta^-}{\text{O}}\!-\!\text{CH}_2\!-\!\text{CH}_3$$

The resulting dipole can be attracted to other molecules of the same or different structure

$$\text{C}_2\text{H}_5\!-\!\text{O}\!-\!\overset{\delta+}{\text{H}} \cdots \overset{\delta-}{\text{O}}\!-\!\text{C}_2\text{H}_5$$

Opposite charges attract each other; H-bond forms.

Hydrogen bonds are dipole–dipole bonds and, although weak (about 5 kcal per mole), affect the physical properties of compounds by virtue of their great numbers. Thus ethyl alcohol, with a molecular weight of only 46, boils at a remarkable high 78°C. Extra energy must be supplied to break the H-bonds before alcohol can boil. Hydrogen bonding is also sometimes called *association*.

Hydrogen bonding is possible in organic molecules in which a hydrogen atom is attached to an electronegative element (O, N, halogen). Compounds containing only carbon and hydrogen cannot undergo hydrogen bonding.

EXAMPLE 10.1

With your knowledge of H-bonding in alcohols, predict possible association between two molecules of an amine (such as CH_3NH_2) or between an amine and an alcohol.

Solution

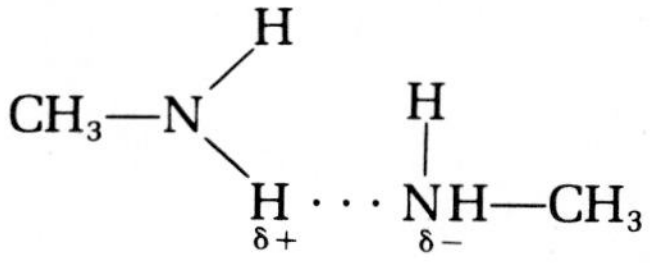

intermolecular H-bonding
in an amine

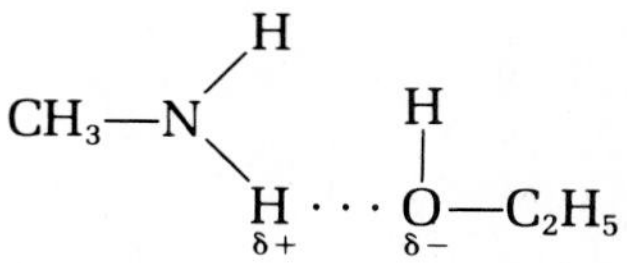

intermolecular H-bonding
between an alcohol
and an amine

IUPAC Naming of Alcohols

The rules for naming an alcohol by the IUPAC system are as follows:

1. Locate the longest, continuous carbon chain *that includes the* OH *group*, and apply the name of the parent hydrocarbon.
2. Drop the alkane *e* suffix and add the suffix *ol*.
3. Number from the end of the chain nearest the OH group.
4. Locate and number the hydroxyl group and all other substituents.
5. Condense the name if two or more of the same substituent are present.

Here are some examples of IUPAC names of alcohols:

CH_3OH $CH_3-CH-CH_3$ $C_3H_7-CH-CH(CH_3)_2$ $HOCH_2CH_2CH_2OH$

methanol   2-propanol 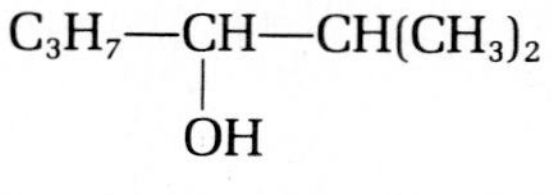   2-methyl-3-hexanol   1,3-propanediol

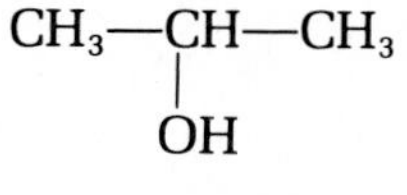

Ethyl Alcohol

In terms of its total impact on world society, ethyl alcohol, C_2H_5OH, might well be considered one of the three most incredible substances known to man. First, it is our oldest known synthetic substance. Archaeological records of the

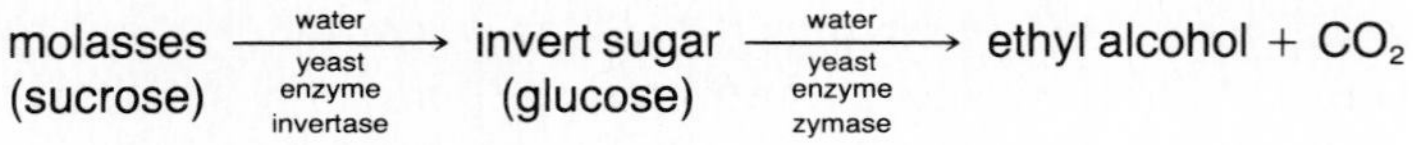

molasses $\xrightarrow[\substack{\text{yeast} \\ \text{enzyme} \\ \text{invertase}}]{\text{water}}$ invert sugar $\xrightarrow[\substack{\text{yeast} \\ \text{enzyme} \\ \text{zymase}}]{\text{water}}$ ethyl alcohol + CO_2
(sucrose)   (glucose)

FIGURE 10.2   **Fermentation as a Source of Ethyl Alcohol.**

oldest civilizations show the use of beer and wine, prepared by fermentation of carbohydrates. The first brewery dates to about 3700 B.C.

Second, as the ingredient of alcoholic beverages, ethyl alcohol is regularly consumed by 100 million Americans; 10 million Americans are problem drinkers or frank alcoholics (that's one in twenty-two).

Third, alcohol is of great importance to our technology as an industrial solvent, chemical intermediate in synthesis, and ingredient in numerous drug preparations.

Alcohol (when used without qualification we take this term to mean *ethyl alcohol*) sold for beverage use is taxed by the federal government at the rate of $10.50 per proof gallon (1 gal of 50% v/v alcohol is a proof gallon). To prevent tax-free alcohol for industrial use from being diverted to beverages, poisons are added. Called *denaturants*, these poisons make the alcohol unfit for human consumption. There is no way the lay person can remove denaturants.

There are two important means of obtaining ethyl alcohol. In fermentation, yeast organisms feed on carbohydrate (molasses, rye, corn, grapes, dandelions, etc.), producing a complex mixture of organic by-products, in which C_2H_5OH is prominent (Figure 10.2).

The hydration of ethylene (Figure 10.3) accounts for most of the industrial alcohol produced in the United States.

Most of the C_2H_5OH that is sold is 95% v/v solution with water. Pure 100% ethanol is known as *absolute alcohol*; it is very hygroscopic (water-loving). For the health professional, a 70% v/v solution of alcohol in water is important as a bactericide. The 70% actually works better than the 100% because the presence of water helps the alcohol to destroy bacterial protein more completely and prevents excessive dehydration of the skin.

Besides acting as an antiseptic, alcohol has some important effects in the human including the following.

- *CNS depressant effect.* Alcohol is a general central nervous system (CNS) depressant drug. It removes inhibitions (which makes some people believe it is a stimulant, though it is not). As the blood alcohol level increases, depression of the CNS increases until coma is reached; death

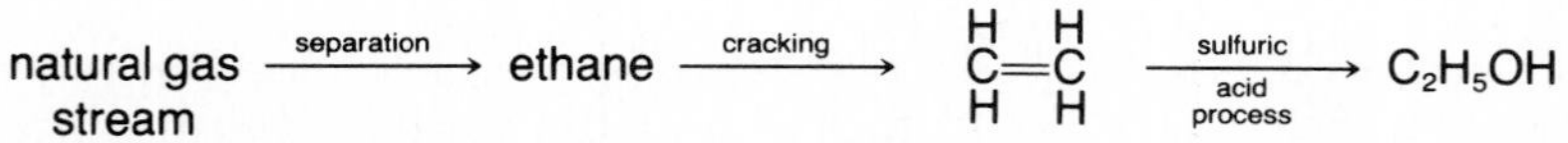

natural gas $\xrightarrow{\text{separation}}$ ethane $\xrightarrow{\text{cracking}}$ $\underset{\text{H}\ \text{H}}{\overset{\text{H}\ \text{H}}{\text{C}=\text{C}}}$ $\xrightarrow[\substack{\text{acid} \\ \text{process}}]{\text{sulfuric}}$ C_2H_5OH
stream

FIGURE 10.3   Chemical synthesis is the major source of industrial alcohol.

is usually due to respiratory depression. In most of the United States, a blood alcohol concentration of 0.1% w/v (0.1 g in each 100 cc of blood) is legal evidence of intoxication. And although some individuals behave as if they were intoxicated with a blood alcohol level of 0.05%, others are driving on our highways with blood alcohol levels of 0.4%! Alcohol induces tolerance in the *chronic* user, permitting him to sustain blood levels that would have killed him originally.

- *Vasodilator effect.* Ethanol acts to dilate (enlarge) arterioles and capillaries, permitting increased blood flow in these vessels. Hence some authorities recommend limited doses of alcoholic beverages in elderly people who have restricted blood flow because of hardening of the arteries. However, chronic (long-term) consumption of alcohol can lead to broken blood vessels in the nose and upper face because of the long-standing, repeated vasodilation in these areas. The full facial flush seen in some people after alcohol consumption is due to massive vasodilation. Alcohol's powerful vasodilator effect appears to give warmth in cold weather, but at the expense of internal heat loss. In the long run, the drinker winds up colder than he would have been without alcohol.

- *Teratogenic effect.* It is now generally accepted that alcohol, consumed during pregnancy, can act as a teratogen, damaging the embryo and resulting in the birth of deformed babies. Even *one* night of heavy drinking during the first trimester is considered capable of causing birth defects. For more on this, consult a modern pharmacology textbook under the heading "fetal alcohol syndrome."

- *Analgesia.* Alcohol is a pain killer (analgesic). It has been used by injection to relieve *tic douloureux*, a painful facial neuralgia.

- *Hypoglycemic effect.* Alcohol acts to lower blood sugar, producing hypoglycemia (*hypo* = low, *glyc* = sugar, *emia* = in the blood). This effect can be profound in the alcoholic. In fact, hypoglycemic coma should be suspected in any chronic drinker found unconscious.

- *Miscellaneous effects.* Ethyl alcohol is a diuretic (increases the flow of urine). It can irritate the lining of the stomach and worsen a peptic ulcer. And recent evidence indicates that moderate doses of alcohol increase the body's level of high density lipoprotein (HDL). See Chapter 13 for more on HDL and its role in reducing atherosclerosis.

Alcohol in Chemical Synthesis

Alcohol occupies a key position in chemical synthesis. It is the starting material for the synthesis of many important classes of compounds and other functional groups (Figure 10.4). For example, from ethyl alcohol we can obtain ethyl ether, a formerly much-used general inhalation anesthetic.

$$2\ C_2H_5OH \xrightarrow[\text{acid}]{140°C} C_2H_5{-}O{-}C_2H_5 + H_2O$$

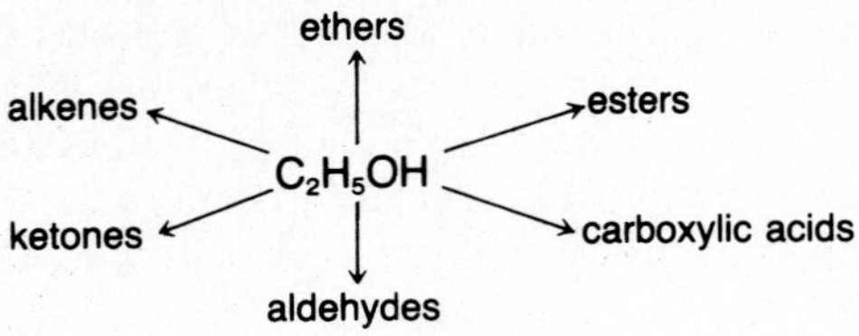

ethers

alkenes ←——— C$_2$H$_5$OH ———→ esters

ketones ←——— ———→ carboxylic acids

aldehydes

FIGURE 10.4 Alcohol occupies a key intermediate position in synthesis.

If, in the same reaction, the temperature is raised to 180°C, an alkene is the chief product.

$$C_2H_5OH \xrightarrow[\text{acid}]{\Delta} CH_2{=}CH_2 + H_2O$$

Referring to Figure 10.4, we see that other important classes of compounds can be obtained from ethanol, including esters, aldehydes, and ketones. We shall have more to say about these other classes of organic compounds later in this chapter.

Polyols

When two OH groups exist in an organic compound, we term the compound a *diol*. More than two OH groups make the compound a *polyol* (*poly* = many). Ethylene glycol (I), used as antifreeze, is a diol; glycerin (II) is a polyol. Meprobamate (Equanil, Miltown) is a famous tranquilizer having a diol structure (shown in bold type in Structure III) that has been converted into a carbamate ester:

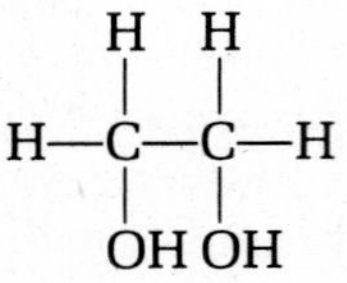

$$
\begin{array}{ccc}
\text{H} \quad \text{H} & \text{H} \quad \text{H} \quad \text{H} & \text{H} \quad \text{C}_3\text{H}_7 \\
\text{H—C—C—H} & \text{H—C—C—C—H} & \text{H—C—C— CH}_2 \\
\text{OH OH} & \text{OH OH OH} & \text{O} \quad \text{CH}_3 \text{ O} \\
& & \text{H}_2\text{NOC} \qquad \text{CONH}_2 \\
\text{I} & \text{II} & \text{III}
\end{array}
$$

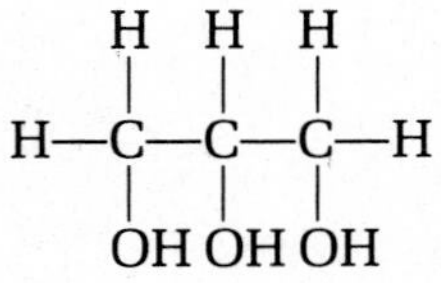

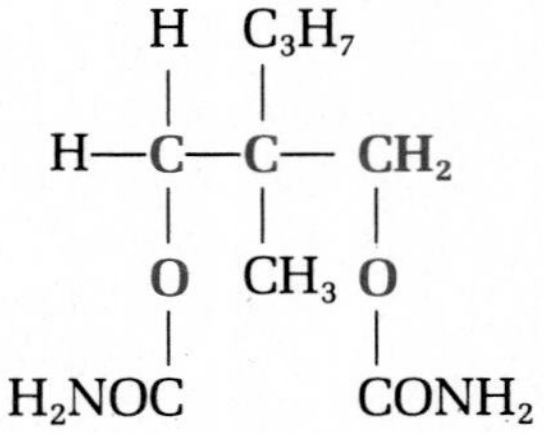

Glycerin, also called glycerol, is of great importance in foods, fats, digestion, medicines, soap making, explosives, industrial applications, and as a preservative, emollient (skin softener), and solvent. It is hygroscopic and is included in various preparations to prevent dehydration. Used in suppositories, glycerin is a mild laxative. We shall study glycerol again in Chapter 13. The polyol glucose is discussed in Chapter 12.

Mannitol USP and sorbitol USP are polyols that have wide use in medicine and the food industry. Sorbitol is metabolized by the body as a source of food; mannitol is not. Both compounds have six carbons and six hydroxyl groups (but different stereochemistry).

CARBOXYLIC ACIDS

If you know that the acid in vinegar is acetic acid, CH_3COOH, you have a head start in understanding **carboxylic acids.** The functional group in acetic acid and in all carboxylic acids is the *carboxyl group,* —COOH. Figure 10.5 shows four ways in which this functional group can be written. In every carboxylic

acid there is an acyl group, $\begin{smallmatrix}O\\\parallel\\R-C-\end{smallmatrix}$. In acetic acid, the acyl group is

$\begin{smallmatrix}O\\\parallel\\CH_3C-\end{smallmatrix}$. We will need to know about acyl groups when we study fats in Chapter 13.

The carboxyl group is classified as an acid group because the hydrogen attached to the oxygen ionizes in water, yielding an acid solution:

$$RCOOH + H_2O \rightleftharpoons RCOO^- + H_3O^+$$

Chemists know that the tendency for the hydrogen atom in a carboxylic acid to ionize and produce an acid solution is not great. In fact, only about 1 to 2 percent of the molecules are ionized in a 0.1-M solution. For this reason, we term carboxylic acids *weak organic acids.* Nonetheless, these compounds are acids. They neutralize bases, form salts, and produce the characteristic tart taste of a solution having an excess of hydrogen ions.

Table 10.3 is one of the most important tables in this book. It lists twelve important aliphatic carboxylic acids by common name, structure, and occurrence. You will encounter many of these acids repeatedly in your study of organic chemistry. Some of them, like the highly unsaturated linoleic acid and linolenic acids, are essential in human nutrition. That means we need them in our diet because our bodies cannot make them at all or at a rate fast enough to meet body demands for them.

The high molecular weight acids in Table 10.3 are often called fatty acids because they can be obtained from fats. The clinical abbreviation *FFA* stands for "free fatty acids." (More on this in Chapter 13.)

In the common system for naming carboxylic acids, the first carbon next to the carboxyl group is referred to as the *alpha* carbon, and substituents

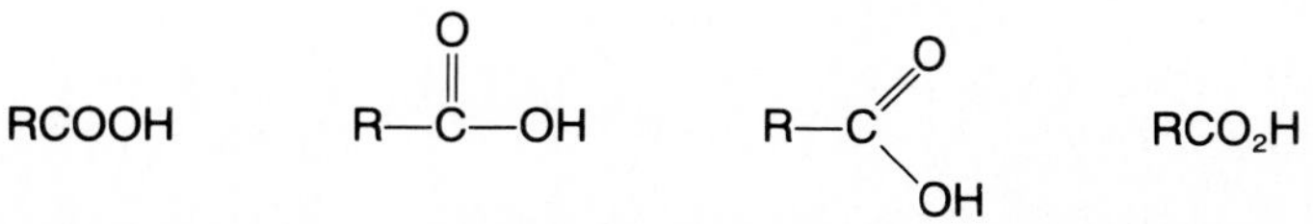

RCOOH $\begin{smallmatrix}O\\\parallel\\R-C-OH\end{smallmatrix}$ $R-C\begin{smallmatrix}O\\\diagup\\\diagdown\\OH\end{smallmatrix}$ RCO_2H

FIGURE 10.5   The Carboxylic Acid Group, Drawn in Four Equivalent Ways.

Table 10.3. Important Aliphatic Carboxylic Acids

Common name	Structure	Occurrence in nature
Formic acid	HCOOH	Venom of red ants and stinging nettle
Acetic acid	CH_3COOH	Destructive distillation of wood; vinegar
Propionic acid	C_2H_5COOH	In small amounts, in dairy products
Butyric acid	$CH_3(CH_2)_2COOH$	In butter as an ester. Free acid = stench!
Valeric acid	$CH_3(CH_2)_3COOH$	Valerian root, Stench!
Caproic acid	$CH_3(CH_2)_4COOH$	Small amounts in milk fat and coconut oil
Lauric acid	$CH_3(CH_2)_{10}COOH$	Vegetable oils, e.g., coconut
Palmitic acid	$CH_3(CH_2)_{14}COOH$	As glyceryl ester in many oils and fats
Stearic acid	$CH_3(CH_2)_{16}COOH$	As glyceryl ester in many animal and vegetable oils (pronounced "steeric")
Oleic acid	$cis\text{-}CH_3(CH_2)_7CH{=}CH(CH_2)_7COOH$	As glyceryl ester in many animal and vegetable oils
Linoleic acid	All $cis\text{-}C_{18}H_{32}O_2$ (2 double bonds)	As a glyceride in many vegetable oils, e.g., cottonseed, soybean, safflower
Arachidonic acid	$cis\text{-}C_{20}H_{32}O_2$ (4 double bonds) all cis	Liver, brain, grandular organs and depot fat of animals

thereon are referred to as alpha substituents. The next carbon is called the *beta* carbon. Examples of naming are

$$ClCH_2COOH$$
alpha-chloroacetic acid

$$CH_3{-}\overset{\overset{\textstyle O}{\|}}{C}{-}CH_2{-}COOH$$
beta-ketobutyric acid

$$C_6H_5CH_2COOH$$
alpha-phenylacetic acid

The widely used symbols for the first four letters of the Greek alphabet (alpha, beta, gamma, delta) are, respectively, α, β, γ, δ.

Important aromatic carboxylic acids are few but nevertheless vital to our study of drugs and biochemistry. Here are some structures and names:

benzoic acid

salicylic acid

phthalic acid

p-aminobenzoic acid (PABA)

IUPAC Names of RCOOH

The rules for naming RCOOH by the IUPAC system are as follows:

1. Locate the longest carbon chain *that includes the carboxyl group;* apply the name of the parent hydrocarbon.
2. Drop the e suffix and add *oic acid.*
3. Begin numbering with the carboxylic acid carbon as number 1.
4. Locate and number all substituents on the chain.
5. Condense the name as needed.

Examples of IUPAC names are given in Figure 10.6.

Synthesis of Carboxylic Acids

Oxidation is one of the important ways by which chemists obtain carboxylic acids. Potassium permanganate, ozone, potassium dichromate, or other agents can be used, depending on the substance to be oxidized. Here are some examples:

$$\text{alcohol oxidation: } CH_3CH_2CH_2OH \xrightarrow[\text{heat}]{KMnO_4} CH_3CH_2COOH$$

$$\text{aldehyde: } CH_3CHO \xrightarrow[\text{heat}]{KMnO_4} CH_3COOH$$

$$\text{alkylbenzene oxidation: } C_6H_5CH_3 \xrightarrow[\text{heat}]{KMnO_4} C_6H_5COOH$$

$$\text{alkene oxidation: } R\text{—}CH{=}CH\text{—}R' \xrightarrow{\text{ozone }(O_3)} RCOOH + HOOC\text{—}R'$$

$$\text{alkyne oxidation: } R\text{—}C{\equiv}C\text{—}R \xrightarrow{O_3} 2\ RCOOH$$

Carboxylic acids can also be obtained by hydrolysis of naturally occurring esters (see next section), by hydrolysis of amides (see Chapter 11) and peptides, and by a considerable variety of other chemical procedures. In their metabolic pathways, plants and animals produce carboxylic acids by enzyme-catalyzed

$HCOOH$	$(CH_3)_2CHCH_2COOH$	$Cl_3C\text{—}COOH$	$CH_3(CH_3)_{16}COOH$
methanoic acid	3-methylbutanoic acid	2,2,2-trichloroethanoic acid	octadecanoic acid

FIGURE 10.6 IUPAC Names of Alkanoic Acids.

oxidations. One of the ways in which the human body rids itself of certain drugs (e.g., barbiturates) is by liver enzyme-catalyzed oxidation.

Acids Used in Medicine

Not many carboxylic acids are used *as such* in medicine, but a few are. Acetylsalicylic acid (ASA) is an analgesic and antipyretic; PABA is a sunscreen agent; benzoic acid is used in athlete's foot products; nicotinic acid is the vitamin, niacin; and 2-propylpentanoic acid is available on prescription for treatment of petit mal seizures (epilepsy).

Salts of Carboxylic Acids

Carboxylic acids react with bases to form water and a salt. This neutralization is quite analogous to the neutralization of a mineral acid with a mineral base:

Mineral acid with base

$$HCl(aq) + NaOH(aq) \rightarrow H_2O + NaCl(aq)$$

Carboxylic acid and base

$$RCOOH + NaOH(aq) \rightarrow H_2O + RCOONa \text{ (a salt)}$$

Although weak by comparison to HCl, carboxylic acids are strong enough to decompose bicarbonate; this is used as a test for organic acids.

$$CH_3COOH(aq) + NaHCO_3(aq) \rightarrow H_2O + CO_2 \uparrow + CH_3COONa$$

To name a salt of a carboxylic acid, first write the name of the positive (usually metal) ion. Then replace the *ic* suffix of the acid name with an *ate* suffix. Hence acetic acid gives *acetate* and butyric acid gives *butyrate*. Some more examples of naming are given in Figure 10.7.

In biological fluids, we do not usually see the unionized form of a carboxylic acid, as there usually is sufficient base present to strip away the acidic proton. Thus, in the metabolism of carbohydrate, we read about "acetate" entering the Kreb's cycle, or we read about a "formate" pool. This simply

CH_3COONa	$HCOONH_4$	$Ca(OOCCH_2CH_3)_2$	$CH_3(CH_2)_{16}COONa$	$Al(OOCCH_3)_3$
sodium acetate	ammonium formate	calcium propionate	sodium stearate	aluminum acetate

FIGURE 10.7 Names and Structures of Salts of Carboxylic Acids.

means that though we are still considering acetic acid or formic acid, we recognize that they may be present as their anions (CH_3COO^- and $HCOO^-$, respectively).

Solubility in water is one important difference between the higher molecular weight carboxylic acids and their sodium or potassium salts. To prepare the sodium or potassium salt usually means to solubilize an otherwise insoluble molecule.

$$\text{stearic acid} + KOH(aq) \rightarrow C_{17}H_{35}COOK + H_2O$$

(water insoluble)   (water soluble)

Likewise,

$$\text{benzoic acid} + NaOH \rightarrow C_6H_5COONa + H_2O$$

(water insoluble)   (water soluble)

Soaps are the potassium or sodium salts of fatty acids, especially stearic, lauric, and palmitic. Their preparation by saponification of fats is described in Chapter 13.

Zinc undecylenate, a zinc salt of an unsaturated 11-carbon fatty acid, is used externally in the treatment of fungal infections. Sodium benzoate is a preservative very commonly found in foods and beverages.

Before we leave the subject of carboxylic acids, here's an opportunity to test your knowledge. Read the descriptions of five biochemically important carboxylic acids and then see if you can match the description with one of the following structures. Look for not-so-subtle clues in the description. The answers are printed after the structures.

1. *Nicotinic Acid.* Chemically related to the alkaloid nicotine, this acid is also known as the vitamin niacin. A deficiency of it results in the serious condition called pellagra. Nicotinic acid contains a nitrogen-heterocycle.
2. *Ascorbic Acid.* Vitamin C is another name for this famous vitamin having one COOH group. It is an alkene with lots of hydroxyl groups. Ascorbic acid is specific for the prevention and cure of scurvy.
3. *Citric Acid.* Found in citrus plants, 1 mol of this acid requires 3 mol of NaOH for neutralization. A famous biochemical pathway is named after this acid. Citric acid is a monohydroxycarboxylic acid.
4. *Lactic Acid.* When we exercise strenuously, lactic acid accumulates in our muscle tissues. At rest, it is oxidized to pyruvic acid. Lactic acid has a molecular weight of 90.

5. *Carbonic Acid.* The combination of CO_2 and water gives carbonic acid. Carbonated beverages contain this acid. It has two ionizable hydrogens.

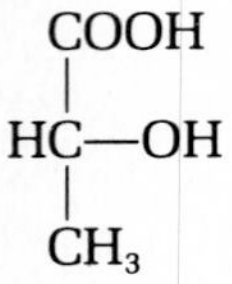

A

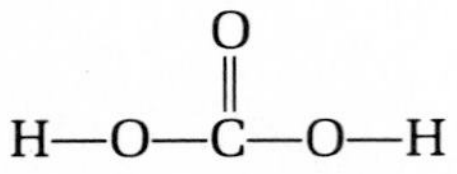

B

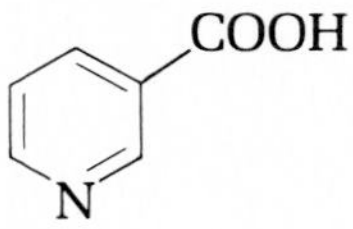

C

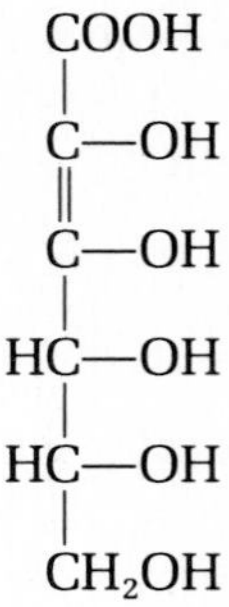

D

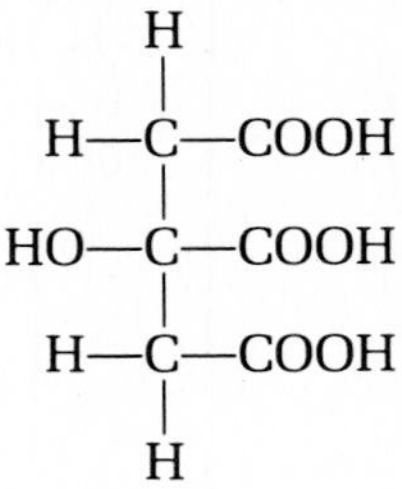

E

Answers: 1-C, 2-D, 3-E, 4-A, 5-B.

ESTERS

The way for you to remember what an ester is chemically is to remember this analogy: You can hitch a trailer to your car with a strong chain and drive around for 6 months in this new vehicular combination. The new combination will behave differently from either of the components that constitute it. At any time, you or someone else can split the chain, freeing your car and the trailer to act independently again.

Likewise, an **ester,** a new chemical combination between a carboxylic acid and an alcohol, behaves uniquely, not at all like the components from which it is constituted. An ester is formed when one molecule of a carboxylic acid hitches up with one molecule of an alcohol; in the process, one molecule of water is eliminated (Figure 10.8).

We can write a general formula for an ester like this:

$$R-\overset{\overset{O}{\|}}{C}-O-R'$$

where R stands for what used to be part of the carboxylic acid and R′ (pronounced "R prime") stands for what used to be part of the alcohol.

Just as your trailer and your car can be split, each returning to its original separate state, esters can be split, regenerating the acid and the alcohol. We term this splitting of the ester *hydrolysis* because it is a *lysis* (splitting) carried out by the addition of water (*hydro*).

Word equation: a carboxylic acid + an alcohol $\xrightarrow{-H_2O}$ ester

Chemical equation:

$$CH_3\overset{O}{\overset{\|}{C}}\!\!\left[\!OH + H\!\right]\!OC_2H_5 \xrightarrow{-H_2O} CH_3-\overset{O}{\overset{\|}{C}}-OC_2H_5$$

FIGURE 10.8   How Esters Are Synthesized.

Let us illustrate ester hydrolysis by using aspirin as our example. Aspirin is acetylsalicylic acid (ASA), an ester formed from a carboxylic acid and a phenol (instead of the usual alcohol, but the situation is analogous).

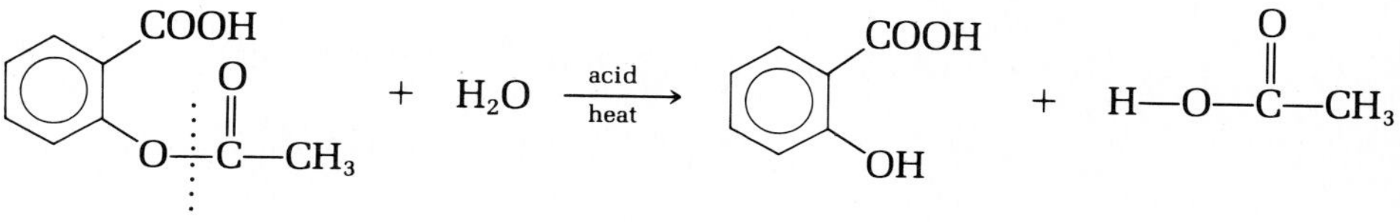

aspirin   salicylic acid   acetic acid

We see that a molecule of water has participated in the splitting of the ester bond, liberating the original phenol and the original carboxylic acid. Old aspirin tablets that have been exposed to moisture sometimes smell like acetic acid. From the preceding equation, you can now understand why. Such samples should, of course, be discarded.

Inside our bodies, esters are hydrolyzed through the catalytic action of enzymes. The anesthetic benzocaine (ethyl ester of PABA) is enzymatically split into PABA and ethyl alcohol, thereby losing its anesthetic potency.

Esters of carboxylic acids are named much like salts of carboxylic acids. Just use the name of the alcohol in place of the metal ion. Table 10.4 lists some esters by name and component parts. Esters have found use as flavoring and fragrance agents. For example, amyl acetate, butyl butyrate, and isobutyl formate smell like banana, pineapple, and raspberry, respectively.

Table 10.4.   Names and Structures of Esters

Name	Structure	Acid component	Alcohol component
Ethyl acetate	$CH_3COOC_2H_5$	Acetic acid	Ethyl alcohol
Isopropyl butyrate	$CH_3(CH_2)_2COOCH(CH_3)_2$	Butyric acid	Isopropyl alcohol
Methyl salicylate (oil of wintergreen)	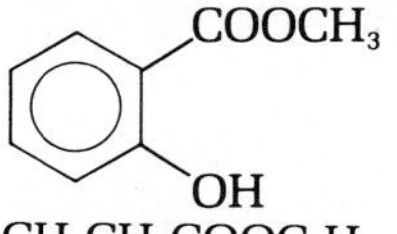 (structure: benzene ring with COOCH$_3$ and OH)	Salicylic acid	Methyl alcohol
Phenyl propionate	$CH_3CH_2COOC_6H_5$	Propionic acid	Phenol (not an alcohol)

Esters Important in Medicine

Cocaine, an ester from the plant *Erythroxylon coca*, has the structure

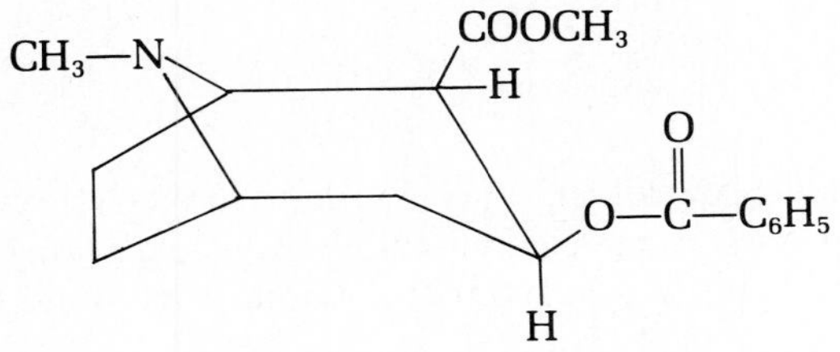

Cocaine's actions in the human include the following:

Local anesthetic   Topically, in 2 to 5% w/v solutions, blocks pain as in surgery of ear, nose, and throat; used as an anesthetic in intubations.

Vasoconstrictor   Constricts (narrows) blood vessels, limiting blood flow in the area of application.

CNS stimulant   Produces mental excitement, euphoria; humans can become psychologically dependent.

Salicylates (aspirin, oil of wintergreen) are valuable as analgesics (pain relievers), antipyretics (fever reducers), and anti-inflammatory agents. Aspirin's ability to upset the stomach, to cause bleeding from the stomach lining, and to interfere with the blood clotting mechanism reduces its usefulness as a drug.

Nitroglycerin (glyceryl trinitrate) is the trinitrate ester of glycerin.

Its formation illustrates the fact that alcohols can form esters with inorganic (i.e., mineral) acids as well as carboxylic acids. Amyl (pentyl) nitrate and amyl nitrite are additional examples of esters made from mineral acids. Nitroglycerin is a powerful vasodilator, used especially in attacks of *angina pectoris*. In this painful and dangerous condition, constricted coronary (heart) arteries prevent adequate oxygen-carrying blood from getting to heart muscle tissue, which results in chest pain that frequently radiates to the left shoulder and then to the left arm. A nitroglycerin tablet held under the tongue (i.e., sublingually) quickly dissolves, is absorbed into the bloodstream, and acts to dilate the coronary arteries. Blood flow is soon returned to adequate levels; oxygenation of coronary musculature is also returned to normal. An attack of angina pectoris can be precipitated by exercise, emotion, or a heavy meal.

Taking too much of a vasodilator drug can cause excessive vasodilation, a profound drop in blood pressure (hypotension) with dizziness and possible fainting.

Acetylcholine is probably the most important ester in the human body. You could not function without it, for it acts as a neurotransmitter substance at synaptic junctions (places where the end of one nerve meets the beginning of the next). A thousandth of a second after acetylcholine has done its job of transmitting a nerve impulse, it is destroyed by enzymatic hydrolysis.

$$\overset{\displaystyle CH_3}{\underset{\displaystyle CH_3}{CH_3 - \overset{\oplus}{N} - CH_2 - CH_2 - O - \overset{\displaystyle O}{\overset{\|}{C}} - CH_3}} \cdot Cl^-$$

acetylcholine chloride

EXAMPLE 10.2

Predict the functional group at which acetylcholine is hydrolyzed and show the structure of hydrolysis products.

Solution
The ester is cleaved as follows:

$$(CH_3)_3\overset{\oplus}{N}CH_2CH_2 - O \vdots \overset{\displaystyle C}{\underset{\displaystyle O}{\overset{\|}{C}}} - CH_3 \quad \xrightarrow[\text{enzyme}]{H_2O} \quad (CH_3)_3\overset{\oplus}{N}CH_2CH_2OH \quad + \quad HO - \overset{\displaystyle C}{\underset{\displaystyle O}{\overset{\|}{C}}} - CH_3$$

choline   acetic acid

Benzyl benzoate, $C_6H_5COOCH_2C_6H_5$, is an effective parasiticide used in the treatment of scabies by local application.

ETHERS

The word *ether* is a general term that includes hundreds of forms. The most famous is ethyl ether (often just called ether), which has the structure

$$\overset{\displaystyle H \quad H \qquad H \quad H}{\underset{\displaystyle H \quad H \qquad H \quad H}{H - \overset{|}{C} - \overset{|}{C} - O - \overset{|}{C} - \overset{|}{C} - H}}$$

All ethers have an oxygen bridge between two carbon atoms. The two carbon atoms can be part of aliphatic groups (R—O—R), aromatic groups (Ar—O—Ar), or both (R—O—Ar). Additional examples of ethers are

CH_3—O—C_2H_5 CH_2=CH—O—CH=CH_2 —OCH_3

methyl ethyl ether   vinyl ether   methyl phenyl   tetrahy-   eugenol
                                                                ether (anisole)   dropyran

When the health professional hears the word *ether*, she or he is inclined to think of general inhalational anesthetics, for ethers constitute some of the most valuable examples of this type of agent. Some of the currently used ether anesthetics are

2-chloro-1,1,
2-trifluoroethyl
difluoromethyl ether
(Ethrane)

2,2-dichloro-1,1-difluoroethyl
methyl ether
(Penthrane)

2,2,2-trifluoroethyl
vinyl ether
(Fluomar)

Ethyl ether, $(C_2H_5)_2O$, once widely used in inhalational anesthesia, is no longer used at most large modern hospitals because of its flammability and tendency to cause postoperative nausea. Ethyl ether was the first anesthetic to be used successfully in a surgical operation (1846). The story of its introduction is fascinating reading (see L. Goodman and A. Gilman, *The Pharmacological Basis of Therapeutics*, 5th ed., New York: Macmillan, 1975, pp. 89 ff.).

The gaseous anesthetics currently in use have the great advantage of being nonexplosive, but each suffers from one disadvantage or another. For example, one may be potent in inducing anesthesia, but also cause delirium. Another may be free of delirium effects but depress respiration or cause irregularity in the heartbeat (arrhythmia). Actually, then, modern inhalation anesthetics are not used alone but in combination with other anesthetics or with preoperative medication. An antisialagogue is a preop medication given to reduce salivation during anesthesia. Atropine is an antisialagogue.

Inhalation anesthetics obtund pain by depressing the CNS, and although we do not know the exact mechanism by which this occurs, we do know that the anesthetic must be fat-soluble if it is to pass the blood–brain barrier. This barrier consists in part of a fatty sheath surrounding blood vessels in the brain.

Fat-soluble chemicals have a better chance of crossing than water-soluble chemicals. An indication of the fat solubility of anesthetic drug is given by its oil/water **partition coefficient.** To determine this coefficient, we shake up the drug with an immiscible mixture of oil and water and then determine how much of the drug has distributed itself (i.e., partitioned) in the oil and how much in the water. For Penthrane (formula given earlier), the partition coefficient is

$$\frac{\text{olive oil}}{\text{water}} = \frac{400}{1} \text{ at } 36°C$$

Such a high solubility in fat compared to water indicates that Penthrane will be able to get to the brain in concentrations great enough to induce anesthesia.

The principle of a drug partitioning itself between two immiscible fluids can be applied to other situations such as drugs at a cell membrane, or gases in the alveolar spaces of the lungs.

A cyclic ether of special interest is ethylene oxide, a

$$CH_2\!\!-\!\!CH_2$$
$$\diagdown O \diagup$$

compound long used to sterilize temperature-sensitive equipment and drugs, as it effectively destroys all forms of microorganisms at room temperature. Plastic IV injection equipment can be sterilized in the shipping carton using ethylene oxide. Ethylene oxide, however, can form explosive mixtures with air; it is also suspected of being a human carcinogen.

A group of ethers of biochemical importance are the *tocopherols* (collectively, vitamin E). Although all green plants contain some tocopherols, rice germ, wheat germ, corn germ, lettuce, soya, and cottonseed oils are especially rich sources.

alpha-tocopherol

We do not know for certain the exact role played by the tocopherols in the human, but evidence indicates that they are antioxidants, that is, they are easily oxidized themselves, thus sparing the oxidative destruction of other important body chemicals such as vitamin C, vitamin A, unsaturated fatty acids, and the fat in red blood cells (RBCs).

Vitamin E has been actively promoted as a treatment for heart disease, as a fertility drug, and as a drug to help wounds heal without scarring. The validity of such claims remains in doubt.

Recommended daily allowances for vitamin E range from 5 International Units (IU) for infants to 30 IU for older males and pregnant women.

ALDEHYDES AND KETONES

We study **aldehydes** so that we can understand what a sugar is. The sugar we have in our blood is glucose; it has the formula

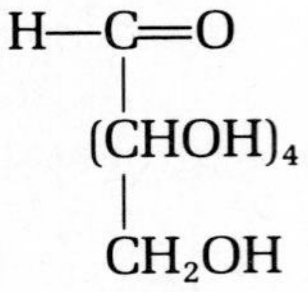

$$H—C\!=\!\!O$$
$$|$$
$$(CHOH)_4$$
$$|$$
$$CH_2OH$$

glucose

That unusual arrangement at the top of the chain is an aldehyde group, RCHO. Glucose is essential for our existence, so in one sense we could not live without aldehyde groups.

We study **ketones** so that, among other things, we can understand that vitamin K is.

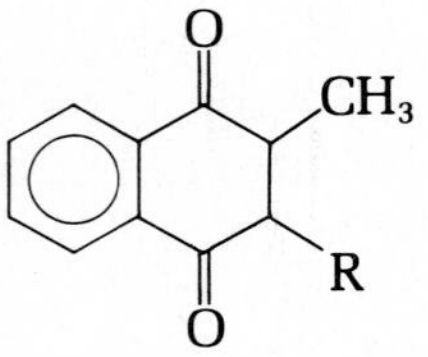

vitamin K (R = long alkene group)

The two C=O groups in vitamin K are ketone groups. Since vitamin K is essential for proper blood clotting, you might say that ketones keep us from exsanguinating (bleeding out).

Collectively, aldehydes and ketones are termed **carbonyl compounds** because they all have the carbonyl group

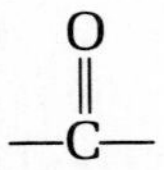

$$O$$
$$\|$$
$$—C—$$

carbonyl group

EXAMPLE 10.3

Is the carbonyl group polarized (i.e., is it a dipole)? If so, specify the charges.

Solution
Oxygen, being much more electronegative than carbon, pulls the bonding electrons closer to itself, creating a partial charge separation.

$$
\begin{array}{c}
\diagdown \\
C \overset{\displaystyle O^{\delta-}}{\underset{\displaystyle C^{\delta+}}{\|}} \\
\diagup \diagdown
\end{array}
$$

This gives the carbonyl carbon atom a partial positive charge, a situation that can greatly affect the chemical behavior of aldehydes and ketones.

Note that every aldehyde has an H atom attached to the carbonyl group, plus the alkyl or aryl group. In a line formula, the aldehyde group is written RCHO. Every ketone has two alkyl or aryl groups, or both, each attached to the carbonyl carbon. Figure 10.9 gives structures and names of some important aldehydes and ketones.

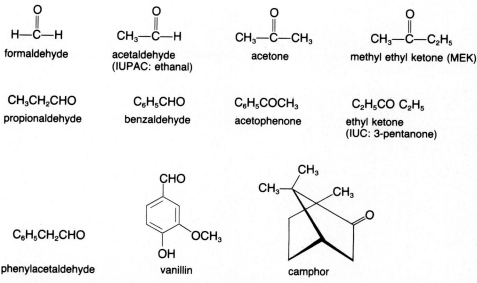

formaldehyde

acetaldehyde
(IUPAC: ethanal)

acetone

methyl ethyl ketone (MEK)

CH_3CH_2CHO
propionaldehyde

C_6H_5CHO
benzaldehyde

$C_6H_5COCH_3$
acetophenone

$C_2H_5CO\ C_2H_5$
ethyl ketone
(IUC: 3-pentanone)

$C_6H_5CH_2CHO$
phenylacetaldehyde

vanillin

camphor

FIGURE 10.9 Some Important Aldehydes and Ketones.

Aldehydes and ketones arise in the laboratory and in living organisms by an oxidation process in which alcohols are stripped of some of their hydrogen atoms. When 1° alcohols are oxidized, aldehydes result.

$$R\text{—}CH_2OH \xrightarrow{\text{oxidation}} R\text{—}CHO$$

a primary alcohol   an aldehyde

When secondary alcohols are oxidized, ketones result.

$$R\text{—}CHOH\text{—}R \xrightarrow{\text{oxidation}} R\text{—}CO\text{—}R$$

a secondary alcohol   a ketone

A good oxidizing agent for both of these reactions is potassium dichromate ($K_2Cr_2O_7$), but when it is used to prepare an aldehyde, care must be taken to prevent overoxidation of the aldehyde to a carboxylic acid.

$$R\text{—}CHO \xrightarrow{\text{oxidation}} RCOOH$$

EXAMPLE 10.4

By oxidation of alcohols, the human body can form (a) formaldehyde, (b) acetaldehyde, (c) HOOC—CHO, (d) HOOC—COOH, and (e) $CH_3COCOOH$. Write the formula of the alcohol that upon oxidation will give each of these.

Solution
(a) CH_3OH, (b) CH_3CH_2OH, (c) $HOCH_2COOH$, (d) $HOCH_2CH_2OH$, and (e) $CH_3CHOHCOOH$ (lactic acid).

CASE HISTORY INVOLVING BIO-OXIDATION OF A DIOL

Oxidation of alcohols in the human body is not just an academic topic, it really happens, as shown in a report by J. Stokes and F. Aueron in the *Journal of the American Medical Association* (243:20, May 1980). A 33-year-old female was brought to the emergency room after drinking 2 L of ethylene glycol ($HOCH_2CH_2OH$). It was known that although ethylene glycol per se is not acutely toxic, it is metabolized in the liver to oxalic acid

oxalic acid

Ca(II) ions in the blood will combine with oxalic acid to form crystals of calcium oxalate that can precipitate in the urine and kidneys—a critical situation. In addition, a metabolic acidosis can occur as the body is forced to handle large quantities of oxalic acid.

Knowing this chemistry, and the possible toxic metabolites, the emergency room staff instituted the following therapy:

1. Gastric lavage removed approximately 1 L of ethylene glycol.
2. Sixty milliliters of absolute ethanol was infused IV stat and a continuous infusion of 10 mL per hour was begun. The rationale here is that the alcohol will compete with the glycol for alcohol dehydrogenase, thus reducing the overall metabolism of the glycol.
3. Large quantities of IV fluids were administered which, in conjunction with hemodialysis, prevented calcium oxalate from crystallizing in the urine.

The patient recovered without any of the damage anticipated from ethylene glycol ingestion.

In Chapter 12 of this book, you will study abnormal carbohydrate metabolism, including the disease diabetes mellitus. Two important chemicals that the body produces in abnormally large quantities in this disease are the ketone bodies

$$
\begin{array}{ccc}
& \quad \overset{\displaystyle O}{\underset{\displaystyle \|}{}} & \qquad\qquad \overset{\displaystyle O}{\underset{\displaystyle \|}{}} \\
CH_3\!-\!C\!-\!CH_3 & \text{and} & CH_3\!-\!C\!-\!CH_2COOH
\end{array}
$$

acetone   acetoacetic acid (beta-ketobutyric acid)

Because the liver has difficulty metabolizing these compounds quickly, they diffuse into the blood, and in the case of acetone, are exhaled on the breath.

Formaldehyde, the simplest aldehyde, is sold as formalin, a 40% solution in water; it finds wide use as an embalming fluid and preservative for specimens in biology laboratories. Formaldehyde is now recognized as toxic to the human; avoid all contact with its liquid or vapor forms.

DRUG ACTIVITY IN CARBONYL COMPOUNDS

Chloral hydrate, $Cl_3CCH(OH)_2$, is produced when a molecule of water attacks the aldehyde chloral, Cl_3CCHO. Chloral hydrate is a reliable and safe hypnotic (sleep inducer), often dispensed in hospitals under the trade name Noctec. With the usual oral dose of 0.5 to 1.0 g, sleep occurs within the hour and lasts for up to 8 hours. Chloral hydrate suppresses REM sleep (rapid eye movement sleep) less than the barbiturates, a clear advantage. In combination with alcohol, chloral hydrate constituted the famous Mickey Finn knockout drops. Although the potency of the mixture was probably overrated, it is true that chloral hydrate and alcohol offer a synergistic combination of CNS depressants that make sleep hard to resist. (In *synergistic action*, two drugs potentiate each other so that their combined action is greater than the sum of their independent actions.)

Methadone (Dolophine) is the much discussed synthetic opioid compound that has had wide use in methadone maintenance clinics in the United States and England. The idea is to get a heroin addict voluntarily to enter a clinic and to receive a near-free daily dose of methadone that substitutes well in satisfying his desire for heroin. *Ideally*, this gets the addict off the street and eliminates theft, trafficking, and prostitution. *Ideally*, the dose of methadone is gradually reduced until the addict is drug-free and can reenter society. In practice, the concept falls short of expectations. For your comparison, here are structures of methadone and heroin.

methadone   heroin (diacetylmorphine)

The structures are sufficiently similar for both to act at the same receptor site in the brain (hence the ability of methadone to satisfy heroin hunger). They induce mutual cross-tolerance (that is, tolerance to one automatically means tolerance to the other).

AMINES

Compounds with this very important functional group contain nitrogen in the form of $R-NH_2$. Examining the periodic chart of the elements, we discover that nitrogen lies in Group Va and needs three electrons to complete its octet. In covalent bonding, nitrogen thus forms three bonds with other elements or with other nitrogen atoms. Figure 10.10 shows examples of aminoid compounds and also shows that the N atom in amines bears a pair of nonbonded electrons. Their presence determines an amine's chemical behavior.

Amines can be aliphatic (A, B, C, and E in Figure 10.10), or aromatic (D in Figure 10.10). They can be primary (1°), that is, bearing one alkyl or aryl

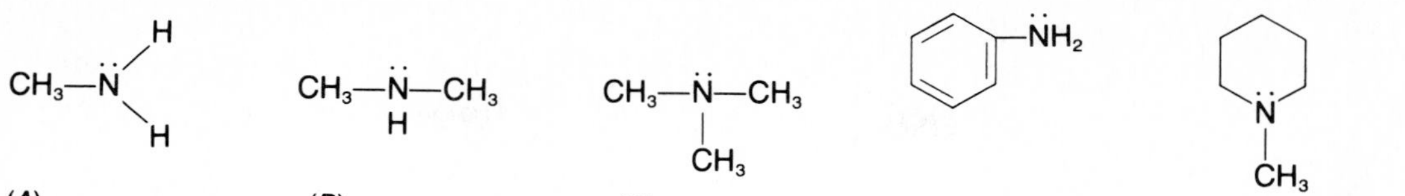

(A) methylamine

(B) dimethylamine

(C) trimethylamine

(D) aniline

(E) N-methylpiperidine

FIGURE 10.10 Examples of Amines. **Note unshared pairs of electrons on N.**

group on the nitrogen (A and D in Figure 10.10), or secondary (2°), that is, bearing two such groups (B) on the nitrogen. A tertiary amine (3°) bears no remaining H atoms (C and E) on the amine group.

All amines are basic compounds; in water they produce an alkaline solution by reacting with water to produce hydroxide ions:

$$RNH_2 + H_2O \rightleftharpoons \overset{\oplus}{RNH_3} + OH^-$$

In this regard they are very much like ammonia, NH_3, from which they can be derived. Aliphatic amines generally are stronger bases than aromatic amines; their solutions in water are more alkaline.

Being basic, amines react with acid to form salts.

$$RNH_2 + HCl \rightarrow \overset{\oplus}{RNH_3} + Cl^-$$

an amine salt

This reaction is of exceptional importance to the health professional because the formation of amine salts is a means of solubilizing an otherwise insoluble amine. It is the pair of electrons on the nitrogen atom that provide the basis for salt formation; they attract and bind the proton of the acid.

$$R-\overset{..}{N}H_2 + H^+Cl^- \rightarrow R-\overset{H}{\overset{|}{N}}H_2^\oplus + Cl^-$$

ionic and water soluble

Before the amine is protonated, it is called a **free base.** After protonation, it is the *amine salt form.* Many different acids can be used to form amine salts: hydrochloric, sulfuric, phosphoric, nitric. This is why you will read about Visine hydrochloride, morphine sulfate, codeine phosphate, physostigmine nitrate. The reverse reaction, the liberation of an amine from its salt, is accomplished by adding a strong base. This neutralization is entirely analogous to the liberation of ammonia from ammonium salts.

Ammonium salts

$$NH_4Cl + NaOH(aq) \rightarrow NH_3 + H_2O + NaCl$$

Amine salts

$$C_2H_5NH_3Cl + NaOH(aq) \rightarrow C_2H_5NH_2 + H_2O + NaCl$$

Morphine is a narcotic analgesic obtained from the opium poppy (Figure 10.11). It is of inestimable value in killing pain (typical dose: 10 to 15 mg). Morphine free base is insoluble in water and cannot be given hypodermically in this state. However, in Figure 10.12 we see how medicinal chemists can prepare a water-soluble sulfate salt suitable for intramuscular (IM) injection.

FIGURE 10.11   Opium is obtained by incising the unripe seed capsule of the opium poppy and collecting the milky exudate. Upon exposure to the air, the exudate turns brown. It is extracted for its content of morphine and codeine.

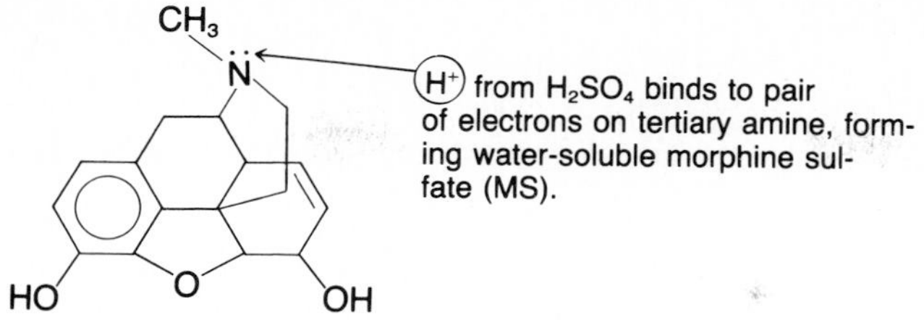

FIGURE 10.12 Morphine-free base can be converted to a water-soluble salt.

Hydrogen bonding can occur in amines, much the same as it occurs in alcohols. The dipole arises because of the greater electronegativity of the nitrogen compared to hydrogen. Hydrogen bonding in amines affects their physical properties: $C_2H_5NH_2$, molecular weight, 45; boiling point, 17°C. Compare to propane, molecular weight, 44; boiling point, -42°C. Amine H-bonding plays a large role in the geometry of DNA.

Naming Amines and Their Salts

To name an amine, write the names of all the aliphatic and aromatic groups attached to the aminoid nitrogen, followed by the word *amine*. If a group appears more than once, condense through the use of *di* or *tri*. In C_3H_7—NH—C_2H_5, the groups are ethyl and propyl, and the name is ethylpropylamine. In $(CH_3)_2NC_6H_5$, two methyl groups and one phenyl group exist; the name is dimethylphenylamine (usually, however, called N,N-dimethylaniline since the parent aromatic amine $C_6H_5NH_2$ is called aniline).

EXAMPLE 10.5

Provide one name each for (a) $H_2NC_4H_9$, (b) $(C_2H_5)_2NH$, (c) $C_3H_7NHC_5H_{11}$, and (d) $C_6H_5NH_2$.

Solution
(a) Butylamine, (b) diethylamine, (c) propylpentylamine, and (d) aniline.

To name salts of amines, use the ammonium ion, $NH_4{}^+$, as your starting point. Then preface names of groups, condensing with *di* or *tri* as necessary. The second word in the name of a salt is the anion from the acid.

EXAMPLE 10.6

Provide one name each for (a) $C_2H_5NH_3Cl$, (b) $(CH_3)_2NH_2I$, and (c) $(C_3H_7NH_3)_2SO_4$.

Solution

(a) Ethylammonium chloride, (b) dimethylammonium iodide, (c) propylammonium sulfate.

IUPAC names of amines follow the usual IUPAC rules and make use of the prefix *amino*. Simply locate the longest unbranched carbon chain that holds the amino group and number from the end of the chain that gives the lowest total of numerical prefixes. Here are examples of IUPAC names of amines:

$H_2N—(CH_2)_3CH_3$ $CH_3CH(NH_2)CH_2CH_2CH_2NH$ $CH_3CH(NH_2)COOH$

1-aminobutane   1,4-diaminopentane   2-aminopropanoic acid

$H_2NCH_2CH_2OH$

2-aminoethanol

Amines of Biochemical and Medicinal Importance

From nature and from the synthetic chemist, we have gotten a massive number of important biomolecules that owe all or most of their pharmacological activity to their amino group. The list is so vast that we can only summarize it here.

1. The **Catecholamines** (cat-eh-call'-ameens). During periods of excitement, anger, or stress, our adrenal glands secrete the two neurotransmitters epinephrine (adrenalin) and norepinephrine (noradrenalin).

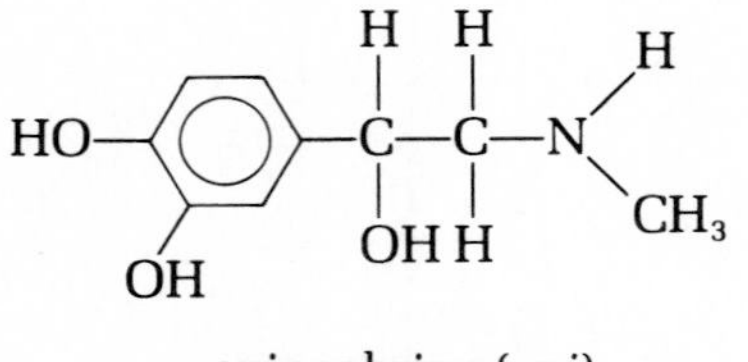

epinephrine (*epi*) 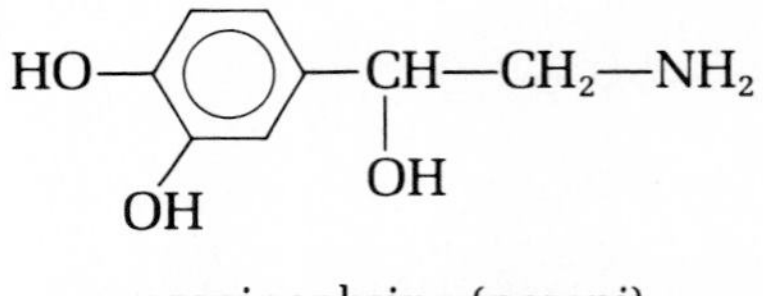   norepinephrine (*norepi*)

The name catecholamine derives from the presence of the dihydroxybenzene (catechol) portion of the molecule plus the fact that they are amines. Norepi is especially important in its role of transmitting impulses in nerves that speed up the heart, dilate the bronchi, constrict peripheral blood vessels, halt food digestion, and excite the CNS. In other words, norepinephrine helps prepare the body for a "fight or flight" type of response. Each catecholamine, superimportant in our biochemistry, is at one and the same time an amine, an alcohol, a diphenol, an aromatic compound, and an aliphatic compound. Can you pick out all of these functional groups in the preceding structures? Dopamine is a catecholamine that functions as a neurotransmitter in the brain. A deficiency of dopamine is associated with the onset of Parkinson's disease. There is increasing evidence that dopamine may be involved in schizophrenia.

2. The **Alkaloids.** Many plants produce alkaline, nitrogen-containing compounds called alkaloids. All alkaloids are amines, either aromatic or aliphatic, carbocyclic or heterocyclic, 1°, 2°, or 3°. Often these alkaloids have potent, useful pharmacological activity in the human.

Morphine and codeine—pain relief.
Ephedrine—bronchodilator in asthma.
Ergotamine—relief of migraine headaches.
Cocaine—local anesthetic.
Atropine—anticholinergic.
Scopolamine—preop medication.
Tubocurarine—muscle relaxant.
Pilocarpine—constricts pupil of the eye.
Reserpine—antipsychotic, antihypertensive.
Caffeine—CNS stimulant.
Theophylline—diuretic.
Quinine—antimalarial.
Quinidine—cardiac antiarrhythmic.

3. The **Amino Acids.** We have studied carboxylic acids, so you know what acetic acid is. Take an NH_2 group and substitute it on the second carbon in acetic acid. The result is an α-amino acid.

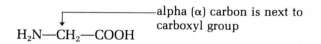

alpha (α) carbon is next to
carboxyl group

$$H_2N—CH_2—COOH$$

We say "alpha" because NH_2 group substitutes on the *first* carbon down the chain from the COOH group. Similarly, $CH_3CH(NH_2)COOH$ is α-aminopropionic acid, which has the special name of alanine. Dozens of amino acids are known to science. About twenty are especially important in the human, for as we shall see in Chapter 11, amino acids are what proteins are made of. Figure 10.13 gives structures of a variety of structural types of amino acids. Although greatly different in some respects, all have the alpha-aminocarboxylic acid arrangement. We learn from Figure 10.13 that some amino acids are aromatic in nature (tyrosine and tryptophan), some are heterocyclic (tryptophan), some have two amino groups (lysine), some have two carboxyl groups (aspartic acid), and some are strictly aliphatic in nature (valine).

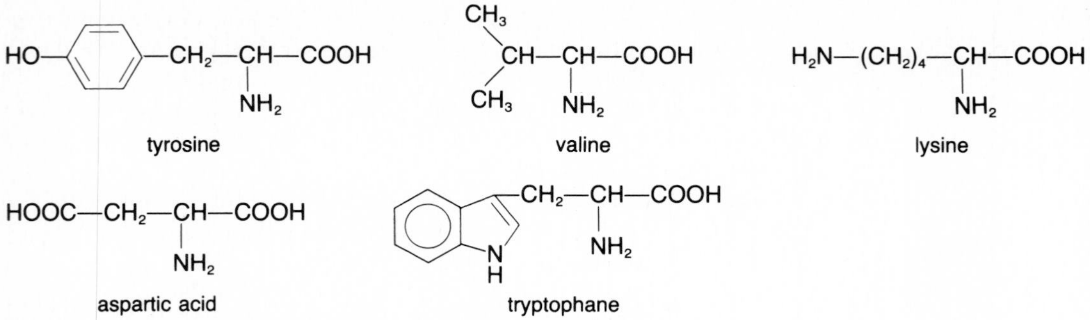

FIGURE 10.13   Some Structural Types of Alpha-Amino Acids.

4. **Aniline Derivatives.** Aniline, $C_6H_5NH_2$, is called a coal tar chemical because originally it was obtained from that source. Aniline is very important because many of its derivatives are highly useful drugs and chemicals. Figure 10.14 lists some important aniline-based compounds.

Acetaminophen (Datril, Tempra, Tylenol, Valadol), shown in Figure 10.14, is currently heavily advertised for the relief of pain and fever. Although a Food and Drug Administration panel rated acetaminophen as "safe and effective" for the relief of pain and the reduction of fever, the wise health professional will exercise caution in its use, for all aniline derivatives are toxic to the liver if used in large enough doses or for a long enough time. In fact, very recent reports suggest that even normal doses of acetaminophen may sometimes cause liver damage. For this reason, the maximum adult daily dose of 4 g should not be exceeded, the duration of use should be limited, and the use of the newer "extra strength" formulations should be made with great caution. Acetaminophen is not useful as an anti-inflammatory in the treatment

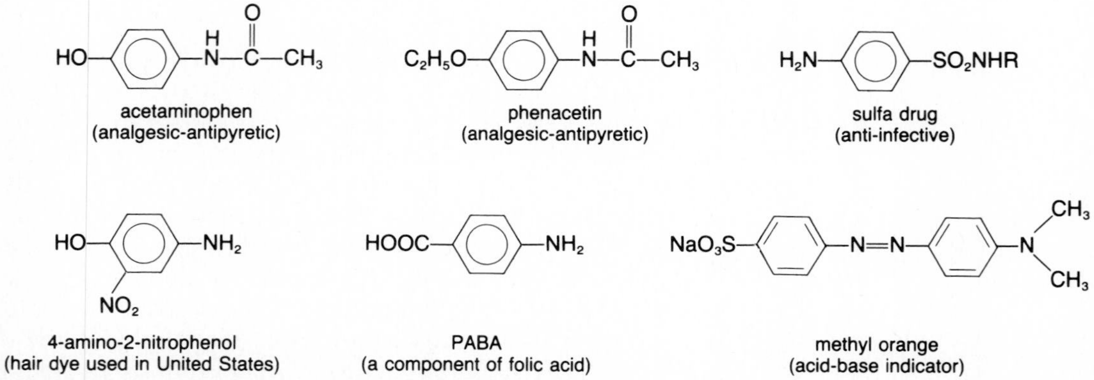

FIGURE 10.14   Aniline-Based Drugs and Chemicals.

of arthritis and rheumatism (aspirin is). Acetaminophen does not increase bleeding tendency in the human (aspirin does).

5. **Amines Used as Drugs.** A very large number of synthetic drugs are amines. This is because medicinal chemists recognize the special properties of an amino group: potential for H-bonding, similarity to the functional groups in body chemicals and neurotransmitters, affinity for receptors, ability to form salts. We cannot examine all major amine drug classes in this book; however, we will identify a few of the very important.

Amphetamines such as methamphetamine (I) (Desoxyn), all possess the phenethylamine skeleton (II) that is found in catecholamines

I   II   III

such as adrenalin (III). Methamphetamine, given in 2.5 to 5.0-mg doses, stimulates the CNS and the heart, increases blood pressure, dilates the bronchi, and reduces appetite. It has been used for years as an appetite suppressant in the treatment of obesity, a practice roundly criticized by the FDA as unwise and dangerous. Heavy amphetamine use can result in the development of tolerance and psychological dependence. One California housewife worked her way up to a daily dose of 250 mg of Desoxyn (that's twenty-five 10-mg tablets a day) before voluntarily entering a clinic for detoxification. Legitimate, reasonable medical uses of amphetamines include increasing blood pressure (i.e., a pressor agent in certain hypotensive conditions), calming hyperactive children (paradoxically, in this situation, they work!), treating narcolepsy.

Antihistamines. Histamine (I), always present in our bodies, is released from its storage sites when we are exposed to pollen, dust, and other allergens; when cell damage occurs; and when we come in contact with chemicals to which we have become sensitized. A powerful agent, upon release into the blood and tissues histamine causes dilation of capillaries, increased capillary permeability, bronchoconstriction, and increased gastric secretion, among other effects. Chemists have synthesized many antihistamine drugs (II and III), which are chemically similar to histamine. This similarity permits the antihistamine molecules to occupy and block receptor sites intended for histamine. This technique is termed competitive antogonism (think of your neighbor's car occupying your garage; it prevents you from parking there and unloading your groceries). If histamine is blocked from the tissue sites where it functions, there will be less sneezing, less tearing, less constricted bronchi, less local irritation. Note, however, that most

I	II	III
histamine	Benadryl (diphenhydramine)	Dimetane, Dimetapp (brompheniramine)

people pay a price for protection against histamine. Drowsiness is a side effect of antihistamines. The ability to operate machinery or drive a car can be impaired after dosing with antihistamines.

Quaternary Amine Salts. When a tertiary amine such as trimethylamine reacts with methyl iodide, a special type of amine salt is produced in which four groups are attached directly to the nitrogen atom. We term these products quaternary ammonium salts.

$(CH_3)_3N$	CH_3I	quaternary ammonium salt
tertiary amine	alkyl halide	

Note that a positive charge resides on the nitrogen atom; an anion such as iodide always accompanies the cation. Two important applications of quaternary salts have been discovered: (1) as germs killers for use on skin and instruments, and (2) as skeletal muscle relaxants.

Zephiran chloride (benzalkonium chloride USP) is a well-known antiseptic quaternary chloride that kills many pathogenic bacteria and fungi on contact. It does this through its detergent action, which destroys the membranes that coat and protect the germs. Zephiran 1:750 w/v is recommended for surgical and gynecological procedures, antisepsis of skin and hands, and sterile storage of sterilized instruments. It must never be mixed with any soap, as soaps cause inactivation. Other well-known quarternary anti-infectives are benzethonium chloride (Phemerol), methylbenzethonium chloride (Bactine, Diaparene), and cetylpyridinium chloride (Ceepryn). Succinylcholine chloride USP

$$(CH_3)_3\overset{\oplus}{N}-CH_2CH_2OOC(CH_2)_2-COOCH_2CH_2\overset{\oplus}{N}(CH_3)_3 \cdot 2\ Cl^-$$

succinylcholine chloride

is a quarternary the idea for which came from the structure of curare, the South American arrow poison. Both compounds cause skeletal muscle paralysis by blocking nerve transmission at the point where muscle meets nerve tissue (the myoneural junction). Actually, they compete with acetylcholine for receptor sites as the junction. Skeletal muscle relaxation is required in surgery such as in the thorax. These drugs are very potent and in larger doses will paralyze the muscles of breathing. If this happens in surgery, respirators keep the patient breathing. Local anesthetics are used topically to relieve pain in a specific, limited body location, or to block the transmission of pain sensation by infiltration of a nerve fiber. The widely used local anesthetics are all amines, often aminoalkyl esters of p-aminobenzoic acid (see procaine type). Table 10.5 gives the names, chemistry, and pharmacological actions of a few of the more important local anesthetics. Cocaine, the structure of which was given earlier in the chapter, occurs

Table 10.5. Some Amines Used as Local Anesthetics

Generic name (trade name)	Chemical structure[9]	Use in medicine
Benzocaine (Americaine)	H_2N—⟨ ⟩—$COOC_2H_5$	Anesthetic on mucous membranes, hemorrhoids, itching; appetite suppressant
Procaine (Novocaine)	H_2N—⟨ ⟩—$COOCH_2CH_2N(C_2H_5)_2$	Anesthetic for minor and major surgery; spinal anesthesia; burns
Tetracaine (Pontocaine)	C_4H_9\N—⟨ ⟩—$COOCH_2CH_2N(CH_3)_2$ H/	Spinal anesthesia, OTC creams; used as spray-on
Lidocaine (Xylocaine)	CH_3 ... N—$COCH_2N(C_2H_5)_2$ ... CH_3	Dental anesthetic, by injection; OTC creams; used to treat heart arrhythmias
Cyclomethycaine (Surfacaine)	CH_3 ... N—$(CH_2)_3$—O—C(=O)—⟨ ⟩—O—⟨ ⟩	Local anesthetic on skin and in rectum; sunburns, abrasions, poison ivy

[9] All of these local anesthetics are supplied as the water-soluble hydrochloride, sulfate, or other salt.

in a plant, and was the chemical inspiration for all of these other synthetic local anesthetics. Can you find the chemical similarities?

Many other examples of important naturally occurring and synthetic amines exist, including enzymes and vitamins. We shall encounter these two type later in this book.

ANTAGONISTS ACT AT RECEPTOR SITES

Our study of amines offers us an opportunity to examine the receptor site theory of drug action and the role of drug **antagonists.** According to the receptor site theory, drug molecules such as morphine or heroin produce their pharmacological effect because they fit exactly a specialized body tissue termed a **receptor site,** much as a key fits a lock. To occupy the receptor site, the drug molecule (called an **agonist**) must have not only the correct three-dimensional size and shape, but also the correct chemical and polar properties to insure high affinity by means of hydrogen bonding or other attractive force. When these requirements are met, the agonist can occupy the receptor site (in the brain, spinal cord, heart, lung, kidney, or wherever) and start a chain of events that culminates in typical pharmacological activity. Your car key will not open your house door because it does not have the correct size and shape; epinephrine does not alleviate pain because it does not fit the brain's receptors for analgesia. The essential feature of the receptor site theory is the specificity of agonist molecules for their own receptor tissue.

Next, consider the naloxone molecule (Figure 10.15). Because this man-made substance is structurally and three-dimensionally so similar to morphine and heroin (see Figure 10.12), it can occupy the brain's opiate receptor site, and with a high degree of affinity. But one profound pharmacological difference exists. Naloxone produces no narcotic effect! Like a key that fits your front door but does not turn and open the door, naloxone is a biochemical key that fits a receptor site, but unlocks no cellular response. Furthermore, its presence at the receptor site prevents agonist molecules from occupying the site and producing their drug action. Naloxone is termed an *antagonist*—in this instance a narcotic antagonist. Under the trade name Narcan, it is used in hospital emergency rooms to treat heroin overdose patients, in whom it acts by virtue

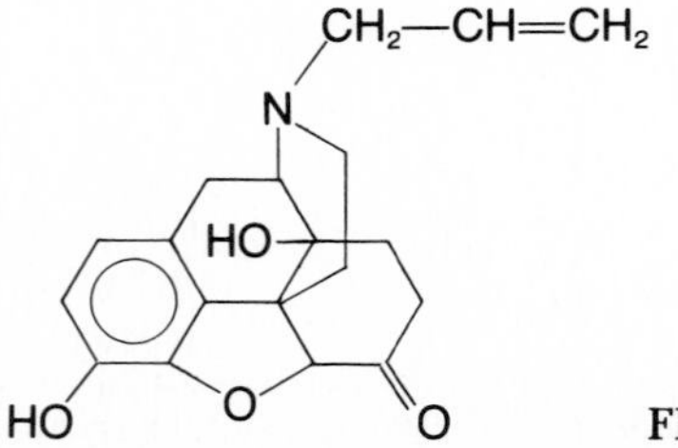

FIGURE 10.15 Naloxone, a Narcotic Antagonist.

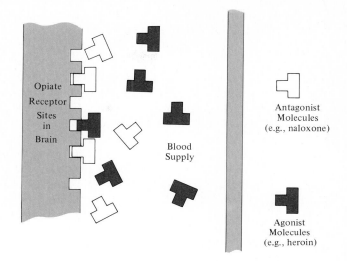

FIGURE 10.16 The Competition between Agonist and Antagonist Molecules for Positions on Receptor Sites. **Which will win out (and therefore which pharmacological action will ensue) is determined by affinity for the receptor site plus concentration of each kind of molecule. A 1-mg intravenous dose of naloxone (antagonist) completely blocks the effects of 25 mg of heroin (agonist), showing the greater affinity of naloxone for the receptor site.**

of its superior affinity for the receptor site, to actually *displace* heroin agonist molecules. Figure 10.16 represents typical competition between agonist and antagonist molecules for a receptor site.

SUMMARY

Functional groups give organic molecules their chemical, physical, and pharmacological characteristics. The seven functional groups that we studied in this chapter are alcohols, ethers, carboxylic acids, esters, aldehydes, ketones, and amines. In Chapter 11 we shall study amides in connection with peptide bonds and proteins.

In this chapter we examined alcohols (ROH) from the standpoint of hydrogen bonding and naming. We concentrated on ethyl alcohol, the alcohol found in alcoholic beverages, and were introduced to its pharmacology. (There are an estimated 10 million alcoholics and problem drinkers in the United States.)

The carboxylic acid, RCOOH, is among the most important of all functional groups. We learned that carboxylic acids are weak acids, although strong enough to decompose bicarbonate. We named carboxylic acids by the common system (e.g., acetic acid) and by the IUPAC system (e.g., ethanoic acid). We studied salt formation in carboxylic acids. Soaps are sodium or potassium salts of carboxylic acids, especially the long chain (fatty acid) type.

Esters (RCOOR′) are produced from carboxylic acids and alcohols by removing the elements of one molecule of water. In the reverse reaction, esters can be hydrolyzed by inserting a molecule of water to regenerate the original carboxylic acid and alcohol. In this chapter we examined three esters important in medicine.

Ethers (ROR) can be made by intermolecular removal of a molecule of water from two molecules of alcohol. We learned in this chapter that some important inhalational anesthetics are ethers and that they must have high fat solubility if they are to pass the blood–brain barrier and enter the central nervous system. Here the concept of partition coefficient became important. We learned that vitamin E is an ether.

Aldehydes and ketones have one structure in common: the carbonyl group ($\diagup$C$=$O). But aldehydes (RCHO) have a hydrogen atom on the carbonyl group whereas ketones (RCOR) have two other carbon atoms attached. Aldehydes and ketones are produced by oxidation of the appropriate alcohols—a reaction that we can perform in the laboratory with oxidizers such as $K_2Cr_2O_7$, and that occurs in our body under the influence of enzymes. We learned that some very important drugs are carbonyl compounds (e.g., chloral hydrate, methadone).

A large portion of this chapter was devoted to a discussion of the chemistry and pharmacology of amines (RNH_2). Amines are basic and form many water-soluble salts, a function used to solubilize certain drugs that are amines. Some of the amines that we examined were catecholamines, certain alkaloids, amino acids, aniline derivatives, amphetamines, antihistamines, and quaternary amine salts.

The last topic in this chapter was the receptor site. We were introduced to the theory of the receptor site and to the concept of agonists and antagonists.

KEY TERMS

Check your understanding of this chapter. Can you explain what is meant by each of these terms?

agonist	ether
alcohol	free base
aldehyde	hydrogen bonding
alkaloid	ketone
amine	partition coefficient
amino acid	phenol
antagonist	quaternary amine
carbonyl compound	receptor site
carboxylic acid	salicylate
catecholamine	soap
ester	

STUDY QUESTIONS

1. Examine each compound in the following list and then assign it to *one or more* of the following categories, based on functional group: alcohol, ether, carboxylic acid, ester, aldehyde, ketone, amine, amine salt, phenol.

   a. $C_2H_5OC_6H_5$
   b. Lactic acid
   c. $C_6H_5NHCH_3$
   d. Camphor
   e. $C_4H_9COOC_4H_9$
   f. HCHO
   g. $CH_3CH_2CHOHCH_3$
   h. $HOOC—COOC_2H_5$
   i. $C_5H_{11}COCH_3$
   j. $CH_3(CH_2)_{16}COOH$
   k. $(CH_3)_3NHCl$
   l. Morphine
   m. o-cresol
   n. ASA
   o. Niacin

2. What structural differences distinguish a phenol from an alcohol?

3. Match the chemical with its well-known medicinal or biological action. (Some have more than one action.)

   A. Phenol
   B. Aspirin
   C. Sodium stearate
   D. Nitroglycerin
   E. Ethyl alcohol
   F. Nicotinic acid
   G. Acetylcholine
   H. Ethyl ether
   I. Morpine
   J. Chloral hydrate
   K. Norepinephrine
   L. Amphetamine
   M. Cocaine

   1. Antipyretic
   2. Vasodilator
   3. Local anesthetic
   4. CNS depressant
   5. Vitamin
   6. Analgesic
   7. Hypnotic
   8. Neurotransmitter
   9. Inhalational anesthetic
   10. Soap (cleanser)
   11. CNS stimulant
   12. Disinfectant

4. Intermolecular hydrogen bonding can occur in wood alcohol. Draw the structures and show the bonds to explain this phenomenon.

5. Explain the processes by which ethyl alcohol is obtained in large quantities in the United States.

6. Define the following:
   a. Absolute alcohol
   b. Rubbing alcohol
   c. Teratogen
   d. FFA
   e. Polyol
   f. Weak organic acid
   g. Unsaturated fatty acid
   h. Ester
   i. CNS stimulant
   j. Partition coefficient
   k. Carbonyl compound
   l. Ketone bodies
   m. Alkaloid
   n. Catecholamine

7. Ethanol is a key intermediate in organic synthesis.
   a. Show the chemical steps by which ethyl ether is synthesized from C_2H_5OH.
   b. Do the same for the synthesis of acetaldehyde.
   c. How can acetic acid be obtained synthetically from ethyl alcohol?

8. Show the chemical steps and reagents necessary to synthesize acetone, starting with any three-carbon alcohol.

9. Provide the IUPAC name for the following:
   a. $CH_3CH(OH)CH_2CH(CH_3)C_2H_5$
   b. $ClCH_2(CH_2)_5COOH$
   c. CH_3CH_2COOK
   d. $(CH_3)_2CHCHOHCH(CH_3)_2$
   e. $C_3H_7COOCH_3$
   f.

   —OH   (ring is saturated)

10. Aspirin is never to be given to postoperative tonsillectomy patients to relieve pain. Why not?

11. What chemical structure is common to all salicylates?

12. Ethyl alcohol, with a molecular weight equal to 46, boils at 78°C. Yet ethyl ether, with a much higher molecular weight (74) boils at a low 38°C. What phenomenon can be used to explain the unexpectedly high boiling point of ethanol? Explain in detail.

13. Draw structures to show the difference between o-, m-, and p-cresol.

14. A drug manufacturer wishes to sterilize a drug for injection but discovers that the drug decomposes if heated above 50°C. Suggest another means of sterilization that does not involve heating.

15. How do aldehydes and ketones differ on the basis of chemical structure? In what respects are they similar?

16. Provide one common name for the following:
   a. CH_3CHO
   b. $CH_3CHOHCH_3$
   c. $CH_3(CH_2)_{16}COOH$
   d. $C_2H_5COOC_2H_5$
   e. $HCOONa$
   f. $C_6H_5COONH_4$
   g. $C_6H_5COOCH_2C_6H_5$
   h. $(C_2H_5)_2NH_2Br$
   i. C_3H_7—CO—$CH(CH_3)_2$
   j. CH_2—CH_2
      \ /
       O

17. Complete and balance the following equations:
   a. $CH_3CH_2COOH + NaHCO_3(aq) \longrightarrow$
   b. $CH_3COOH + CH_3OH \xrightarrow[\text{catalyst}]{\text{acid}}$
   c. $CH_3CHOHC_2H_5 \xrightarrow[\text{oxidation}]{K_2Cr_2O_7}$
   d. n-Butyl alcohol $\xrightarrow[\text{oxidation}]{K_2Cr_2O_7}$ (an aldehyde)
   e. Trimethylamine + $H_2O \longrightarrow$

   f. Trimethylamine + HNO_3(aq) $\longrightarrow$

   g. $CH_3(CH_2)_8COOCH_3$ + H_2O $\xrightarrow[\text{hydrolysis}]{\text{acid}}$

18. Eugenol comes from oil of cloves. There are several different functional groups in eugenol (see text for structure). Identify each.

19. Compare the structure of methadone with that of heroin.
    a. What functional groups do they have in common?
    b. What functional groups do they have but not share?
    c. How do we know methadone can occupy the same brain receptor sites that heroin occupies?

20. What is the difference between a primary, a secondary, and a tertiary amine?

21. The toxic alkaloid nicotine has the structure:

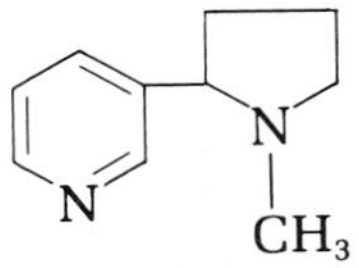

    a. Show the structure of the hydrochloride salt of nicotine.
    b. Show the structure of the quaternary salt formed with methyl iodide.
    c. What is meant by the expression "nicotine free base"?

22. Approximately 20 amino acids are found in human protein tissue. They differ greatly in some respects, but what chemical structure do they all have in common?

23. How do antihistamines actually function to protect us against the effects of released histamine?

24. 1-Phenyl-2-aminopropane can exist in two enantiomeric forms. Give the trade name for one of them. (*Hint:* the racemic mixture is termed Benzedrine.)

ADVANCED STUDY QUESTIONS

25. Starting from ethylene only, show all of the steps, reagents, and conditions necessary for the synthesis of ethyl acetate.

26. In this chapter you read the case history of a person who ingested ethylene glycol. You learned how the human body metabolizes this type of compound. Predict the metabolic fate of ethyl alcohol in the human. Show structures and write equations in your explanation.

27. An article in the June 12, 1981, issue of the *Journal of the American Medical Association* discussed the drug action of isoflurane (1-chloro-2,2,2-trifluoroethyl difluoromethyl ether). From its structure, predict the use of this substance in medicine.

11

Biochemistry of the Amide Group: Peptides and Proteins

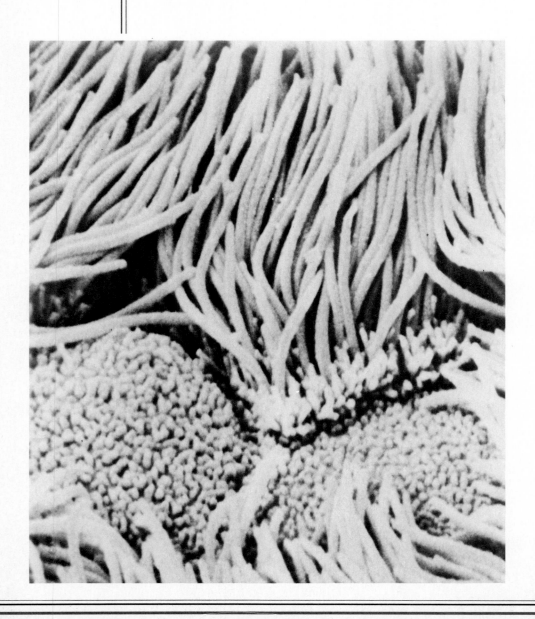

Proteins occupy a central position in the architecture and functioning of living matter. They are intimately associated with all phases of biochemical and physical activity of living cells since all enzymes and many hormones are proteins. Hair, wool, and collagen (a part of connective tissue) are protein, as are muscle, antibodies, and the globin in hemoglobin.

Proteins are associated with tissues and structures throughout the body. Our photo shows cilia lining the surface of a bronchiole in horse lung (the cilia are the long, wavy tubular extensions at the top of the photo). These cilia are almost totally protein in nature; through their movement, mucous and trapped foreign particles ae moved up and out of air passageways.

In this chapter you will learn that all protein is composed of amino acids, joined through a special kind of amide bond termed a peptide bond. We emphasize the geometry of proteins because their three-dimensional aspects govern their ability to act as enzymes, receptor sites for drugs, or specialized tissues like hemoglobin. In Chapter 17 we discuss the role of protein in diet and body chemistry. (Photo courtesy of American Lung Association.)

INTRODUCTION

We have saved the amide functional group for this chapter because we want you to associate amides with their most important application, *proteins*. In this chapter you will learn that the peptide bond in proteins is actually a special kind of amide bond.

You will learn what an amino acid is and how nature links many amino acids through amide bonds to form the "polymers" we call proteins.

After you have completed your study of this chapter, you will understand the very important role played by proteins in body chemistry, including their function as enzymes, hormones, and brain chemicals.

In Chapters 16, 17, and 18, you will apply your knowledge of proteins as you study, respectively, biotransformations, essential food factors, and the chemistry of heredity.

CHEMISTRY OF AMIDES

One of the ways to remember what **amides** are is to remember how they can be formed. Think of a carboxylic acid and ammonia (or an amine) combining with the loss of a molecule of a water. Draw a lasso around the elements of water, pull them out, and join what is left:

$$R-\overset{\overset{\textstyle O}{\|}}{C}-OH \quad + H-\overset{\overset{\textstyle H}{|}}{N}-H \quad \rightarrow \quad R-\overset{\overset{\textstyle O}{\|}}{C}-NH_2 + H_2O$$

→ pull out H_2O

a carboxylic acid   ammonia   an amide

or

$$CH_3-\overset{\overset{\textstyle O}{\|}}{C}-OH \quad + H-\overset{\overset{\textstyle H}{|}}{N}-H \quad \rightarrow \quad CH_3-\overset{\overset{\textstyle O}{\|}}{C}-NH_2 + H_2O$$

→ pull out H_2O

acetic acid   ammonia   acetamide

Even though there are better chemical ways of forming amides, our lasso method emphasizes that in every amide there is a direct carbon-to-nitrogen bond:

$$R-\overset{\overset{\textstyle O}{\|}}{C}-N\overset{\textstyle R}{\underset{\textstyle R}{}}$$

$$RCONR_2$$

amide bond   the amide bond in a
line formula

It also emphasizes that in every amide there is the remnant of a former carboxylic acid. This carboxylic acid will be regenerated when and if the amide bond is broken by hydrolysis. Here is an example:

$$CH_3-CH_2-\overset{\overset{\textstyle O}{\|}}{C}-NH_2 \quad \xrightarrow[\text{means of } H_2O)]{\overset{\text{hydrolysis}}{\text{(splitting by}}} \quad CH_3-CH_2-COOH + NH_3$$

propionamide   propionic acid

We can think of the carboxylic acids in amides as being latent, waiting to be freed by hydrolysis.

Because this type of amide is formed from a carboxylic acid, it is termed a *carboxamide*. This distinguishes it from *sulfonamide* (RSO_2NH_2), as found in sulfa drugs. Figure 11.1 lists names and structures of some important amides.

Naming Amides

Two systems of naming, the common and the IUPAC, exist for amides. Common system names for amides are easy: drop the *ic* or *oic* suffix of the common name of the corresponding carboxylic acid, and add the suffix *amide*. Table

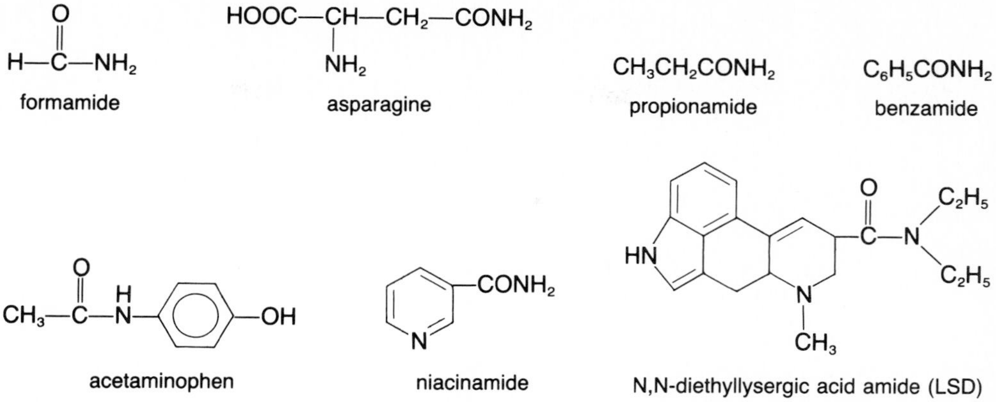

FIGURE 11.1 Some Important Amides.

11.1 gives examples. For IUPAC names, drop the *oic* suffix of the IUPAC name of the corresponding carboxylic acid, and add the suffix *amide*. Table 11.1 gives examples.

EXAMPLE 11.1

Draw the structures for benzamide, 2-bromopropanamide, and salicylamide.

Solution

benzamide $CH_3CHBrCONH_2$ salicylamide

2-bromopropanamide

Polymers

The topic of amides introduces us to **polymers**. Polymers are macromolecules— large collections of atoms sometimes having molecular weights in the millions.

Table 11.1. Common and IUPAC Names of Amides

Amide	Common name	IUPAC name
$HCONH_2$	Formamide	Methanamide
CH_3CONH_2	Acetamide	Ethanamide
$C_3H_7CONH_2$	Butyramide	Butanamide
$CH_3CH_2CHClCONH_2$	α-Chlorobutyramide	2-Chlorobutanamide
$(CH_3)_2CH\!-\!CONH_2$	Isobutyramide	2-Methylpropanamide

Polymers occur in nature and are also man-made; we encounter them in such forms as cellulose, glycogen, starch, natural rubber, polyethylene, polypropylene, silicone rubber, Orlon, Teflon, and Styrofoam.

More precisely, polymers are big compounds that are made up of small chemical units repeated over and over. By themselves, these small units are called *monomers*. For example, plants make the polymer starch by hooking together hundreds or thousands of glucose molecules. This hooking together is called *polymerization*. Hence glucose is the repeating unit in starch. For natural rubber, nature uses the isoprene (C_5H_8) building block as the monomer to be polymerized. For Styrofoam, chemists use styrene, $C_6H_5CH{=}CH_2$, as the building block.

Amides as Polymers

Nylon is a very important man-made polymer that is based on the amide linkage. Each molecule that participates in the polyamide bond in nylon has *two* functional groups; in this way, a long chain can form:

$$NH_2{-}(CH_2)_6{-}NH_2 \quad HO{-}\overset{O}{\overset{\|}{C}}{-}(CH_2)_4{-}\overset{O}{\overset{\|}{C}}{-}OH \quad H_2N{-}(CH_2)_6{-}NH_2 \quad HO{-}\overset{O}{\overset{\|}{C}}{-}(CH_2)_4{-}\overset{O}{\overset{\|}{C}}{-}OH$$

a diamine · a di-acid · a diamine · a di-acid · links up to next NH_2 group (−H₂O steps)

Water molecules split out, amide bonds form, and polymerization is complete, to give

$$-NH{-}(CH_2)_6{-}NH{-}\overset{O}{\overset{\|}{C}}{-}(CH_2)_4{-}\overset{O}{\overset{\|}{C}}{-}NH{-}(CH_2)_6{-}NH{-}\overset{O}{\overset{\|}{C}}{-}(CH_2)_4{-}\overset{O}{\overset{\|}{C}}-$$

When writing the structure of the polymer, it is customary to show only the portion that repeats:

$$-\!\!\left(NH{-}(CH_2)_6{-}NH{-}\overset{O}{\overset{\|}{C}}{-}(CH_2)_4{-}\overset{O}{\overset{\|}{C}}\right)\!\!-$$

the repeating unit in the polymer Nylon-66

Quite correctly, then, we can look at nylon as a polyamide, subject to the same chemistry as any amide. Nylon can be hydrolyzed by heating with strong acid or strong base. It undergoes all of the chemical reactions we would expect of any amide. But it, like amides generally, is relatively inert and resistant to oils, water, mildew, sunlight, and bacteria. These properties make nylon especially suitable for rope and cord manufacture, and for bearings, valves, and machine parts. Nylon can be formed into fibers, bristles, sheets, rods, coatings, and tubes. Nylon was the first synthetic suture material.

We must examine one more type of polymer that is based on the amide bond. Nature has selected this chemical link for the creation of peptides and proteins.

The Peptide Bond

One of the most important topics we can study in this book is amide bond formation between amino acids. This type of bond forms the basis of peptides and thus of proteins. Since enzymes, hemoglobin, viruses, antibodies, and certain hormones are proteins, the amide bond becomes involved in the most profound aspects of human biochemistry.

Recall that alpha-amino acids have a minimum of two functional groups, the —NH$_2$ and the —COOH.

$$R—CH—COOH$$
$$|$$
$$NH_2$$

an α-amino acid

Additional functional groups may occur in the remainder of the molecule (R in our formula), including another NH$_2$ or COOH group, an aromatic ring, an alcohol group, and so on.

Two amino acids, the same or different, can combine in an amide bond involving the NH$_2$ group of one amino acid and the COOH group of the other:

$$R'—CH—N—H \ + \ HO—C—CH—R'' \rightarrow \ R'—CH—N—C—CH—R'' + H_2O$$

first amino acid   second amino acid   amide (peptide) bond

An amide bond formed between two amino acids is termed a **peptide bond.**

When two amino acids join in a peptide bond, we call the product a *dipeptide.* Three combine to form a *tripeptide;* four, a *tetrapeptide,* and so forth. In nature, there is almost no limit to how many amino acids can be joined. Macromolecular peptides (i.e., proteins) having thousands of amino acids joined in peptide bonds are known. On the other hand, the dipeptide sweetener and sugar substitute, aspartame, consists of only two amino acids, L-aspartic acid and L-phenylalanine. The term *polypeptide* is applied to the higher molecular weight amino acid combinations.

Let us consider three specific amino acids and examine one of the six possible tripeptide arrangements. The amino acids are phenylalanine, valine, and aspartic acid. These three happen to be three of the amino acids that constitute a small segment of insulin from cows. The sequence we will illustrate here is identical to that in bovine insulin.

$$CH_2-CH-COOH \qquad CH-CH-COOH \qquad HOOC-CH_2-CH-COOH$$

phenylalanine   valine   aspartic acid

To prepare the peptide, we must create a total of two peptide bonds. We begin with phenylalanine's COOH group and join it to the amino group of valine by removal of a molecule of water:

phenylalanine   valine   a dipeptide

Next we join the free COOH of the dipeptide to the NH_2 group of aspartic acid:

a dipeptide   aspartic acid   a tripeptide

  Examine the structure of the tripeptide we have prepared and you will note two peptide (i.e., amide) bonds, three amino acid "residues," one free amino group (on the phenylalanine), and two free carboxyl groups. There are two free COOH groups because aspartic acid was a dicarboxylic acid to begin with. You can now also appreciate the fact that two other amino acid sequences are possible with the three amino acids we used. (They could have been joined Phe-Asp-Val or Asp-Phe-Val instead of Phe-Val-Asp.) (See the table on page 297 for an explanation of these symbols.) Also, the peptide bond arrangements *within* this sequence could have been different, that is, the NH_2 group of phenylalanine could have participated in the peptide bond instead of the COOH group of phenylalanine. Thus, altogether there are six possible tripeptide arrangements.

For polypeptides created from 10, 50, or even 100 amino acids, the number of possible arrangements becomes staggering. Remember, not only can the amino acid sequence be different in the peptide, but each amino acid has the possibility of using either its NH_2 group or its COOH group in forming a peptide link.

We can now understand how it is possible that millions of animals, insects, and plants can exist in this universe, each with different polypeptide cell or protein constituents.

▬▬▬ EXAMPLE 11.2 ▬▬▬

Consider the sequence of amino acids: tyrosine-isoleucine-glutamine-asparagine, as found in a part of oxytocin, the pituitary hormone that causes the uterus (womb) to contract at childbirth. If the amino group of tyrosine is free (not part of the peptide bond), what is the structure of the tetrapeptide?

Solution

Individual amino acids can join together in chains by forming amide (peptide) bonds between their amino and carboxyl groups. Shorter chain are termed peptides; very long chains are termed polypeptides or proteins. Some proteins can have molecular weights in the millions and be very complex. The possibility exists for enormous variation in the sequence of amino acids in the polymer chain.

Hydrolysis of Peptide Bonds

The hydrolysis (chemical breakdown using water) of proteins in our food supplies us with the amino acids our bodies require to synthesize our own proteins. Whatever the source of our protein—soybeans, red meat, fish, or dairy products—our bodies handle it in the reverse of the process we described earlier. Through the action of digestive enzymes, peptide bonds are broken and individual amino acids are liberated. In the laboratory we can accomplish hydrolysis of amide bonds using either acid or base:

$$\text{amide} + \text{water} \xrightarrow[\text{hydrolysis}]{\text{acid}} \text{carboxylic acid} + \text{salt of base}$$

$$\underset{\overset{\|}{\text{O}}}{\text{R}-\text{C}}-\text{NHR}' + H_2O \xrightarrow[\text{heat}]{H_2SO_4} \text{RCOOH} + \text{R}'NH_2 \text{ (as the salt)}$$

EXAMPLE 11.3

A certain tetrapeptide was known to contain glycine, methionine, serine, and alanine (not necessarily in that order). Partial digestion at various times gave the fragments: Gly-Ala, Ala-Met, and Ser-Gly. What was the correct sequence of amino acids in the tetrapeptide?

Solution
Ser-Gly-Ala-Met.

MORE ABOUT AMINO ACIDS

If we are to understand proteins, we must further examine **amino acids** because amino acids are the building blocks of proteins. All alpha amino acids are identical in the first two carbons:

$$\overset{2 \quad\ 1}{\text{R}-\text{CH}-\text{COOH}}$$
$$\underset{\text{NH}_2}{|}$$

It is the nature of the **R group** that gives amino acids different physical and chemical properties—and which therefore gives polypeptides and proteins different physical and chemical properties. Here is a classification of the twenty amino acids that make up most of the world's simple protein. The classification is based on the types of R groups in the amino acids.

I. Nonpolar R Groups

R group	Name	Recommended symbol
—H	Glycine	Gly
—CH$_3$	Alanine	Ala
—CH(CH$_3$)$_2$	Valine	Val

R group	Name	Recommended symbol
—CH₂CH(CH₃)₂	Leucine	Leu
—CH(CH₃)CH₂CH₃	Isoleucine	Ile
	Phenylalanine	Phe
	Tryptophan	Trp

II. Polar R Groups

R group	Name	Recommended symbol
A. R group contains sulfur		
—CH₂SH	Cysteine	Cys
—CH₂CH₂SCH₃	Methionine	Met
B. R group contains alcoholic hydroxyl group		
—CH₂OH	Serine	Ser
—CH(OH)CH₃	Threonine	Thr
C. R group contains phenolic OH group		
	Tyrosine	Tyr
D. R group contains a COOH (free or as its amide derivative)		
—CH₂COOH	Aspartic acid	Asp
—CH₂CONH₂	Asparagine	Ans
—CH₂CH₂COOH	Glutamic acid	Glu
—CH₂CH₂CONH₂	Glutamine	Gln
E. R group contains an amine group		
—(CH₂)₄NH₂	Lysine	Lys
	Arginine	Arg
	Histidine	His

III.   Proline[a]

Group	Name	Recommended symbol
HOOC—⬠(N—H)	Proline	Pro

[a] Belongs in Group I since it is essentially nonpolar, but we have listed it separately because its alpha-amino group is not free but is part of a heterocycle.

The amino acids in our classification deserve some further comment:

- The twenty amino acids in the list constitute almost all of the world's simple protein, such as enzymes, hemoglobin, viruses, antibodies, certain hormones, muscle, and connective tissue. Over 200 other amino acids exist in nature but are not found in proteins.
- Functional groups in the amino acids allow for special physical and chemical behavior, such as nonpolar groups giving a peptide hydrophobic (water-repelling) properties. The —SH groups in cysteine allow for chains of peptides to hook up to each other through disulfide bridge (R—S—S—R) formation.
- The three-letter symbols for the amino acids are used in a special shorthand that denotes sequences of amino acids in peptides. For example, the complete sequence of amino acids in the nonapeptide (*nona* = nine) oxytocin is

$$\underset{\text{Cys—Tyr—Ile—Gln—Asn—Cys—Pro—Leu—Gly—NH}_2}{\overset{\displaystyle \text{S————————S}}{\big|\big|}}$$

Note the presence of a disulfide bridge in oxytocin.
- Nine of the twenty amino acids cannot be synthesized by the adult body at all or at rates fast enough to meet adult body demands. These are termed *essential* amino acids. Much more on this can be found in Chapter 17.
- The pH of the biological fluid in which the amino acid exists determines the electrical charge on the amino acid:

$$\underset{\text{NH}_2}{\overset{}{\text{R—CH—COOH}}} + \text{H}^+ \rightarrow \underset{\text{NH}_3{}^+}{\overset{}{\text{R—CH—COOH}}}$$

    if acid combines   a positive ion results

$$\underset{\text{NH}_2}{\overset{}{\text{R—CH—COOH}}} + \text{OH}^- \rightarrow \underset{\text{NH}_2}{\overset{}{\text{R—CH—COO}^-}}$$

    if base combines   a negative ion results

For each amino acid there is a specific pH, called the **isoelectric point**, at which the dipolar ion (also called *di-ion*) form exists:

$$R—CH—COO^-$$
$$|$$
$$NH_3{}^+$$

dipolar ion form of amino acid

- All amino acids (except glycine) fit our "four-different-group rule" (Chapter 9) and therefore possess a chiral center; they can exist in optically active enantiomeric forms. The amino acids that constitute living protein show optical activity.

A recently developed technique for establishng the age of artifiacts is based on the time-dependent process of racemization of amino acids. When a living creature dies, the amino acids in its protein begin the slow process of racemization (into a 50–50 mixture of enantiomers). Dr. Jeffrey Bada, a researcher at the Scripps Institute of Oceanography, has been able to correlate the extent of racemization in a sample to the length of time the sample has existed, thus establishing its age. Of course, the sample must have a protein content for this technique to be applicable (hair can be used for this). At 20°C and a pH of 7, racemization of a typical amino acid has a half-life for reaction of about 200,000 years. Storage temperature and structure of the amino acid can greatly influence this.

PROTEINS

Proteins are polymers—with amino acids as the building blocks. When only a few amino acids join together, the products are called *peptides*. But when very large numbers of amino acids polymerize, the products are called proteins. How large is "very large"? Here are a few proteins, selected to demonstrate a range of molecular weights.

Protein	Molecular weight	Amino acid residues
Insulin	5,734	51
Cytochrome C	15,600	104
Growth hormone	49,000	191
Hexokinase	96,000	730 (est)
Gamma-globulin	176,000	1,320
Myosin	800,000+	6,100 (est)
Tobacco mosaic virus	59 million	348,000

Proteins hold first place (Greek: *proteios*) in living organisms because nothing exceeds them in importance. Proteins, as enzymes, control the rate of

Table 11.2.   A Classification of Proteins Based on Function

Function	Example of this type of protein
Defense	Antibodies (gamma-globulin), certain antibiotics (bacitracin)
Digestion	Digestive enzymes (pepsin, amylase, fumarase)
Catalysis	Approximately 2000 enzymes have been identified that catalyze a myriad of biochemical reactions
Support and connective tissue	Cartilage, tendons
Motion	Muscle tissue
Communication	Nerve tissue
Protection	Hair, nails, skin, feathers
Regulation	Hormones (gastrin, glucagon, ACTH, FSH), pituitary releasing factors
Regeneration	Nucleoproteins
Receptor sites	Biological response to neurotransmitters, mediators, and drugs; the receptor site for a naturally occurring tranquilizer has been identified
Transport	Albumins carry fats and metal ions otherwise insoluble in the blood; hemoglobin transports oxygen
Pathogenesis	Viruses; diphtheria toxin

biological processes; as hormones, they regulate biochemical relationships; as building blocks, they constitute organic structure; and as contractile tissue, they accomplish motion. A classification of proteins based on function (Table 11.2) illustrates how widespread and important proteins are.

We can consider another classification of proteins based on composition. **Simple proteins,** when broken down by hydrolysis, yield nothing but their amino acid content. Egg albumin (from egg whites) hydrolyzes to yield only amino acids, as does the lactalbumin of milk. The globulins are a group of simple proteins of wide distribution in animals and plants. The gamma-globulin fraction of human plasma protein contains antibodies we have manufactured in response to invasion of antigens. Gamma-globulin is used, for example, when a pregnant woman who has never had German measles (rubella) has been exposed to the disease. She is given a dose of the gamma-globulin and thus receives "passive" immunity, which temporarily protects her developing embryo from the possible teratogenic effects of the measles virus.

Conjugated proteins consist of simple proteins plus some other nonprotein substance, referred to as the **prosthetic group.** In nucleoproteins, nucleic acids are the prosthetic groups. The chromoproteins hemoglobin and cytochrome (involved in metabolism) contain protein plus an iron-containing complex that gives pigment (color) to the entire conjugated protein. In lipoproteins, lipids such as cholesterol and glycerides are combined with protein. High-density lipoprotein (HDL, Chapter 13), is currently of great interest in theories regarding hardening of the arteries. Egg yolk contains lecithoprotein, in which lecithin is the prosthetic group.

Glycoproteins are a very important category of conjugated proteins in which the protein is covalently bound to a sugar such as mannose or glucose by means of a carbohydrate–peptide bond (carbohydrates are discussed in Chapter 12). Researchers have found that most body proteins are actually glycoproteins, as are many polysaccharides such as glycogen and starch. The carbohydrate portion of the glycoprotein can play important biological roles such as stabilization of protein conformation, specification of human blood types, control of the lifetime of glycoproteins in the circulatory system, and control of glycoprotein uptake by cells.

The Total Geometry of Proteins

It is not enough to know the sequence of amino acids in a protein, for proteins are much more complex. For a protein to function as a carrier of oxygen, as a biological catalyst, as a receptor site for drugs, or as a bearer of genetic information, it must have a three-dimensional structure, a configuration in space, a geometry that helps carry out its function.

There are four steps in defining the total geometry of a protein. In the first step, we look at the **primary structure.** You are already familiar with this aspect, for this is simply the sequence of amino acids in the protein, bound through peptide bonds. To their great credit, biochemists have been able to determine the exact primary structure of proteins having as many as 1320 amino acids! The primary structure of bovine insulin is shown in Figure 11.2.

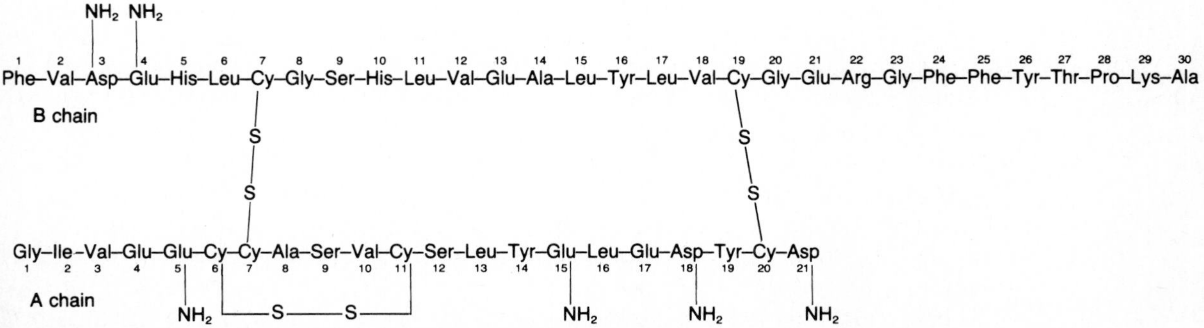

FIGURE 11.2 The Primary Structure of Bovine Insulin.

FIGURE 11.3 An Illustration of the α-helical Secondary Structure of a Protein. **These coiled fragments represent portions of the actual conformation of the 293-amino acid cytochrome c peroxidase enzyme. The coils are themselves part of the total three-dimensional structure of the enzyme molecule.** (Courtesy of Thomas Poulos, Department of Chemistry, University of California, San Diego, La Jolla, CA.)

The **secondary structure** of proteins specifies the ways in which a polypeptide chain can fold or spiral as it forms the backbone of a particular protein. Linus Pauling and Robert Corey in 1951 made the great discovery that many proteins have, in part, a spiral shape (much like a spiral staircase), and that the coil shape in the spiral is maintained by hydrogen bonding between functional groups in the polypeptide chain. Protein backbones with this type of spiral structure are said to have the *alpha-helical conformation* (or, *α-helix*). Note that a helix of this type can turn either clockwise or counterclockwise when viewed from one end (Figure 11.3). The α-helix in proteins is held together all along its length by hydrogen bonds between amide hydrogens and the carboxamide carbonyl oxygen (Figure 11.4*a*). Note again, the α-helix is the protein backbone made up of the peptide bonds between amino acids. In Figure 11.4*a* we note that the R groups in the amino acid residues stick out from the α-helix. These R groups, sometimes hydrophobic, sometimes hydrophilic, can influence the overall shape of the protein molecule (more about these R groups in the next section).

The α-helix, though occurring in many proteins—including those of hair, wool, and muscle myosin—is not the only secondary structure assumed by proteins. A side-by-side or sheetlike arrangement can also be assumed. However, biochemists have theorized that in order to accommodate the R groups of the amino acid residues, the sheets must be "pleated" or buckled (Figure 11.4*b*). The protein found in silk is arranged predominantly in the *pleated sheet conformation*. Proteins can have a combination of helical, pleated sheet, and other types of secondary structures.

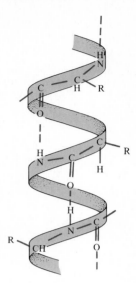

FIGURE 11.4*a* **The α-helix backbone in a protein's second-ary structure is held together by hydrogen bonds between the amide hydrogen of one peptide bond and the carboxamide carbonyl oxygen of another peptide bond in the next turn of the helix. Hydrogen bonds are denoted by broken lines. There are 3.6 amino acids per turn of the helix.**

FIGURE 11.4*b* **The pleated sheet secondary structure is found in some proteins. Two parallel strands of polypeptides are held together by hydrogen bonds.**

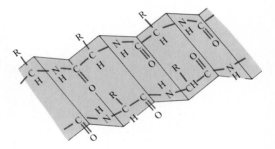

In **tertiary** (or *ternary*) **structures** of proteins is where the real excitement lies. Here we consider the total geometry of a protein, that is, how all the amino acid sequences, α-helix, and pleated sheets are arranged in a total three-dimensional structure that gives a protein its specific character. Imagine a long sring of different-sized beads, arranged in a coil. But the coil itself is looping, twisting, flattening out, and turning back upon itself, to give a specific, special arrangement in space.

Protein chains, whether α-helical or pleated, can also loop, twist, flatten, and fold back to assume a final, three-dimensinal shape. It is the differing chemical natures of the R groups—some with polar R groups like —OH or —NH_2, others with nonpolar alkyl groups—that help determine the final shape of a protein. Polar groups can influence the shape of a protein by hydrogen bonding. Nonpolar groups (as in valine, isoleucine, phenylalanine) tend to cluster together in the interior of a protein, giving a special shape to the molecule as well as providing an area in which nonpolar agents can dissolve.

In a globular protein (like hemoglobin), nonpolar R groups cluster together in the interior while polar, hydrophilic (water-loving) groups are situated on the outside of the sphere, imparting an overall solubility to the protein. In addition, disulfide bridges (R—S—S—R) can form between appropriately spaced cysteine residues; such bridges hold long polypeptide chains in the correct orientation (see prior structure of insulin molecule). In summary, then, the tertiary structure of a protein is very special because it depends on the special kinds and arrangements of the amino acid residues in the protein.

And how important the final protein shape is! An enzyme's shape must be just right for a food molecule or a drug to fit it and be metabolized. The coiled shape of collagen is necessary for its role in connective and supportive tissue. And in hemoglobin, an improper shape can give rise to sickle cell anemia. People who suffer from this condition produce a protein (globin) in which only one out of 146 amino acids is "wrong," yet the total geometry of the protein is affected seriously, as is the red blood cell in which it occurs. For more on sickle cell disease, see Chapter 15.

In Figure 11.5, you can examine a remarkable schematic drawing of the tertiary structure of an enzyme obtained from the bacterium *Escherichia coli*. The ribbons you see represent the backbone of the protein, sometimes coiled but mostly flattened out into an eight-stranded form. This arrangement is not accidental. It is specific and fairly well fixed by the chemical nature of the 162 amino acid residues in the protein.

The tertiary nature of the enzyme in Figure 11.5 allows it to act as an enzyme catalyzing the synthesis of tetrahydrofolate—a key factor required by normal *and cancer cells* if they are to grow and multiply. In cancer chemotherapy, if an enzyme like this dihydrofolate reductase can be inhibited, cancer cell proliferation can be arrested. Methotrexate is a currently highly useful chemotherapeutic agent used as an inhibitor of cancer cell growth in the treatment of leukemias in children. Examine the right center portion of the enzyme backbone and note a deep cavity cutting across one face of the enzyme. Bound in this cavity is a molecule of methotrexate that was added to the enzyme before X-ray crystallographic analysis was begun. Here we have graphic evidence of how an anti-cancer drug works at the *moleclar level* to inhibit cancer cell growth! The foreign methotrexate molecule can fit the receptor site in the enzyme, inhibiting the normal action of the enzyme in catalyzing synthesis of tetrahydrofolate. We also have an exciting view of the receptor site in the tertiary structure of an important enzyme.

Summary

The tertiary structure of a protein describes its three-dimensional geometry or arrangement in space. The tertiary structure arises when the polypeptide backbone of the protein folds and twists into a specific conformation that is required for the protein to carry out its function. The protein is held in its conformation by noncovalent attractions arising from the R groups.

The **quaternary structure** of a protein describes the joining together of

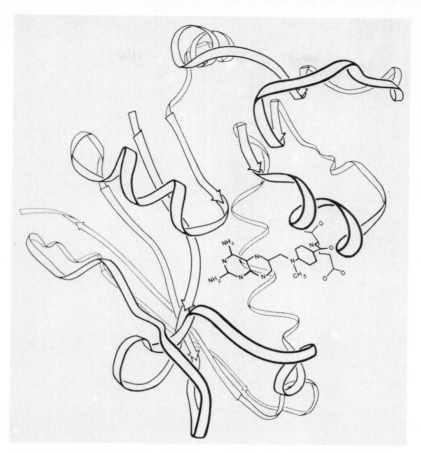

FIGURE 11.5   The tertiary structure of *E. coli* dihydrofolate reductase with bound methotrexate, solved at 2.5 A resolution. See text for discussion. (Courtesy David Matthews, Dept. Chemistry, Univ. California San Diego, La Jolla, CA.)

two or more polypeptide subunits to form a giant protein. These subunits can be the same or different, a few or as many as several thousand. In other words, some proteins are really multiproteins, constructed of many smaller protein units. A study of the quaternary structure of phosphorylase shows that it is comprised of two polypeptide subunits, neither of which is catalytic alone; the polio virus (a multiprotein) consists of 130 subunits. There are many proteins that have a quaternary structure. Figure 11.6 shows the quaternary structure of hemoglobin. In hemoglobin, four peptide subunits join to form a large protein, with molecular weight of 64,450. Each of the four subunits brings a *heme* group to the protein; it is the heme that actually holds the oxygen during transport in the blood. Heme is a nonprotein, iron-containing molecule that is made up of four pyrrole rings.

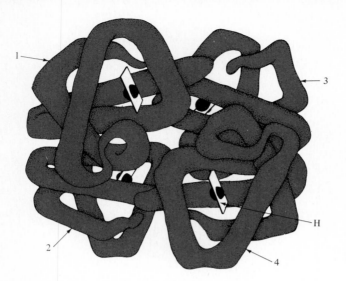

FIGURE 11.6   As shown, the quaternary structure of hemoglobin consists of four separate polypeptide chains, numbered 1 through 4, joined to form one large, complex protein. Also shown are the four home units (H), one from each of the contributing subunits. (Modified from "The Structure and Action of Proteins" by R. E. Dickerson and I. Geis, W. A. Benjamin, Inc., Menlo Park, CA, p. 56, 1969.)

Denaturation of Proteins

The health professional has a special interest in denaturation of proteins because the health professional has a special interest in *asepsis* (*a* = without, *sepsis* = infection). Pathogenic (disease-causing) bacteria, yeasts, and viruses can be killed if the protein that makes up these organisms can be forced out of its normal three-dimensional geometry into an altered state. For example, if a coiled protein can be uncoiled, or if a pleated sheet can be pulled apart, the microorganism will lose its functional integrity and die. The process of drastically changing the secondary tertiary, or quaternary structure of a protein is termed **denaturation.** It can occur without the breaking of any peptide bonds but often involves the disruption of hydrogen bonds.

Heat or chemicals can act as denaturants. Heating an egg denatures the albumins in the egg. A hot curling iron alters the geometry of hair protein. Stomach acid curdles milk. Certain detergents so alter the protein sheath protecting bacteria and viruses that the sheath breaks apart, permitting cell contents to spill out. Hence some detergents are germicidal.

Certain heavy metal ions, especially Hg(II), Pb(II), and Ag(I), denature protein generally; this is the basis for their poisonous character if taken internally. Mercury(II) chloride, $HgCl_2$, has long been used in water solution to sterilize inanimate objects. Containers of mercury solutions must have *Poison!* labels.

EXAMPLE 11.4

Select an appropriate antidote to be given to a person who has accidentally swallowed a heavy metal poison such as mercuric chloride.

Solution

Administer a protein such as milk or egg, which will preferentially react with the metal ion in the stomach, preventing denaturation of body proteins. (*Note:* the contents of the stomach must then be removed by gastric lavage.)

Denaturation of proteins can also be effected by exposure to ultraviolet (UV) light. Recall that UV light is of short wavelength but of high energy. Bacteria exposed to UV light for long enough periods will be killed, a fact put to use in some sterilization laboratories and on some fancy toilet seats.

The geometry of proteins can also be disrupted through the use of certain organic solvents. Ethyl alcohol, 70 percent in water, is an effective denaturant, as are isopropyl alcohol and acetone.

AMIDES AND PEPTIDES IN MEDICINE

Insulin, a polypeptide containing fifty-one amino acid residues, is made by the beta cells of the islets of Langerhans in the pancreas and is essential to life for it promotes uptake and metabolism of glucose in body tissues. A deficiency of insulin, as in *diabetes mellitus,* results in abnormal carbohydrate and fat metabolism with consequent high blood sugar levels and acid–base imbalance (see Chapter 12). The insulin made by cattle, goats, sheep, horses, pigs, and whales is not identical to human insulin. In beef insulin, for example, four amino acids out of fifty-one are "wrong." Nevertheless, many of these animal insulins can substitute well in the human and thereby prevent diabetic debilitation. Beef and pig insulin are most often used, being readily available at the slaughterhouse (Figure 11.7). It takes almost 8000 g of pancreas tissue to make 1 g insulin. One complication of the use of a **foreign protein** such as beef insulin in humans is that the human body develops antibodies to it. In fact, almost all humans who have received commercial beef insulin for longer than 2 months have developed antibodies, some of which are so intense they nullify the beneficial effects of the insulin. Fortunately, these individuals can switch to pork or goat or sheep insulin, at least until such time as antibodies to these alternatives develop. The ultimate solution to the insulin supply problem appears to be the discovery that the human gene that codes for insulin can be inserted into the genetic material of the bacterium *E. coli.* Eli Lilly and Company announced in 1981 that it had used recombinant DNA techniques to produce synthetic insulin and had successfully tested it in humans. Lilly has received FDA approval to market their new insulin under the name Humulin. See Chapter 18 for a more complete discussion of recombinant DNA techniques and their application to other biomolecules.

Human growth hormone, a protein composed of 191 amino acids, binds to receptor sites at the ends of the long bones (the epiphyses) stimulating bone

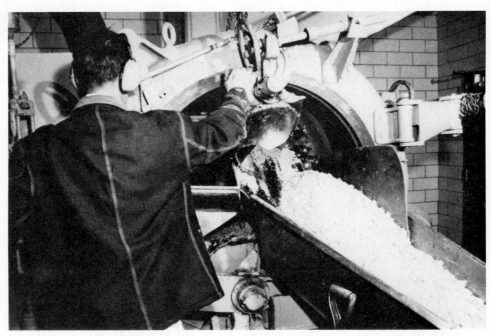

FIGURE 11.7   Grinding frozen pancreases in the preparation of insulin. Almost 8000 g of pancrease tissue is required to prepare 1 g of insulin. (Photo courtesy of Eli Lilly and Company, Indianapolis, Indiana 46206.)

growth. Human growth hormone has been synthesized, but the process is not commercially economical, and supplies of the hormone must come from the pituitary glands of human cadavers. No mammal except the monkey produces a growth hormone structurally similar enough to humans' for it to substitute in humans. Supplies of human growth hormone for use in treating incipient pituitary dwarfism are obtained from the National Pituitary Agency, 210 West Fayette Street, Baltimore, Md. 21201.

Certain *polypeptide antibiotics* have assumed importance as medicinal agents, among them the tyrocidines, gramicidins, polymyxins, and bacitracin USP.

Oxytocin and **vasopressin** are two peptide posterior pituitary hormones that are made up of nine amino acids each:

$$\overset{1}{\text{Cys}}\text{—}\overset{2}{\text{Tyr}}\text{—}\overset{3}{\text{Ile}}\text{—}\overset{4}{\text{Gln}}\text{—}\overset{5}{\text{Asn}}\text{—}\overset{6}{\text{Cys}}\text{—}\overset{7}{\text{Pro}}\text{—}\overset{8}{\text{Leu}}\text{—}\overset{9}{\text{Gly}}\text{—NH}_2$$

$$\text{|——————S————S——|}$$

oxytocin

$$\overset{1}{\text{Cys}}-\overset{2}{\text{Tyr}}-\overset{3}{\text{Phe}}-\overset{4}{\text{Gln}}-\overset{5}{\text{Asn}}-\overset{6}{\text{Cys}}-\overset{7}{\text{Pro}}-\overset{8}{\text{Arg}}-\overset{9}{\text{Gly}}-\text{NH}_2$$

$$\llcorner\underline{\quad\quad}S\underline{\quad\quad}S\underline{\quad\quad}\lrcorner$$

arginine vasopressin

The disulfide bridge (R—S—S—R) arises from two molecules of cysteine by intramolecular oxidation.

Oxytocin causes contraction of smooth muscle in (a) the uterus (labor) and (b) the lactating breast (milk ejection). Stimulation of the genitals in coitus releases oxytocin, and some believe that the resulting uterine contractions facilitate sperm transport. Pitocin (oxytocin injection USP) is a synthetic form of oxytocin marketed as a uterine stimulant for the induction of labor or for the reinforcement of labor.

Vasopressin is also known as **antidiuretic hormone (ADH)** because of its action in reducing the volume of urine flow, thus conserving body water. Vasopressin acts in the kidney. Pitressin is a commercial preparation of vasopressin injection USP.

EXAMPLE 11.5

If vasopressin is an antidiuretic (reduces urine volume), what is a diuretic? Give several examples of diuretics.

Solution
Diuretics such as alcohol, caffeine, Dyazide and Diuril increase the flow of urine and facilitate loss of body water.

The kidneys are the site of the production of a powerful vasoconstrictor hormone termed **angiotensin**.

$$\overset{1}{\text{Asp}}-\overset{2}{\text{Arg}}-\overset{3}{\text{Val}}-\overset{4}{\text{Tyr}}-\overset{5}{\text{Ile}}-\overset{6}{\text{His}}-\overset{7}{\text{Pro}}-\overset{8}{\text{Phe}}$$

angiotensin II

This octapeptide can constrict arteries and raise blood pressure; it can also act directly on the adrenal cortex to stimulate hormone production at that site. Angiotensin is *pressor agent*. Pressor agents are used in cases of hypotension (low blood pressure) caused by hemorrhage or associated with shock.

Have you had a head cold or a fever blister recently? If so, you probably had a viral infection and your body responded by manufacturing a protein called **interferon.** Interferon is produced by our bodies only when we need it, that is, when we are invaded by a virus. Although interferon itself does not attack viruses, it establishes the *antiviral state* (AVS) in tissues adjacent to the infection site. As long as interferon is present, the antiviral state will be maintained and the infecting virus will be attacked. The ability to produce interferon is not limited to humans: monkeys, cows, mice, birds, and even fish can also manufacture it. But interferons are generally species specific, and interferon produced in the rat will not work in the human. Currently a great research effort is under way in the world to prepare larger quantities of interferon (again, see Chapter 18 for recombinant DNA discussion) and to study its possible application in the treatment of hepatitis, shingles, and cancer. If you are betting on a drug of the future, put your money on interferon.

Recently Discovered Brain Peptides

Luteinizing hormone-releasing factor (LHRF) is a brain peptide (10 amino acid residues) that controls the rate of secretion of luteinizing hormone (LH) from the pituitary gland. As you may know, ovulation and therefore fertility depend on the release of luteinizing hormone and thus also depend on LHRF. Recently organic chemists have prepared synthetic analogs of LHRF (these are man-made chemical relatives of the real thing) that antagonize (block) the action of LHRF. These analogs have been shown to be good prospects in a new approach to birth control. We may hear a lot more about this novel approach to contraception.

Paradoxcally, some of the LHRF analogs are *more* effective than LHRF itself in stimlating the release of LH. These superactive analogs have been successfully administered experimentally to nonfertile women by nasal spray. Altogether, more than 1000 analogs of LHRF have been prepared. The intensity and complexity of biochemical research in this area are phenomenal.

The human brain contains countless numbers of **receptor sites** (specialized tissues to which drugs attach). We know, however, that there are two receptor sites in the brain that bind morphine and other opiates like codeine, the first step in starting their pain-killing action. One of the great recent discoveries in brain chemistry has been that our brain *itself* manufactures small peptides called *enkephalins* that bind to the same receptors as the opiates and also have a pain-killing effect. The lay press calls these naturally occurring compounds the "brain's own opiates." They are not opiates, but pentapeptides, such as

Tyr-Gly-Gly-Phe-Met

an enkephalin

The enkephalins are part of the amino acid sequence of larger peptides called *endorphins*. And the amino acid sequence in endorphin is, in turn, found in a 91-amino-acid peptide called *beta-lipotropin* that is secreted by the anterior lobe of the pituitary gland. These discoveries have not been put to clinical use yet, but they have great implications for mankind in the areas of pain relief, treatment of narcotic addiction, and brain biochemistry.

SUMMARY

The thrust of this chapter was toward understanding proteins—those very important biological polymers that play a variety of major roles in our bodies.

We began the chapter with a discussion of the amide bond, for proteins are amides formed from amino acid molecules. The special term *peptide bond* is applied to the amide bond in proteins. When amides (i.e., peptides) are hydrolyzed (cleaved with water) a carboxylic acid is liberated. We learned to name amides by the common and IUPAC systems.

With a view toward understanding proteins, we examined the concept of polymers—high molecular weight compounds made up by joining together a large number of individual molecules called monomers. Nylon is a man-made polymer obtained from a diacid and a diamine by means of polyamide formation. Proteins are polymers obtained by condensing many amino acids by polyamide formation. We specified a dozen types of proteins, based on their function.

Before we took up a detailed study of proteins, we examined the twenty amino acids that make up most of nature's protein. These twenty amino acids all have an alpha-amino group and a carboxylic acid group, but they differ in the nature of the remainder of the molecule, the "R" group. We categorized the twenty amino acids as nonpolar or polar, and if polar, as sulfur-, alcoholic OH-, or phenolic OH-containing, or as diacid or diamino. We emphasized the importance of R groups in helping to determine the shape and behavior of the proteins they comprise.

Proteins can be simple or conjugated (that is, containing a prosthetic group). Proteins can be defined on the basis of their primary, secondary, tertiary, or quaternary structure. We spent considerable time defining these four structures, emphasizing the α-helical and pleated sheet secondary structures, and the factors that contribute to a protein's tertiary structure. It is the tertiary structure that gives a protein its unique three-dimensional qualities and this is the key factor in enzyme activity and receptor site binding capabilities.

Agents for and applications of denaturation of proteins were examined. Finally, we studied amides and proteins used in medicine or found occurring in the human. These included insulin, growth hormone, antidiuretic hormone, angiotensin, interferon, and a number of brain peptides that act as releasing factors for other hormones.

KEY TERMS

Check your understanding of this chapter. Can you explain what is meant by each of these terms?

ADH	amino acid
amide	angiotensin

denaturation
foreign protein
interferon
isoelectric point
oxytocin
peptide bond
polymer
primary structure
prosthetic group

protein, conjugated
protein, simple
quaternary structure
receptor site
R group
secondary structure
tertiary structure
vasopressin

STUDY QUESTIONS

1. Write the structure of the amide that can be formed from the following:
   a. Propionic acid and ammonia.
   b. Benzoic acid and ammonia.
   c. Salicylic acid and ammonia.
   d. Formic acid and diethylamine.
   e. Two moles of glycine.
   f. Stearic acid and ammonia.

2. Provide either a common name or an IUPAC name for the following:
   a. CH_3CONH_2
   b. $CH_3(CH_2)_8CONH_2$
   c. $HCONH_2$
   d. $(CH_3)_2CHCH_2CONH_2$
   e. $CH_3(CH_2)_{10}CCONH_2$
   f. Cl_3CCONH_2

3. Chemically speaking, what are polymers? What are monomers?

4. You have probably heard the word *polyester* in connection with clothing. Show how a polyester (a polymeric ester) could form between ethylene glycol ($HOCH_2CH_2OH$) and malonic acid ($HOOC—CH_2—COOH$). Show the structure of the repeating unit.

5. The amine that copolymerizes to form nylon (see text) is called hexamethylenediamine. Is that an ambiguous name? (*Hint:* Methylene is $—CH_2—$).

6. When is R—CONHR′ a plain amide and when is it a peptide bond?

7. Draw the structures of the two possible dipeptides that you can form between leucine and phenylalanine.

8. Draw the structure of the tripeptide Met-Ser-Gly. Assume that the free amino group is on the methionine.

9. A pentapeptide contains Ala, Leu, Trp, Lys, and Glu, not necessarily in that order. Partial hydrolysis at various times gives the fragments Lys-Ala, Glu-Leu, Ala-Glu, and Lys-Trp. What is the amino acid sequence in the pentapeptide?

10. Answer the following questions.
   a. How many different ways can three entities, △, □, and ○, be hooked together in an open chain? Show the possible sequences.
   b. How many different ways can the same three entities be hooked together if they each have two different hooks for joining up? The hooks are as follows:

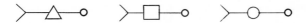

   The bonds can be only of the ——○⟩—— type.

11. Would you expect the amino acid tryptophan, as found in proteins, to be optically active? Why?

12. Show the products of acid hydrolysis of amides by completing and balancing the following equations:

   a. $CH_3(CH_2)_4CONH_2 + H_2O \xrightarrow[\text{heat}]{\text{HCl}}$

   b. $H_2N—CH_2—CONH—\underset{\underset{CH_3}{|}}{CH}—COOH + H_2O \xrightarrow[\text{heat}]{\text{HCl}}$

13. Examine the text's list of twenty amino acids and identify those that (a) are branched chain, (b) show aromatic properties, (c) have two acid groups, (d) have two amine groups, (e) have more than one chiral center.

14. We consume fats, carbohydrates, and proteins in our diet. Generally speaking, of those three, proteins cost the most per gram. From what you have read in this chapter, explain protein's "first place" in cost.

15. Table 11.2 lists a dozen functions of proteins. Which one of these categories probably accounts for the greatest mass of protein on the earth?

16. How do conjugated proteins differ from simple proteins?

17. What possible effects can the R groups (on the amino acid residues) in a peptide have on its structure and properties?

18. How does digestion of a protein differ from denaturation of a protein?

19. Show how the dipolar ion form of alanine behaves chemically toward addition of base and toward addition of acid.

$$\xleftarrow{+OH^-} CH_3—\underset{\underset{NH_3^+}{|}}{CH}—COO^- \xrightarrow{+H^+}$$

alanine

20. What are the names of the amino acids that have alcoholic hydroxyl groups? That contain sulfur? That are abbreviated Asn, Glu, Gln, Arg?

21. Define (a) essential amino acid, (b) isoelectric point, (c) denaturation, (d) epiphyses, (e) HDL, (f) enkephalin, (g) protein.

22. Predict what would happen if insulin were administered by the oral route instead of parenterally in the treatment of diabetes mellitus.

23. There are four steps in defining the total geometry of a protein: primary, secondary, tertiary, and quaternary. To which of these does the α-helix structure pertain? The pleated sheet? The receptor site? The structure of hemoglobin? The amino acid sequence in insulin?

24. Does the word *denatured* mean the same in denatured alcohol as it does in denatured protein?

25. It is hydrogen bonding in the backbone of a peptide that helps hold the α-helix conformation together. What atoms in the backbone are actually involved in the hydrogen bonding? Draw the hydrogen bonds.

26. Which of the following processes will be effective in denaturing a bacterial protein and which will be essentially ineffective?
    a. Heating the bacterium in water at 38°C for 1 hour.
    b. Treating with 5 percent silver nitrate solution.
    c. Heating in air at 90°C for 1 hour.
    d. Treating with 5 percent $HgCl_2$ aqueous solution.
    e. Treating with 5 percent $FeCl_2$ aqueous solution.
    f. Exposure to sunlight for 1 hour.
    g. Exposure to infrared light for 1 hour.
    h. Treatment with pH 6.5 water.
    i. Treatment with pH 1.5 water.

27. What pharmacological (drug) effects in the human are expected following administration of the following:
    a. Pitocin   b. Vasopressin
    c. Angiotensin   d. Interferon
    e. LHRF   f. Insulin

28. Bradykinin (a pain-causing nonapeptide) is released by globulins in the blood as a result of wasp stings. Partial hydrolysis of bradykinin gives the following tripeptides: Gly-Phe-Ser, Arg-Pro-Pro, Ser-Pro-Phe, Pro-Gly-Phe, Pro-Pro-Gly, Pro-Phe-Arg, and Phe-Ser-Pro. Give the complete amino acid sequence in bradykinin.

ADVANCED STUDY QUESTIONS

29. Tobacco mosaic virus (TMV) has a molecular weight of 59 million. Using Avogadro's number, calculate how many actual molecules of TMV would be present in 1.0 mg of this very large protein.

30. Wool is a protein, composed of amino acids; it therefore has the amide bond in its primary structure. Nylon is a polymer that also has the amide bond in its backbone. Propose an explanation for the fact that moths can attack wool and digest it but cannot digest Nylon, even though both have the amide bond.

31. A protein was found to contain 0.32 percent sulfur by weight, all of it from six cysteine residues. What was the molecular weight of the protein?

12 Carbohydrates

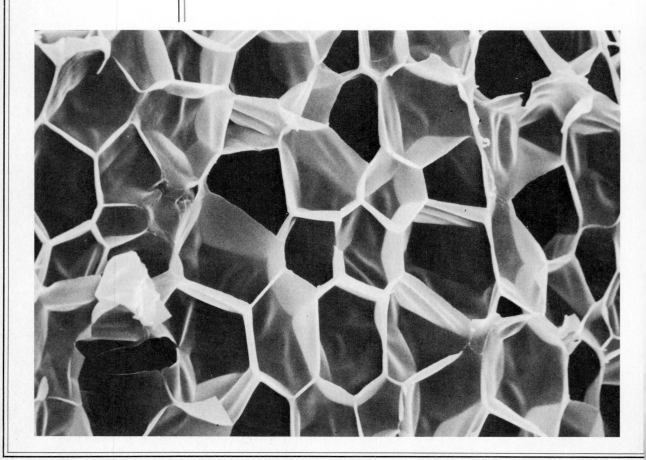

This is an electron microscope photo of a carbohydrate familiar to everyone, popcorn. The compartments shown in the photo result from the explosive burst of water as the kernel of corn is heated. We classify popcorn as a carbohydrate because of its high starch and cellulose content.

Carbohydrates have come in for a lot of criticism. Picture a family sitting down to a breakfast of pancakes, syrup, toast, and jam, followed by sugared donuts washed down by a cola drink! Of course, such over-indulgence in carbohydrates—to the exclusion of fiber, protein, and fresh fruits—it to be deplored. Obesity, tooth decay, and diabetes are statistically related to overindulgence in carbohydrates.

On the other hand, most people receive most of their daily calories from carbohydrates. They are the backbone of our diet. They should be consumed with a *variety* of foods, to give nutritional balance to our diet. Counsel your patients to select from the four basic food groups: cereals, meat, vegetables, and dairy products, with a good measure of fiber thrown in.

They key compound in this chapter on carbohydrates is glucose. It is the sugar in our blood and the building block for starch, cellulose, and glycogen.

INTRODUCTION

The glucose that courses through our blood vessels is a carbohydrate. So is the corn starch (Figure 12.1) in the tacos we eat and the cellulose in the cotton swabs we use. Glycogen in the liver and xylans in straw and wood are carbohydrates, too. Carbohydrates comprise one of the three main food groups (along with fats and proteins); we humans obtain the majority of our calories from carbohydrates. Cellulose derivatives such as acetate are important commercial products of wide application. Rayon and cellophane are regenerated cellulose. Tragacanth, agar, chondrus, pectin, and guar gum are all carbohydrates, used often as thickeners for water solutions.

In this chapter we will examine sources of carbohydrates, their chemical structure and names, and how we recognize their formulas. More important, we will examine the role of carbohydrates in human biochemistry, in disease states, and in therapy. We will take a close look at a disease that sooner or later afflicts one American in twenty—diabetes mellitus. And we will examine briefly the metabolism of carbohydrates as a source of energy.

To the health professional, the study of carbohydrates is important for the following reasons:

1. Glucose, the most important carbohydrate, is human blood sugar.
2. Sugars, such as ribose, are part of important biomolecules (RNA, glycoproteins, riboflavin, immunoglobulin).

FIGURE 12.1 Scanning Electron Micrograph of Corn Starch Granules. (Courtesy Kraft, Inc., Glenview, IL.)

3. The carbohydrate cellulose constitutes most of the indigestible bulk in our diet.
4. The human body manufactures glycogen, a storage form of carbohydrate.
5. The carbohydrate inulin can be used to study kidney function (glomerular membrane filtration rates).
6. Dextran, a carbohydrate derivative, is used as a plasma substitute and in eye products to adjust tonicity.
7. Oxidized cellulose is used in surgery as a hemostatic (helps stop persistent bleeding).
8. Heparin occurs naturally in humans; this carbohydrate is a blood anticoagulant.

DEFINITION AND CLASSIFICATION OF CARBOHYDRATES

Carbohydrates all contain C, H, and O, most of them in the ratio $C_1:H_2:O_1$, or one H_2O for each C. Recognizing this, the French gave them the name *hydrate de carbone*, from which we get *carbohydrate*. Glucose and arabinose are carbohydrates that have the formulas $C_6H_{12}O_6$ and $C_5H_{10}O_5$, respectively.

Chemically, carbohydrates are polyhydroxy aldehydes, polyhydroxy ketones, or substances that can be hydrolyzed to these compounds. That sounds complicated, but it is easy to understand if we remember that aldehydes and ketones are carbonyl compounds, RCHO or R—CO—R. If we add many (poly) alcohol groups to an aldehyde or a ketone, we get a carbohydrate:

$$
\begin{array}{cc}
\text{H—C=O} & \text{CH}_2\text{OH} \\
| & | \\
\text{HC—OH} & \text{C=O} \\
| & | \\
\text{HC—OH} & \text{HC—OH} \\
| & | \\
\text{HC—OH} & \text{HC—OH} \\
| & | \\
\text{HC—OH} & \text{HC—OH} \\
| & | \\
\text{CH}_2\text{OH} & \text{CH}_2\text{OH}
\end{array}
$$

a polyhydroxy   a polyhydroxy
aldehyde   ketone

A **monosaccharide** is a carbohydrate that cannot be broken down by hydrolysis into any simpler compounds. A **disaccharide** is a carbohydrate that, upon hydrolysis, gives two monosaccharides. A *trisaccharide* upon hydrolysis yields three monosaccharides, and so forth. Collectively, disaccharides, trisaccharides, and tetrasaccharides are known as *oligosaccharides (oligo = a few, little)*, that is, carbohydrates that yield a few molecules of monosaccharide on hydrolysis. A carbohydrate that can be hydrolyzed to yield many monosaccharides is termed a **polysaccharide**. Table 12.1 gives some examples.

An imprecise category of carbohydrates includes those crystalline compounds referred to as **sugars.** We think of sugars as sweet, and though most of them are, some are tasteless. Most are monosaccharides, but some, like sucrose, are disaccharides. The suffix *ose* is commonly used to denote a sugar, and "simple sugar" usually refers to a monosaccharide. Actually, the only reasonable basis for defining sugars is their chemical behavior. Typically, they form crystalline derivatives called osazones, can be reduced to sugar alcohols, can be oxidized to sugar acids, and so on.

Table 12.1. Some Categories of Carbohydrates

Category	Example	Products of hydrolysis
Monosaccharide	Glucose	Cannot be hydrolyzed
Disaccharide	Maltose	Two glucoses
Disaccharide	Lactose	Glucose plus galactose
Disaccharide	Sucrose	Glucose plus fructose
Trisaccharide	Raffinose	Glucose, fructose, and galactose
Polysaccharide	Starch	Many glucoses

Sources of Carbohydrates

In the process of photosynthesis plants use the sun's energy to convert carbon dioxide and water into carbohydrates—most often into the monosaccharide glucose. This process is the source of all of the world's food. The plant may combine its glucose with other monosaccharides to make a disaccharide. We obtain sucrose (table sugar) in vast quantities from sugar cane and sugar beets. Plants also combine many molecules of glucose to form starch, the plant's storage form of energy. Mankind uses this starch in the form of potatoes, rice, beans, and other foods. Some sources of carbohydrates are as follows.

Source	Carbohydrate (approximate percent by weight)	Source	Carbohydrate (approximate percent by weight)
Sugar (sucrose)	100	Lima bean flour	66
Rice	90	White bread	53
Honey	82	Whole wheat bread	50
Prunes	73	Kidney beans	21
Molasses (cane)	69	Green beans	5
Oatmeal	67		

Plants can also convert the glucose they make into structural material such as cellulose. Humans cannot digest cellulose and obtain food energy (cows can).

Naming Carbohydrates

We use the common system in naming carbohydrates and there is little systematization to it. It is mostly a matter of memorizing the names, such as those given in Figure 12.2. However, to describe the kind of sugar and the length of its carbon chain, there is a systematic approach. It has three aspects:

1. General carbohydrate suffix: -ose.
2. Length of monosaccharide chain: *tri* (three), *tetr* (four), *pent* (five), and *hex* (six).
3. Chemical type: *aldo* (aldehyde), and *ket* (ketone).

Examples:
D-glucose ($C_6H_{12}O_6$) is an aldohexose.
D-ribose ($C_5H_{10}O_5$) is an aldopentose.
D-fructose ($C_6H_{12}O_6$) is a ketohexose.
D-threose ($C_4H_8O_4$) is an aldotetrose.

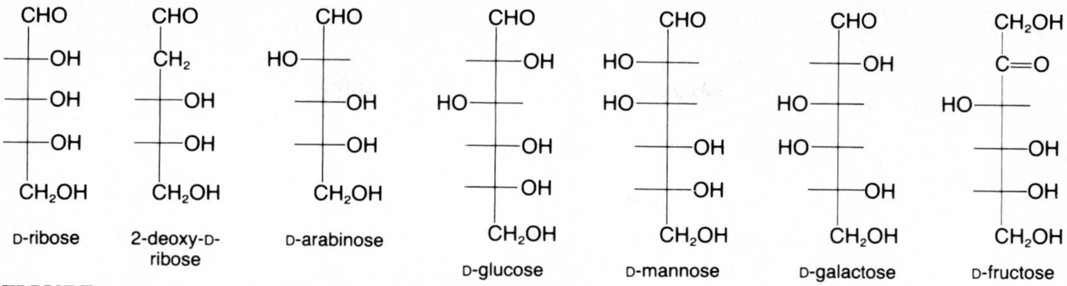

```
  CHO         CHO          CHO          CHO          CHO          CHO          CH₂OH
 —OH          CH₂     HO—            —OH      HO—            —OH          C=O
 —OH         —OH          —OH     HO—        HO—        HO—          HO—
 —OH         —OH          —OH          —OH          —OH          —OH          —OH
  CH₂OH       CH₂OH        CH₂OH       —OH          —OH          —OH          —OH
                                        CH₂OH        CH₂OH        CH₂OH        CH₂OH
 D-ribose   2-deoxy-D-   D-arabinose
             ribose                   D-glucose    D-mannose    D-galactose   D-fructose
```

FIGURE 12.2 Important Monosaccharide Structures.

Glucose is also known as **dextrose** and as **blood sugar.** Table sugar is sucrose and is implied by the term *sugar.* Lactose is milk sugar. Molasses is a thick, sweet, sticky syrup obtained as a by-product in the manufacture of sugar from sugar cane.

MONOSACCHARIDES

Monosaccharides can be thought of as the building blocks of polysaccharides and also as constituents of important biomolecules (DNA, RNA, the heart drug digitalis, and the citrus bioflavonoid hesperidin). Monosaccharides are the simplest of the carbohydrates, for there is no possibility of hydrolyzing them into smaller carbohydrate fragments. More than 200 different monosaccharides have been found in nature.

For us, the most important monosaccharides are those containing five or six carbon atoms. Combining this with our requirement of a polyhydroxy aldehyde or ketone, we get the following structures for three key monosaccharides:

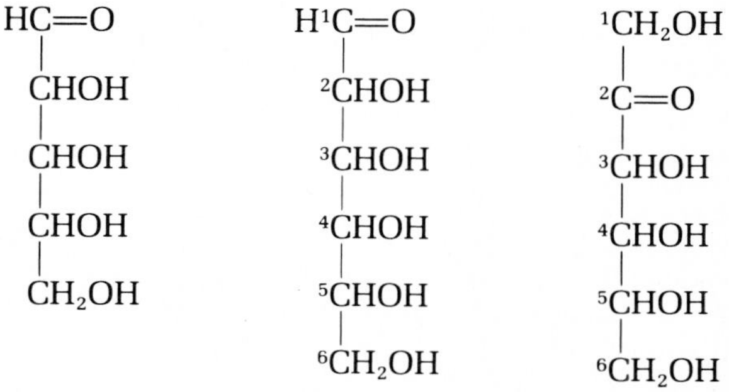

```
   HC=O             H¹C=O            ¹CH₂OH
   CHOH             ²CHOH            ²C=O
   CHOH             ³CHOH            ³CHOH
   CHOH             ⁴CHOH            ⁴CHOH
   CH₂OH            ⁵CHOH            ⁵CHOH
                    ⁶CH₂OH           ⁶CH₂OH

   ribose           glucose          fructose
(a five-carbon   (a six-carbon     (a six-carbon
  aldehyde)        aldehyde)          ketone)
```

All three of these compounds are properly referred to as sugars. Each is important in its own right plus as a part of a di- or polysaccharide or as a component of a biomolecule.

In an aldohexose such as glucose, four chiral centers exist and sixteen stereoisomers are possible. Glucose is but one of these! We neeed some kind of shorthand system to represent so many stereoisomers of this type, and a man named Emil Fischer[1] invented such a system.

In the **Fischer projection system** for drawing the structures of sugars, we let a two-dimensional drawing represent a three-dimensional molecule (think of a flashlight projecting an image of your right hand onto a movie screen). We begin by placing all of the carbons on a vertical line with the carbonyl carbon as close to the top as we can get it (I). Second, eliminate the carbons

and let the intersection of two lines represent each chiral center (II). Third, draw the OH group on each chiral center to the right or to the left, depending on its arrangement in the stereoisomer (III is glucose). Once we draw a Fischer projection formula like structure III , we must not take any part of it out of the plane in which it is written. This is because a projection formula like III stands for the actual three-dimensional structure we see in structure IV. Figure 12.2 gives the Fischer projection formulas for sugars that are important to discussions in this book.

D- and L-Configurations of Sugars

Glucose (blood sugar) is more correctly termed D-glucose. The D stands for D-**configurational series,** a classification of sugars having a common stereochemical ancestor. There is a companion L-**configurational series** that has its ancestor, too.

[1] Fischer (1852–1919), the greatest organic chemist of his time, was the first to explain the structures of the sugars, the polypeptides, and the purines. He won the Nobel prize in 1902.

For our purposes, it is enough to know that if the OH group on the *next-to-the-bottom* carbon in the Fischer projection formula is drawn to the *right*, the sugar belongs to the D-configurational series. If it is drawn to the *left*, the sugar belongs to the L-series. Here are some examples.

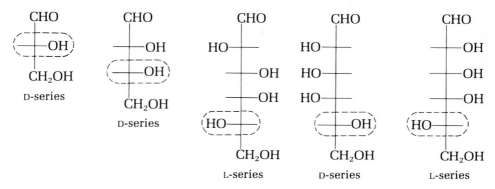

D-Glucose and L-glucose are enantiomers (a pair of mirror images). Although the differences between them are only stereochemical, there is still a great difference in their biological activity. This is because their three-dimensional structures are specific for the enzymes with which they react.

One reason we need to know about D- and L-sugars is the recent proposal to market foods containing only L-configurational sugars as diet aids. Reportedly, these L-sugar foods look and taste like the real thing but provide no calories because our enzymes are not capable of metabolizing them.

Almost all sugars that are biochemically imported are of the D-series. (Amino acids have their configurational series, too. Almost all of the amino acids that occur in peptides are of the L-series.)

Summary

The monosaccharides that are important in biochemistry are polyhydroxy aldehydes or polyhydroxyketones that can have up to four chiral centers. This means that up to sixteen stereoisomers are possible for a given six-carbon aldehyde structure. Blood sugar is one of sixteen isomers. A good shorthand system for drawing two-dimensional pictures of sugars is the Fischer projection system. It has specific rules that must not be violated. All monosaccharides can be classified into one of two series: the D-configurational series or the L-configurational series; the basis for assignment is the configuration of the OH group on the carbon atom adjacent to the terminal —CH_2OH group.

The Cyclic Structure of Sugars; Haworth Formulas

Up to now, we have depicted glucose as an open-chain, polyhydroxy aldehyde. This, however, is not the form in which 99 percent of glucose in blood or water exists. Actually, D-glucose forms a cyclic etherlike structure wherein its No. 5 hydroxyl has reacted with the aldehyde group:

glucose in open chain form   glucose in cyclic form

The cyclic form is termed a *hemiacetal* because it has the structure

hemiacetals have an alcohol and an ether group on
the same carbon

All of the aldohexoses and aldopentoses we have studied thus far can and do form similar cyclic structures. The six-membered ring is most common, but five-membered rings are also known (as in fructose).

The Englishman W. N. Haworth (1883–1950) devised a system for depicting monosaccharides and disaccharides in their cyclic forms. You may encounter **Haworth formulas** in the professional literature you read.

In a Haworth formula, we draw D-glucose as follows. Note the numbering.

down because it is to the *right*
in the Fischer formula

Haworth drawing of D-glucose

In proceeding from a Fischer projection formula to a Haworth, the rule is if an OH is drawn to the *right* in the Fischer formula, it is drawn *down* in the Haworth. If an OH is to the *left* in a Fischer formula, it is drawn *up* in the

Haworth. Check back to our earlier Fischer projection formula for D-glucose to confirm that the rule has been followed in the preceding Haworth drawing.

Anomeric Carbons

When the open chain form of a monosaccharide closes to form a cyclic hemiacetal, carbon number 1 becomes a chiral center and two new stereoisomers become possible. These are termed the *alpha-anomer* and the *beta-anomer*. In the α-anomer, the OH on carbon number 1 is drawn *down*. In the β-anomer, the OH on carbon number 1 is drawn *up*. We will illustrate using D-mannose:

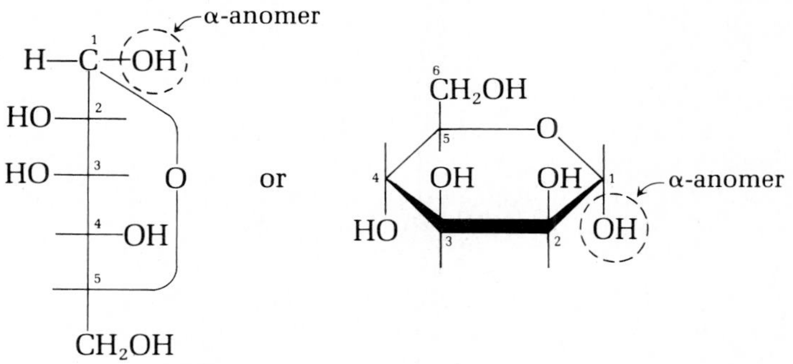

the alpha-anomer of D-mannose

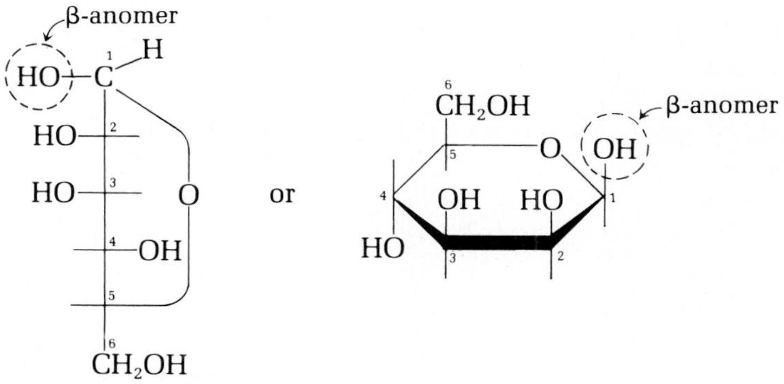

the beta-anomer of D-mannose

Summary

To represent the cyclic hemiacetal structure of sugars, we use the Haworth formula. The sugar ring is represented as a hexagon (considered to be flat) with the oxygen atom in the right rear. Carbon number 1 is always placed to the extreme right. If an OH group is drawn to the right in the sugar's Fischer formula, it is drawn down (below the ring) in the Haworth, and vice versa. The OH group on carbon number 1 is either beta (above the plane of the hexagon) or alpha (below the plane).

DISACCHARIDES

Both plants and animals can join two monosaccharide molecules together, splitting out a molecule of water, and forming a disaccharide:

$$\text{monosaccharide} + \text{monosaccharide} \xrightarrow[\text{catalyzed}]{\text{enzyme-}} \text{disaccharide} + H_2O$$

Disaccharides can also result when polysaccharides are broken down (by chemical hydrolysis or by digestion), step by step, ultimately to a two-saccharide unit. Thus, maltose is a disaccharide that is an end-product resulting from the action of diastase enzyme on starch.

Some of the disaccharides so far discovered are of tremendous importance to mankind. Table 12.2 lists various disaccharides.

Sucrose is by far the most abundantly distributed of the sugars. It appears openly as table sugar, cake frostings, and in other forms, and not so obviously in many of our foods (breakfast cereals, baby foods, vitamin C drinks, salad dressings) where it is a factor in customer acceptance and therefore sales volume. Per capita consumption of sucrose in America is close to 100 lb per year, a disturbingly high level. Hydrolysis of sucrose gives a 50–50 mixture of glucose and fructose. Honey (invert sugar) is already a 50–50 mixture of glucose and fructose. It is on this basis that some ads claim honey is easier to digest than sucrose. In the sense of saving the energy of hydrolysis, the ads are correct. Invert sugar (Travert) is a hydrolyzed product of sucrose prepared for IV use.

Lactose is found in cow's milk and in higher concentrations in human milk. Lactose helps promote absorption of calcium in the infant, a fact that partially accounts for the excellent growth associated with consumption of

Table 12.2. Some Important Disaccharides

Name	Formula	Monosaccharide constituents	Source
Maltose	$C_{12}H_{22}O_{11}$	Two glucoses	Hydrolysis of starch
Cellobiose	"	Two glucoses	Hydrolysis of cellulose
Gentiobiose	"	Two glucoses	Hydrolysis of amygdalin (laetrile, from bitter almond seeds)
Melibiose	"	Two glucoses	Hydrolysis of raffinose
Lactose	"	Glucose plus galactose	Mammalian milk
Sucrose	"	Glucose plus fructose	Cane and beet sugar

human milk. Some newborns, especially prematures, have not yet fully developed the capacity to synthesize the enzyme lactase and are prone to lactose intolerance, manifested by diarrhea and cramping for the first weeks of life. Hydrolysis of 1 mol of lactose yields 1 mol each of glucose and galactose.

Maltose is formed as an intermediate product in the acid hydrolysis of starch; it is a constituent of corn syrups that are prepared from starch by acid hydrolysis. We can summarize the progressive hydrolysis of starch in the following display:

$$\text{starch} \xrightarrow[\text{H}_2\text{O}]{\text{H}^+} \text{dextrins} \xrightarrow[\text{H}_2\text{O}]{\text{H}^+} \text{maltose} \xrightarrow[\text{H}_2\text{O}]{\text{H}^+} \text{glucose}$$

| a polymer of glucose | a complex mixture of intermediate-weight hydrolysis products |

Haworth formulas work very well when used to depict the structures of disaccharides. You will recall that a disaccharide forms when a molecule of water splits out from two monosaccharides. The union between the two former monosaccharides can be 1,4-, 1,6-, 1,2-, and so on. It can also be alpha or beta, depending on which anomer joined in the bond. Here are some examples:

maltose has the α-1,4-link between the two D-glucose residues

cellobiose has the β-1,4-link between the two D-glucose residues

There is a fascinating contrast between maltose (from starch) and callobiose (from cellulose). They both contain D-glucose and appear almost identical, yet humans can digest and obtain food value only from the maltose. Cotton, newspapers, and celery stalks are mostly cellulose and contain D-glucose, but they would only provide bulk, not calories, if we ate them. Can you spot the explanation? It is the beta-1,4-link in cellobiose, for which we humans have no digestive enzyme. We simply cannot hydrolyze the disaccharide to free its glucose content. Fortunately, cows can.

The Haworth structure of sucrose shows that the D-fructose and the D-glucose are both linked through their anomeric carbons:

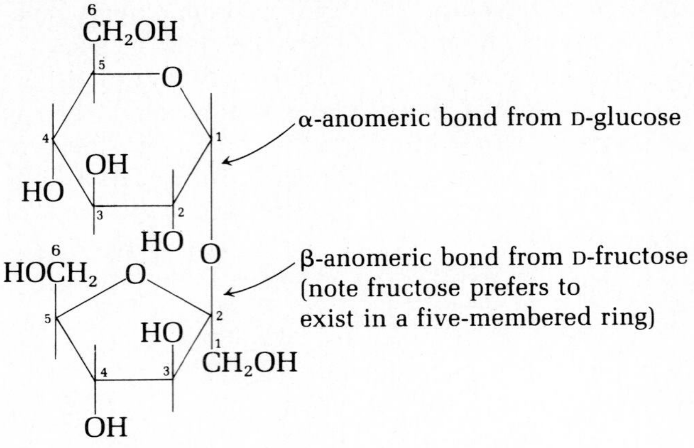

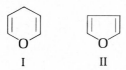

sucrose

Naming Cyclic Sugar Structures

The six-membered cyclic sugar structure is termed the *pyranose* form because it is related to pyran (I); the five-membered cyclic sugar structure is termed the *furanose* form because it is related to furan (II).

I   II

In nature, glucose exists predominantly in the pyranose form; fructose exists predominantly in the furanose form.

REDUCING SUGARS

Later in this chapter, in our discussion of diabetes mellitus, we will mention tests for the presence of sugar in the urine. Some of these tests depend chemically on the sugar acting as a *reducing agent*. Hence these sugars, glucose included, have become known as **reducing sugars**.

A reducing sugar is defined as a mono- or disaccharide that, without hydrolysis, reacts to give a positive test with *Benedict's solution*, *Tollens' solution*, or *Fehling's solution*. It is the easily oxidized free aldehyde group in a reducing sugar that makes it a reducing agent. But, you say, we learned earlier that 99 percent of a sugar like glucose exists in the cyclic hemiacetal. So where is the free aldehyde group to react? The explanation lies in the equilibrium that exists between free aldehyde and anomeric forms of a sugar:

$$\alpha\text{-anomer} \rightleftharpoons \text{R-CHO} \rightleftharpoons \beta\text{-anomer}$$

Table 12.3. Chemical Tests for Reducing Sugars

Name of test	*Reagents in test*	*Tests for*	*Results of a positive test*
Benedict's	A solution of Cu^{2+} ion in citrate	Reducing sugars or any aliphatic (but not aromatic) aldehyde	Appearance of yellow, red-orange, or brick-red precipitate of cuprous oxide Cu_2O
Tollens'	A solution of Ag^+ in ammonia water (i.e., $Ag[NH_3]_2^+$	Reducing sugars or any aliphatic or aromatic aldehyde	Appearance of Ag metal (as a silver mirror if glassware is very clean)
Fehling's	A solution of Cu^{+2} in tartrate	Reducing sugars	Appearance of brick-red precipitate of Cu_2O

When some of the R-CHO form of the sugar is removed by chemical reaction, the equilibrium shifts to restore it. Hence the reduction "funnels" through the free aldehyde form. The reducing sugar itself is oxidized to a carboxylic acid (the OH groups are not changed chemically). Table 12.3 summarizes tests for reducing sugars.

Sucrose is not a reducing sugar. No free or potentially free aldehyde group exists in sucrose because *both* anomeric carbons are tied up in the formation of the disaccharide link.

Even though it has no aldehyde group, fructose *is* a reducing sugar. This is because the alpha-hydroxyketone structure in fructose is oxidized just as easily as an aldehyde group by the alkaline test solutions listed previously.

POLYSACCHARIDES

For health professionals, the three most important polysaccharides are

1. Starch
2. Glycogen
3. Cellulose

Other polysaccharides you may encounter are inulin (a poly-D-fructose used to study kidney function), mucopolysaccharides (such as heparin and chondroitin), and chitin (a polymer of an amino sugar, found in invertebrates). The blood-group polysaccharides, found in red blood cells, are polymers of an amino sugar; they play a role in the typing of blood. Dextrans (not to be confused with dextrins[2]) are linear polysaccharides of glucose units joined

---

[2] Dextrins are fragments composed of several connected glucose units formed in the early stages of starch hydrolysis.

together by alpha-1,6-linkages and synthesized by various microorganisms. Dextrans have been used in medicine as plasma substitutes and blood volume expanders in the treatment of shock; they were used extensively in World War II. Dextrans, which can be hydrolyzed down to intermediate molecular weights, are also used to adjust tonicity (osmotic pressure) in ophthalmic preparations.

Starches are the reserve carbohydrate in plants. They occur in grains such as wheat, rye, oats, rice, and corn, and in roots and tubers such as potato (Figure 12.3) and cassava. D-glucose is the "monomer" from which starch is synthesized in the plant. The D-glucose units are linked α-1,4- in the fraction of starch known as amylose, and α-1,6- and α-1,4- in the branched fraction called amylopectin. If we concentrate on the amylose fraction, we can picture the following chain in starch:

α-1,4-glucose chain in starch

It is obvious from the amylose chain that maltose will be the disaccharide arising from partial hydrolysis. Amylases are the enzymes that catalyze hydrolysis of starch (see Chapter 9). Studies have shown that a single, giant polymeric molecule of amylose may contain as many as 12,000 glucose units.

When treated with iodine (I_2), starches with the amylose (unbranched) fraction give a blue color. The amylopectin (branched) fraction of starch gives a red-violet color with iodine. High molecular weight dextrins give a red color. Maltose gives no color with iodine.

The nurse may be called upon to prepare and administer a *starch bath*, used as an emollient to relieve skin irritation. Mix 1 lb of cornstarch with cold water to make a paste. Add hot water and boil until thick. Add this to water in a bathtub and have the patient sit for up to one-half hour. Afterwards, the film of starch is allowed to remain on the skin.

Glycogen is a very important substance in the human body. Formed and broken down in the liver, this polymer has been referred to as "animal starch." And, like starch, glycogen uses D-glucose as the building block, hooking up the monosaccharide units mostly α-1,4-, with some α-1,6-branching. Thus the glycogen chain looks much like the chain we drew a few paragraphs back for starch. Glycogen is even more highly branched than amylopectin; its moleclar weight averages in the 3 to 15 million range.

Glycogen is the primary carbohydrate of tissues, where it serves as a storage form of reserve carbohydrate. Most of the glucose in our bodies is present as glycogen, and when tissue supplies of glucose become low, as in fasting, liver-glycogen is broken down to replenish blood glucose levels.

FIGURE 12.3 This South American woman is proud of her harvest. The potato, with 85% of its dry weight starch, is a staple of the diet world wide. (Courtesy World Health Organization.)

Conversely, when the blood and tissues are rich in glucose, as after a rich carbohydrate meal, glycogen is formed.

Dozens of different dietary factors can be converted to glycogen, providing there is need for it and other conditions are right. Serving as precursors to glycogen can be D-glucose, D-mannose, D-fructose, and also glycerin, citric acid, lactic acid, mannitol, and at least a dozen amino acids. Glycogen is constantly present in heart muscle. Liver tissue contains the highest percentage of glycogen (about 5 percent); our skeletal muscle averages much less (about 0.5 percent). The process of glycogen formation is called *glycogenesis*; glycogen breakdown is called *glycogenolysis*.

Insulin, glucagon, and **epinephrine** (adrenalin) are three hormones that play important roles in regulating body glycogen. Human glucagon is a peptide hormone (twenty-nine amino acids) and is identical to pig glucagon. Table 12.4 summarizes some of the effects of these hormones on glycogen levels.

Table 12.4.   Hormonal Control of Glycogen

Hormone	Source	Effect on glycogen	Related effects
Insulin	β-Cells of pancrease	Promotes formation	Lowers blood sugar levels
Glucagon	α-Cells of pancreas	Accelerates breakdown	Acts only on liver glycogen, not muscle; raises blood sugar levels
Epinephrine	Adrenal medulla	Accelerates breakdown	Raises blood sugar levels

It can be seen (Figure 12.4) that insulin and glucagon work in opposition to each other, and that blood sugar levels at any given time depend, in part, on the biochemical balance between these two hormones. Blood levels of glucose are also influenced to some degree by growth hormone, sympathomimetic amines, and adrenal cortex steroids.

Cellulose and its derivatives have already been discussed at various times in this chapter. Here is a summary of the important facts on cellulose:

- Cellulose is a polymer consisting of D-glucose units linked β-1,4-.
- Purified cotton USP is essentially 100 percent pure cellulose.
- Methylcellulose USP (Methocel) is a methyl ether derivative available in various molecular weights; its ability to swell in water giving bulk or viscous solutions has led to its use as a laxative (Hydrolase), appetite suppressant (many trade names), and a cushioning agent in contact lens wetting solutions.
- In oxidized cellulose USP (Oxycel, Hemo-Pak), some of the terminal

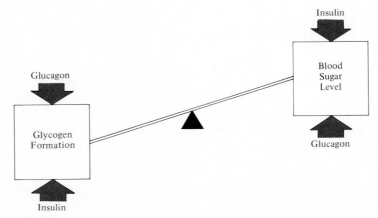

FIGURE 12.4   Effects of the Biochemical Balance between the Two Mutually Antagonistic Hormones, Insulin and Glucagon.

CH$_2$OH groups have been oxidized to COOH, giving a substance that is useful as a local hemostatic (*hemo* = blood, *static* = not flowing). If left in the wound, oxidized cellulose eventually is biodegraded and absorbed.

- Sodium carboxymethylcellulose USP (CMC) produces thick, viscous solutions when mixed with water. It is sold in antidiarrheal products (Kaolin Pectin Suspension) and as a laxative [bulk former].

THE IMPORTANCE OF BLOOD SUGAR

You could not live without glucose. In fact, if your blood glucose level fell much below its normal 70–100 mg % level, say, down to 30 mg % or so, you could fall into a coma and possibly suffer brain damage. And levels in the other direction are bad, too. One patient could not understand why she felt so terribly ill—until she walked into a doctor's office and discovered she had a blood sugar level of nearly 500 mg %! She was a severe diabetic and didn't know it.

Glucose (blood sugar, dextrose) is essential to normal brain function. In **hypoglycemia** (*hypo* = below normal, *gly* = glucose, *emia* = of the blood), the blood sugar level falls below normal for that individual. He or she can become edgy, irritable, and hard to get along with. Chronic, mild hypoglycemia can cause lack of coordination and slurred speech. Individuals with this condition sometimes hallucinate or experience convulsions; they have been mistaken for public drunks. Ironically, one of the pharmacological effects of alcohol ingestion is hypoglycemia. The alcohol is believed either to stimulate insulin secretion or to inhibit release of glucose from the liver. In any event, hypoglycemic crisis is not uncommon in alcoholism. Hypoglycemics may have to eat more often to maintain a sufficiently high blood sugar level. *Hyperglycemia* is an abnormally high blood sugar level. Diabetics are hyperglycemic.

Glucose is not to be confused with table sugar (sucrose). Glucose is dextrose USP and is usually obtained by acid hydrolysis of cornstarch or other starches. One gram of glucose dissolves in about 1 mL of water and in about 100 mL of alcohol. Aqueous solutions of glucose can be sterilized by autoclaving. When dextrose is administered IV, its solutions (5 to 50% w/v) are made up in physiologic (isotonic) salt solution or in Ringer's solution. Glucose's aldehyde group can easily be oxidized to a carboxyl group, giving *gluconic acid*. Two salts, calcium gluconate USP and ferrous gluconate NF, are used, the former as a calcium replenisher and the latter as a source of iron.

The normal *fasting* venous blood level of glucose is 60 to 80 mg per 100 mL of blood. That is 3.4 to 4.5 mmol/L in SI units. The level in *arterial* blood is about 15 to 30 percent higher than in venous blood. To cover the general case of a person in between meals, a range of 70 to 100 mg % is usually stated as normal.

Diabetes Mellitus[3] (or just plain diabetes) is a chronic disease characterized by abnormal carbohydrate and fat metabolism due to a deficiency in the body's production of insulin or to an inability of the body to utilize the insulin produced. In the healthy person, insulin acts to promote tissue uptake of glucose with resultant glucose metabolism (Figure 12.5).

In the diabetic, glucose oxidation in the tissues (skeletal muscle, cardiac and smooth muscle, fat, mammary gland) is diminished because of a lack of insulin. As a result, blood glucose levels rise. (Levels as high as 700 mg % have been measured in some cases.) When the blood sugar level increases to about 180 mg % glucose "spills over" into the urine and the disease is now detectable by urinalysis. In the healthy person, glucose is normally never found in the urine.

Because calories are not coming from glucose anymore, fat metabolism greatly increases in the diabetic. As a consequence, there is much greater than normal liver synthesis of three chemicals that are by-products of fat metabolism: acetoacetic acid (CH_3COCH_2COOH), β-hydroxybutyric acid ($CH_3CHOHCH_2COOH$), and acetone (CH_3COCH_3). See our discussion of fat metabolism in Chapter 13 for the mechanisms by which these three compounds arise. Collectively, these three compounds are known as **ketone bodies**. In severe diabetics, blood levels of acetone become so high that acetone is detectable on the breath. Excess ketone body formation is also observed in starvation and in high-fat/low-carbohydrate diets.

Two severe complications of diabetes are often observed. First *ketoacidosis* can occur. This is an abnormal acidification of the body from the high production of acetoacetic acid and β-hydroxybutyric acid. Second, long-standing diabetes can damage capillary beds, especially in the retina of the eye. The leading cause of new blindness in the United States is diabetes mellitus.

Additionally, diabetics can suffer from slow-healing of wounds, from an increased thirst (polydipsia), and from an increased frequency of urination (polyuria). They may eat more but continue to lose weight and may suffer from itching and weakness.

Causes of Diabetes Mellitus

The prevalence of diabetes is about 10 percent of the population, diagnosed and undiagnosed. About 50 million Americans either have diabetes, will develop it, or have a diabetic relative. Diabetes is the third leading cause of death; it kills about 340,000 Americans each year, either outright or because of side effects. Women are 50 percent more likely to have diabetes than men. The number of diabetics in the United States is increasing at a rate of 5 to 6 percent a year.

---

[3] Not to be confused with diabetes insipidus, in which enormous volumes of urine are excreted, with no sugar content. Common in the young, diabetes insipidus can be caused by tumors of the pituitary gland or head injury.

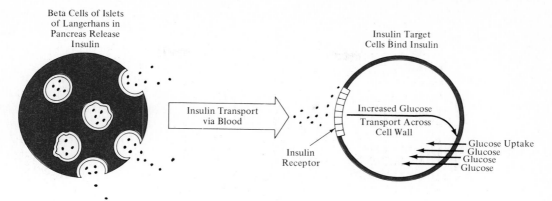

FIGURE 12.5 **Insulin secreted in the pancreas is transported to cells throughout the body where it acts to promote cellular uptake of glucose with resultant glucose metabolism. In the diabetic, cellular uptake of glucose is diminished either because of lack of insulin or because of an inability to utilize secreted insulin. Hence, glucose is not metabolized.**

It is believed that three major causes of diabetes exist:

1. Hereditary tendency (genetic predisposition).
2. Viral infection or obesity (environmental factors).
3. Autoimmune response of the body.

The sometimes abrupt onset of the juvenile form has given credence to the theory that a viral infection involving the pancreas may be the cause in some cases. The isolation of coxsackievirus B4 from the damaged pancreatic cells of a 10-year-old boy who died of diabetic ketoacidosis adds to the growing evidence that implicates viruses in some forms of insulin-dependent diabetes mellitus. In any event, the beta cells of the islets of Langerhans of the pancreas are permanently damaged and cease synthesizing insulin, or greatly reduce synthesis.

Here are some helpful definitions for key terms used in discussing diabetes mellitus:

- *Insulinopenic diabetes*—(Greek: *penia*, poverty or need) Diabetes occurring because of an insufficiency of insulin.
- **Juvenile diabetes**—The severe form, usually having an abrupt onset before age 25 and tending to be difficult to control and unstable ("*brittle*"). Requires insulin injections to control; pancreatic activity is lost.
- *Maturity-onset diabetes*—A mild, often asymptomatic form with onset usually after age 40, most often in overweight persons; diet control is possible.

- *Diabetic coma*—A medical crisis that threatens when blood sugar levels rise above 500 mg %; requires prompt injection of insulin; diabetics carry cards identifying their disease so that, if unconscious, they can be promptly and correctly treated.
- Insulin shock—A dangerous too-low blood sugar due to overadministration of insulin; occurs accidentally in diabetics or deliberately in the mentally ill as one means of treating schizophrenia (an approach analogous to electroshock therapy).
- **Oral hyopglycemic drugs**—Tolbutamide (Orinase), acetochexamide (Dymelor), and chlorpropamide (Diabenese); do not substitute for insulin but rather stimulate the pancreas to secrete more insulin; require a functionally intact pancreas; are possible teratogens, and should not be used in pregnancy; are often of use in maturity-onset diabetes.

Medical Treatment of Diabetes

The treatment of diabetes mellitus is fourfold:

1. Injection of insulin for the severe insulinopenic.
2. Control of diet for the mild to moderate diabetic.
3. Use of oral hypoglycemic drugs, especially in the maturity-onset case.
4. Any combination of the above.

Exogenous insulin substitutes very well in the insulinopenic individual. It must, of course, be injected, as insulin by the oral route would be destroyed by digestive enzymes. The physician will establish the correct dose and the most appropriate form. At least twenty-one different forms and brands of insulin are marketed. Here is a summary of just a few.

Type	*Onset (hours)*	*Peak (hours)*	*Duration (hours)*
Fast-acting			
Regular iletin	1	4–6	6–8
Semilente	1	4–6	12–14
Intermediate			
NPH iletin	2–4	8–12	20–24
Lente iletin	2–4	8–12	18–28
Globin zinc	1–3	6–10	12–18
Long-acting			
Protamine zinc	3–4	14–20	24–36
Ultralente	3–4	16–19	30–36

Figure 12.6 shows how various insulin preparations differ in their onset and duration of action in lowering too-high blood glucose levels.

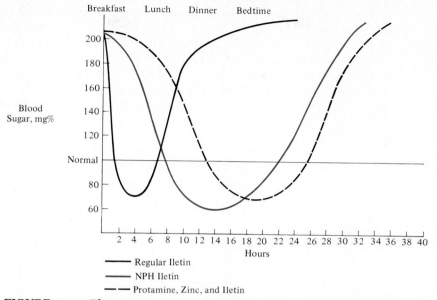

FIGURE 12.6   The comparative onset, peak action, and duration of action of one dose of three different types of insulin preparations.

The Oral Glucose Tolerance Test

To screen for diabetes, the physician may administer an *oral glucose tolerance test* (OGTT) (Figure 12.7), in which the patient's *insulin-producing capacity is challenged* by the rapid ingestion of up to 100 g of glucose. A diabetic will have a poor ability to handle that much sugar, and his blood sugar level will soar. The normal response is shown in Figure 12.7. According to the American Diabetic Association's standards, the test indicates diabetes

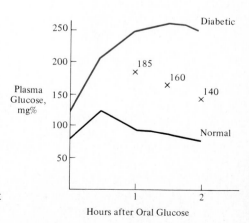

FIGURE 12.7   The glucose tolerance test is used to screen for diabetes.

if fasting glucose is 115 mg % or if the 1-, 1.5-, and 2-hour postglucose values are 185, 165, and 140 mg %, respectively, with all three values abnormal (Figure 12.7). Other diagnostic criteria are applied, including those of the University Group Diabetes Program, the Fajans-Conn Criteria, and the U.S. Public Health Service's Wilkerson Point System. The person undergoing the test should restrict carbohydrate intake to no more than 150 g per day for 3 days prior to the test and should fast for at least 8 hours but less than 16 hours before the test. Significant exercise should be avoided before the test, as should a variety of drugs (caffeine, nicotine, diuretics, oral contraceptives).

Home Monitoring of Diabetes Mellitus

Handy, easy-to-use tests for detection of glucose and ketone bodies in the urine have been developed and are available for home use. Their use tells the diabetic how well he or she is controlling the disease. At least fourteen products are available commercially, and we can only summarize here. Clinitest utilizes copper reduction as the basis for glucose detection, whereas TesTape and Diastix use glucose oxidase enzyme. Both are widely used to test for spilling of sugar but are not 100 percent reliable (either false negative or false positive test results are noted in a small percentage of users). Ketodiastix is useful for both glucose and ketone urine detection by the dip and read method; there are some false negatives, and it is used mainly in the more severe cases.

Readings of the amounts of glucose in the urine are reported as O, trace, +1, +2, +3, or +4, and must be carefully interpreted according to the comparative readings given in Table 12.5.

Normal Metabolism of Carbohydrates

Because carbohydrates constitute the largest mass of foods we consume and are the source of the majority of the calories we burn, the metabolism of carbohydrates to energy, carbon dioxide, and water is of special consequence to the health professional. The overall reaction representing carbohydrate metabolism is

$$\text{carbohydrate} \xrightarrow[\text{tissue cells}]{\text{oxidation in}} CO_2 + H_2O + \text{energy (both heat and potential)}$$

The energy produced by this process is used, in part, to keep the process going, but also to supply all of the energy-demanding processes that occur in the human. These include digestion and absorption of foods, muscle contraction, growth, kidney function, and the synthesis of nucleic acids, protein, fat, and thousands of biochemical compounds.

The biochemist recognizes two types of metabolic change in the body. **Catabolism** is the breaking down of complex molecules such as starch, protein, or fat into simpler ones through hydrolysis, oxidation, deamination, or some

Table 12.5. Comparative Values for Urine Glucose Test Readings

Test	0	Trace	+	++	+++	++++
Glucose oxidase tests						
TesTape	0%	—	0.1 %	0.25%	0.5%	2%
Diastix	0%	0.1 %	0.25%	0.5%	1.0%	2%
Clinistix	0%	—	0.25%	Present	0.5%	—
Copper reduction tests						
Clinitest	0%	0.25%	0.5%	0.75%	1.0%	2%
Clinitest 2-Drop*	0%	Trace	0.5% 1.0%	2%	3%	5%

* The two-drop method has seven possible readings; other tests have six possible readings.

Source: Handbook of Nonprescription Drugs, Sixth Edition, American Pharmaceutical Association, 2215 Constitution Ave., Washington, D.C. Used with permission.

such process. **Anabolism** is the building up (the synthesis) of body chemicals needed in growth, repair, and functioning. In anabolic changes, ions and simple molecules are incorporated into the living cytoplasm of the cell for the synthesis of more complex molecules.

The sites of catabolism of carbohydrates are mainly cells of the liver and muscle, but significant oxidation occurs in fat tissue, white blood cells, pituitary gland, and mammary gland. More specifically, glucose metabolism occurs partly in the cytoplasm of a cell and partly in the **mitochondria**—small sausage-shaped granules in cells, bound to cell membranes and packed with many kinds of enzymes (Figure 12.8). The mitochondria are the power-generating units of a cell; they are found in cells and areas where energy-requiring processes occur.

The energy produced by oxidation of glucose is packed into a high-energy storage molecule called *adenosine triphosphate (ATP)* (Figure 12.9). Occurring everywhere in the body, ATP is the main source of energy for energy-requiring reactions (anabolic and catabolic). For example, if a phosphate ester is required in some process, ATP can supply it by breaking down to yield a phosphate and ADP (adenosine diphosphate).

$$\text{glucose} + \text{ATP} \longrightarrow \text{glucose-6-phosphate} + \text{ADP}$$

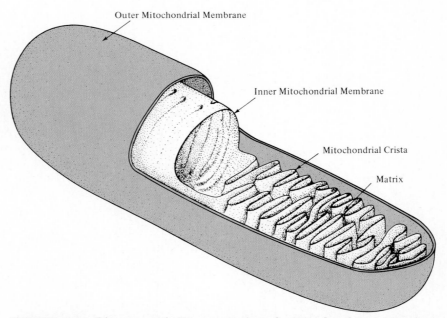

Outer Mitochondrial Membrane

Inner Mitochondrial Membrane

Mitochondrial Crista

Matrix

FIGURE 12.8   Diagrammatic Representation of a Mitochondrion. Mitochondria, found in the cytoplasm of cells, have double membranes. They contain the enzymes of the Krebs and fatty acid cycles and the respiratory pathway. They are the principal sites of the generation of energy in the form of ATP.

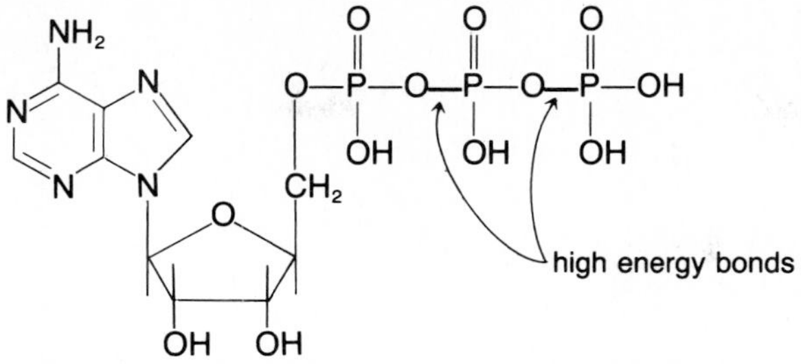

FIGURE 12.9   Adenosine Triphosphate (ATP).

Adenosine triphosphate is the immediate source of energy for many other mechanical and chemical body processes, including contraction of muscle, protein and fat synthesis, and transport of wanted substances across cell membranes into cells and transport of unwanted substances out of cells. The energy required for reforming ATP comes from the catabolism of glucose.

Glycolysis

The metabolism of glucose can be considered as occurring along two consecutive pathways:

1. The *glycolytic pathway* (or **glycolysis** or the **Embden-Meyerhof pathway**).
2. The **Krebs Cycle** (or **citric acid cycle**).

The glycolytic pathway can be summarized as shown in Figure 12.10. Note that in glycolysis, two pyruvate molecules are produced from each glucose molecule metabolized. Although it requires ATP to keep it operating, the glycolytic pathway yields a net gain of ATP. Of course, this is the main goal in metabolism—the production of energy from foodstuffs.

The Embden-Meyerhof pathway functions typically in the absence of oxygen (**anaerobic metabolism**) but can also operate in the presence of oxygen (**aerobic metabolism**). Under extended anaerobic conditions, as in prolonged muscle exertion, the end product of glycolysis will not be pyruvic acid, but rather lactic acid:

$$CH_3-\overset{\overset{\displaystyle O}{\|}}{C}-COOH \qquad\qquad CH_3-\overset{\overset{\displaystyle OH}{|}}{CH}-COOH$$

pyruvic acid—formed
when oxygen supply is
adequate

lactic acid—formed
when oxygen supply is
inadequate

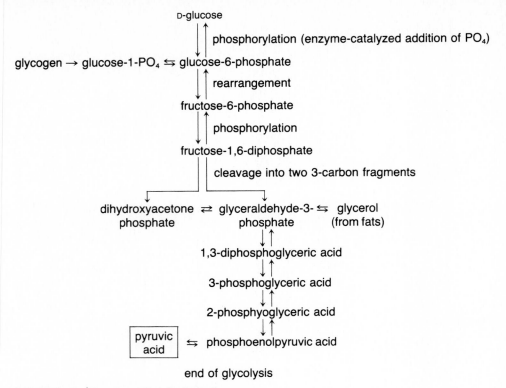

FIGURE 12.10 The Glycolytic Pathway.

Any lactic acid that might temporarily accumulate in muscle tissue is converted back to pyruvate when the oxygen supply in the muscle is restored through muscle rest with adequate ventilation.

(It is of interest to note here that yeast organisms are able to ferment carbohydrate to ethyl alcohol by an anaerobic process that, in its first part, is identical to the Embden-Meyerhof pathway.)

The Krebs Cycle

In the mitochondria, pyruvic acid molecules next enter a metabolic cycle called the *citric acid cycle* or *Krebs' cycle*, named after Hans Krebs (1900–1981), winner of the 1953 Nobel prize for physiology and medicine (Figure 12.11).

The citric acid cycle produces much energy in the form of ATP. If we consider pyruvate as the starting point, each "turn" of the cycle produces 15 ATPs. That is equivalent to 30 ATPs from each glucose. Molecular oxygen is required for the Krebs cycle (hence it is termed aerobic in contrast to the

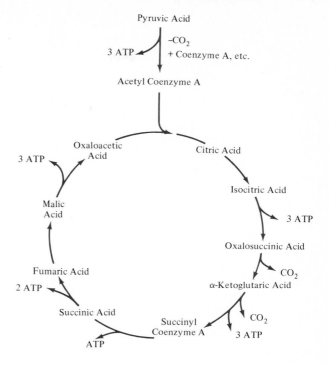

FIGURE 12.11 The Krebs Cycle (citric acid cycle). Starting with pyruvate, each turn of the cycle produces fifteen ATPs.

typically anaerobic glycolysis pathway). We note that the chemical compound that actually enters the Krebs cycle is acetate, CH_3COO^- (in the form of the high-energy compound, *acetyl Coenzyme A*, shown in Figure 12.11). Acetate can arise from fat metabolism as well as from carbohydrate metabolism, and so the Krebs cycle is not exclusively a source of glucose energy.

Last, the cell must provide a chemical means of forming water, which, with carbon dioxide, is the end-product of carbohydrate metabolism. To be more specific, the cell must have a pathway for transferring electrons and protons from chemical species in the Krebs cycle to molecular oxygen to form water, simultaneously producing ATP. All of the details of such a pathway, called the **respiratory chain** or **electron transport system,** have been worked out. All reactions occur in the mitochondria and involve a complex series of enzymes that can exist reversibly in reduced and oxidized forms. These enzymes are as follows:

1. Nicotinamide adenine dinucleotide (NAD); the need for the vitamin nicotinamide[4] (niacinamide) now becomes obvious.

[4] Both nicotinic acid and its amide are physiologically active in the human. Nicotinic acid per se is also termed niacin.

2. Flavin adenine dinucleotide (FAD), which incorporates a molecule of the vitamin riboflavin.
3. Coenzyme Q, which incorporates a quinone molecule; quinones are easily reversibly oxidized and reduced.
4. The cytochromes, b, c_1, c, a, and a_3, which are conjugated protein enzymes; iron is required for their functioning, reversibly shifting between the ferrous and ferric states.

Electrons generated in the Krebs cycle are passed on down the respiratory chain until they are finally united with H^+ and O_2 to yield water (Figure 12.12).

Summary

Body cells metabolize glucose to produce heat plus potential chemical energy stored in ATP. Each step in the oxidation of $C_6H_{12}O_6$ to CO_2 and H_2O requires an enzyme; many steps require cofactors such as vitamins or metal ions. At certain points in the pathways, other metabolic pathways (such as for fats) can intermingle. The metabolic pathways for glucose are described in the following table.

Pathway	Oxygen requirements	Scope	ATP's produced
Glycolysis (Embden-Meyerhof)	Anaerobic or aerobic	Glucose → pyruvate	
Citric acid cycle (Krebs cycle, tricarboxylic acid cycle)	Aerobic	Pyruvate → CO_2 + hydrogen atoms	
Respiratory chain (electron transport system)	Aerobic	$2H^+ + \frac{1}{2}O_2 + 2e^- \rightarrow H_2O$	36 per glucose* (6 via glycolysis and 30 via Krebs cycle)

* This quantity can vary depending on the source of glucose and aerobic or anaerobic conditions.

FRUCTOSE AND GALACTOSE METABOLISM

We obtain fructose in our diets from the digestion of sucrose or from honey. Some food faddist promoters would have us believe there are special health benefits from foods high in fructose, but no scientific evidence has been

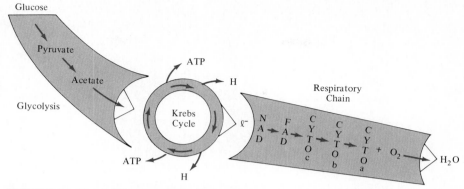

FIGURE 12.12   A Diagrammatic Representation of Glucose Metabolism Showing the Relationship between Glycolysis, the Krebs Cycle, and the Respiratory Chain.

presented to substantiate such claims. Fructose, as the 6-phosphate, is a normal constituent in the glycolytic pathway we have just studied, and can thus serve as a source of ATP. Fructose has been of only limited value in restoring carbohydrate reserve in the diabetic.

Digestion of the lactose in milk gives galactose, which is then absorbed from the GI tract and enters the general circulation. After cellular phosphorylation galactose is enzymatically converted to glucose, providing calories and ATP. In the **inborn error of metabolism** called *galactosemia*, the enzyme responsible for the conversion of galactose to glucose is missing or deficient, and a buildup of galactose in the blood occurs. Children with this genetic error may consequently suffer stunted physical and mental growth (mental retardation). They must be identified early in life so that all dietary sources of galactose can be eliminated. For a more complete discussion of inborn errors of metabolism, see Chapter 18.

PROTEIN-SPARING EFFECT OF GLUCOSE

When carbohydrate intake and liver glycogen levels are high, the body relies on carbohydrate as a primary source of energy, and thus the rate of protein catabolism is significantly diminished. Deamination of amino acids is reduced and the amino acids are saved for other body uses. This is termed the *protein-sparing effect* of glucose. High body glucose levels also suppress ketone body formation but increase the body's ability to excrete some chemicals by acetylation and glucuronide formation (glucuronides are inactive forms of phenols, alcohols, and carboxylic acids made in the liver by condensation with *glucuronic acid*, $HOOCH-(CHOH)_4CHO$, made in the body by oxidation of glucose).

A high carbohydrate intake should be maintained during the course of treatment of infectious hepatitis because a glycogen-rich liver is more resistant to toxic agents and pathologic processes.

SUMMARY

Carbohydrates are polyhydroxyaldehydes and ketones occurring in animals and plants. They can be classified as mono-, di-, tri-, and polysaccharides. Sugars are an ill-defined class of carbohydrates, typically crystalline mono- and di-saccharides. Sugars are described on the basis of the number of carbons and presence of either an aldehyde or ketone group; *ose* is the suffix used. Glucose is an aldohexose.

We learned to use two systems for drawing mono- and disaccharides: the Fischer projection formula and the Haworth formula. The Fischer projection is a two-dimensional representation of the three-dimensional stereochemistry of sugars; there are special rules concerning its use. Fischer formulas permit us to compare the stereochemistry of sugars and to specify D- or L-configurations. The Haworth formula is derived from the Fischer; it corresponds to the cyclic hemiacetal form in which sugars mostly exist.

When a monosaccharide cyclizes to a hemiacetal, a new chiral center is created and two new stereoisomers become possible at the former aldehyde carbon. They are termed the α-anomer and the β-anomer. We learned the rules for drawing anomeric forms of monosaccharides.

We studied four important disaccharides in the chapter: sucrose, lactose, maltose, and cellobiose. All contain at least one molecule of D-glucose. Reducing sugars are those mono and disaccharides having a free or potentially free aldehyde group; they give positive tests with Benedict's, Tollens', and Fehling's solutions.

For the health professional, the three most important polysaccharides are starch, glycogen, and cellulose. All three, upon complete hydrolysis, yield only D-glucose. Levels of body glycogen are regulated mainly by three hormones: insulin, glucagon, and epinephrine. Insulin and glucagon work in opposition to each other in regulating glycogen levels.

Blood sugar is D-glucose. In this chapter we learned its normal blood levels and how to recognize pathologic blood levels (hypoglycemia and hyperglycemia).

Diabetes mellitus was discussed in this chapter from the standpoint of cause, clinical signs and symptoms, ketone bodies, medical treatment, diagnosis, and home monitoring. We define key terms used in discussing diabetes mellitus.

A brief introduction to normal metabolism of carbohydrates was given in this chapter. The concepts of anabolism and catabolism were introduced and we learned that ATP is the bodywide storage form for the energy we derive from our metabolism of foods. Normal body metabolism of carbohydrate is divided into (a) glycolysis, (b) the Krebs cycle, and (c) the respiratory chain. We learned that each cycle has its special constituents, aerobic or anaerobic demands, and production of ATP. Thus we were able to follow the metabolism of glucose through pryruvate, acetate, and electron transport, to yield CO_2 and water.

KEY TERMS

Check your understanding of this chapter. Can you explain what is meant by each of these terms?

aerobic metabolism	glycogen
anabolism	glycolysis
anaerobic metabolism	Haworth formula
blood sugar	hypoglycemia
carbohydrate	inborn error of metabolism
catabolism	insulin
cellulose	juvenile diabetic
citric acid cycle	ketone bodies
D-configurational series	Krebs cycle
L-configurational series	mitochondria
dextrose	monosaccharide
diabetes mellitus	oral hypoglycemics
disaccharide	polysaccharide
electron transport system	reducing sugar
Embden-Meyerhof pathway	respiratory chain
epinephrine	starch
Fischer projection formula	sugar
glucagon	

STUDY QUESTIONS

1. Define the terms (a) carbohydrate, (b) monosaccharide, (c) disaccharide, and (d) polysaccharide.

2. Give five different examples of (a) monosaccharides, (b) disaccharides, and (c) polysaccharides.

3. What are the chemical differences between an aldopentose and a keto-hexose?

4. Fill in the blank spaces:

Carbohydrate	Products of complete hydrolysis	Occurrence
Maltose		
	D-glucose and D-fructose	
		Human and cow's milk
Cellulose		
		Liver, as the body's storage form of carbohydrate

5. If D-glucose, $C_6H_{12}O_6$, is an aldohexose, what would be the classification of the following:

    a. CH_2OH
         |
        $C=O$
         |
       $(CHOH)_2$
         |
      CH_2OH

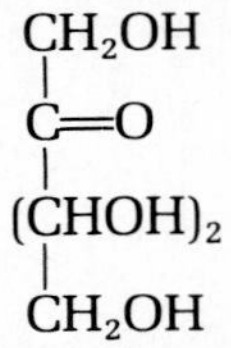

    b. $HOCH_2CHOHCHOHCHO$

    c. D-galactose   **d.** $CH_2OH-CHOH-CHO$

6. Identify the configurational series, D- or L-, to which each sugar belongs:

    a.   **b.**   **c.**   **d.**   **e.**

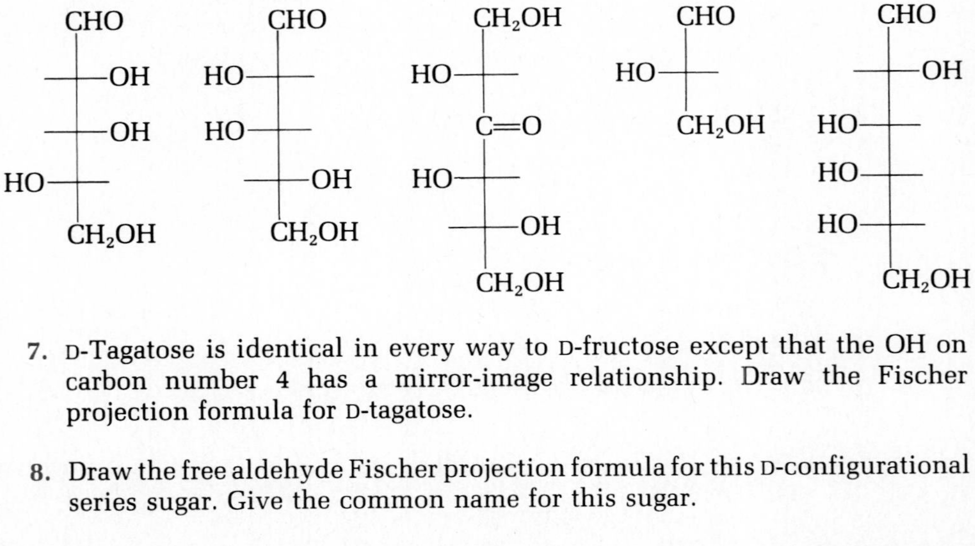

    a. CHO, —OH, —OH, HO—, CH_2OH

    b. CHO, HO—, HO—, —OH, CH_2OH

    c. CH_2OH, HO—, $C=O$, HO—, —OH, CH_2OH

    d. CHO, HO—, CH_2OH

    e. CHO, —OH, HO—, HO—, HO—, CH_2OH

7. D-Tagatose is identical in every way to D-fructose except that the OH on carbon number 4 has a mirror-image relationship. Draw the Fischer projection formula for D-tagatose.

8. Draw the free aldehyde Fischer projection formula for this D-configurational series sugar. Give the common name for this sugar.

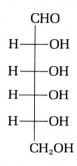

9. Using a Haworth formula, draw the alpha-anomeric isomer of the pyranose form of this sugar:

```
        CHO
   H ——|—— OH
   H ——|—— OH
   H ——|—— OH
   H ——|—— OH
       CH₂OH
```

10. True or false.
    a. Dextrose is a synonym for sucrose.
    b. Milk sugar is a synonym for dextrose.
    c. Blood sugar is a synonym for lactose.
    d. Mannose is a disaccharide.
    e. Fructose is a ketopentose.

11. Which of these formulas correctly represent a hemiacetal? (*a*) $RCH(OCH_3)_2$, (*b*) $RCH(OH)OC_2H_5$, (*c*) RCO_2R, (*d*) $RCH(OH)_2$, (*e*) $RCH(OCH_3)OH$, (*f*) R_2CHOH.

12. This is lactose:

```
    CH₂OH            CH₂OH
  HO⎺⎺⎺⎺O          ⎺⎺⎺⎺O  OH
    ⎛  OH  ⎞⎺⎺O⎺⎺⎛  OH  ⎞
         OH              OH
         A                B
```

    a. Identify the monosaccharide residue marked A and that marked B.
    b. What are the numbers of the carbon atoms involved in the disaccharide link? (Are they 1,1- or 1,6- or ?)
    c. Is the bond between the sugar residues α or β?

13. What is a "reducing sugar"? Is the sugar depicted in Question 12 a reducing sugar?

14. What is the "monomer" that plants use to form starch? What is the relationship or significance of each of the following with respect to starch?
    a. Amylases   b. Iodine
    c. Dextrins   d. Maltose

15. I have this funny friend Fred who, when hungry, chews on trees. Why is Fred wasting his time? Can any good come from such a diet?

16. What is the source and biochemical role of glycogen in the human?

17. Describe the mutually antagonistic roles played by insulin and glucagon in regulating glycogen levels in the human.

18. a. Will the secretion of glucagon be increased or inhibited by the administration of D-glucose? (*Hint:* Examine the role of glucagon without administration of glucose.)
    b. Will exercise increase or inhibit glucagon secretion?

19. The thyroid hormone *thyroxin* acts to accelerate breakdown of glycogen in the liver. Why does this seem logical, as compared to glycogen formation? (*Hint:* Check the physiological role of thyroxin.)

20. Name three synthetic derivatives of cellulose and state how they are used in medicine.

21. How do blood sugar levels in hypoglycemia compare to the normal range? (Give actual values.) How does the normal fasting *venous* blood sugar level compare with that in arterial blood? (Give actual values.)

22. Briefly, what is the role of insulin in the human organism?

23. What clinical signs and symptoms are associated with untreated diabetes?

24. Contrast the juvenile diabetic with the maturity-onset type from the standpoint of age of onset, integrity of pancreatic cells, requirement for injected insulin, and success of control by diet regulation.

25. What is meant by "ketone bodies" and where are they produced?

26. Diagnosis of the disease diabetes mellitus is made by blood analysis after challenging the system with a glucose tolerance test. What tests are available for urinalysis and how are they useful to the diabetic?

27. Are oral antidiabetic drugs such as Orinase insulin substitutes? How do they function and in what type of diabetes are they most likely to succeed?

28. Name the pathway in the metabolism of glucose that can operate in the *absence* of molecular oxygen. What is the chemical end-product of this

pathway and how many ATPs are produced per molecule of glucose metabolized?

29. Name the cycle in the metabolism of glucose that operates in the mitochondria and that is aerobic. How many ATPs are produced in this cycle per molecule of glucose oxidized?

30. The two-carbon unit, acetate (CH_3COO^-), in the form of acetyl Coenzyme A, enters the citric acid cycle. Where does it go? That is, chemically, where does it wind up?

31. The respiratory chain is sometimes referred to as the *electron transport system*. What electrons are transported and to where?

ADVANCED STUDY QUESTIONS

32. Coenzyme A is vital in the conversion of pyruvate to acetylcoenzyme A (and in many other situations). Describe the nature of this enzyme and the high-energy bond it forms.

33. Diabetes is increasing at a 6 percent annual rate. List five factors that could account for this increase.

34. Low-income people (less than $5000 a year) are three times more likely to have diabetes than middle- or upper-income people. Speculate on reasons for this.

35. English researchers have reported (*British Medical Journal*, 280:604, 1980) that giving the morning insulin injection before washing and dressing rather than afterwards can significantly reduce postprandial hyperglycemia. Preprandial hypoglycemia did not occur with the early injection regimen and blood glucose levels were stable or rising before breakfast. Explain how timing can make this difference.

36. Mathematically, show how to convert 85 mg of insulin per 100 mL to mmol/L.

13

Fats and Other Lipids

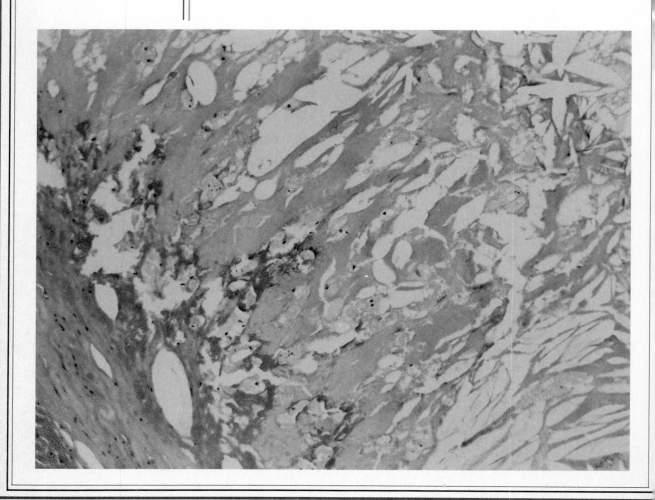

Cardiovascular disease (CVD) accounts for about 52 percent of the yearly deaths in this country. In our photo you can see the cause of death of one of the individuals who contributed to this statistic. This is a high magnification photo of a thin cross section of the plaque (the "plug") in a coronary artery taken on autopsy from a patient who succumbed to a myocardial infarct.

All along the right-hand side of our photo one can see the elongated white outlines of cholesterol clefts. These are deposits of solid cholesterol in the mass of material that plugged up the coronary artery.

Cholesterol is a lipid—as are the free fatty acids, saturated fats, and polyunsaturated fats from our diet. All of these lipids have been under investigation for many years, but as you will read in this chapter there is still no general agreement among the experts on the roles played by the various lipids in the cause (or prevention) of heart disease.

Whatever our position in the lipid—heart disease controversy, we must agree that fats and other lipids play prominent roles in human biochemistry. In this chapter you will have the opportunity to examine these roles. (Photo courtesy Heinz R. Hoenecke, M.D., Sharp Hospital, San Diego.)

INTRODUCTION

Everybody knows what fat is—it's that stuff that accumulates around the belly area of overweight humans and animals. It is what we as a nation spend millions of dollars trying to lose while the Madison Avenue hucksters spend billions of dollars trying to get us to put it on (by eating and drinking their products). Take a look around and you can see who is winning the battle.

Students of biochemistry learn to view fat in a scientific and health-oriented way. They ask questions like, What is the chemical nature of fat? How is fat metabolized normally (and abnormally in disease states)? How does it act as a storage depot for drugs like marijuana and the barbiturates? and What is the relationship, if any, between dietary fat and cardiovascular disease such as hardening of the arteries?

We will examine all of these topics in this chapter, plus others. We will examine a new theory about the process of atherosclerosis and study some biologically important molecules that are known by the collective name lipids. In Chapter 17 we will apply some of the principles we learn here to dietary, health, and professional situations.

DEFINITION AND CLASSIFICATION

By tradition, "fat" has usually meant fatty acid esters of glycerol such as the lard from animals, or vegetable oils such as corn, safflower, or peanut. We, however, need a term that is more inclusive, one that recognizes a half dozen different kinds of fatlike compounds. **Lipids** are all of those low-polarity, greasy, water-insoluble biochemical substances that dissolve in the so-called organic solvents such as chloroform, carbon tetrachloride, ether, or benzene. Lipids serve special functions in the human: storage of energy, insulation against cold, padding of organs against shock, acting as chemical intermediates in biochemical pathways, serving as constituents of cell membranes, and nerve transmission.

Among lipids, we recognize the following types of biomolecules:

1. Fatty acids—free or esterified.
2. Fats and oils—fatty acid esters of glycerol.
3. Waxes—fatty acid esters excluding those of glycerol.
4. Phopholipids—glycerol esters containing phosphorus.
5. Sphingolipids—derivatives of an aminoalcohol, containing phosphorus.
6. Sterols—such as cholesterol.

Although all six types of compounds are biochemically important, the types that impact most strongly on our lives are found in categories 1, 2, and 6.

FATS AND OILS

Chemically, there is not a great deal of difference between the larded fat in a steak and the free-flowing vegetable oils we cook with. Both are esters. The acids that make up the ester links are fatty acids (see our discussion in Chapter 10). The alcohol that participates in the ester link is the polyol *glycerol* (synonym: glycerin). Thus the general formula for a **fat** or **oil** is as follows:

$$
\begin{array}{c}
\text{H} \quad\quad\quad \overset{\displaystyle O}{\overset{\|}{}} \\
\text{HC}-\text{O}-\text{C}-\text{R} \\
| \quad\quad\quad \overset{\displaystyle O}{\overset{\|}{}} \\
\text{HC}-\text{O}-\text{C}-\text{R} \\
| \quad\quad\quad \overset{\displaystyle O}{\overset{\|}{}} \\
\text{HC}-\text{O}-\text{C}-\text{R} \\
\text{H}
\end{array}
$$

from glycerol · from the fatty acid

any fat or oil

This chemical structure is termed a **triacylglycerol** (formerly known as

$$\overset{O}{\overset{\|}{}}$$

triglyceride). Recall from Chapter 10 that an acyl group is R—C—. A fat molecule, then, has three acyl groups linked to glycerol.

The fatty acid residues in one glycerol ester can all be the same or all different. Most natural fats contain many different fatty acids. Altogether, about fifty different fatty acids have been found in nature, and all but one of these are unbranched compounds containing an even number of carbon atoms. The only odd-numbered fatty acid is isovaleric acid, from dolphin and porpoise oils.

Figure 13.1 provides some examples of simple glyceryl esters (all the same fatty acid) and mixed esters (a mixture of fatty acids).

You can see that some fats have unsaturated fatty acid residues in them. And some are polyunsaturated. Since unsaturated fat is an important factor in current theories of the cause of hardening of the arteries, let us specify the chemical structures we are discussing. The biochemically important unsaturated fatty acids are shown here.

$$CH_3(CH_2)_7CH{=}CH(CH_2)_7COOH$$

oleic acid *(cis)*
(9-octadecenoic acid)

$$CH_3(CH_2)_4CH{=}CH{-}CH_2CH{=}CH(CH_2)_7COOH$$

linoleic acid (all *cis*)
(9,12-octadecadienoic acid)

$$CH_3CH_2CH{=}CHCH_2CH{=}CHCH_2CH{=}CH(CH_2)_7COOH$$

linolenic acid (all *cis*)
(9,12,15-octadecatrienoic acid)

$$CH_3(CH_2)_4CH{=}CHCH_2CH{=}CH_2CH_2CH{=}CHCH_2CH{=}CH(CH_2)_3COOH$$

arachidonic acid (all *cis*)

In these fatty acids, all double bonds are *cis* and methylene (—CH$_2$—) interrupted.

Currently, in some circles a special system of notation is used to describe long-chain polyunsaturated fatty acids. For example, linoleic acid is 18:2ω6 where ω is the Greek letter omega. Here the 18 represents the total number of carbon atoms in the fatty acid, the 2 gives the number of double bonds in the molecule, and the 6 tells us the number of carbon atoms beyond the last double bond. In this system of notation, arachidonic acid becomes 20:3ω6.

For a discussion of dietary essential fatty acids, see Chapter 17.

In human fat, almost 25 percent of the fatty acid residues are palmitic acid. Palmitic acid is $C_{15}H_{31}COOH$—that is sixteen carbon atoms altogether.

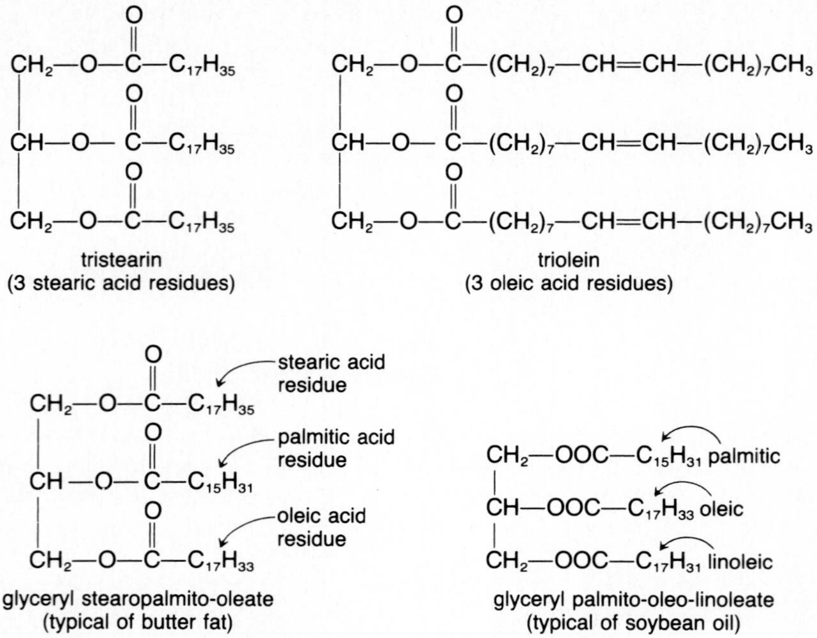
tristearin
(3 stearic acid residues)
triolein
(3 oleic acid residues)
stearic acid residue
palmitic acid residue
oleic acid residue
palmitic
oleic
linoleic
glyceryl stearopalmito-oleate
(typical of butter fat)
glyceryl palmito-oleo-linoleate
(typical of soybean oil)

FIGURE 13.1 The Chemical Nature of Some Fats.

Another 40 percent is oleic acid. Other well-represented human fatty acids are linoleic acid, stearic acid (C_{18}, saturated), and a C_{16} acid with one double bond, 9-hexadecenoic acid. Figures for the cow and goat are not much different from those for the human. Table 13.1 gives the saturated and unsaturated fat content of selected lipids.

Fats from vegetable sources often have lower melting points than animal fats and are typically liquid at room temperature. Hence we refer to them as vegetable oils. Their lower melting point is correlated to their degree of unsaturation: the greater the number of double bonds, the lower the melting point. Food chemists routinely hydrogenate liquid fats (such as cottonseed oil)

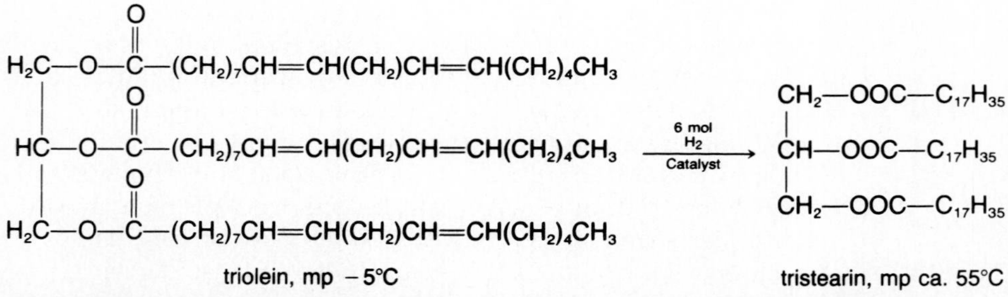
6 mol
H2
Catalyst
triolein, mp −5°C
tristearin, mp ca. 55°C

FIGURE 13.2 Complete Hydrogenation of a Liquid Fat Gives a Solid Fat.

Table 13.1. Saturated and Unsaturated Fat Content of Selected Lipids

Fat or oil	Total fat (g)	Total saturated fatty acids		Oleic acid		Linoleic acid	
		Amount in 100 g of edible portion (g)					
Coconut oil	32.2	28	(87%)	2	(6%)	Trace	
Butter	81	46	(57%)	27	(33%)	2	(2.5%)
Milk, cow (whole)	3.7	2	(54%)	1	(27%)	Trace	
Lard	100	38	(38%)	46	(46%)	10	(10%)
Eggs, raw (fresh)	11.5	4	(35%)	5	(43%)	1	(9.7%)
Margarine	81	18	(22%)	47	(58%)	14	(17%)
Peanuts, shelled	47.5	10	(21%)	20	(42%)	14	(29%)
Safflower oil	100	18	(18%)	15	(15%)	72	(72%)
Olive oil	100	11	(11%)	76	(76%)	7	(7%)
Corn oil	100	10	(10%)	28	(28%)	53	(53%)

Source: U.S. Department of Agriculture Handbook No. 8, *Composition of Foods* (Washington, D.C., U.S. Department of Agriculture, 1975).

to obtain the solid or semisolid products desired as margarine or cooking fat. The reduction of double bonds is only partial, for if it were complete a solid with too high a melting point would be obtained (Figure 13.2). Nonetheless, in preparing margarine from cottonseed oil, polyunsaturation disappears, and regrettably so, say many nutrition experts (see later in this chapter). To indicate how much unsaturation exists in a fat, the *iodine number* is used. The higher the iodine number, the greater the degree of unsaturation.

	Iodine Number
Butter	26–28
Lard	46–70
Peanut oil	84–102
Cottonseed oil	105–114
Safflower oil	140–156

The iodine number represents the number of grams of iodine (or equivalent) absorbed by the carbon-to-carbon double bonds in 100 g of a fat. The chemical reaction is of this type:

$$R-O-\overset{O}{\overset{\|}{C}}-(CH_2)_7CH=CH-(CH_2)_7-CH_3 + I_2 \rightarrow R-O-\overset{O}{\overset{\|}{C}}-(CH_2)_7-\overset{I}{\overset{|}{CH}}-\overset{I}{\overset{|}{CH}}-(CH_2)_7CH$$

Iodized oils, prepared in very much the same manner, have been injected into body areas as x-ray contrast media. Iodized poppyseed oil is used to visualize intradural (spinal cord) tumors.

Pure fats of the glyceryl ester type have no color, odor, or taste. Any pigment in them, as in butter, is due to extraneous matter. Current law permits a yellow dye to be incorporated into margarine before it is sold to the customer. Early laws required that the consumer add the dye himself after purchase (if he wanted yellow margarine). In the 1930s a yellow dye was proposed for margarine that would have been sold to millions of Americans. Very fortunately, use of the dye was stopped at the last minute after an examination revealed it to be one of the most potent carcinogens known. Our federal Food, Drug, and Cosmetic Act is designed to prevent just this kind of potential catastrophe.

Exposed to air, fats and oils will eventually become rancid as they undergo hydrolysis or oxidation to form products with unpleasant tastes or odors. For example, butter (which contains 3 to 4 percent butyric acid esters of glycerol) will eventually hydrolyze to yield free butyric acid (stench odor!). The $C{=}C$ double bonds in unsaturated fats can become oxidized by molecular oxygen to yield bad-smelling, short-chain fatty acids (butyric, caproic, caprylic), also contributing to the rancidity. Food chemists routinely add antioxidants such as butylated hydroxytoluene (BHT) or ascorbic acid (vitamin C) to prevent or delay development of rancidity by air oxidation of vegetable oils or prepared foods. Vitamin E is considered to be a natural antioxidant.

Medically Important Fats and Oils

Theobroma oil USP (cocoa butter, cacao butter) is the lipid obtained from the roasted seed of *Theobroma cacao*. Containing tristearin, tripalmitin, and triolein, this substance melts between 30° and 35°C (body temperature is 38°) and is therefore useful as a suppository base.

Castor oil USP comes from the seed of the castor bean. It is classified as a cathartic, dose 15 mL (1 tablespoonful). Castor oil is one of the few vegetable oils soluble in ethyl alcohol.

Cod liver oil NF is obtained from the livers of cod fish. It is a very good source of vitamins A and D and is used internally for that purpose. Cod liver oil is available in ointment and suppository form (as various "A and D" products), and is believed to be of value in treating burns and other injuries. Halibut liver oil is another source of vitamins A and D.

Peanut oil USP contains triolein, trilinolein, and other fatty acid esters of glycerin and is important because of its use as a solvent (vehicle) for long-acting drugs such as testosterone, given by deep intramuscular injection. Sesame oil is also used for this purpose. Oils are more slowly absorbed from the site of injection than aqueous solutions, and this delay is desirable for hormones and antibiotics the actions of which should be spread out over a long period. Penicillin oil for injection has lost favor, however, since it tends to cause fatty abscesses at the site of injection.

Olive oil is rich in triolein (70 percent) and is used as a food and in soap manufacture. Olive oil is useful as an emollient (skin softener).

Saponification of Fats and Oils

The Latin word for soap is *sapo*, and is the basis for our word for soap making, **saponification.** Modern soaps are made by alkaline hydrolysis (saponification) of fats and oils. Chemically, that can be represented as

$$
\begin{array}{c}
\underset{\text{fat}}{
\begin{array}{l}
CH_2\!-\!O\!-\!\overset{\displaystyle O}{\overset{\|}{C}}\!-\!R \\[4pt]
CH\!-\!O\!-\!\overset{\displaystyle O}{\overset{\|}{C}}\!-\!R \\[4pt]
CH_2\!-\!O\!-\!\overset{\displaystyle O}{\overset{\|}{C}}\!-\!R
\end{array}}
\quad 3\ KOH \quad \xrightarrow{\text{heat}} \quad
\underset{\text{glycerin}}{
\begin{array}{l}
CH_2OH \\[4pt]
CHOH \\[4pt]
CH_2OH
\end{array}}
\quad + \quad \underset{\text{soap}}{3\ R\!-\!COOK}
\end{array}
$$

In saponification, each molecule of triacylglycerol provides one molecule of glycerol and three molecules of the *potassium or sodium salt of a fatty acid.* That is our definition of a **soap.** It is restricted to K^+ or Na^+ because these salts are soluble. Calcium and magnesium salts of fatty acids are insoluble in water and would never function as cleansing agents.

Before the soap is compressed into a cake and sold, it is washed to remove the glycerol and the unreacted alkali. Air bubbles are incorporated into some soaps so that they will float. Perfume or bactericidal agents are added to others for sales appeal. These are all extraneous to the real function of the soap—to cleanse.

Soap cleansing action is really one of **emulsification.** An emulsion is a suspension of fine droplets of oil in water (o/w). Sometimes the opposite forms, water in oil (w/o). Emulsions must be stabilized so that the tiny oil droplets will not coalesce (clump together). Well-stabilized emulsions will last a long time before eventually "breaking," or separating into the two phases, oil and water.

Many agents can stabilize o/w emulsions. Soap is one of them. Let us consider a specific soap, sodium oleate (castile soap), and determine exactly how it acts to emulsify fat and, therefore, to cleanse. A nonpolar end and a polar end can be seen in the oleate ion:

$$
\underset{\text{nonpolar}}{\underbrace{C\!-\!C\!-\!C\!-\!C\!-\!C\!-\!C\!-\!C\!-\!C\!=\!C\!-\!C\!-\!C\!-\!C\!-\!C\!-\!C\!-\!C\!-\!C}}\!-\!\underset{\text{polar}}{\underbrace{COO^-}}
$$

Let us represent this bipolar ion by the symbol ⸺○ in which the circle is the COO^- group and the tail is the nonpolar hydrocarbon chain. As we wash something greasy, the force of our washing breaks up the oil or grease into droplets that tend to suspend in the wash water. As the droplets form,

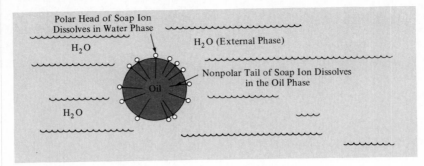

FIGURE 13.3 How Soaps Act to Stabilize Emulsions.

they dissolve the nonpolar end of the soap (like dissolves like), but reject the polar carboxylate anion (opposites repel). Therefore, in each tiny oil droplet— also termed a *micelle*—the structure looks like that in Figure 13.3. The lipophilic (lipid loving) end of the soap molecule has dissolved in the oil. The lipophobic (lipid rejecting) ionic end is sticking out in the water. In effect, this places a negative charge all over the surface of each oil droplet (Figure 13.4). Since like charges repel each other, the micelles will repel each other, and the emulsion will be stabilized. The oil droplets will remain suspended in the water and can be rinsed off and flushed away. Without the emulsifying agent, the cleaning action would be much less effective.

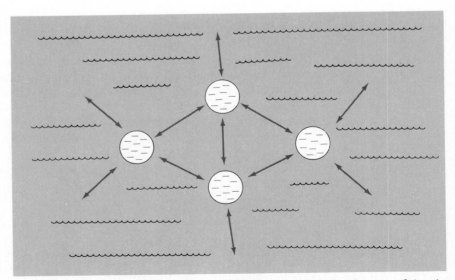

FIGURE 13.4 Negatively charged oil droplets repel each other; emulsion is stabilized.

Soaps are also surface active agents. As we learned in Chapter 7, surfactants lower the surface tension of water and increase its penetrating power. Thus soaps cleanse in two ways: by emulsifying oils and greases and by increasing the wetting or penetrating ability of the aqueous phase.

Hard water is "hard" partly because of its Ca(II) and Mg(II) ion content. In hard water, soaps lose their effectiveness because they are converted to water-insoluble Ca and Mg salts, which precipitate out of solution. This is what constitutes the ring around the bathtub.

Medicinally, *green soap* NF is a well-known potassium soap manufactured with the glycerin left in. It appears in green soap tincture NF and is used in the hospital to cleanse the skin of calloused tissue and in the treatment of skin diseases.

Sodium stearate NF, $CH_3(CH_2)_{16}COONa$, is an emulsifying and stiffening agent used in ointments, suppositories, creams, and cosmetics.

Detergents, though cleansing by a mechanism much like that of soap, have different chemical structures, as shown in I and II:

$$CH_3-(CH_2)_3-CH-(CH_2)_6CH_3$$

$$CH_3(CH_2)_{11}SO_3Na$$

$$SO_3Na$$

I
a detergent of the sodium alkysulfonate type

II
a detergent of the sodium alkylbenzenesulfonate type

Can you spot the lipophilic (grease-loving) and lipophobic (grease-hating) structures in formulas I and II?

One great advantage of detergents over soaps is that the Ca^{2+} and Mg^{2+} salts of the former are water soluble; hence there is no ring around the bathtub (i.e., no formation of insoluble salts in hard water) with detergents.

HUMAN FAT METABOLISM

Fat in the diet and fat in the body are frequent topics for discussion in our society. Obesity is a medical problem. It places the person at greater risk of cardiovascular disease and of diabetes. In the nonobese person, fat, especially saturated fat, is suspected of increasing the risk of vascular problems such as hardening of the arteries. It is worthwhile to take a look, therefore, at how the human body handles dietary fat and the theories of fat as a factor in disease.

If you are consuming 100 to 125 g of fat a day, you are about average, because the average American gets approximately 40 percent of his calories from fat. (West Germans get an average of 70 percent of their calories from fat.) Through the action of pancreatic lipase enzyme, your dietary triacylglycerols are hydrolyzed to glycerol, free fatty acids, monoacylglycerols and diacylglycerols.

```
G—O—fatty acid                    G—OH
L                                 L
Y                                 Y
C—O—fatty acid     lipase         C—O—fatty acid   +   fatty acids
E              ——————→            E
R                                 R
O                                 O
L—O—fatty acid                    L—OH

 triacylglycerol                  monoacylglycerol
```

In the small intestine, bile from the gallbladder helps emulsify these lipids into suspensions of fat micelles less than 100 Å (one-millionth of a centimeter) in diameter. In this size, the saponified fat can be absorbed from the lumen of the gut into cells within the intestinal wall. Here triacylglycerols are re-formed, coated with a layer of lipoprotein, transported by the lymph system to the blood and then to the liver.

Lipids in the blood are not transported in the free state. "Free" fatty acids (FFA) are always bound to albumin. And triacylglycerols and cholesterol are transported in the form of *lipoprotein complexes* such as *low-density lipoprotein (LDL)* and **high-density lipoprotein (HDL).** More will be said on this a little later.

We can summarize as follows:

1. Dietary fats are enzymatically hydrolyzed to fatty acid and glycerol.
2. Bile salts help emulsify the fat, forming micelles.
3. Micelles diffuse to cells in the intestinal wall.
4. Here, triacylglycerols are resynthesized and transported to the lymph system in combination with a lipoprotein.

After a heavy fat meal, the human lymphatics can be observed to be milky white with emulsified fat.

Fats can enter the circulation from another source: **depot fat.** Here we are referring to the estimated 1 *billion* lb of mainly excess lipid that saddles the United States population. Depot fat is an active tissue, undergoing continual breakdown to FFA and glycerol, and recombination to triacylglycerol. In depot fat, glucose can be metabolized and converted to lipid. In fact, it is believed that 30 percent of our carbohydrate intake is typically converted to adipose tissue. Protein, too, can end up as adipose tissue, provided it is not required somewhere else.

If we are fasting or engaged in prolonged exercise, depot fat is acting as the prime source of energy. Triacylglycerols are hydrolyzed and FFA are liberated in the bloodstream, along with glycerol. We saw in Chapter 12 that glycerol and fats can both be fed into the metabolic pathways of glucose; hence depot fat can serve as a source of heat and ATP energy.

Free fatty acids do not last long in the circulation; they are quickly metabolized in the liver, heart, brain, and in almost all other tissues. The normal fasting FFA range in humans is 190 to 420 mg per 100 mL. Levels consistently higher than this suggest to the clinician that the person may be at increased risk of cardiovascular disease.

Just how do free fatty acids get metabolized to ATP? The steps have all been worked out and the **beta-oxidation** pathway is summarized in Figure 13.5. It is obvious from this pathway that a fatty acid is metabolized piecemeal, with two-carbon fragments being chopped off one at a time:

$$CH_3CH_2 \!-\! CH_2CH_2 \!-\! CH_2CH_2 \!-\! CH_2CH_2 \!-\! CH_2COOH$$

They enter the Krebs cycle, which is also operating in the mitochondria. The beta-oxidation is actually a dehydrogenation followed by hydration and oxidation of the resultant 2° alcohol to a ketone. For each acetyl unit that is oxidized off of the fatty acid, approximately sixteen ATPs are produced, most via the Krebs cycle—quite a large gain, showing that fats are a very good source of energy.

ABNORMAL LIPID METABOLISM

In *diabetes mellitus*, inadequate insulin secretion (or inability to utilize what is secreted) results in diminished carbohydrate metabolism. The burden for calories then falls on lipids, and their catabolism is greatly increased. Ketone bodies (for structure, see Chapter 12), which are always produced in small amounts, now are produced faster than the tissues can handle them; they spill over into the urine and acetone is exhaled on the breath. The breakdown of fat in fat depots results in greatly increased FFA blood levels. Blood levels of triacylglycerols increase because of diminished removal of these compounds into fat depots. The net result is that the system becomes flooded with FFA and triacylglycerols. A ketoacidosis develops as ketone bodies such as β-ketobutyric *acid* and β-hydroxybutyric *acid* are formed in the liver and released to the circulation.

In the *alcoholic*, long-duration consumption of alcohol plus poor nutrition can damage the liver to the point where tissues die and scars form. This fibrotic change in liver structure is termed *cirrhosis*. Accompanying it is diminished liver function, including reduced capacity to handle the metabolism of lipids. The liver can become infiltrated with fat. Hence the term "fatty liver."

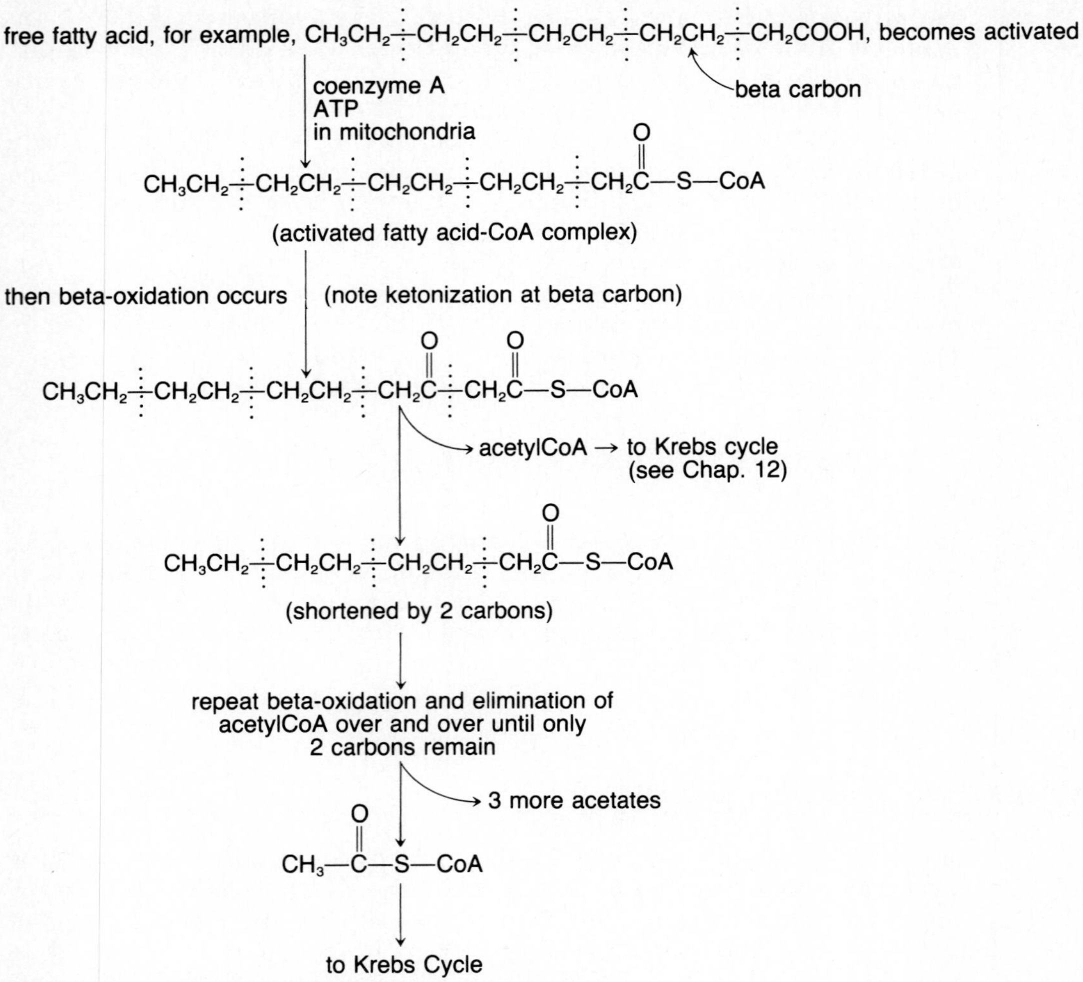

free fatty acid, for example, $CH_3CH_2\!\vdots\!CH_2CH_2\!\vdots\!CH_2CH_2\!\vdots\!CH_2CH_2\!\vdots\!CH_2COOH$, becomes activated

coenzyme A
ATP
in mitochondria

beta carbon

$$CH_3CH_2\!\vdots\!CH_2CH_2\!\vdots\!CH_2CH_2\!\vdots\!CH_2CH_2\!\vdots\!CH_2\overset{\displaystyle O}{\overset{\|}{C}}\!-\!S\!-\!CoA$$

(activated fatty acid-CoA complex)

then beta-oxidation occurs   (note ketonization at beta carbon)

$$CH_3CH_2\!\vdots\!CH_2CH_2\!\vdots\!CH_2CH_2\!\vdots\!CH_2\overset{\displaystyle O}{\overset{\|}{C}}\!\vdots\!CH_2\overset{\displaystyle O}{\overset{\|}{C}}\!-\!S\!-\!CoA$$

acetylCoA → to Krebs cycle
(see Chap. 12)

$$CH_3CH_2\!\vdots\!CH_2CH_2\!\vdots\!CH_2CH_2\!\vdots\!CH_2\overset{\displaystyle O}{\overset{\|}{C}}\!-\!S\!-\!CoA$$

(shortened by 2 carbons)

repeat beta-oxidation and elimination of
acetylCoA over and over until only
2 carbons remain

3 more acetates

$$CH_3\!-\!\overset{\displaystyle O}{\overset{\|}{C}}\!-\!S\!-\!CoA$$

to Krebs Cycle

FIGURE 13.5   Beta-Oxidation of Fatty Acids.

Inborn errors of metabolism, as the result of genetic defects, can give rise to abnormal lipid metabolism. One of the better known of these is **hypercholesterolemia,** or elevated serum cholesterol. The steroidal nature of cholesterol is discussed in Chapter 14. The normal fasting blood level of cholesterol is 120 to 220 mg %, or 3.10 to 5.69 mmol/L; however, 12-year-old children with this genetic defect can show serum cholesterol levels of 350 mg %. The prognosis for individuals suffering from hypercholesterolemia is poor. Atherosclerotic heart disease is an early manifestation. Another, similar, disease of lipid metabolism is hypertriacylglycerolemia (or, hypertriglyceridemia). In this common disorder, triacylglycerol levels in the serum are consistently elevated above the normal 40 to 150 mg per 100 mL and are associated with very low density lipoproteins. These individuals have a high predisposition to atherosclerosis.

CHOLESTEROL, HARDENING OF THE ARTERIES, AND HDLC

Postmortem examinations of what were considered healthy, physically fit American men have revealed shockingly advanced **atherosclerosis,**[1] that is, degenerative changes in the walls of arteries accompanied by hardening and deposition of fatty plaques, or, in other words, hardening of the arteries. The *plaque* (or deposit) in the arteries is found to contain cholesterol and other lipid material such as phospholipid, triacylglycerols, and carotene. In many cases, calcium salts deposit, making the plaque more brittle. This has the effect of making the artery less flexible—more rigid—which results in an increase in blood pressure. As the disease advances, the lumen of the artery narrows. In the coronary arteries, this can mean greatly reduced blood flow, diminished oxygen supply, and pain in the heart muscle. The name for this condition is *angina pectoris.* The danger of a blood clot increases as the plaque builds up (Figures 13.6 and 13.7). Sudden occlusion of the artery by clot formation is

[1] Arteriosclerosis is a generic term covering a variety of vascular diseases, of which the most important is atherosclerosis.

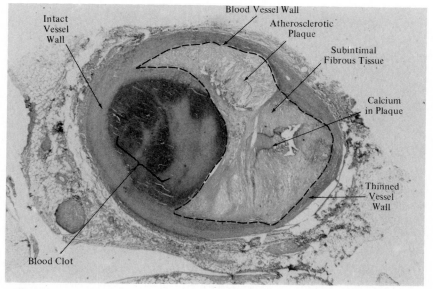

FIGURE 13.6 Cross-Section of Occluded Coronary Artery of M.I. Patient.

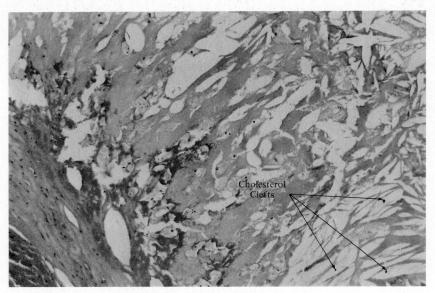

Cholesterol
Clefts

FIGURE 13.7   A Close-Up of the Plaque in Figure 13.6.

termed *myocardial infarct*, if it occurs in the heart. The lay term is *heart attack*. If it occurs in the brain, it is a stroke.

Arteries that are most likely to be damaged by atherosclerotic plaque are the coronaries, cerebral, and arteries of the lower extremities.

Cardiovascular disease now accounts for more than half of the yearly deaths in America. Coronary occlusion (myocardial infarct), cerebrovascular accidents (CVAs, stroke), and hypertension are all related to hardening of the arteries. Atherosclerosis is primarily a disease of the aged but can occur in young people, which suggests that genetic predisposition is important in some cases. A college woman told us that all of her male relatives above the age of 43 had died of heart attacks. The wipeout was so complete that a team of medical investigators had been sent to examine the survivors. Atherosclerosis is also a disease of the well-to-do nations.

At Framingham Union Hospital in Massachusetts, in 1948, the National Heart Institute began a long-range study of 5209 men and women to determine what effects, if any, diet, smoking, alcohol use, and a sedentary life had on the development of cardiovascular disease. From this study, which lasted 22 years (and followed 98 percent of the participants), and from others that lasted 6 and 7 years (see, "The Anticoronary Club, A Seven-Year Report," *American Journal of Public Health,* 56:226, 1966), plus studies of populations around the world, researchers reached the following conclusions:

1. Diet appears to play a role in the development of cardiovascular disease.
2. High intake of saturated fat is directly related to prevalence of coronary heart disease.
3. High risk populations tend to have higher cholesterol plasma levels.
4. Increased cholesterol and saturated triacylglycerols in the diet cause an increase of the same in the blood; and the substitution of polyunsaturated for saturated fat reduced plasma cholesterol levels that are elevated.
5. Reduction of dietary cholesterol reduces plasma cholesterol levels.
6. Cardiovascular disease is directly related to the number of cigarettes smoked daily.

As a consequence of the Framingham and other studies, a national campaign was begun to educate the American people to avoid foods high in cholesterol and saturated fat. Butter, cream, lard, and fatty meats were out; margarine and fish were in. Egg yolks with their relatively high cholesterol content were viewed with alarm; soybean with zero cholesterol was encouraged (cholesterol is associated with animal products, not with vegetable products). Saturated fats, that is, triacylglycerols having fatty acid residues with no C$=$C double bonds, were considered risky as they predisposed to plaque formation. Polyunsaturated oils such as safflower, corn, and soybean retarded plaque formation.

On this basis, physicians counseled their patients, some of whom changed life-long dietary habits. The American dairy association and the egg lobby fought back with ads proclaiming the healthfulness of their products. Researchers countered with reports of low rates of heart disease in Japan, where fat consumption is only half of what it is in America.

There are opponents to the theory that dietary cholesterol and saturated fat lead to atherosclerosis. A June 1980 report issued by the National Research Council's Food and Nutrition Board challenged the idea that eating less cholesterol and saturated fat is good for your health. The report questions the scientific validity of advising healthy people to reduce consumption of these foods. The National Research Council is a private organization, and the report we are referring to here was financed by the food industry.

In January of 1981 scientists at Chicago's Rush-Presbyterian-St. Luke's medical center published the results of a 20-year study of the diets of 1900 middle-aged American men. They concluded that men who had consumed large amounts of foods high in cholesterol and saturated fat suffered upwards of a third more deaths from heart disease than men who consumed relatively small amounts (*New England Journal of Medicine*, January 8, 1981). A similar conclusion was reached in 1982 from the results of a Norwegian study of 1200 healthy men who had high blood levels of cholesterol and who faced a high risk of heart disease. Their regular diets were high in saturated fats and in cholesterol (butter, sausage, high-fat cheese, eggs, and whole milk). Half of the men were induced to stop smoking and to follow a cholesterol-lowering diet. In this group, after 5 years, there was a 47 percent lower rate of heart attacks and sudden deaths than in the control group that did not change dietary habits. Those men who experienced the greatest drop in blood cholesterol levels were those who had adhered most closely to the dietary recommendations. Further, triacylglycerol levels in these men had dropped substantially and their blood levels of high-density lipoprotein-cholesterol (see later) had risen. Many researchers have concluded from this study that heart attacks and diet are persuasively linked.

As it now stands, there is a lot of statistical evidence implicating dietary cholesterol and saturated fat in the etiology of atherosclerosis. When these dietary factors are reduced, and when polyunsaturated fat replaces saturated fat, the progress of atherosclerosis is halted. But scientific proof is still lacking that low-cholesterol/low-saturated-fat diets will *prevent* the occurrence of hardening of the arteries in healthy individuals. For that proof, we will need the results of large-scale studies now under way.

It is interesting that whatever Americans are doing in response to the threat of atherosclerosis, morbidity and mortality are decreasing. Figure 13.8 shows that cardiovascular disease and stroke in the United States population have declined rather dramatically in the decade of the seventies. Based on the

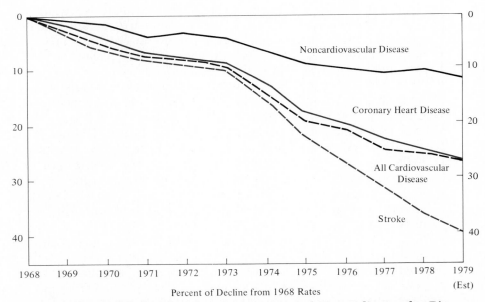

Percent of Decline from 1968 Rates

FIGURE 13.8   **Trends in Cardiovascular Disease and Noncardiovascular Disease: Decline by Age-Adjusted Death Rates, 1968–1979** (*Source: National Heart, Lung and Blood Institute's Fact Book for Fiscal Year, 1980,* U.S. Department of Health and Human Services).

1980 census, the Metropolitan Life Insurance Company has calculated that the average life expectancy for persons in the United States born in 1981 reached an all-time high of 70.3 years for men and 77.9 years for women.

It is important to note that many other theories have been proposed to explain the cause of atherosclerosis. An enzyme in cow's milk (bovine milk xanthine oxidase) has been blamed as the agent that initiates plaque formation. The herpes virus has been called the cause; another theory proposes that protein is the villain in combination with a lack of vitamin B6 (destroyed by overcooking foods). The most recent and perhaps the most cogent of the alternative theories involves HDLC.

HDL stands for *high-density lipoprotein.* You will recall from our earlier discussion that a combination of lipid plus protein is involved in transport of fat in the blood. These lipoproteins come in different molecular weights and in different lipid contents. This gives them different densities in the ultracentrifuge. Lipoproteins containing the most lipid and the least protein are least dense; those containing the least lipid and the most protein are most dense. There are, accordingly

1. Very low density lipoproteins (VLDLs).
2. Low-density lipoproteins (LDLs).
3. High-density lipoproteins (HDLs).

The HDLs consist of 50 percent protein, twice that of LDLs. The lipoproteins, especially HDLs, bind and transport blood lipids such as cholesterol from sites in the tissues and arterial walls to the liver where they can be metabolized and eliminated. The combination of HDL and cholesterol is abbreviated *HDLC.* Evidence is accumulating that the prime factor in regulating atherosclerosis is not dietary lipid intake, but how much circulating HDL a person has. If HDLC levels are high, cholesterol will be transported and metabolized effectively; it may even be removed from its deposit in a plaque. But if LDLC predominates, with its poor ability to transport cholesterol (and its poor ability to activate lipoprotein lipase), atherosclerosis will not be halted (Figure 13.9). One researcher has said that the single most powerful predictor of coronary heart disease in people over 50 is HDL. Risk can be determined by comparing the cholesterol in HDLC to the total cholesterol. The more present as HDLC, he says, the less the risk of plaque formation. Low serum HDL is a common antecedent to coronary heart disease, according to this theory.

Additional research reports tend to substantiate the HDLC theory. Indeed, some groups have found no correlation between how much cholesterol there is in an animal's diet and how much there is in his circulation. Also, we know that the body synthesizes its own cholesterol—1.0 to 1.5 g per day, in the process of making steroid hormones and bile salts.

Well, then, how does one get and maintain lots of HDLC? The answer appears to be threefold: (1) be born a woman, (2) exercise diligently, and (3)

FIGURE 13.9   One new theory of the cause and prevention of hardening of the arteries suggests that cholesterol is transported to the liver (for elimination) by high-density lipoprotein (HDL), and that low-density lipoprotein (LDL) and very low density lipoprotein (VLDL) are ineffective in such transport. Thus the higher the serum HDL levels, the less chance for cholesterol and plaque buildup.

do not smoke. Women generally have higher HDL levels than men. Numerous studies show an apparent direct relationship between serious exercise and increased HDL levels. However, a University of Wisconsin study (1980) found that at least 4 weeks of sustained endurance exercise is necessary to get a significant increase in the level of HDL. For example, sixteen college students averaged 54 mg % of HDLC at the start of their program, 52 mg % after 2 weeks, 55 mg % at 4 weeks, 56 mg % at 6 weeks, and 58 mg % after 8 weeks. Figure 13.10 shows the relationship between different types of physical activity and serum levels of HDLC. Cigarette smoking has been associated with morbidity for a long time and in many ways, and we can now add another debit to its ledger. Smoking diminishes circulating levels of HDL. Alcohol, incidentally, tends to increase HDL (only *moderate* amounts are needed for this effect). The average (normal) man in the United States has approximately 45 mg % of HDL in his serum; he has a one-in-four chance of developing coronary heart disease by age 60. Some authorities believe that a 45 mg % level of HDLC is not a healthy level.

We emphasize that the HDL-lowered cholesterol-reduced-atherosclerosis concept is still theoretical. Much more remains to be discovered about lipoproteins themselves and about their biochemical roles. This discussion has omitted many of the details of the roles played by enzymes and by chemical intermediates. Nonetheless, HDL has many believers and research on it continues.

Summary

Atherosclerosis is a killer disease. We Americans appear to be especially vulnerable to it. One school of thought points the accusing finger at dietary cholesterol and dietary saturated fat, and urges us to reduce our intake of these factors and of calories in general. A second school claims that diet has little to do with how much plaque is deposited in an artery; they claim the key role is the protective role played by high-density lipoprotein (HDL). To build up effective serum HDL levels, it is important that we exercise seriously and eliminate smoking.

Eicosapentaenoic Acid

Greenland eskimos have superb cardiovascular health (only a 3.5 percent annual mortality from heart disease), in spite of the fact that they have the highest fat intake in the world. A major component of the eskimo's diet of seal and fish is eicosapentaenoic acid (EPA), a twenty-carbon penta-unsaturated fatty acid found in fish oil. EPA serves as a precursor for body synthesis of prostacyclin, a powerful vasodilator and inhibitor of platelet aggregation. Platelet clumping is, of course, a major factor in coronary heart disease.

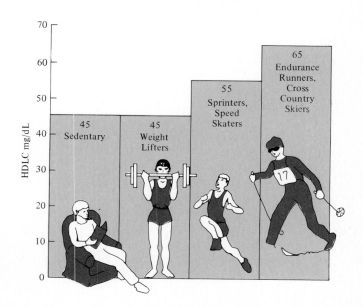

FIGURE 13.10 Exercise and HDLC. **Higher HDLC levels are associated with exercises in which the cardiovascular system is seriously challenged.**

A dietary supplement called MaxEPA, available commercially, contains the omega-3-polyunsaturated fatty acids, including EPA, which are obtained only from fish and marine predators who live on fish. Like fish, MaxEPA reportedly elevates HDL and reduces triacylglycerols, cholesterol, and the tendency for blood to clot.

OBESITY AND HEART DISEASE

Health professionals may have the opportunity to counsel the obese patient on the risks associated with the overweight condition. The Framingham study showed that the risk of heart disease increases with body weight even if no increase in blood pressure is initially evident. Dr. S. M. Grundy of the University of Texas Health Center at Dallas states, "(obesity) should probably take its place with other big factors" in causing heart disease. He adds, "obesity (also) brings out a tendency for diabetes." The actual mechanism of action by which obesity contributes to heart disease is not understood.

DEPOT FAT IN DRUG STORAGE

According to Theodore B. Van Itallie, special adviser to the surgeon general on human nutrition, at the time of our Civil War the average weight of a 5-ft-8-in. man between 30 and 34 years was 137 lb. Today, with muscular exertion decreased and with food more plentiful, the average 5-ft-8-in. man between 30 and 34 is likely to weight 170 pounds. And among Americans between 20 and 74 years, 14 percent of men and 23.8 percent of women are 20 percent or more over their optimum weights. About 5 percent of men and 7 percent of women are severely obese. The news media recently reported on a hospitalized man who once weighed 1400 lb and who had gained as much as 200 lb in 1 week. During his hospital stay, this man was limited to 1200 calories per day, and lost 900 lb!

If you are a college-aged male, your body probably consists of 15 percent fat. College females average 25 percent fat, whereas marathon runners can get down to 5 or 6 percent fat. In one study of twenty obese people, the mass of body fat ranged from 92 to 227 lb. The average American eats 115 lb of fat a year—40 percent in the saturated form.

Putting all of this together indicates there is a great deal of depot fat in humans, more in some, less in others. Many commonly used drugs, prescription and recreational, are essentially nonpolar and will dissolve in nonpolar solvents such as depot fat. Marijuana is one of these. Psychiatrist D. H. Powelson used radioactive tracer techniques to follow a dose of marijuana in human volunteers. He concluded that only half a dose of pot leaves the body after 1 week (marijuana is 600 times more soluble in body fat than in water). Barbiturates are another example of high-fat-soluble/low-water-soluble drugs. The *half-life* of a drug is the time it takes for the body to get rid of 50 percent of an ingested dose. The half-life for the oft-used barbiturate Seconal is more than 20 hours. Ether, after its use as an anesthetic, remains dissolved in fat depots for a considerable period of time; its fat/water solubility ratio is at least 12:1.

The point of this discussion is that drug life, and therefore drug activity, is governed partly by the patient's physical composition, and this should be taken into account in selection of dose size and frequency of administration.

There is some evidence to implicate body fat as a reservoir for hormones that have been found to be carcinogenic.

Another "depot" for drugs in the body is protein. Serum albumin binds many drugs, prolonging their half-lives.

PHOSPHOLIPIDS

The **phospholipids** are a diverse category of cell constituents that are found especially in the nervous system and that have phosphorus in their chemical makeup. They can be obtained from egg yolk, brain, certain seeds, as well as liver, heart mucle, and red blood cells. Their main role is in the formation of cell membranes, but they also are involved in fat transport.

All phospholipids are derivatives of glycerol—with two of the glycerol's hydroxyls esterified with fatty acid and the third esterified with phosphoric acid (Figure 13.11). Variations on this structure occur in the nature of the three R groups. When R''' is choline, $HOCH_2CH_2\overset{\oplus}{N}-(CH_3)_3$, the compound is a **lecithin** (Figure 13.12). Variations in the fatty acids (R' and R'') give rise to a variety of lecithins.

Other alcohols can be esterified and occupy the R''' position. Ethanolamine ($HOCH_2CH_2NH_2$) and serine ($HOCH_2CH(NH_2)COOH$) in the R''' position give phospholipids known as *cephalins*.

The lecithins take part in transport of fatty acids and triacylglycerols in the human. You can see from structure 13.12 that the lecithins carry a + and a − charge (actually in a dipolar ion arrangement) and also a long, nonpolar

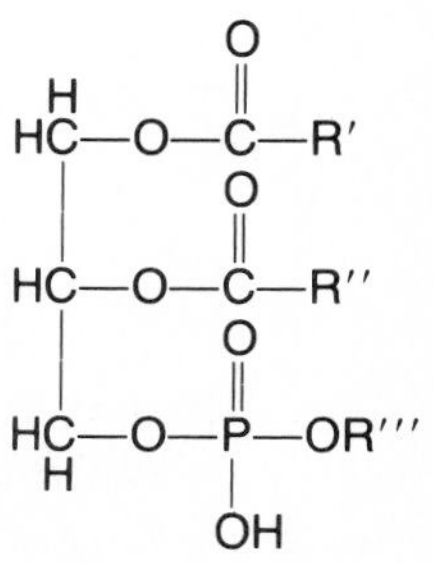

FIGURE 13.11 The Structure Common to all Phospholipids.

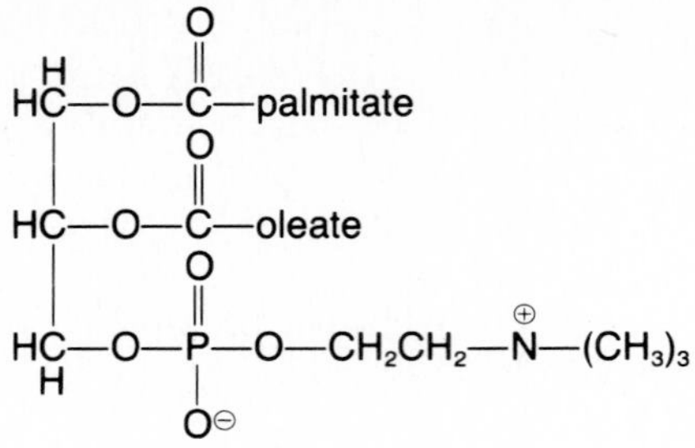

$$
\begin{array}{l}
\overset{\text{H}}{\text{H}}\text{C}-\text{O}-\overset{\overset{\displaystyle O}{\|}}{\text{C}}-\text{palmitate}\\[6pt]
\text{HC}-\text{O}-\overset{\overset{\displaystyle O}{\|}}{\text{C}}-\text{oleate}\\[6pt]
\overset{}{\underset{\text{H}}{\text{HC}}}-\text{O}-\overset{\overset{}{}}{\underset{\underset{\displaystyle O^{\ominus}}{|}}{\text{P}}}-\text{O}-\text{CH}_2\text{CH}_2-\overset{\displaystyle\oplus}{\text{N}}-(\text{CH}_3)_3
\end{array}
$$

FIGURE 13.12   Choline, a Lecithin.

fatty acid ester "tail." This makes them good emulsifying agents, and they are so used in the food industry. Lecithins are used in vegetable cooking sprays that are meant to keep pans from sticking. Food faddists maintain that lecithins are effective cholesterol reducers and antihyperlipidemics, and lecithins are sold as such in health food stores. Nationally known nutritionists Frederick Stare and Margaret McWilliams have rejected this view in their book, *Living Nutrition* (1973, p. 267).

SPHINGOLIPIDS

In **sphingolipids,** the glycerol is replaced by a long-chain aminoalcohol called sphingosine. To it are bound a fatty acid and choline, the former in an amide bond and the latter as a phosphoryl ester:

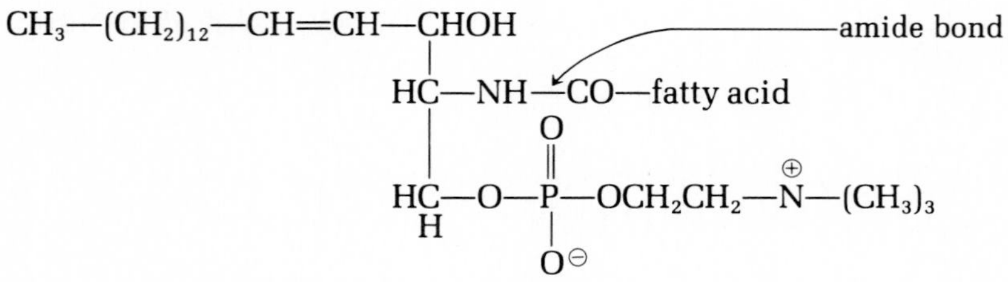

$$
\text{CH}_3-(\text{CH}_2)_{12}-\text{CH}=\text{CH}-\text{CHOH} \qquad\qquad\qquad\text{amide bond}
$$

$$
\text{HC}-\text{NH}-\text{CO}-\text{fatty acid}
$$

$$
\overset{}{\underset{\text{H}}{\text{HC}}}-\text{O}-\overset{\overset{\displaystyle O}{\|}}{\underset{\underset{\displaystyle O^{\ominus}}{|}}{\text{P}}}-\text{OCH}_2\text{CH}_2-\overset{\displaystyle\oplus}{\text{N}}-(\text{CH}_3)_3
$$

sphingomyelin, a sphingosine-type lipid

Sphingolipids are found in cell membranes and in large amounts in brain and nerve tissue. In Nieman-Pick disease, appearing chiefly in Jewish families, an insufficiency of the enzyme that catalyzes the hydrolysis of sphingomyelin results in a buildup of this compound in the brain, with subsequent mental retardation. Nieman-Pick disease is another example of an inborn error of metabolism, also called a "familial disorder."

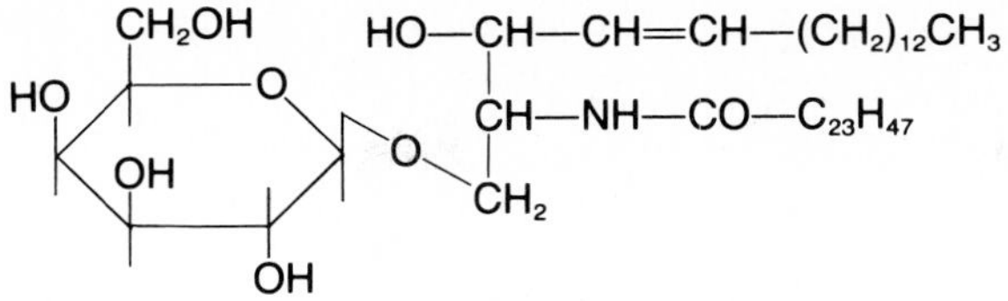

FIGURE 13.13 Typical Cerebroside Structure.

GLYCOLIPIDS

The glycolipids (Greek *glykys* = sweet) are carbohydrate–lipid combinations. Each contains a sugar, often galactose, but also glucose or inositol. Phosphate may or may not be present and glycerol or sphingosine may occur. The cerebrosides are brain glycolipids similar to sphingosine except that the fatty acids they contain are up to twenty-four carbons long and there is no phosphate. Figure 13.13 gives the structure of a typical cerebroside.

▰▰▰▰▰ **EXAMPLE 13.1** ▰▰▰▰▰

Complete hydrolysis of the cerebroside in Figure 13.13 yields three products. Identify them.

Solution
β-D-galactose, a twenty-four-carbon fatty acid (tetracosanoic acid), and sphingosine (2-amino-4-octadecen-1,2-diol).

WAXES

The functional group in **waxes** is the ester group, just as in fats. The acid residues in waxes are long-chain fatty acids, just as in fats. However, the alcohol residues in waxes are not glycerol, but long-chain alcohols called *fatty alcohols*. Beeswax is an insect wax mixture one component of which is shown in Figure 13.14. We can see that the fatty alcohols in waxes are long chain, indeed. In some cases, alcohols are found up to thirty-four carbons in length.

Anhydrous lanolin USP (wool fat), obtained from the wool of sheep, is a very complex mixture of dozens of fatty acids and fatty alcohols in ester

$$CH_{15}H_{31}—COO—C_{30}H_{61}$$
palmitic acid myricyl alcohol   **FIGURE 13.14** An Ester Component of Beeswax.

linkage. Some of the alcohols in lanolin are sterols, which we will discuss in Chapter 14. Lanolin USP is wool fat containing 25 to 30 percent of water. Lanolin has a great capacity for absorbing water and is used as an ointment base. It is stable and tenacious.

Waxes are useful in the form of candles, polishes, lubricants, and ointments.

SUMMARY

Fats, along with fatty acids, oils, waxes, phopholipids, sphingolipids, and sterols, are examples of the general category of lipids. Lipids are low-polarity, greasy, water-insoluble compounds that tend to dissolve in low-polarity solvents like ether.

Chemically, fats and oils are esters formed between glycerol and fatty acids. If the fatty acid portion has one or more C=C double bonds, the fat is said to be unsaturated.

In soap making (saponification), a fat is hydrolyzed in the presence of KOH or NaOH to yield glycerol and a K- or Na-salt of a fatty acid (or mixture of fatty acids). Soaps possess a hydrophilic "head" and a hydrophobic "tail" and can act as surfactants to help sabilize oil-in-water emulsions.

In fat metabolism, dietary fat is enzymatically cleaved to free fatty acid and glycerol. Bile aids in the process by emulsifying the fat into very tiny globules called micelles.

Free fatty acids are metabolized in almost all body tissues to heat energy and ATP. This process is termed beta-oxidation; it was summarized in this chapter.

The subject of atherosclerosis (hardderning of the arteries) and the role of dietary factors in it were discussed in this chapter. Evidence was presented to support the theory that a diet high in saturated fat and cholesterol contributes to early atherosclerosis, but this theory has been discredited by other authorities. A second theory was presented, namely, that the key factor in preventing atherosclerosis is HDLC (high-density lipoprotein-cholesterol). High blood levels of HDLC are believed to prevent or retard plaque buildup. We made the point that all of these theories are based on statistical information and are not considered to be the final word on the subject. Indeed, there is still much disagreement.

The role of depot fat in drug storage was discussed. Here, the concept of half-life was introduced.

We concluded the chapter with definitions of four additional classes of lipids: phospholipids, sphingolipids, glycolipids, and waxes.

KEY TERMS

Check your understanding of this chapter. Can you exlain what is meant by each of these terms?

atherosclerosis
beta-oxidation
depot fat
detergent
emulsification
fat
high-density lipoprotein (HDL)
hypercholesterolemia
lecithin

lipid
oil
phospholipid
saponification
soap
sphingolipid
triacylglycerol
waxes

STUDY QUESTIONS

1. What is the chemical difference between a fatty acid, a fat, and an oil?

2. What functions do fats serve in the body?

3. Draw the general type of structure of a triacylglycerol. Identify the ester bonds and the fatty acid residues.

4. Draw the structure of the human fat that contains one palmitic acid, one oleic acid, and one linoleic acid residue.

5. From Table 13.1, find the vegetable product and the animal product that have the highest percentage of saturated fat. What oil has the highest percentage of linoleic acid?

6. Chemically, what is the difference between a saturated fat and an unsaturated fat?

7. Write and balance the equation that represents the lipase-catalyzed hydrolysis of triolein, as it might occur in the small intestine of a human.

8. Hydrogenation of a fat or oil involves chemical alteration of a functional group; so does determining the iodine number. State which functional group is involved and how it is altered in each case.

9. Why would an oil (such as Theobroma oil) that melts at 35°C make an especially suitable suppository base?

10. Define the following:
    a. Lipid   b. Soap
    c. Green soap   d. Micelle
    e. Lipophilic   f. Emulsification
    g. Depot fat   h. Half-life (of a drug)

11. Often it is desirable for a hormone like testosterone or an antibiotic like penicillin to be absorbed very slowly from an injection site so that the drug effect is prolonged. Which of the following factors would facilitate a slow absorption and therefore prolonged action?

a. IV injection.   **b.** Subcutaneous injection.
c. Deep IM injection.   **d.** Injection of a suspension.
e. Injection of a solution.   **f.** Use of an aqueous solvent.
g. Use of an oil solvent.

12. Write and balance the chemical equation to represent the saponification of tristearin using KOH(aq).

13. Explain how soaps work to remove grease and oils.

14. How do free fatty acids get into the bloodstream? (*Hint:* Consider diet and depot fat.)

15. Lipoproteins play important roles in human biochemistry. What are lipoproteins, what are some of their types, and how do they function in fat metabolism?

16. In the mitochondria, fats are metabolized by beta-oxidation. What is the length of the carbon fragments that are serially oxidized off? Into what cycle do they enter for further oxidation? How many of these units would be provided by one molecule of stearic acid?

17. Hypercholesterolemia can be the result of an inborn error of metabolism. What is the normal blood level of cholesterol, and what would be an example of an elevated blood cholesterol?

18. What pathological changes occur in atherosclerosis?

19. There is a school of thought that advocates dietary control of impending atherosclerosis. What dietary factors are to be avoided in this approach? Be specific.

20. What are LDL, HDL, and HDLC? How do some believe that HDL is involved in the prevention of hardening of the arteries?

21. **a.** Write the chemical structure that is common to all phospholipids.
    b. Write the structure that is common to all lecithins.

22. Like fats, waxes are esters of long-chain fatty acids. But what significant chemical difference exists between them which warrants placing waxes in their own category?

ADVANCED STUDY QUESTION

23. Write a chemical scheme showing all of the chemical intermediates, enzymes, transport mechanisms between an ingested fat and its eventual deposit in depot fat.

14 | Steroids

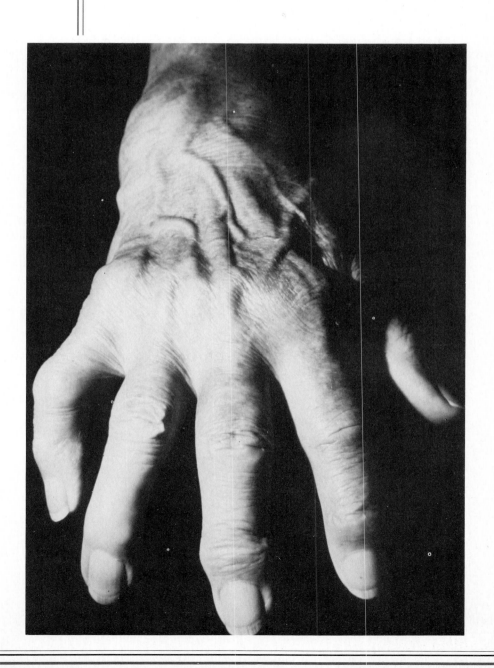

T he patient whose hand is shown in our photo is suffering from rheumatoid arthritis, an inflammatory disease that involves body joints (especially wrist, fingers, feet) and that causes changes in joint fluid, cartilage, and skeletal muscle. Women comprise 75 percent of rheumatoid arthritis patients, and in 80 percent of the cases, onset is before 40 years of age.

Of the half-dozen different chemotherapeutic approaches to the treatment of rheumatoid arthritis, steroid-type drugs are the most effective as regards anti-inflammatory activity. But anti-inflammatory activity in arthritis is only one of a host of important actions we ascribe to steroids. Indeed, there are so many vital uses of steroids that we have devoted this entire chapter to them.

When we discuss cholesterol, vitamin D precursors, and the sex hormones, we are considering steroid-type compounds. The steroid nucleus is also found in bile acids, in hormones from the adrenal cortex, and in the heart stimulant drug digitalis.

With a knowledge of steroid chemistry, organic chemists have synthesized a great many new steroid compounds; some are used as anti-inflammatory agents and certain others in birth control pills and as anabolic steroids. Steroidal drugs generally are potent substances, active in milligram quantities. This is a reflection of their relationship to the hormones, which we recognize to be highly active in tiny doses.

INTRODUCTION

Nature has produced a very interesting, unique hydrocarbon with four carbon rings and has chosen it to be the nucleus for some of the most potent and biologically significant molecules in the human body. After all, what could be more important than the sex hormones, the adrenal cortex hormones, or cholesterol!

We call this four-ring system the **steroid** *nucleus,* and it looks like this:

or in shorthand

steroid nucleus

You can see that each carbon has a number—very important for locating substituents in specific compounds. And each ring has a letter designation, A, B, C, or D.

Here's a little exercise to test your knowledge of how carbon rings are joined to each other (and to help you learn the steroid nucleus better). Which of the following four structures is or are identical to the steroid nucleus?

I

II

III

IV

As you work on this exercise, remember that although we draw structures a certain way on paper, in reality molecules are free to move about and assume any orientation in space. It will also help you to letter structures I through IV as we did in the previous steroid nucleus; lettering will show you how ring C is fused to ring D. The answer: structures I and III are the steroid nucleus drawn in a slightly different way.

In this chapter we are going to examine steroids that play biological roles in humans—including those that regulate salt balance and act as anti-inflammatory agents, sex hormones, bile salts, and contraceptives. Those steroids that are alcohols (such as choles*terol*) are sometimes termed *sterols*. We owe much of our knowledge of steroids to the great biochemists of the 1930s era, including Kendall, Windaus, Allen, Doisy, and Reichstein. More recently, Carl Djerassi has made immense contributions to the steroid field.

In the last chapter we classified steroids as lipids, and indeed they are. However, they are nonsaponifiable lipids—one major difference from the triacylglycerols.

The hundreds of steroids that are known differ from each other in the kind and position of substituents on the steroid nucleus. Common substituents are hydroxy (both alcoholic and phenolic) and keto. Unsaturation is frequently encountered; alkyl groups are found on carbon number 17. Figure 14.1 shows a make-believe steroid that illustrates typical substitution.

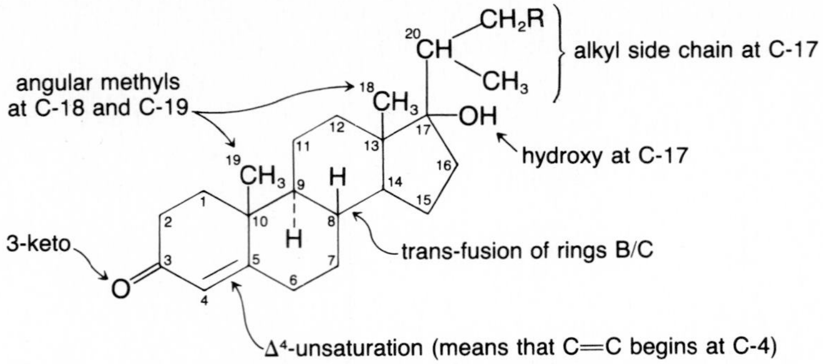

alkyl side chain at C-17
angular methyls at C-18 and C-19
hydroxy at C-17
3-keto
trans-fusion of rings B/C
Δ⁴-unsaturation (means that C=C begins at C-4)

FIGURE 14.1 An Imaginary Steroid Illustrating Substitution.

CLASSIFICATION AND FUNCTION OF STEROIDS

The *adrenal corticosteroids* are secreted from the adrenal cortex. To put this in anatomical perspective, Figure 14.2 pictures an adrenal gland perched atop a kidney and distinguishes between the adrenal medulla and the adrenal cortex.

If we surgically remove an animal's adrenal glands, the animal will die in a few days. In the human, diminished adrenal cortex activity is termed *Addison's disease* (discovered by Thomas Addison in 1855), and is characterized by disturbance in the levels of K^+, Na^+, Cl^-, and water, by changes in carbohydrate metabolism, and by reduced ability to handle stress. This tells us a lot about the functions of the adrenal cortex hormones (of which over fifty have been isolated crystalline).

There are five principal, naturally occurring **adrenocorticosteroids:** hy-

Adrenal Gland
Medulla
Cortex
Adrenal Gland
Kidney
Abdominal Aorta

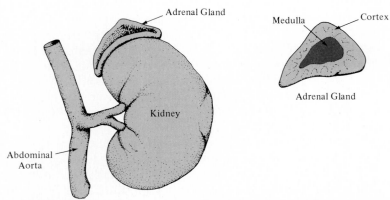

FIGURE 14.2 The Anatomical Location and Composition of the Adrenal Gland.

Table 14.1. Structures and Roles of Important Corticosteroids

Name	Structure	Biological roles in human
Hydrocortisone USP (cortisol)		A *glucocorticoid*. Controls glucose production in the liver; increases glycogen stores and elevates blood sugar levels. Also controls electrolyte balance but not as powerfully as aldosterone. Has anti-inflammatory activity in trauma, asthma, allergy, arthritis, etc.
Aldosterone		The major *mineralocorticoid*; regulates Na^+ absorption and excretion in the kidney; thus it acts to retain Na^+ and conserve water.

drocortisone, corticosterone, aldosterone, deoxycorticosterone, and dehydro-epiandrosterone. Table 14.1 summarizes the structures and biological roles of two representative adrenocorticosteroids. Carefully read the third column in the table and note the distinction between a *glucocorticoid* and *mineralo-corticoid*.

When hydrocortisone is used to control inflammation in arthritis, it has the undesirable side effect of upsetting Na^+, K^+, and water balance. Recognizing this, pharmaceutical chemists designed and synthesized chemical analogs of hydrocortisone that have an increased anti-inflammatory activity but little or no effect on electrolyte and water balance. Some of these synthetic analogs have become quite well known: prednisone, prednisolone, methylprednisolone, triamcinolone, dexamethasone, and betamethasone (all generic names).

=========== **EXAMPLE 14.1** ===========

What are Kenalog and Aristocort and how are they used medicinally? (*Hint:* Check your library for a reference book called the *Physician's Desk Reference.*)

Solution

These two trade products have the generic name triamcinolone. They are synthetic anti-inflammatory steroids, injected into bone joints in 2.5- to 15-mg doses (for arthritis) and used externally in ointments, creams, or lotions for dermatitis, psoriasis, allergic conditions, and so on.

Adrenocorticosteroids occur naturally in the adrenal cortex and are also available synthetically. The naturally occurring steroids perform a vital role in regulating electrolyte and water balance and in helping us cope with stress. The synthetic analogs are very useful especially as anti-inflammatory drugs.

The *sex hormones* can be classified as **estrogens** (female sex hormones) and **androgens** (male sex hormones). All of the secondary sex characteristics that accompany the change from a prepubertal boy or girl to a postadolescent man or woman are regulated in large measure by the sex hormones. These substances are powerfully effective in microgram amounts.

Three female sex hormones are recognized in the human: **estradiol, estriol,** and **estrone:**

estradiol   estrone   estriol

The three estrogens are chemically unique because ring A is aromatic and that makes the OH group on C-3 phenolic instead of alcoholic. Equine (horse) estrogens are chemically similar to the human except that they have extra unsaturation in the 6,7- and 8,9- positions. Conjugated estrogens USP (Premarin) is a very widely prescribed product that is obtained from pregnant mares' urine. Human pregnancy urine also contains large quantities of estrogenic material.

Estradiol is the primary estrogen in the human and is produced in the ovaries. At menopause, ovarian secretory function slows and then ceases, and estrogen production diminishes. The signs and symptoms of the menopause are too many to list here, but they can be profound. Ovariectomy (*ectomy* = surgical removal) is a form of surgical menopause. Males secrete small quantities of estrogen, too, and it can be detected in their blood. But the great preponderance of androgen in males produces an overriding masculinity.

Estrogens maintain the physical and emotional qualities of femininity. Their presence in the circulation helps rebuild the uterine endometrium (lining of the womb) after menstruation and stimulates the growth of the ovarian follicle. Estrogens maintain the life of the corpus luteum; they stimulate uterine growth, activity, and excitability. Estrous in animals is dependent on estrogen; in humans, estrogens increase the libido. Breast enlargement at puberty is effected principally by estrogen. Dermatologists believe that estrogens are valuable in the treatment of acne. Estrogens exhibit a significant plasma cholesterol-lowering activity; they diminish FSH secretion and can act to inhibit ovulation. Estrogen use in the Pill is discussed later in this chapter.

Estrogens are heavily prescribed to American women. In 1977, approximately 5 percent of women in the United States, or 4.1 million, were taking estrogen therapy, many for the treatment of the menopause. It is estimated that more than one-fifth of the women reaching the age of 60 have taken the drug, many in the hope of retaining youthfulness. They should be aware of the risks of estrogen therapy as well as the benefits. After a 4-year study involving 1339 women, University of Pennsylvania School of Medicine researchers concluded that women who use estrogen for 5 years or more are fifteen times more likely to develop uterine cancer than nonusers of estrogen.

EXAMPLE 14.2

We know that the prostate gland in the male develops in response to androgens. What is the explanation of the use of *estrogen* in men with inoperable *prostatic* cancer?

Solution
The feminizing effect of estrogen in the male extends to the prostate gland. Estrogen causes this gland to atrophy (shrink); the cancerous growth in the gland does likewise.

DES stands for **diethylstilbestrol,** a synthetic *trans*-alkene that the liver biotransforms into a high-potency estrogen. DES has a long and not-so-illustrious history.

DES

Thirty to 40 years ago it was given to pregnant women who were threatening to abort. Any success it had then was overshadowed by the discovery years later that daughters born to the women who took DES had an increased incidence of cancer of the cervix. Today we know that DES and all other estrogenic substances, including those in oral contraceptives, are *never* to be given to pregnant women.

The use of DES as a "morning after" pill, that is, as an abortifacient given the morning after sexual intercourse, is strongly discouraged by the FDA, which has removed from the market the 25-mg tablets formerly used for that purpose. The FDA still approves of the use of DES as a morning after pill in the case of rape or incest. The newest "morning after" pills are the prostaglandins (Chapter 17).

Testosterone,[1] the most important androgen (male sex hormone), is synthesized in the cells of Leydig in the testes. It is necessary for the normal development of the male gonads and for the continued production of sperm. The deep voice of the male, the pattern of hair growth, and skeletal muscular development are all dependent on the presence of testosterone.

testosterone

Males castrated before they reach puberty never develop secondary sex characteristics. They often grow tall and have slender bodies with little muscular development. Their genitals are small and their voice high pitched.

Females secrete small quantities of androgens. Androgens are chemical intermediates in the biosynthesis of estrogens from acetate and cholesterol.

Anabolic steroids are synthetic compounds patterned after testosterone and having androgenic activity. They also have the effect of stimulating protein synthesis (anabolism) and hence of supposedly increasing muscle mass. Anabolic steroids have become a point of contention with professional and amateur athletes who sometime use them to build muscle and therefore presumably increase athletic performance. Given to female athletes, anabolic steroids are capable of causing maculinization and of terminating menstruation. They are also not without potentially serious side effects such as liver damage. Scientific opinion is that the anabolic steroids do not increase athletic performance. That view is held by Syntex Puerto Rico, Inc., a company that specializes in the development and marketing of steroidal hormones. Their product, Anadrol-50, is an anabolic steroid that enhances red blood cell

[1] Research has shown that dihydrotestosterone, and not testosterone per se, is the physiologically active form.

formation in the treatment of anemia. The company cautions that their product should never be used during pregnancy as it can masculinize the fetus. They also state that anabolic steroids do not enhance athletic ability. In spite of this, there are reports of widespread use of anabolic steroids in sports.

Bile salts are found in the bile of man and a number of animals. Bile is made in the liver, stored in the gallbladder, and ejected into the small intestine under the stimulus of food in the alimentary tract. As we learned in Chapter 13, bile functions to emulsify fat, forming micelles, and aiding in fat digestion and absorption. About half a liter of bile is secreted per day.

Bile salts consists of steroids having a side chain at C-17 that ends in a carboxyl group. The carboxyl group is present as the amide of either glycine or the rare amino acid taurine. Figure 14.3 shows a bile salt and the steroid acid from which it is formed.

Bile salts have an ionic "head" (the COO⁻ group) and a generally nonpolar, hydrophobic hydrocarbon tail. This fits our structural requirement for an emulsifying agent and explains the ability of bile salts to emulsify intestinal lipids.

Gall stones afflict some 10 to 20 percent of our population. They occur when a constituent of bile precipitates out of solution in the gallbladder. In advanced cases, the entire bladder can be packed with stones, wedged in side to side. The condition is very painful and requires surgical intervention. The great majority of stones in gallbladders of Americans consist of cholesterol. Cholesterol was first isolated from gall stones in 1770 by de la Salle.

The role of cholesterol in atherosclerosis was discussed in Chapter 13. Here we will examine its chemistry, how it is formed in the body, and its biochemical relationship to steroidal hormones and bile acids.

Figure 14.4 presents the chemical structure of **cholesterol.** The groups at C-18 and C-19 are termed angular methyl groups; the side chain at C-17 has eight carbons. It is through the alcohol at C-3 that cholesterol can be esterified with various acids (e.g., fatty acids). As a whole, the cholesterol molecule is clearly nonpolar and lipidlike.

Cholesterol is synthesized in the body in many tissues, especially in the liver, skin, and adrenal glands. Its ultimate precursor is acetyl CoA with a

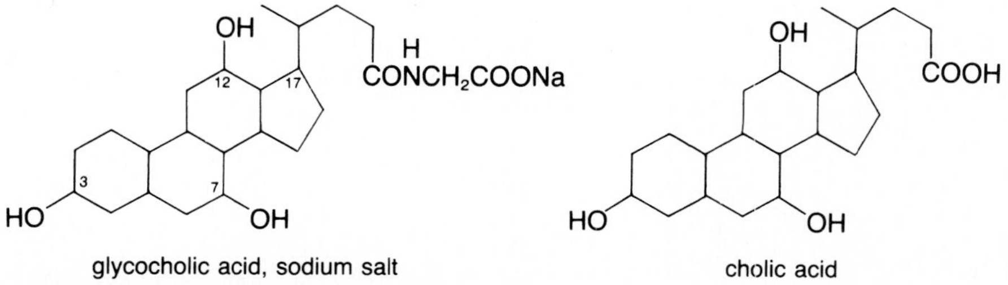

glycocholic acid, sodium salt   cholic acid

FIGURE 14.3 The Chemistry of Bile Salts.

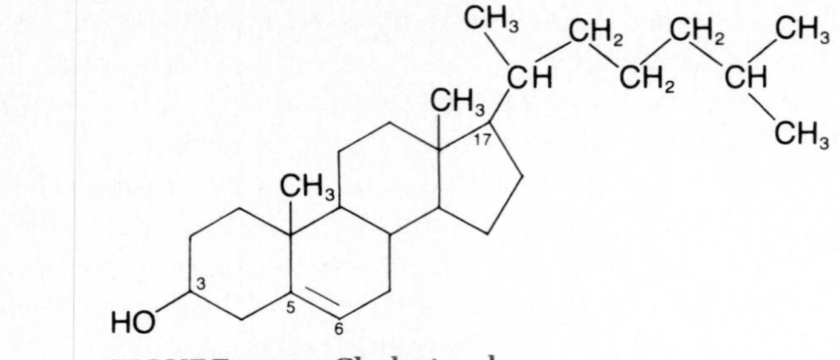

FIGURE 14.4 Cholesterol.

total of eighteen acetyls required to make one cholesterol, $C_{27}H_{46}O$. The several dozen steps in the total biosynthesis of cholesterol are known, but we will describe only the steps in which the cholesterol nucleus is formed:

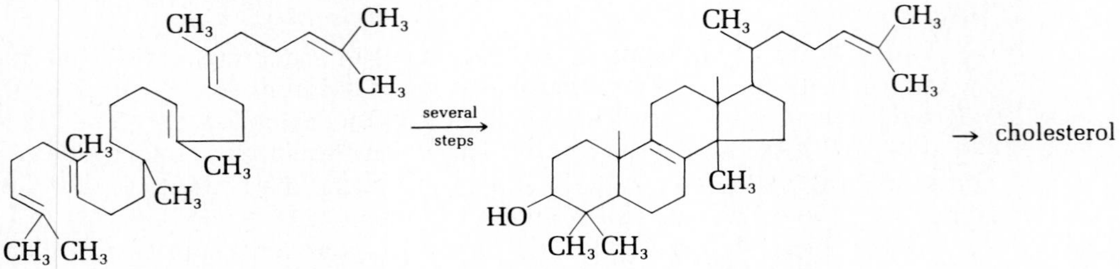

squalene (open chain) lanosterol (a steroid)

The final steps from lanosterol to cholesterol involve demethylations and changes in the unsaturation. All of these steps are enzyme-catalyzed. About 1.0 to 1.5 g of cholesterol is synthesized in the human body by this route daily. The remainder is obtained from the diet. For example, there is 0.550 g of cholesterol in each 100 g of whole egg and 0.011 g of cholesterol in each 100 g of whole milk.

Cholesterol, once formed, is the starting point for the body's synthesis of sex hormones, corticosteroids, and bile salts. It is through bile acids, actually, that the body metabolizes and excretes cholesterol. In summary,

$18\ CH_3COO^- \xrightarrow{\text{many steps}}$ cholesterol → adrenocorticosteroids

cholesterol → progesterone → testosterone

cholesterol → estrogenic hormones

cholesterol ↓ (oxidation) bile acids ↓ excretion in feces

STEROIDS FROM PLANT SOURCES

As you will learn in Chapter 17, vitamin D (which is actually a group of compounds, D_2, D_3, and D_4) is necessary to prevent the development of rickets in children. In this section we will examine only the chemistry of the steroid precursors of vitamin D and the role played by high-energy radiation.

Humans can obtain vitamin D from fish liver oils such as cod liver oil. Egg yolks provide limited amounts. However, excluding vitamin pills, the way in which most of us obtain our vitamin D is through the action of sunlight on vitamin D precursors (called provitamins) that we can get in our diet and that are present in our skin. One such provitamin D compound is 7-dehydrocholesterol (Figure 14.5), which, when irradiated with high-energy (UV) light, is chemically transformed into physiologically active vitamin D_3; the reaction occurs in human skin, as shown in Figure 14.5. The essential feature of the process is that the high-energy radiation must get to the precursor in the skin; protective clothing, cloudy weather, and smoky skies hinder vitamin D formation. Black people, especially those living in the industrialized cities of the north (with smog, clouds, and cold weather), are at greater risk of rickets because the melanin pigment in their skin acts to filter out sunlight.

7-Dehydrocholesterol is a steroid; vitamin D_3 has lost its steroid character through rupture of ring B. However, this is what is required for creation of antirachitic activity. Ergosterol is another provitamin D steroid. You may be familiar with it as irradiated ergosterol, or vitamin D_2 (calciferol), often added to milk. The usual daily prophylactic dose of vitamin D_2 is 10 μg. Since 1 μg equals 40 USP units, that is 400 USP units a day. The usual *therapeutic* dose of D_3 is 30 μg.

Health professionals should have a knowledge of the relationship between wavelength and energy of radiation because they may need such knowledge in professional settings (irradiation of neonates who suffer from jaundice, Chapter 15). Briefly, UV light is part of the *electromagnetic spectrum* of radiation, as shown in Figure 14.6. A nanometer (nm, formerly millimicron) is 1×10^{-9} (one-billionth) m.

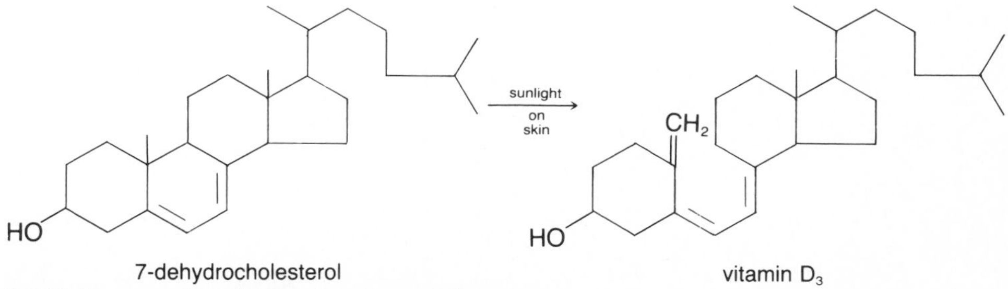

7-dehydrocholesterol   vitamin D_3

FIGURE 14.5   Irradiation Produces Vitamin D.

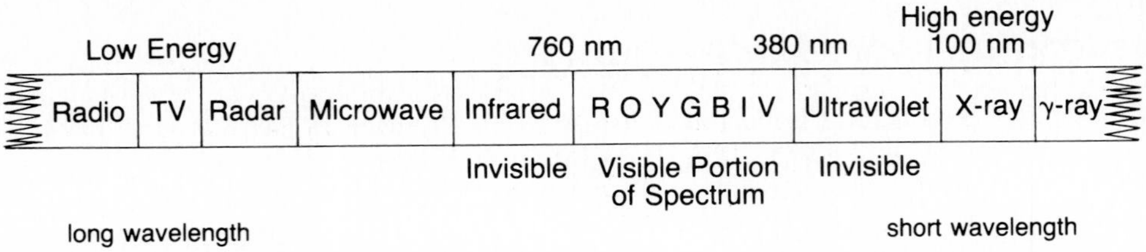

FIGURE 14.6   **Part of the Electromagnetic Spectrum Showing the Relationship Between Wavelength and Energy.**

Ultraviolet light is high-energy radiation. It can tan the skin, kill bacteria, damage the retina of the eye, and induce chemical changes, as in the formation of vitamins D_2 and D_3. True UV light is invisible; its wavelength range is from about 350 to about 100 nm. In the irradiation of ergosterol, light of about 300 to 280 nm is utilized.

Digitoxin USP is a powerful stimulant of heart muscle, a life-saving drug that is used to treat patients with heart failure. It and other digitalis glycosides come from the plants *Digitalis purpurea* and *D. lanata*. Figure 14.7 shows that digitoxin is a steroid glycoside, that is, its C-3 hydroxyl is bound in a hemiacetal link to three sugars (appropriately called digitoxose).

EXAMPLE 14.3

Recall question: identify the functional groups that comprise the fifth ring in digitoxin.

Solution
The link R—CO—O—R is an ester. Thus this compound is an ester of an α,β-unsaturated acid. Actually, intramolecular esters of this type are given the special name, *lactones*.

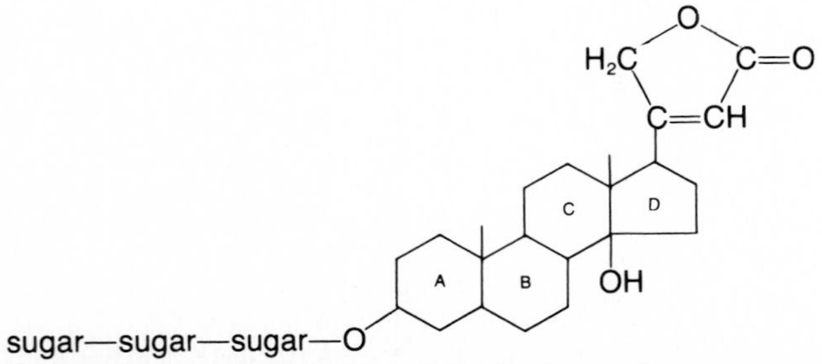

FIGURE 14.7   Digitoxin, a Steroid Glycoside.

The digitalis drugs are unusually potent. A daily dose of only 100 to 200 µg is needed to maintain a heart patient. Classified as a cardiotonic, digitoxin makes the heart beat more efficiently, with increased stroke volume. More effective blood circulation reduces edema and diminishes fluid accumulation in the lung. The toxic dose of digitalis is very close to the therapeutic dose, a fact that must never be forgotten.

ORAL CONTRACEPTIVES

There are about 50 million American women of childbearing age, and about 10 million of these take **oral contraceptives,** or the Pill. That constitutes a tremendous social, economic, and even political impact on our society. But use of these oral anovulatory agents is not without some degree of risk, especially in certain categories of users, and the health professional should become familiar with Pill action, use, and risk/benefit ratios.

When we say "Pill" we mean a synthetic steroidal compound patterned sructurally and physiologically after **progesterone.** That is why the name *progestin* is applied to these agents. Progesterone occurs naturally as a hormone of pregnancy. When a woman is pregnant, her circulating progesterone,

CH_3
$C=O$

progesterone

OH
$C\equiv CH$

norethindrone

synthesized in the corpus luteum and in the placenta, acts by way of a negative feedback mechanism to suppress pituitary production of follicle stimulating hormone (FSH) and luteinizing hormone (LH). We know that FSH and LH are necessary for ovulation to occur. They stimulate the primordial egg follicle to develop and mature and eventually to be ejected from the surface of the ovary into the Fallopian tubes. This is the process of ovulation. All of this is held in abeyance during pregnancy by the action of circulating progesterone.

If you got the idea just now—why not give synthetic, oral progesterone to a nonpregnant woman and fool her pituitary into recognizing her as pregnant— you are on the right track. Except that progesterone by the oral route is destroyed by digestive juices, and it is too painful to inject it daily. Here is where the synthetic progestins come in. These steroids (norethindrone is one of the most widely used) can be taken orally, are not destroyed by digestive

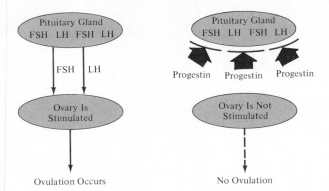

FIGURE 14.8   How the Pill Works as a Contraceptive.

juices, cost only about $4 to $5 for a month's supply, and *do* trick the pituitary into perceiving the uterus as gravid (pregnant). Daily progestin use results in suppression of FSH and LH; ovulation ceases. Hence the name *anovulatory agents*. Progestins are also believed to have an effect on the uterus directly, making it less receptive to a fertilized egg. Figure 14.8 explains the contraceptive action of progestins.

Because the use of progestins alone resulted in spotting, or breakthrough bleeding, it was decided to add an estrogen to the Pill, resulting in what we now call the combination Pill. The estrogen commonly used in today's Pill is a synthetic derivative of estradiol called mestranol, and it provides a product that gives good control over ovulation with minimal spotting.

Use of the Pill makes pregnancy impossible because ovulation is inhibited. Today's combination Pill contains a synthetic progestin and smaller amounts of estrogen. Taken once a day, these two chemicals act to suppress the release of the hormones that are needed to start the process of ovulation.

Some three dozen oral contraceptive products are marketed currently; they differ in progestin used, its strength, and the estrogen used and its strength. Reflecting concern over the possible deleterious effects of the estrogen component of the Pill, drug companies have introduced the "minipill" in which progestin alone, with no estrogen, is used. Minipills are taken every day of the year without interruption, a major difference from the combination products. Minipills are less effective in suppressing ovulation, but this is a trade-off for greater safety in use.

A woman who is using the Pill wants to know how safe it is. She should be informed that concern over these agents is in two main areas:

1. Blood clots (thromboemboli).
2. Cancer of the uterus.

Within 3 years of their introduction in 1960, oral contraceptives were linked to increased incidence of blood clotting with resultant strokes, phlebitis, and clots in the lungs. These are potentially life-threatening episodes. The studies that have been conducted on the relationship of the Pill to clotting have given conflicting results. Retrospective studies tend to establish a cause-and-effect relationship, whereas one large prospective study found no increase in incidence of thromboembolism in oral contraceptive users. Despite the conflicting evidence, the consensus today is that use of the Pill does increase a woman's risk of experiencing a blood clot some four to ten times over what she could expect without the use of the Pill. *Smokers, women over 40, and women with a history of blood clots are at especially high risk* when on the Pill, as shown in Figure 14.9.

A large British study, reported in 1977, estimated the excess annual death rate from circulatory diseases for oral contraceptive users to be as follows.

Ages	Excess death annually (per 100,000)
15–34	5
35–44	33
45–49	140

In summary, the role of the Pill in causing thromboembolism is uncertain. Healthy young women who do not smoke and have no history of blood clots stand a low risk of experiencing clot formation because of oral contraceptive use. Such women may decide that the benefits of Pill use clearly outweigh the risks. On the other hand, risk/benefit ratios are high in women over 40 who smoke. They would be well advised to investigate other contraceptive technology.

If the causal role played by the Pill in blood clotting is uncertain, its role in causing cancer is tenuous at best. The suspicions we have are all statistical in nature, based on retrospective studies, but that does not mean we should ignore them. The combination Pill contains estrogen. From all information available, the FDA has concluded that women who take estrogen are four to seven times more likely to experience cancer of the uterus than nonusers of estrogen. Earlier we cited a University of Pennsylvania study that concluded estrogen use for 5 years or longer resulted in a uterine cancer rate fifteen times greater than in the nonuser. To repeat, these figures do not *prove* anything. They do give a woman reason to pause and evaluate the risk and the benefits associated with her use of oral contraceptives.

Skin cancer and the Pill were linked satistically in a 10-year study of nearly 18,000 patients on a prepaid health plan in California. Data *suggested* that women who use birth control pills for longer than 4 years expose themselves to almost twice the risk of developing the often fatal skin cancer, malignant melanoma. The report cautioned that suggestive data are not definitive data.

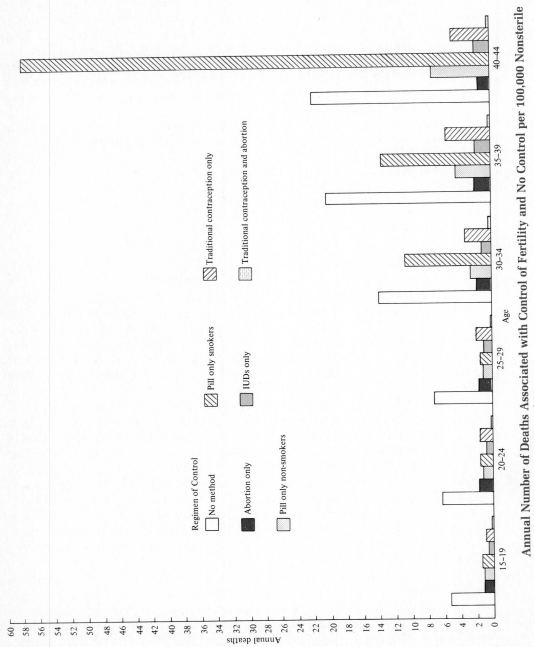

Annual Number of Deaths Associated with Control of Fertility and No Control per 100,000 Nonsterile Women, by Regimen of Control and Age of Women.

SUMMARY

The steroid nucleus (seventeen carbons in four rings) is the chemical basis of the adrenal cortex hormones, sex hormones, bile salts, cholesterol, and oral contraceptives.

Adrenal corticosteroids are essential for life. The glucocorticoids increase blood sugar levels and act as anti-inflammatory agents. Aldosterone, the major mineralocorticoid, acts to retain Na^+ and conserve body water. Synthetic analogs of hydrocortisone have been prepared as anti-inflammatory agents that have little or no side effects on electrolyte balance.

Estrogens are female sex hormones; androgens are male sex hormones. All are steroidal. The three female sex hormones are estradiol, estriol, and estrone. DES is a nonsteroidal synthetic estrogen; it must never be used during pregnancy.

Anabolic steroids are prescribed for the anemic, the debilitated patient, or those who need an increase in protein synthesis. Anabolic steroids have masculinizing side effects and should not be used as muscle builders in athletes.

Bile from the liver contains bile salts. These steroidal acid salts are good emulsifying agents and help in the digestion of fats by emulsifying intestinal lipids.

In this chapter we learned the chemical structure of cholesterol, how it is biosynthesized from acetyl CoA, and its role as an intermediate in the body's synthesis of bile acids, progesterone, adrenocorticosteroids, and sex hormones.

The natural way to obtain vitamin D is through the effect of sunlight on vitamin D precursors in the skin. UV irradiation of 7-dehydrocholesterol yields vitamin D_3. In this chapter we examined the electromagnetic spectrum and learned which portions of it exhibit high energy.

The steroidal hormone progesterone introduced us to the principles of the oral contraceptives (the Pill). Progestins are synthetic steroids having progesteronelike activity; they fool a woman's pituitary into believing the woman is pregnant, that is, they supress release of FSH and LH.

The Pill is a very effective means of contraception, but there are risks associated with its use. These were discussed in this chapter.

KEY TERMS

Check your understanding of this chapter. Can you explain what is meant by each of these terms?

adrenocorticosteroid   androgen
anabolic steroid   bile salt

cholesterol
diethylstilbesterol
estradiol
estriol
estrogen

estrone
oral contraceptive
progesterone
steroid
testosterone

STUDY QUESTIONS

1. Draw the steroid nucleus and number each carbon atom through C-17. Where are positions 18 and 19?

2. **a.** A steroid has a 3-keto, 11,17-dihydroxy substitution pattern with a Δ^4-double bond. Draw the structure of this steroid.
   b. Another steroid has the same structure as estradiol except that ring A is saturated and the C-3 hydroxyl has been oxidized to a ketone. Draw the structure of this steroid.

3. Our text identifies two types of adrenal cortex hormones. What are they? Give the name and chemical structure of one example of each type. What are their biochemical roles in the body?

4. What deleterious side effects can be expected if a drug like hydrocortisone is administered to treat arthritis?

5. If a person has a known sensitivity to bumble bee stings, his physician may prescribe a corticosteroid to keep on hand at all times. What is the rationale behind this? (*Hint:* See Table 14.1.)

6. Explain the statement in the *Merck Manual* that "hyperkalemia is a feature of addisonian crisis." (*Hint:* The Latin word for potassium is *kalium*.)

7. What structural differences exist between aldosterone and hydrocortisone? Predict how these differences would affect differences in metabolism. (See Table 14.1.)

8. Out of all the bioactive compounds mentioned in this chapter, one type is administered by intraarticular injection. What type would that be? (*Hint:* Obtain a medical dictionary and look up the definition of "articulation.")

9. Give names of the three primary human estrogenic hormones. What five chemical features are common to all three?

10. Estrogens appear naturally in the blood of males. In fact, the male blood level of estradiol averages 2 ng/dL. Express that as grams per liter. (*Hint:* ng stands for nanogram and a deciliter is 1/10th of a liter.)

11. Chemically, how would DES have to be altered to bring its structure closer to that of the naturally occurring estrogens?

12. Define the following:
    a. Anabolic steroid
    b. Androgen
    c. Estrogen
    d. Gonad
    e. Irradiate
    f. Cardiac glycoside
    g. MiniPill
    h. Rickets
    i. Nanogram
    j. Morning after pill
    k. Anovulatory agent
    l. Thromboembolism

13. What is the explanation for the use of testosterone or other androgen in women with inoperable breast cancer?

14. Summarize the arguments that have been presented *against* using anabolic steroids to increase athletic ability.

15. Match the drug or hormone with its gland source, if any.
    A. Deoxycorticosterone
    B. Estradiol
    C. Sodium glycocholate
    D. Progesterone
    E. Norethindrone
    F. FSH
    G. Testosterone
    H. DES
    I. Digitoxin

    1. Corpus luteum
    2. Adrenal cortex
    3. Adrenal medulla
    4. Testes
    5. Liver
    6. Ovary
    7. Pituitary
    8. Not from a human gland

16. We know that the body synthesizes estrogens from cholesterol. How many carbons would cholesterol have to lose to be converted to estradiol? What other features or functional group changes would be necessary?

17. What is the difference between a prophylactic dose of a drug and a therapeutic dose of a drug? (*Hint:* See discussion of vitamin D.)

18. a. If one wanted the highest energy, shortest wavelength radiation, which one of the following should be selected: microwave, visible light, x-ray, infrared, UV?
    b. If one wanted the highest energy, shortest wavelength radiation, which one of the following should be selected: green, red, blue?

19. What precautions must be observed in the use of high-energy radiation such as ultraviolet?

20. What is the role of sunlight in vitamin D production?

21. What are progestins and how do they work as contraceptives?

22. Compare the steroid structure of estradiol with that of testosterone. What is the most striking and obvious difference?

23. Cholesterol, whether from the diet or synthesized in the liver, is the chemical precursor for many hormones. Name five steroid hormones that the body makes from cholesterol.

24. Assume that data from a study of many women show a statistical relationship between oral contraceptive use and uterine cancer. Does this prove that oral contraceptives cause cancer? Explain your reasoning.

ADVANCED STUDY QUESTIONS

25. How many chiral carbons exist in cholesterol, and therefore how many stereoisomers are possible?

26. The generic names of six widely used types of anti-inflammatory synthetic steroids are prednisone, prednisolone, methyl prednisolone, triamcinolone, dexamethasone, and betamethasone. Using your library's copy of the *Physician's Desk Reference* (PDR), make a list of all of the trade names under which these generics are sold.

15 Body Fluids

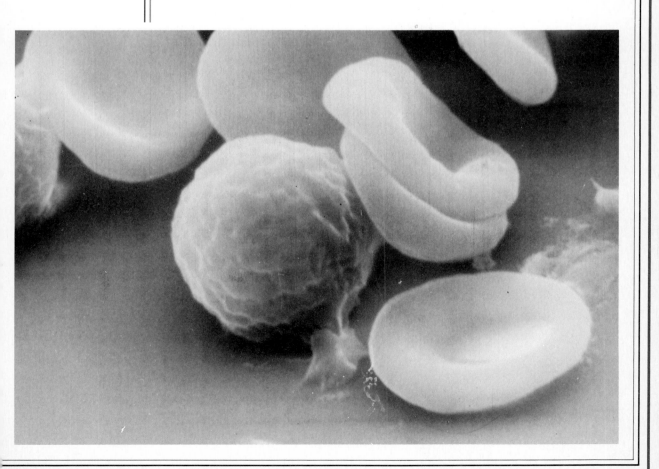

Just as the earth's surface is more than half water, we humans by weight are about 50 to 60 percent water. In this chapter you will study the compartments into which body water is divided, and the biochemical role of fluids in each compartment.

By far, the most important body fluid is blood. Our photo shows two of the formed elements of blood, red blood cells (the discs) and white blood cells (the spheres) at a magnification of about 10,000. The delicate nature of the red cells is a reflection of the delicate balance in which body water, dissolved chemicals, gases, and pH are maintained. Our maintenance of this balance in spite of environmental and disease threats to it, is termed *homeostasis.*

Much of this chapter deals with homeostatic mechanisms which protect us against threatened acidosis, alkalosis, electrolyte imbalance, fall in blood pressure, overheating, or dehydration.

In this chapter, you will be introduced to some very important clinical procedures that affect body fluids. These include hemodialysis, blood typing, urinalysis, and blood gas analysis. In medical jargon, the word "clinical" refers to medical procedures that are performed on patients or their specimens in the hospital, doctor's office, or doctors' clinic, in order to diagnose or treat disease.

As nurses, we are concerned with assisting our patients to attain their highest level of functioning. To achieve this goal, we wish to guide the therapy of our patients in a manner that supports homeostatic mechanisms. This involves an understanding of what makes up our cellular environment, how it functions, and how it is maintained.
–Margaret Wallhagen, RN, MSN, California State University, Chico

INTRODUCTION

Earlier in this book we learned that human adults are about 50 to 60 percent water by weight. This means that we exist in large part in an internal aqueous environment. The transport of thousands of different molecules occurs in body water. The cellular manufacture of many of these same molecules occurs in water. Metabolism of foods and other biotransformations such as hydrolysis, esterification, neutralization, acetylation, methylation, oxidation, and reduction take place in a watery environment.

The entire renal (kidney) process of filtration of blood, selective reabsorption of needed constituents, and excretion of wastes occurs in an aqueous medium. Drugs and medicines are administered in water and transported to their active sites in the blood—a water-based body fluid. The brain is cushioned in a special fluid that is more than 90 percent water.

Abnormal or disease-caused chemical reactions occur in aqueous media, too, and the health professional relies on analysis of blood, urine, and other body fluids for clues in making diagnoses or for following the progress of therapy.

It is our goal in this chapter to learn the composition and important functions of five body fluids: blood, urine, bile, cerebrospinal fluid, and synovial fluid. We will concentrate on blood as the most important of the body fluids, examining its composition, its role in oxygen and carbon dioxide transport, and its function in healthy and diseased states.

CIRCULATING AND NONCIRCULATING BODY FLUIDS

The blood consists of the formed elements, **red blood cells (RBCs), white blood cells (WBCs),** and **platelets,** suspended in water. It circulates in the cardiovascular system: arteries, arterioles, capillaries, venules, and veins.

The blood and the fluid bathing the cells constitute the **extracellular fluid** of the body. The fluid bathing the cells is literally just that—tissue fluid in which cells are nourished and cushioned. Extracellular fluid is generally synonymous with **interstitial fluid** except that the latter does not include the blood. Extracellular fluid is the source of lymph (discussed later in this chapter).

Fluid that exists inside cell membranes is termed **intracellular fluid.** It supports the cell nucleus, numerous cell inclusions such as the mitochondria, and provides a medium in which intracellular chemical reactions can occur. These relationships are summarized in Figure 15.1.

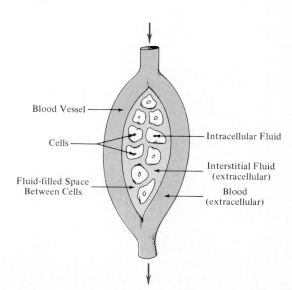

Blood Vessel

Cells

Fluid-filled Space
Between Cells

Intracellular Fluid

Interstitial Fluid
(extracellular)

Blood
(extracellular)

FIGURE 15.1 Circulating and Noncirculating Body Fluids.

Table 15.1. Body Fluid Compartments

Compartment	Percent of all body water	Percent of body weight (approximate)
Fluid inside cells (intracellular fluid)	66	40
Fluid bathing cells (interstitial fluid)	26	15
Blood plasma	8	5

There are other noncirculating body fluids, but they do not add up to much volume and we shall only mention them in passing. Fluid in the eye is either *aqueous humor* or *vitreous humor*, depending on its compartment. Tears are composed of *lachrymal fluid. Seminal fluid,* or semen, is the ejaculated fluid that carries the sperm. *Amnionic fluid* of pregnancy can vary from 100 to 1800 cc in volume. It protects the fetus from shock, allows freedom of fetal movement, supplies water for the fetus, and helps dilate the cervix during contractions. Saliva and gastric, intestinal, and pancreatic juices are digestive fluids.

We can think of the water that is in the body as existing in three compartments, as shown in Table 15.1.

GAIN AND LOSS OF BODY WATER

If we assume that the water content in a 70-kg adult male is 42 L, how much of that is lost and replaced each day? The answer is that with an average daily urine output of 1200 mL, plus 200 mL water lost in the stool (feces), plus loss of water as perspiration and in exhaled breath (calculate as 50 mL/hour or 1200 mL/day), we lose about 2600 mL of water a day. This is about 6 percent of our total body water, and it must be replaced daily. Our calculation does not take into account possible diarrhea, vomiting, hemorrhage, heavy exercise, or other stress that would result in extra water loss with greater requirement for replacement. Remember, loss of body water almost always means loss of body electrolytes, which also must be replaced.

How long can a human live with zero intake of water? A study was made of the efforts of a small group of downed U.S. airmen (World War II) who attempted to walk out of the North African desert. It was estimated that they lived for a maximum of 8 days in that hot environment with no fluid intake before the last of them succumbed.

HOMEOSTASIS

Homeostasis is the condition of maintaining a healthy internal environment; that is to say, the body resists changes in any important function or constituent-changes that tend to threaten good health or mental functioning. When the normal state of any part of our system is disturbed, we have biochemical processes and regulatory mechanisms that can "come on line" to resist the change. Humans are admirably equipped with homeostatic mechanisms. Here are a few:

1. *Stress:* Heavy exercise increases CO_2 production in tissues with buildup of H_2CO_3 and a tendency toward acidosis.

    Reaction: Reflex stimulation of the breathing center results in increased respiration. Excess CO_2 is blown off in the lungs; tissue levels of carbonic acid fall; the kidney acts to neutralize and excrete acid.

2. *Stress:* Working in a hot environment causes significant loss of body fluid. Because of limited intake of water, body dehydration threatens. (The same state can sometimes be seen in persistent nausea with pernicious vomiting.)

    Reaction: Urine output falls dramatically as Na^+ and other electrolytes are conserved. There is redistribution of water in body compartments, until oral intake of water is again possible.

3. *Stress:* Blood pressure drops dramatically in conditions of shock, hemorrhage, or sudden environmental changes, threatening consciousness or life itself.

    Reaction: Compensatory mechanisms come into play, constricting blood vessels (thus raising blood pressure) and redistributing body fluids. Pressor agents are secreted; body water is conserved.

4. *Stress:* During periods of voluntary or forced fasting, blood sugar levels fall below the normal range; the individual becomes weak and irritable.

    Reaction: Through a feedback effect, the pancreas senses the drop in blood sugar, reduces its output of insulin, and increases its secretion of the hormone glucagon. When glucagon enters the hepatic and general circulation, it causes the breakdown of glycogen to glucose, the synthesis of new glucose from available amino acids, and an increase in fat metabolism. As a consequence, blood sugar levels are raised to the normal range.

As we study blood chemistry, urine chemistry, and body fluids in the remainder of this chapter, we will examine these and other homeostatic mechanisms that maintain us as biochemically sound in the face of daily stresses.

BLOOD BIOCHEMISTRY

In areas such as critical care and hemodialysis, excellent patient care demands a working knowledge of biochemistry. The nurse is the person who first receives lab reports, blood gas data, etc., and s/he must be able to accurately interpret a maze of numbers.
–Sally McLeish, RN, MN, California Lutheran College Nurse Practitioner, Conejo Valley Renal Center

Composition of Blood and Plasma

Whole blood is water with dissolved chemicals (called the **plasma**) and suspended **formed elements.** The formed elements are RBCs, WBCs, and platelets (thrombocytes). **Serum** is the straw-colored, watery fluid that remains after a clot forms. These relationships can be summarized as follows:

Whole Blood
      Remove formed elements
  ↓
Plasma
      Remove fibrinogen and clotting factors
  ↓
Serum

A 70-kg adult has about 5 to 6 L of blood; a 36-kg child has about 2.5 L of blood. When stored at 4 to 6°C, whole blood has a "shelf life" of about 3 weeks; plasma can be stored for much longer periods.

Plasma makes up about 55 percent of the volume of the blood. In it are dissolved hundreds of different chemicals that are always present, plus hundreds of others present only after a meal, after drug use, or after exposure to environmental substances. The osmotic pressure of plasma is very close to that of a 0.9% w/v sodium chloride solution (isotonic saline). Its pH is 7.4.

Plasma proteins are those proteins that remain in the blood after the formed elements are removed. They fall into three categories:

1. The *albumins*, the most abundant of the blood proteins, make important contributions to the osmotic pressure of the blood; in the human, they average a molecular weight of 68,000. As we have learned, many drugs and other biomolecules bind to albumin, rather than float free in the blood. The albumins also help buffer against pH changes (a homeostatic mechanism). Blood albumins cannot pass membranes and hence are confined to the cardiovascular tree (except in burns where leakage is common).
2. The *globulins* consist of the alpha, beta, and gamma classes, depending on molecular weight. Gamma-globulin, with a molecular weight of around 156,000, contains the antibodies we have formed against all of the antigens that have threatened us in the course of daily existence (viral disease, foreign protein, pollen, etc.). We have already discussed the use of gamma-globulin injection (immunoglobulin) to confer passive immunity to a person who has been exposed to some infectious disease. The alpha- and beta-globulins function in the transport of lipids in the blood.
3. *Fibrinogen* makes up about 5 percent of the plasma proteins. With a

molecular weight of 340,000, it is the protein that coagulates to form the matrix in blood clots. This matrix consists of fibrin, and its formation can be summarized as follows:

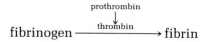

$$\text{fibrinogen} \xrightarrow[\text{thrombin}]{\text{prothrombin}} \text{fibrin}$$

Figure 15.2 shows a red blood cell enmeshed in fibrin fibrils—a part of a typical clot.

Plasma electrolytes consist mainly of NaCl, with lesser amounts of K^+, Ca^{2+}, phosphate, Mg^{2+}, and HCO_3^- (bicarbonate). In Figure 15.3, you can

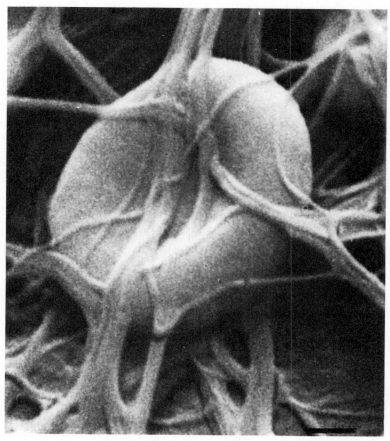

FIGURE 15.2 **Scanning electron micrograph of an erthrocyte enmeshed in fibrin fibrils—a part of a typical blood clot. Fibrin is an insoluble protein formed from fibrinogen. (Bernstein, E. and Kairinen, E., *Science* Vol. 173, Cover, 27 August 1971.)**

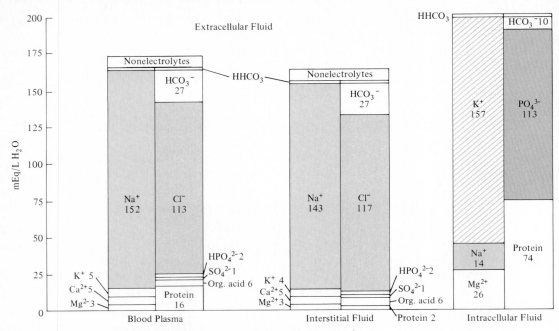

FIGURE 15.3 **Comparison of the Composition of Body Fluids.**

compare the electrolyte composition of plasma to that of other body fluids. It can be seen that

1. The principal cation of plasma is Na^+.
2. The principal cation of intracellular fluid is K^+ (actually, about 98 percent of the body's potassium is found inside cells).
3. While Cl^- is the principal anion of extracellular fluid, phosphate is the main anion inside cells.
4. There is more than 4.5 times as much protein in intracellular fluid as there is in plasma.
5. There is remarkably little protein in interstitial fluid.

EXAMPLE 15.1

Figure 15.3 shows the presence of "nonelectrolytes" in blood plasma, exclusive of proteins. What are these nonelectrolytes?

Solution
Glucose and other nonionized carbohydrates.

The electrolytes and the nonelectrolytes dissolved in the plasma contribute to its osmotic pressure. They therefore influence water volume in the various fluid compartments, diffusion of ions across membranes, and maintenance of homeostasis in the organism. Another factor in the control of blood protein and blood volume is the lymph.

Lymph

The interstitial fluid (surrounding the cells) originates mostly from the blood in the capillaries and contains many of the constituents found in blood. Interstitial fluid contains cells and proteins that cannot diffuse back into the blood; hence a drainage system is required to collect these and other substances and return them to the bloodstream. This drainage system, called the *lymphatic system,* is a very complex network of vessels and nodes; the fluid inside it is called the **lymph.**

The flow of lymph is very slow because there is no pumping force, only the squeezing effect of nearby muscles in action. Lymph eventually empties into the venous blood, mainly via the thoracic duct in the neck, carrying with it proteins and electrolytes picked up in the capillary beds. The importance of this is shown by the fact that each day up to 60 percent of plasma volume and 50 percent of the total quantity of protein circulating in the blood is lost from the capillaries and returned to the bloodstream by the lymphatics.

The lymphatic systems plays other roles:

- Uptake of bacteria, dead cell parts, and other foreign bodies and transport to lymph nodes for digestion by white cells. (Unfortunately, this is also the machinery by which metastasis, or spread, of cancer cells can occur.)
- Transport of tiny fat globules from the intestines to the blood via the thoracic duct; after a high-fat meal, the lymph changes from a transparent yellow to a milky white emulsion.

Summary
Blood consists of two parts, the formed elements and the plasma. Blood plasma contains three important types of proteins: albumins, globulins, and fibrinogen. Plasma also contains dissolved ions, chiefly Na^+ and Cl^-, that play a dominant role in maintaining homeostasis. The lymphatic system drains proteins, fats, and miscellaneous debris.

Ranges of Normal Values for Blood Constituents

The health professional will have occasion to encounter many additional naturally occurring constituents of blood. From years of analysis of the blood of healthy and morbid patients, biochemists have compiled lists of what are considered normal ranges of blood components. A few of these are listed here; a more complete list will be found in Appendix IV.

Constituent	Normal level Conventional	SI
Ammonia	80–110 µg/100 mL	47–65 µmol/L
Bicarbonate	24–30 mEq/L	24–30 mmol/L
Iron	50–150 µg/100 mL (higher in males)	9.0–26.9 µmol/L
Cholesterol	120–220 mg/100 mL	3.10–5.69 mmol/L
Total fatty acids	190–420 mg/100 mL	1.9–4.2 g/L
pH	7.35–7.45	
Potassium	3.5–5.0 mEq/L	3.5–5.0 mmol/L
Sodium	135–145 mEq/L	135–145 mmol/L
Protein (total)	6.0–8.4 g/100 mL	60–84 g/L
Transaminase (SGOT)	10–40 U/ml*	—
Urea nitrogen (BUN)	8–25 mg/100 mL	2.9–8.9 mmol/L

* Instead of using mass units to measure drug activity, we can use biological response. That is, one "unit" of a hormone or enzyme is the quantity that will bring about some desired result. Insulin activity, for example, is measured in units. The SGOT unit is established by the laboratory assay for SGOT.

Clinical Significance of Too-Low Blood Ion Levels

Inorganic ions in the blood play important roles in blood clotting, the maintenance of water balance, cell formation, bone development, energy transfer within cells, and muscle and nerve actions. When blood levels of key electrolytes fall below normal values, serious clinical signs and symptoms may be seen.

A too-low blood level of sodium, called *hyponatremia*, can occur through the use of diuretics, through excessive diarrhea or sweating, from burns, and through urine loss associated with acidosis. Recall that sodium ion plays the key role in maintaining body water volume because it is the most abundant extracellular cation. Hyponatremia is characterized by headache, tachycardia, muscular weakness, and hypotension (too-low blood pressure), leading to circulatory shock and possibly coma.

Changes in blood chloride ion are usually secondary to Na^+ changes. Too little blood chloride ion can be caused by excessive vomiting. Such a *hypo-chloremia* can cause muscle spasms, alkalosis, and depressed respirations.

Health professionals have become aware of a serious condition termed *hypokalemia* (*hypo* = too little; *kalium* = potassium). Diuretics, given to high blood pressure patients, cause loss of water but can also cause loss of electrolytes such as potassium and sodium. High sodium intake and kidney disease can also cause potassium ion loss. Since the electrical activity of heart muscle fibers depends (in part) on the local K^+ concentration, we can expect to see a change in the electrical conductivity of the heart in hypokalemia. Such changes in heart muscle electrical conductivity are measured by taking the patient's electrocardiogram (ECG, or sometimes, EKG). Figure 15.4 shows a normal ECG

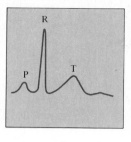
a

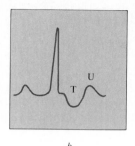
b

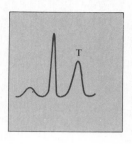
c

FIGURE 15.4 **The important effect of potassium ion concentration on heart muscle contractility is shown by taking the patient's electrocardiogram (ECG). With a normal blood K⁺ level, the ECG (a) is normal. Note the position of the T wave. In severe hypokalemia (b, plasma K⁺ less than 2.5 mEq/L), the PR segment is lengthened, the T wave is inverted, and a new U wave is seen. In hyperkalemia (c), the T wave has increased.**

tracing, a heart tracing in a patient suffering from severe hypokalemia, and for comparison a tracing of a patient with hyperkalemia. Obviously, blood potassium levels can have a very important effect on healthy heart activity.

The Formed Elements of the Blood

A view of a whole blood sample as it might appear through the microscope is given in Figure 15.5. White blood cells (WBC) are the largest of the formed elements; they appear in a variety of forms, depending on the constitution of their cell nucleus.

Certain white cells are capable of killing bacteria; they constitute our first

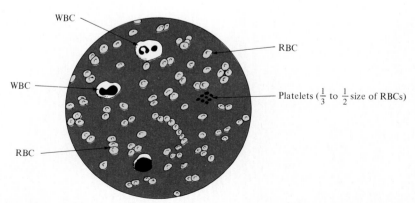

FIGURE 15.5 **In whole blood, WBCs are the largest in size. The RBCs are smaller, doughnut-shaped, but far more numerous.**

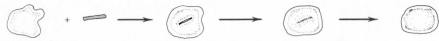

FIGURE 15.6   Phagocytosis of a Bacterium by a White Cell.

line of defense against invading pathogens in wounds or infections. Many billions of leukocytes are produced each day in a person's bone marrow (and thymus and spleen) to replace the billions that die off each day. During an infection, many extra white cells are produced in the bone marrow to help fight invading bacteria. For example, elevated white cell counts are seen in appendicitis and are used in the diagnosis of this condition. The normal range for white blood cells is 5000 to 10,000 cells/µL;[1] in cases of severe infection like appendicitis, the range increases to 16,000 to 20,000 cells/µL.

Phagocytosis (or, "cell eating") is the process in which a leukocyte envelops, ingests, and consumes a bacterium, foreign protein, or dead tissue (Figure 15.6). Phagocytosis is possible because leukocytes have lysosomal enzymes (i.e., ezymes that can *lyse* or split bacterial membranes).

In **leukemia,** a form of cancer, there is uncontrolled growth of white blood cells. Further, the cells are immature and lack the cohesiveness of normal WBCs. Drug treatment modalities for certain forms of leukemia have been

[1] Microliter; identical with cubic millimeter (mm³).

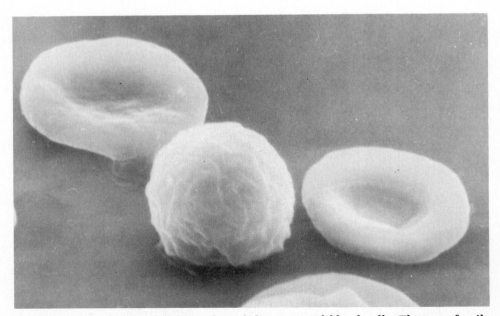

FIGURE 15.7   The two doughnut-shaped discs are red blood cells. They are fragile and, if roughly handled, can rupture, spilling out their hemoglobin. (Courtesy of Benjamin Zweifach, Dept. of Bioengineering, Univ. of Calif. San Diego.)

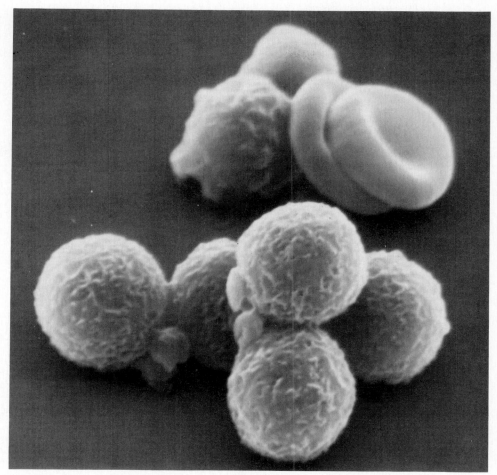

FIGURE 15.8 Two red blood cells accompanied by seven white cells. (Courtesy Benjamin Zweifach, Dept. of Bioengineering, Univ. of Calif. San Diego.)

remarkably successful. For example, vincristine, given with prednisone, is the treatment of choice in childhood acute lymphoblastic leukemia, where it produces complete remission (i.e., return to normal) in about 90 percent of children on their first course of treatment. When some of these children relapse and are given a second course of vincristine therapy, 70 to 80 percent again enter into remission.

Red blood cells (RBCs, **erythrocytes**) are shaped like disks, with two concave surfaces. Figure 15.7 shows two red blood cells, along with one, spherical, white blood cell. Figure 15.8 shows two RBCs accompanied by seven WBCs. Both photos were obtained through scanning electron micrograph techniques and are at about 10,000 magnification.

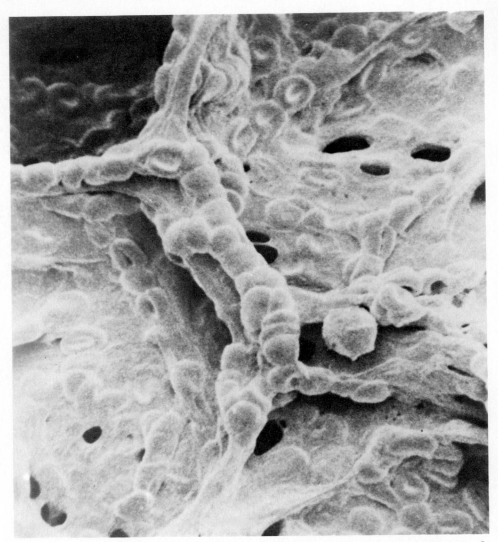

FIGURE 15.9   Outlines of erythrocytes (red cells) in capillaries underlying interalveolar septa are clearly visible due to the epithelial shrinkage produced by air drying. (Field width 64 microns; hamster lung.) (Photo courtesy American Lung Association.)

Erythrocytes are floppy, easily molded, flexible cells that bend to fit the countour and size of the vessel containing them (Figure 15.9). They are easily distorted under the pressure of blood in an arteriole or capillary, where they sometimes can be seen to be streamlined in shape.

In *sickle cell anemia,* some red cells assume a variety of abnormal shapes, including a sickled shape. It is believed that this disease is due to a hereditary defect that results in the incorrect substitution of valine for glutamic acid in the hemoglobin of the afflicted person. (That's only one incorrect amino acid out of 146 in a heme unit!) Sickled cells hemolyze easily and cannot carry adequate amounts of oxygen; the resultant severe anemia can be accompanied by breathing difficulties, abdominal pain, vomiting, fever, and ulcers about the ankles. Few patients live beyond age 40. Patients with sicklemia should be cautioned against high altitudes or oxygen-deficient areas since lowered oxygen tension increases the sickling tendency. The sickle cell gene originated in black peoples of Africa where it confers resistance to one type of malaria. Today 1 in 600 black Americans suffers with sickle cell disease and more than 2 million others carry the trait without showing symptoms. This makes sickle cell anemia the most prevalent inherited disorder known to medical science. Carriers who marry have a 1-in-4 chance of giving the disease to their children, even though the carriers have no sign of the disease themselves.

Red cells are about 7 μm in diameter and about 1 μm thick at their center; they are nonnucleated (have no nucleus). Their hemoglobin content gives them the capacity to transport oxygen, as well as their red color. Normal human red blood cells ranges are 4.6 to 6.2 million per cubic millimeter for men, and 4.2 to 5.4 million per cubic millimeter for women. Since one estimate puts the total number of erythrocytes in an adult human at 10^{12}, and since about one-third of the mass of RBCs is hemoglobin, an adult's blood contains about 950 g of hemoglobin.

The red bone marrow of the ribs, sternum, clavicles, vertebrae, and pelvis is the site of RBC synthesis. Besides protein for hemoglobin production, the process of red cell formation (called *hemopoiesis*) requires iron, trace quantities of copper, and vitamin B_{12}. On the average, a red blood cell functions in the human for about 125 days and then is destroyed in the spleen. Much of the iron content of a destroyed red cell is recovered and reused for additional hemopoiesis. The hemoglobulin liberated is metabolized in the liver and appears as bile pigment. This will be considered further in our discussion of bile later in this chapter. Bone marrow transplants are necessary in certain leukemia patients who have had their own bone marrow deliberately destroyed by chemotherapy or radiation. The marrow transplant renews the patient's ability to synthesize red cells and normal white cells.

Hemolysis (rupture) of red cells is accompanied by leakage into surrounding fluids, imparting a dark red color. Hemolysis can occur when RBCs are placed in hypotonic solutions or in cases of Rh-incompatibility (more on this later).

The **hematocrit** is the percentage of the blood that consists of the formed elements. It is given as a percentage (say, 45 percent) and is used in diagnosis of conditions such as anemia. To determine the hemocrit, a volume of whole blood is centrifuged, a process in which the formed elements are packed by spinning into the bottom of the centrifuge tube, leaving the clear plasma as the supernatant solution (Figure 15.10). The centrifuge tube is graduated and the height of the packed column can be read.

A hematocrit of 47 percent is taken to be normal for adult males. That

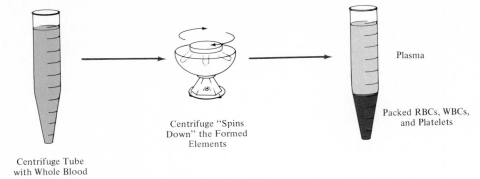

Centrifuge Tube
with Whole Blood

Centrifuge "Spins
Down" the Formed
Elements

Plasma

Packed RBCs, WBCs,
and Platelets

FIGURE 15.10   Determining the Hematocrit.

means that in good health, 47 percent of the volume of whole blood of males is formed elements. For women, the average is 42 percent. Anemic patients may show hematocrits of 30 percent or lower. Burn victims may show a hematocrit of over 50 percent (from protein leaking into the circulation). In the more convenient microhematocrit method, a much smaller volume of blood is required.

Table 15.2.   Anticoagulants and Their Uses

Anticoagulant	*Uses and mechanism of action*
Na heparin USP	Normally found in the body, it inhibits prothrombin conversion to thrombin; clinically, is administered IV, IM, or SC to prevent the formation of a blood clot, to prevent the enlargement of an existing clot, or to prevent clotting of blood being circulated in a heart-lung machine
Coumarin derivatives such as dicumarol USP and acenocoumarol	These "prothrombinopenic" anticoagulants work by lowering prothombin blood levels—a result of an inhibition of vitamin K; coumarin derivatives are used for the prophylaxis and treatment of venous and pulmonary emboli
Oxalate ion	Chelating agents such as oxalate and citrate are able to scavange and bind metal ions like Ca^{2+}, in effect removing them from solution; used in the laboratory analysis of blood to prevent clotting of the blood sample
Citrate ion	Same as for oxalate; in some cases, phosphate, dextrose, and adenosine are added to the citrate; the phosphate acts as a buffer and the dextrose and adenosine help maintain ATP levels so that the blood can be stored for up to 35 days

One method for estimating a person's red cell level (and therefore hemoglobin level) is to pinch a fingernail and note the contrast between the pinkness of the capillary bed of the nail and the blanched area pinched; the more contrast, the richer the hemoglobin levels. Anemic persons exhibit hardly any contrast.

Platelets, shaped like oval disks, occur to the extent of approximately 300,000 per mm³ (microliter) of blood, and contain no hemoglobin. Platelets (also called thrombocytes) are key factors in initiating the blood clotting process. At the site of rupture or trauma of the circulatory system, platelets collect and form a sticky mass. They then release their contents—Ca(II) ion, ADP, and serotonin, helping to begin what ultimately is the formation of fibrin from fibrinogen. Red cells are trapped in the network of fibrin fibrils, and a clot has formed.

Anticoagulants are agents that act to inhibit blood clot formation by a variety of mechanisms. They are used prophylactically to prevent the formation of an embolism (clot), or therapeutically to prevent an existing embolism from increasing. Anticoagulants such as oxalate ion and citrate ion are used *in vitro* (literally "in glass"; compare with *in vivo*, "in body") to inhibit clot formation in blood withdrawn for laboratory analysis.

One of the key steps in the multistep process of blood clotting is the formation of thrombin:

Thromboplastin → Acts on prothrombin → Results in formation
(released at point (in the presence of thrombin
of injury) of Ca^{2+}) ↓

To give fibrin ← Acts on fibrinogen
(spongelike mass of
interlacing fibers)

Table 15.2 summarizes the clinical and laboratory uses of various anticoagulants.

The Functions of the Blood

Adults have about 5 to 6 L of blood coursing through 100,000 miles of arteries, veins, and capillaries. The heart, beating 40 million times a year, pumps the 5 to 6 L of blood through its chambers every 60 seconds.

Six biochemical functions are accomplished as the blood passes through the cardiovascular tree. Let us list them first and then discuss them.

1. Transport of oxygen from the lungs to the tissues.
2. Transport of carbon dioxide from the tissues to the lungs.
3. Provision of a buffering action against threatened changes in pH.
4. Distribution of glucose and other nutrients from depots and the GI tract to all tissues.

5. Conveyance of wastes to the kidneys for elimination.
6. Distribution of heat to all parts of the body.

Blood Gases

Adequate oxygen transport to the tissues would not be possible without the hemoglobin (Hb) in red blood cells. Oxygen forms a relatively strong, reversible bond with the Fe(II) ions in Hb. Actually, a total of four O_2 molecules bind to each hemoglobin (remember, each Hb molecule contains four heme units), but by convention the equation is simplified as follows:

$$O_2 + HHb \rightarrow HbO_2^- + H^+$$

This reaction occurs in the blood in the capillaries of the lung which are an integral part of the approximately 300 million alveolar spaces. Figure 15.11 shows alveoli under high magnification.

When blood is exposed to various drugs and oxidizing agents such as nitrates and nitrites, the Fe(II) in the hemoglobin is oxidized to Fe(III), forming *methemoglobin*. This dark-colored product is toxic because it is incapable of transporting oxygen. Research has shown that shortly after a pregnant woman inhales the smoke of a cigarette, methemoglobin can be detected in fetal blood.

The oxygenation of Hb in the alveolar spaces takes place very rapidly. How much of the Hb is oxygenated depends on the pressure of oxygen gas in the lungs. If we open our mouth and inhale, 760 torr[2] of air pressure is filling alveolar spaces. If oxygen accounts for 150 torr of this, the oxygen's **partial pressure** (P_{O_2}) is 150 torr. Of course, at 20,000 ft, a lot less oxygen is present; its P_{O_2} is smaller, and there will be considerably less oxygenation of Hb. At 20,000 ft, an individual would suffer from *hypoxia*. Conversely, a cardiac patient breathing pure oxygen (P_{O_2} = 760 torr) will have superoxygenation of his Hb, with maximal transport of oxygen to the tissues. Oxygenation of Hb also depends on the temperature and the pH of tissue environment.

Once in the tissues, the Hb-oxygen complex (HbO_2^-) dissociates, liberating oxygen for uptake by oxygen-hungry tissues. The P_{O_2} in tissues is only about 40 torr. This fact, plus the greater acidity of tissues, helps the Hb-oxygen complex to dissociate. Yet another factor promoting release of oxygen in the tissues is the superior affinity of myoglobin for O_2 as compared to hemoglobin. *Myoglobin*, a muscle protein that can bind to, and store, oxygen, consists of 153 amino acids arranged in a single chain.

Not all of the oxygen in the Hb-oxygen complex is transferred to the tissues. In fact, in an individual at rest, more than half of the oxygen is still bound to Hb in venous blood on its way back to the heart. With heavy exercise, this figure falls. The color of blood is related to its oxygen content. Arterial blood is crimson; venous blood is a darker red.

---

[2] One torr is equal to 1 mm Hg pressure.

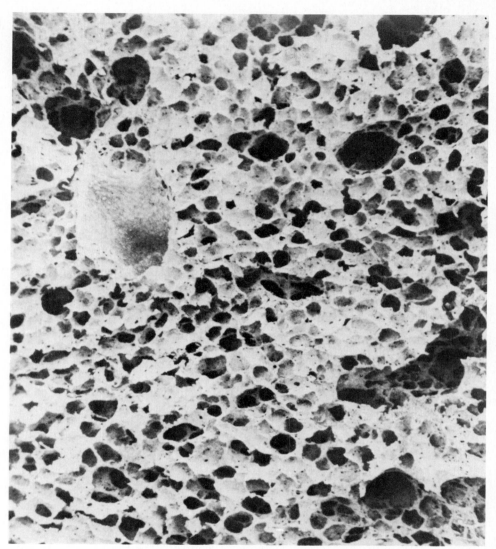

FIGURE 15.11   Aveoli in Hamster Lung. A section of a respiratory bronchiole is seen at upper left; alveolar ducts and alveoli occupy remainder of field. It is through the walls of alveoli that O_2 and CO_2 gas exchange takes place between alveolar gas and capillary blood. Field width 1.11 mm. (Photo courtesy American Lung Association.)

Carbon dioxide transport from tissues to lungs is intimately associated with oxygen transport. CO_2 arises in muscle and other tissues as the result of carbohydrate metabolism (Chapter 12, Krebs cycle CO_2 production). We can speak of a partial pressure of CO_2 (P_{CO_2}) in the tissues, just as we spoke of a partial pressure of oxygen. The P_{CO_2} in tissues, however, is greater than the

Table 15.3. Blood Gas Tensions in the Vascular System in Torr (approximate)

	Tissues	*Veins*	*Arteries*
P_{CO_2}	55	46	40
P_{O_2}	40	40	100

P_{CO_2} in the lungs, as indicated in Table 15.3. Note that the word *tension* means "partial pressure."

In the tissues, carbon dioxide combines with the Hb, which has just given up its oxygen.

$$CO_2 + HHb \rightarrow HbCO_2^- + H^+$$

Note that in this equation we have used HHb for hemoglobin to emphasize its ability to act as an acid, which in this case it does. Additional CO_2 in the tissues dissolves in blood, forming carbonic acid, which then ionizes:

$$CO_2 + H_2O \rightleftharpoons H_2CO_3 \rightleftharpoons H^+ + HCO_3^-$$
$$\text{carbonic}$$
$$\text{acid}$$

The reaction of CO_2 with water to form carbonic acid, and the reverse reaction, too, are catalyzed by the enzyme *carbonic anhydrase*. Diamox is a diuretic drug used to treat cardiac edema; it owes its activity to its ability to inhibit the action of carbonic anhydrase. Inhibition of carbonic anhydrase means that the reversible kidney reaction $H_2CO_3 \rightleftharpoons H_2O + CO_2$ is also inhibited. Consequently, less H_2CO_3 is free to ionize to HCO_3^- and H^+. The result is kidney loss of Na^+ owing to failure to exchange H^+ for Na^+. Alkalinization of the urine and diuresis are thus effected.

Most of the CO_2 we exhale has been transported to the lungs dissolved in the blood as bicarbonate ion; the remainder is bound as $HbCO_2^-$.

In the lung, the processes are reversed. The $HbCO_2^-$ complex breaks down

$$HbCO_2^- + H^+ \rightarrow HHb + CO_2$$

and bicarbonate combines with acid to reform H_2CO_3:

$$H^+ + HCO_3^- \rightarrow H_2CO_3$$

Carbonic anhydrase in lung tissue catalyzes the necessarily rapid decomposition of H_2CO_3 to CO_2 and water. Since the P_{CO_2} of the blood in the lungs is greater than the P_{CO_2} or air in the lungs, carbon dioxide will pass into the alveolar spaces and be exhaled. Note that bicarbonate ion is a key factor in determining blood pH and tissue pH. Normally in extracellular body fluids, the ratio of HCO_3^- to H_2CO_3 is 20 to 1. If the tissues are too acidic, the body will compensate by increasing its bicarbonate levels (and vice versa). Since oxygen diffuses into alveoli and body cells normally at 250 mL/minute, and since cellular production of CO_2 is approximately 200 mL/minute, we can see that oxygen uptake and CO_2 production are nearly equal.

A summary of blood gas transport is given in Figure 15.12.

Transcutaneous Monitoring of Blood Oxygen

With the present state of biomedical science, it is possible for the nurse to monitor blood P_{O_2} noninvasively and continuously. Apparatus are now commercially available (Figure 15.13) that provide a digital readout (and permanent printout) of approximate blood oxygen tension by a painless through-the-skin (transcutaneous) technique.

This type of noninvasive P_{O_2} monitoring can be very useful in coronary care, surgical patients, OB/GYN patients, inhalation therapy, and especially neonatal care. In neonates (newborn), particularly the premature, the circulating level of oxygen can be critical. In fact, both *hypoxemic* and *hyperoxemic* episodes can pose a threat to the sick neonate with respiratory distress.

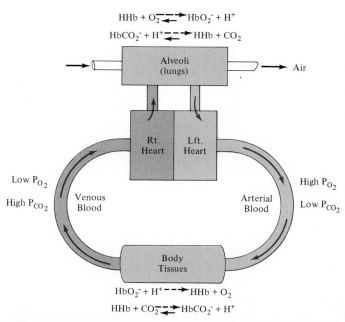

FIGURE 15.12   A Summary of Blood Gas Transport.

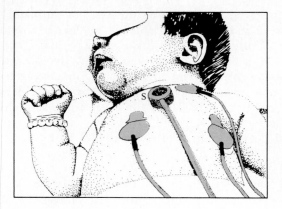

FIGURE 15.13   **Blood oxygen tension (P_{O_2}) can be measured without puncturing the skin. The sensor, S, normally placed over the sternum, continuously measures the amount of oxygen diffusing through the skin into the sensor (which is heated). Other sensors, left and right, monitor breathing and heart rates.**

Transcutaneous oxygen (tcP_{O_2}) monitoring does not do away with arterial blood gas analysis. Blood samples must still be drawn to determine CO_2 tension and pH and to calibrate the instrument. But the tcP_{O_2} technique reduces the need for blood sampling and has the advantage of producing a permanent, continuous printout.

On the subject of respiratory therapy, the nurse should be aware that a too-high oxygen concentration in environmental air can damage the eyes of newborns.

Carbon Monoxide

As a colorless and odorless gas, carbon monoxide is a potentially silent and stealthy killer in the human. When inhaled, CO and O_2 compete for binding sites on hemoglobin. But CO wins out, as it can bind 240 times more strongly than oxygen. The complex of CO and hemoglobin is called *carbonmonoxyhemoglobin* (or carboxyhemoglobin) and is cherry red in color. Its formation reduces the oxygen-carrying and O_2 delivering capacity of the blood. The more CO in the air that is inhaled, the less oxygen will get to the brain and other tissues.

The following are symptoms of CO poisoning, in progression of severity:

1. Slight headache, dizziness, lassitude.
2. Headache, throbbing temples.
3. Weakness, mental confusion, nausea.
4. Rapid pulse and respiration, fainting.
5. Possibly fatal coma.

Where are we exposed to carbon monoxide in our society? We inhale it with automobile exhausts, space heater exhausts, cigarettes, marijuana joints, and industrial and natural air pollution. In Tokyo, traffic policemen are exposed to so much CO that they must take time out every hour to inhale oxygen. In the United States, guards in auto tunnels must use specially ventilated rooms for resuscitation.

Respiratory Acidosis and Alkalosis

The *extreme* range in pH of the extracellular fluid in the human is 7.70 to 7.00; outside of this range life is not possible. In large measure, the acid–base status of the body depends on respiratory function. If ventilation is reduced through disease or drugs (hypoventilation), CO_2 will not be adequately removed from the body, and **respiratory acidosis** may develop (plasma pH < 7.4). On the other hand, a hysterical patient may breathe much too rapidly (hyperventilation), and a **respiratory alkalosis** may develop as an excessive amount of CO_2 is eliminated (blood pH > 7.4). Here is a summary of causes of respiratory acidosis and alkalosis and the clinical signs associated therewith.

Acid–base disorder	Possible causes	Clinical signs
Respiratory acidosis	Impaired ventilation as in pneumonia, emphysema, severe asthma; an excess of CO_2 in inhaled air; depression of brain's respiratory center (as with drugs such as heroin, morphine, barbiturates, anesthetics)	Plasma pH less than 7.4; increased P_{CO_2}; to compensate for the acidosis, the kidney conserves HCO_3^- and thus eventually HCO_3^- levels tend to rise
Respiratory alkalosis	Hyperventilation as in hysteria, neurotic hyperventilation, salicylate (ASA) poisoning, CNS damage, fever, hypoxia	Plasma pH greater than 7.4; a decreased P_{CO_2}, hyperventilation; to compensate for respiratory alkalosis, the kidneys conserve H^+ and reject HCO_3^-, and thus plasma bicarbonate tends to fall

▰▰▰▰▰ EXAMPLE 15.2 ▰▰▰▰▰

Respiratory alkalosis caused by neurotic hyperventilation is sometimes relieved by having the patient rebreathe his own expired CO_2 from a paper bag—but not a plastic bag. Why not a plastic bag?

Solution

Plastic bags offer too great a danger of collapsing upon the patient's face, causing asphyxia.

Table 15.4. How to Diagnose Acid–Base Disorders from Data on Analysis of Blood Gases

---

Look at pH. For this maneuver, only a pH of 7.4 is considered to be normal, even though the range of normal is actually 7.35 to 7.45.

A pH under 7.4 indicates acidemia.

A pH over 7.4 indicates alkalemia.

Look at $PaCO_2$ (range of normal: 35 to 45 mm Hg). If $PaCO_2$ is normal, there is no primary respiratory problem and no respiratory compensation for a metabolic problem.

Abnormal $PaCO_2$ values are interpreted in relation to pH:

↑ $PaCO_2$ plus ↓ pH: acidosis of respiratory origin.

↑ $PaCO_2$ plus ↑ pH: respiratory retention of CO_2 to compensate for metabolic alkalosis.

↓ $PaCO_2$ plus ↑ pH: alkalosis of respiratory origin.

↓ $PaCO_2$ plus ↓ pH: respiratory elimination of CO_2 to compensate for metabolic acidosis.

Look at HCO_3^- (range of normal: 22 to 26 mEq per liter). If HCO_3^- is normal, there is no primary metabolic problem and no metabolic compensation for a respiratory problem.

Abnormal HCO_3^- values are interpreted in relation to pH:

↓ HCO_3^- plus ↓ pH: acidosis of metabolic origin.

↓ HCO_3^- plus ↑ pH: renal retention of H^+ or elimination of HCO_3^- to compensate for respiratory alkalosis.

↑ HCO_3^- plus ↑ pH: alkalosis of metabolic origin.

↑ HCO_3^- plus ↓ pH: renal retention of HCO_3^- or elimination of H^+ to compensate for respiratory acidosis.

Use above findings to diagnose acid-base status. Possible disorders include compensated or uncompensated respiratory acidosis (synonym "hypoventilation"), compensated or uncompensated respiratory alkalosis (synonym "hyperventilation"), compensated or uncompensated metabolic acidosis, and compensated or uncompensated metabolic alkalosis. Simultaneous respiratory and metabolic disorders are also possible. If $PaCO_2$, HCO_3^-, and pH are *all* within their normal ranges, acid-base status is normal.

Look at PaO_2 (normal: 80 mm Hg for elderly adults at sea level; 100 mm Hg for young adults at sea level). A PaO_2 below normal for age indicates hypoxemia.

Source: Joanne Schnaidt Rokosky, *The Nursing Clinics of North America* 16:2 (June 1981). Copyright Saunders Publishing Company. Used with permission.

Chemical Application of Blood Gas Analysis Data

The health professional should be familiar with data obtained in the analysis of blood gases and how these data are interpreted clinically in the diagnosis of acidemia or alkalemias. As can be seen in Table 15.4, the pH of arterial blood tells us if the patient is acidemic or alkalemic, and the HCO_3^- and

P_{CO_2} values tell us if the acid–base problem is of respiratory or metabolic origin. You should examine Table 15.4 thoroughly and carefully.

Metabolic Acidosis and Alkalosis

We want to be able to distinguish acid–base disorders caused by errant *metabolic* processes from those caused by respiratory malfunctions. For example, in the diabetic, increased reliance on fats for energy production results in large-scale biosynthesis of acetoacetic acid (I) and beta-hydroxybutyric acid (II).

$$CH_3COCH_2COOH \qquad\qquad CH_3CH(OH)CH_2COOH$$
$$I \qquad\qquad\qquad\qquad II$$

These ketone bodies act as sources of hydrogen ion (H^+), and though buffering occurs to some extent, severe **metabolic acidosis** can still develop. Other common causes of metabolic acidosis are strenuous exercise (heavy pyruvic acid, lactic acid, and CO_2 production), kidney failure to excrete acid (as in diseased kidney) and ingestion of acid-forming salts such as ammonium chloride. Severe dehydration and severe diarrhea can also contribute to acidosis.

To compensate for a metabolic acidosis, the respiratory center in the brain signals for increased respiration (increased loss of CO_2) and the kidneys increase H^+ excretion (in the form of ammonium ion). Clinically, metabolic acidosis is treated by IV administration of bicarbonate, but if kidney failure is diagnosed, hemodialysis would be the treatment of choice.

Metabolic alkalosis can result when acidic gastric juice is lost, as occurs in pernicious vomiting or in gastric lavage (stomach pumping). Some people ill-advisedly use large amounts of bicarbonate of soda ($NaHCO_3$) to treat an acid stomach, resulting in a metabolic alkalosis. Fruits are the principal dietary sources of alkali. Metabolic alkalosis can also develop as the result of diuretic therapy, from a too-active adrenal cortex, or from the administration of corticosteroids.

Treatment of metabolic alkalosis consists in reducing the body's level of bicarbonate. This can be accomplished by the administration of NaCl, which causes increased HCO_3^- excretion:

$$NaCl + H_2CO_3 \rightarrow NaHCO_3 + H^+ + Cl^-$$
$$\text{in the kidney} \qquad \text{excreted}$$

Acidosis or alkalosis is of special concern in infants who typically exchange more than 50 percent of their extracellular fluid each day (adults exchange less than one-fifth).

Blood buffers provide another chemical means of maintaining the pH between 7.35 and 7.45. Buffers, you will recall, are chemicals that tend to resist incipient changes in pH. A combination of a potential acid and a potential base, a buffer will react with added base to neutralize it; it will react with added acid to neutralize it, too.

The buffer systems of the blood are as follows:

- HCO_3^-/H_2CO_3.
- Hemoglobin.
- Plasma proteins.
- $HPO_4^{2-}/H_2PO_4^-$ (of greater significance in cells).

Let's take a look at how the bicarbonate–carbonic acid system can act as a buffer.

1. If acid is dumped into the blood, the buffer reaction will be

$$H^+ + HCO_3^- \rightleftharpoons H_2CO_3$$

acid   weakly dissociated

In effect, the hydrogen ions have been gobbled up and tucked away in the form of weakly dissociated acid, H_2CO_3.

2. If base is dumped into the blood, the reaction will be

$$OH^- + H_2CO_3 \rightleftharpoons H_2O + HCO_3^-$$

base   weakly ionized

The base has been neutralized to water. The other product, HCO_3^-, is a part of the original buffer system.

The buffering actions of hemoglobin and of plasma protein are similar. However, in the case of Hb and protein the groups that buffer against addition of acid and base are found in the same molecule (dipolar ions).

Other Functions of Blood

As a distribution channel, the blood acts to send glucose, proteins, amino acids, cholesterol, fatty acids, lipids, vitamins, metal ions, iodine, and epinephrine and other hormones to depots, organs, receptor sites, and target tissues where these substances are required. Hormones, secreted from ductless glands, enter the blood directly and are distributed generally to all body organs, but specifically to their target tissues. In the case of growth hormone, the target tissues are the epiphyses (ends of the long bones); with ACTH, the adrenal cortex; with thyroid stimulating hormone, the thyroid gland.

The blood functions to transport wastes to the kidney where they may be filtered out and excreted. This is discussed later in the section on urine chemistry.

Finally, the blood acts to distribute heat from the areas of heat production to the areas of heat loss, maintaining a "normal" body temperature of 37°C (98.6°F) orally, and one-half to one degree higher than that, rectally. We recognize, however, that significant variation in "normal" body temperature can occur depending on the time of the day, environmental conditions, exercise, ovulation, or ingestion of hot or cold liquids (Figure 15.14).

Pathways for heat production and heat loss can be summarized as follows.

Heat production	*Heat loss*
1. Body metabolism (e.g., carbohydrate metabolism)	1. Radiant heat loss and conduction of heat to object we touch
2. Food intake (heat is released as food is assimilated)	2. Evaporation of body surface water (perspiration)
3. Muscular activity	3. Moisture lost on exhaled breath
4. Infants and animals have a special type of *brown fat* not found in adults; it has a high rate of metabolism, and produces much heat	4. Urine and fecal heat loss

The body's thermostat, or heat regulatory center, is located in the hypothalamus. Heat loss is mediated through this center and is brought about by perspiration, peripheral vasodilation (increased blood flow in skin and extremities), and increased respiration. Conversely, to conserve heat, the thermostat acts to diminish peripheral circulation and reduce water loss. We urinate more often in cold weather because the water that would ordinarily be lost by skin evaporation is now lost through the kidneys as the body conserves heat.

Alcohol acts to dilate peripheral blood vessels, giving a feeling of warmth in cold weather but causing a net *loss* of heat over the long run.

The *febrile state* is another name for *pyrexia*, or fever. Thus a patient can be febrile or afebrile. Drugs used to treat a fever are termed *antipyretics*.

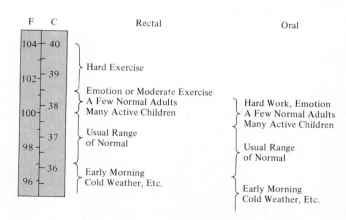

FIGURE 15.14 Normal body temperatures of healthy people can vary widely depending on time, age, activity, and environment.

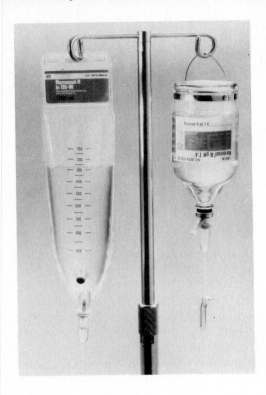

FIGURE 15.15 Water to be used for intravenous injection must have the correct pH and osmolarity, and be sterile and pyrogen free. (Photo courtesy Abbott Laboratories, North Chicago, Illinois.)

Aspirin, in 325 to 650-mg doses, and acetaminophen, in the same dosages, are the two most often used antipyretics. It is important to control a fever, for brain damage can result if the rectal temperature remains above 41°C for prolonged periods. Fevers as high as 41.5°C are not uncommon in encephalitis.

Substances that cause fever, or that "reset the thermostat," are termed **pyrogens.** The toxins (poisons) liberated by pathogenic organisms infecting the body are often potent pyrogens. Scarlet fever toxin is an example. External pyrogens can also stimulate the body to produce endogenous pyrogens. Tiny pieces of foreign protein in water for injection can be pyrogenic. That is why we specify pyrogen-free water for injection (Figure 15.15).

In **heat stroke,** the temperature regulatory center in the brain becomes malfunctional because of prolonged exposure to high temperatures or sun. Ingestion of alcohol makes the condition worse. Early symptoms are dizziness and weakness, progressing to spots before the eyes and ringing in the ears. Bright red, dry skin is seen, as is a strong pulse, later becoming weak. Unconsciousness usually follows as temperatures rise to 108 to 112°F. The condition may be fatal; it is not to be confused with heat exhaustion.

Many apparently normal people have a higher than normal body temperature (rectally, up to 37.8°C). This is termed constitutional *hyperthermia.* Hard exercise, as in running, can raise body temperatures to near 40°C (see Figure 15.14). *Hypothermia* can develop in persons exposed to very cold weather; it can be a serious threat to life. On the other hand, hypothermia is deliberately induced in some patients prior to brain or heart surgery. Cooling reduces body needs for oxygen, and circulation can be stopped for relatively long periods.

Blood Types

To understand the following discussion of **blood types,** you must understand what an **antigen** is and what an **antibody** is. An antigen is any protein substance invading the body or formed within the body which stimulates our immunological response system to manufacture antibodies. The antibodies that are manufactured are specifically tailored protein molecules that can bind to the antigen and knock it out of commission. Body tissues that make up the immunologic response system are the lymph, bone marrow, and spleen. The key point in this is that specific antibodies are produced in response to specific antigens. The antibody for measles virus, for example, cannot knock out the antigen for polio virus, because it does not recognize the polio virus and cannot bind to it. Antibodies have a very high degree of specificity.

Animals are capable of producing a tremendous variety of antibodies. All vertebrates that have been studied so far can produce 100,000 to 1 million antibodies.

Figure 15.16 helps explain the relationship between antigen and antibody.

Why is all of this necessary? It provides protection against the many antigenic bacteria, viruses, foreign proteins, and so on that invade our systems daily. Our immunological response system is essential for our existence if we are going to handle and resist the pathogens that attack us. After an organ transplant, the immunologic response system is temporarily knocked out of commission with drugs so that it cannot reject the organ transplanted (which

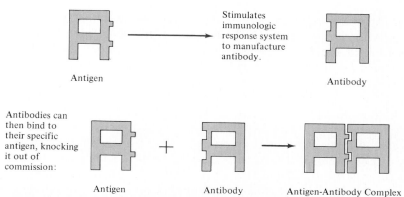

FIGURE 15.16   The Immunological Response to Antigen.

Table 15.5. The Four Blood Types and Their Antigen–Antibody Relationships

	RBC antigen	*Plasma antibodies*	*Percent in U.S.*
People with type A blood have	Type A antigen	Antibodies to type B blood	41
People with type B blood have	Type B antigen	Antibodies to type A blood	10
People with type AB blood have	Type A and B antigens	None	4
People with type O blood have	None	Antibodies to both type A and B blood	45

it perceives as foreign protein). During this period of a nonfunctioning response system, the patient is at high risk of death from overwhelming infection.

Next, consider the red blood cells that exist in our bodies. On the surface membranes of these cells, a variety of antigens exist. They are glycoproteins which are capable of inducing the production of antibodies. They are termed *agglutinogens*. There are many hundreds of types on each red cell surface, but the most important agglutinogens are of the A and B types.

Let's assume something at this point. Let us assume you already know that humans have *one* of four possible blood types:

1. Type A blood
2. Type B blood
3. Type AB blood
4. Type O blood

How can we explain the possibility of four blood types from only two types of agglutinogens, A and B? The answer is given in Table 15.5, which shows that each of the four blood types has its own antigen and antibody profile. Note that each blood type contains its own antibodies, called *agglutinins*, either inherited or induced by exposure to another type of blood. Type AB blood, however, contains no antibodies (no agglutinins). A person with type AB blood can receive a transfusion of any other blood type safely because no antigen–antibody reaction (termed *agglutination*, or *clumping* of cells) can occur. But if type B blood is given to a type A person, a dangerous and possibly fatal agglutination reaction can occur as the anti-B bodies in the type A blood clump with the B antigens in the type B blood. Such clumping destroys cells by causing hemolysis. Likewise, type B blood is incompatible with type A blood.

EXAMPLE 15.3

To what type of blood group can type O blood be safely transfused?

Solution

Type O blood can safely be donated to *any* type. It is the universal donor. This is possible because type O blood has no A or B antigens. When mixed, say, with type A blood, the antibodies in type A blood have nothing with which to agglutinate. (If you are wondering why the anti-A bodies and the anti-B bodies in type O blood are not a factor in clumping, it is because they are so diluted in the large volume of blood in the recipient that they cannot cause significant agglutination.)

There is one more RBC antigen that you must study. First identified in the *rhesus* monkey, this agglutinogen is termed the **Rh factor.** Only about 10 to 15 percent of Americans do *not* carry this antigen on their RBC membranes. They are identified as Rh-negative (Rh −), and they do not normally have anti-Rh bodies in their blood. In fact, there is no naturally occurring anti-Rh antibody. But consider this real possibility. An Rh- woman becomes pregnant. Her fetus has Rh positive (Rh +) blood, inherited from the father. During delivery, some fetal blood gets into the mother's circulation and induces her immune response system to form a significant level of anti-Rh bodies. Years later, in a *subsequent pregnancy* with another Rh + baby, the mother's acquired anti-Rh agglutinins cross the placenta and clump with fetal blood in utero, hemolyzing and destroying fetal RBCs. This condition is termed *hemolytic disease of the newborn.* If the fetus dies, it is termed *erythroblastosis fetalis.*

This type of hemolytic disease of the neonate can be prevented if the mother-to-be, at the time of her *first* pregnancy with a Rh + baby, is given a dose of an immune globulin which prevents her immune response system from manufacturing anti-Rh bodies. The dose must be given immediately postpartum. One of the commercial products for this purpose is RhoGAM. A product such as RhoGAM is necessary only if the fetus is Rh + and the mother is Rh −. It is given after the first pregnancy with an Rh + baby in order to prevent agglutination in a subsequent pregnancy with an Rh + baby.

Summary

Blood types arise because RBCs carry different antigenic proteins on their surfaces. In addition, blood plasma has antibodies to RBC antigens of the other groups. Hence potentially serious incompatibilities can arise if the wrong blood is tranfused to a patient. The antigen–antibody (or agglutinogen–agglutinin) reaction is termed clumping (or agglutination) and hemolysis of red cells can occur. Type AB blood is the universal recipient because it has no agglutinins; type O blood is the universal donor because it has no agglutinogens. Childbearing women who are Rh− can sometimes manufacture agglutinins to the Rh+ fetus; in subsequent pregnancies this can cause hemolytic diseases in the newborn.

Hemodialysis

Visiting a **hemodialysis** unit in a large modern hospital, one gets the impression that not a lot is going on. Patients enter in their street clothes, recline in large, padded, dentist-type chairs, quickly get hooked up to a plastic tube, and sit there for 3 to 4 hours, reading a book or eating their lunch. What one does not see is a great deal of sophisticated biochemistry going on in the background, the result of a great deal of biomedical research.

The hemodialysis patient has kidneys that are functioning poorly or not at all. As a result, his blood is not being cleansed of toxic wastes such as urea, creatinine, and uric acid (which can add up to 55 to 70 g every 24 hours). If these wastes are permitted to accumulate uncontrolled, death will result from uremia.

The principle behind hemodialysis is simple: pass the patient's blood through tiny tubes with walls as permeable as a membrane so that the toxic wastes can pass through the tubes (dialyze) and be excreted while the formed elements and the desired blood constituents are retained (Figure 15.17). Note that there is a difference between an osmotic membrane and a dialyzing membrane. An osmotic membrane allows the passage of solvent only whereas a dialyzing membrane allows the passage of a number of *small* ions and molecules. A modern hemodialysis unit is shown in Figure 15.18 along with the heart of the apparatus, the hollow fiber dialyzer. In the dialyzer, blood passes through thousands of tiny tubes (the hollow, thin-wall fibers), which present a large, effective surface area (greater than 1 m²) for membrane passage. Externally, the tubes are bathed in the dialysate fluid.

The challenge in hemodialysis is to design a system that will selectively remove wastes and conserve the rest. Formed elements are easy; they are too large to pass through the membrane. Chemicals in solution are a different matter. They are *in effect* retained by bathing the entire tube apparatus in a dialyzing solution (the dialysate) that has the desired chemicals already

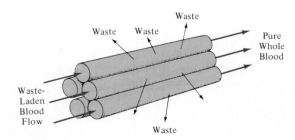

Waste

Waste   Waste

Pure Whole Blood

Waste-Laden Blood Flow

Waste

FIGURE 15.17   The Principle Behind Hemodialysis. **The membranes permit waste to pass out but retain formed elements.**

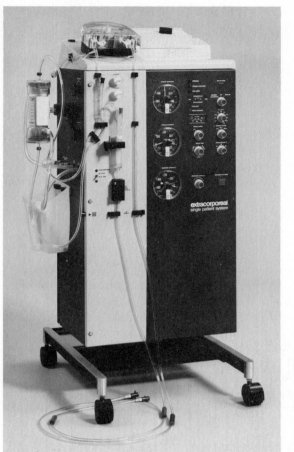

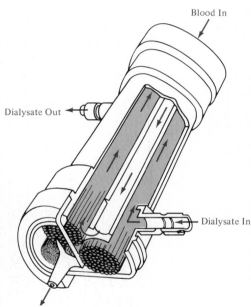

Blood In

Dialysate Out

Dialysate In

FIGURE 15.18 **(Left)** A Modern Hemodialysis Unit. **(Right)** Hollow Fiber Dialyzer. **The heart of the artificial kidney is the hollow fiber dialyzer, shown here. Blood passes through the thousands of hollow fibers while dialysate fluid (in separate compartments) flows outside the fibers. The resulting hemodialysis clears the blood of impurities. (Photo courtesy Extracorporeal Inc., King of Prussia, PA. 19406.)**

dissolved in it at the same concentrations as found in whole blood. There will be no net membrane transport of Na^+, K^+, Ca^{2+}, HCO_3^-, Mg^{2+}, and so on across the membrane if the concentrations of electrolytes on both sides are the same. By regulating the composition of the dialysate, the physician can regulate the net flow of waste material out of the blood and also the net flow of needed electrolytes into the blood.

Typically, the "artificial kidney machine" patient will have to be dialyzed two or three times a week for 4 hours per visit to maintain his health. Although it is true that in recent years, 30,000 kidney transplants have been performed, it is also true that large numbers of people continue to undergo routine kidney dialysis. In 1979, about $1 billion was spent on such dialysis in the United States, with the cost per patient per year in hospitals being more than $30,000. (For an article describing the cost of hospital and home dialysis treatment, survival rates of hemodialysis patients, and the government's share of the expense of uremia therapy, see the *Journal of the American Medical Association*, July 17, 1981, p. 230.)

A recent innovation in hemodialysis is continuous ambulatory peritoneal dialysis (CAPD) in which 2 L of dialysate is injected through the patient's abdominal wall to mix with fluid bathing the intestines. Wastes dialyze into the fluid, which, after a time, is drained out of the peritoneum and discarded. The procedure is repeated about four times a day, forever. It has the great advantage of allowing mobility. Patients can be taught to carry out the procedure at home.

Occult Blood

The health professional should be familiar with the meaning and significance of the term **occult blood**. Blood that has entered the intestinal tract—from disease, hemorrhage, trauma, or drug use—in quantities so minute as to be recognized only by microscope or special chemical test, is termed occult blood (*occult* = hidden). The *guaiac* test for occult blood is a do-it-yourself test for hidden blood in the stool that might be a sign of cancer of the bowel. The patient is instructed to take stool smears at home using a special slide. If laboratory analysis then shows signs of blood, further testing to discover its source can be ordered.

EXAMPLE 15.4

Why should the patient who is to take the guaiac test be advised to avoid meats and aspirin for 4 days prior to taking the fecal smear?

Solution
Meat may have animal blood in it; aspirin is notorious for causing a small amount of blood loss into the intestine.

URINE CHEMISTRY

Two healthy, adult kidneys are capable of filtering 7.5 L of blood an hour, or almost 50 gal a day. By "filtering" we mean that the blood flows through the kidney, wastes are removed as urine, valuable constituents are recovered and placed back into the blood, and acid–base and water balance are maintained.

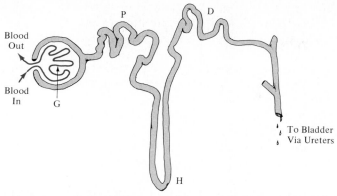

FIGURE 15.19   A highly diagrammatic representation of a human nephron showing glomerulus (*G*), proximal convoluted tubule (*P*), loop of Henle (*H*), distal convoluted tubule (*D*), and ducts collecting urine.

All of this occurs in anatomical units of the kidney called *nephrons*, of which there are over 1 million in each kidney (Figure 15.19). Blood from a renal artery enters the glomerulus, is filtered into the tubules as plasma, containing all blood constituents except the formed elements and blood proteins. As this glomerular filtrate passes through the proximal convoluted tubules, the loop of Henle, and the distal tubules, water, electrolytes, and other desired constituents are reabsorbed into the general circulation. The effectiveness of this reabsorption process is shown by the fact that of 20 mL of fluid reaching the loop of Henle, only 1 mL is eventually collected as urine. Energy as ATP is required for reabsorption since chemicals are being absorbed across membranes against their concentration gradients. Reabsorbed nutrients include Na^+, K^+, Cl^-, HCO_3^-, and glucose.

Many substances, both natural to the body and foreign, are actively secreted into the fluid of the tubules and thus end up in the urine. Urea, the major end-product of protein metabolism in man, and uric acid, an end-product of nucleic acid metabolism, are excreted in the urine.

urea   uric acid

Hydrogen ions (H^+) are actively secreted by the cells of the proximal and distal tubules. In addition, a great and diverse number of prescription and recreational drugs and accidentally ingested environmental agents (unchanged, or as metabolic breakdown products) can be eliminated from the body as solutions in the urine. The kidney is thus seen to play a vital role in maintaining good health.

Note that even though glucose is always present in the blood, it is not a normal constituent of urine. This is because glucose is completely reabsorbed in the tubules. Glucose, of course, can be present in the urine in diabetes mellitus, along with ketone bodies. Tests for glucose and ketone bodies in the urine are discussed in Chapter 12.

Blood and proteins are not found normally in significant quantities in the urine but in pathological conditions may be so (**hematuria** and *proteinuria*, respectively). Urine, as excreted, is ordinarily sterile. The presence of bacteria in the urine is evidence of a urinary tract infection (UTI) or of a sloppy technique in collecting the specimen.

Analysis of the urine (urinalysis) is one of the most frequently performed of all body fluid analyses, for the information it provides can be highly useful in diagnosis. Renal disease, for example, is indicated when a protein such as albumin is found in the urine (**albuminuria).** In clinical laboratories, many tests are performed for a great variety of urine constituents. Home test kits are available, too, for ketone bodies in the urine and for blood, protein, or bilirubin in the urine. These tests consist of dipping a specially prepared, reagent-impregnated test paper into a urine sample and then noting the color change. Some degree of accuracy is sacrificed to convenience, with the multitest strips being the least accurate of all. For a discussion of urinalysis in diabetes mellitus, see Chapter 12.

Urea, $H_2N—CO—NH_2$, which results from human liver metabolism of amino acids, is reported clinically as blood urea nitrogen (BUN), and is normally found in the range of 8 to 25 mg %. BUN values can be used to diagnose disease states.

Acid—Base Balance in the Urine

Healthy urine can range in pH from 4.5 to 8.0, with 6.6 being a reasonable average for a person on an ordinary diet. Many common fruits and vegetables contain salts of organic acids (e.g., calcium citrate, potassium tartrate, sodium fumarate), which tend to make the urine alkaline. On the other hand, high protein foods with their sulfur and phosphorus content are acid-residue foods and tend to make the urine acidic.

Through the urine, the body can reject excess acid or excess base, in its attempt to maintain homeostasis. Thus, if the urine is very acid, we know that the body is removing excess acid. If a person is taking large quantities of sodium bicarbonate or other antacid for "acid stomach," his urine pH can easily rise to 7.5 to 8.0 as his body rejects the excess alkali.

Just as we observed buffers in the blood, there are buffers in the urine.

The most important is the phosphate system $HPO_4^{2-}/H_2PO_4^{-}$. Here is how this buffer can resist threatened changes in pH.

If acid is added

$H^+ + HPO_4^{2-} \rightarrow H_2PO_4^{-}$

If base is added

$OH^- + H_2PO_4^{-} \rightarrow HPO_4^{2-} + H_2O$

The kidneys also handle acid by making ammonia, which can react with acid.

$NH_3 + H^+ \rightarrow NH_4^{+}$

The NH_4^{+} ion, once formed, can be excreted instead of Na^+, conserving Na^+. Actually, the system NH_3/NH_4^{+} constitutes a buffer.

The third buffer system in the urine is HCO_3^{-}/H_2CO_3. We have already discussed the chemistry of this buffer system under blood buffers.

Urine buffers can play an important role. They can gobble up acid or base, keeping the pH of the urine from too quickly getting to its limit of 4.5 on the acid side or 8.0 on the alkaline side.

The kidneys can also help control the acid–base balance of the body by a process that involves active exchange of sodium ions for hydrogen ions across tubule walls. For example, to rid the body of excess acid, the nephrons can "secrete" H^+ ions from blood into the glomerular filtrate and simultaneously exchange them for Na^+ (and accompanying HCO_3^{-}). This acidification of the urine provides a means of controlling excess systemic acidity, for it effectively removes H^+ from blood and retains the base bicarbonate.

Diuretics

Diuretics are agents that increase the volume of urine, thereby reducing total body water. The process is called *diuresis*. In heart failure, poor circulation can result in water accumulating in the legs and lungs. We call this condition *edema*. A diuretic will promote loss of body water and help reduce the edematous condition. Diuretics are often given to *hypertensives* (patients with high blood pressure) in order to cause the loss of Na^+ and water, both of which can contribute to the elevated blood pressure.

The diuretics that are most often used clinically produce water loss by disrupting normal biochemical events in the nephron. For example, the heavily prescribed diuretic furosemide (Lasix) acts to inhibit Cl^- ion reabsorption in the loop of Henle, resulting in extra loss of water.

The kidneys are the target tissue for certain hormones that help regulate the volume of body water. *Antidiuretic hormone* (ADH, or vasopressin) is secreted by the pituitary gland and acts on the kidney to cause retention of water by stimulating its reabsorption in the tubules. ADH is released when body receptors sense that the osmotic pressure of the blood is too high. Conversely, ADH release is inhibited when blood osmotic pressure is too low. Some individuals suffer from diabetes insipidus (a lack of ADH) and pass huge volumes of urine daily; this is accompanied by a great thirst and often a voracious appetite.

Renin is a proteolytic enzyme synthesized, stored, and secreted in kidney cells. Renin plays a role in regulation of blood pressure by catalyzing the conversion of angiotensinogen to angiotensin I, which then is converted in the lungs and elsewhere to angiotensin II. It has been established that angiotensin II is the most potent vasoconstrictor known.

BILE CHEMISTRY

We learned in Chapter 13 that bile acts to emulsify fats in the small intestine, promoting their digestion and absorption. We also discussed the chemistry of bile salts. In this section, we will discuss the anatomical features involved in bile secretion, the formation of bile pigments, and the role of bile pigments in the condition called jaundice.

Bile, secreted continuously by cells in the liver, is stored in the gallbladder and ejected into the small intestine when foods are present. A diagrammatic representation of the anatomy involved is given in Figure 15.20. Like pancreatic juice, bile is alkaline, and helps to neutralize the acidic stomach contents as they squirt into the small intestine. Bile is almost 90 percent water, with small amounts of bile salts, the highly colored bile pigments, cholesterol and other lipids, and inorganic ions. From 500 to 1000 mL of bile can be secreted per day.

Bile pigments are important to the health professional because they are the link to jaundice in adults and in newborns. Let us see how this comes about.

Like everything else, red blood cells get old and die (125-day life span). The hemoglobin in these cells is split into globin plus heme. (Four hemes from each hemoglobin, remember?) In the bone marrow and in the liver, the heme molecules are oxidized, with ring splitting, to a substance called *biliverdin*, which in turn is reduced to the bile pigment *bilirubin*. Bilirubin is excreted with the bile into the small intestine; it gives bile its yellow color. Actually, then, this route constitutes a pathway for metabolism and excretion of hemoglobin (Figure 15.21). It is seen that the brown color of fecal matter is due to the original secretion of bilirubin to the feces by means of bile. Patients with any disease that obstructs the normal flow of bile will perforce have gray-colored feces, a fact that helps the physician diagnose the obstruction.

Jaundice occurs when bilirubin, yellow in color, builds to excessive levels in the blood (more than 2.5 mg %) and is deposited in the skin and sclera

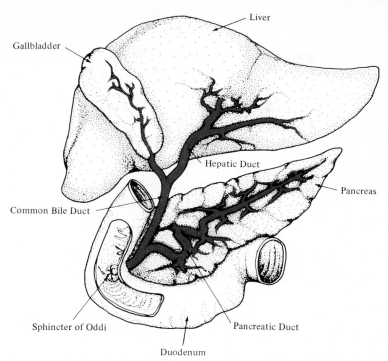

FIGURE 15.20   Bile is made in the liver, stored in the gallbladder, and delivered to the duodenum via the common bile duct.

(whites of the eyes). In serious cases, the skin and sclera are distinctly and persistently yellow. Jaundice in the newborn is not uncommon; the babies are distinctly yellow. Another name for jaundice is *icterus*.

Bilirubin buildup in the blood can result from the following:

1. Hepatic or other liver disease in which liver function is impaired; circulating bilirubin is not excreted and blood levels increase.
2. A blocked bile duct that prevents bilirubin excretion into the intestines; it backs up into the blood.
3. A lack of enzymes needed for the metabolism of bilirubin (in newborns).

Jaundice occurs in up to 70 percent of neonates shortly after birth because of insufficient production of the liver enzyme that catalyzes the biotransformation of bilirubin to water-soluble conjugates for urine excretion. Already at birth, RBCs are breaking down and releasing hemoglobin, and each gram of Hb released can give rise to 35 mg of bilirubin. Almost all neonatal jaundice is mild and disappears spontaneously, but when it is severe it must be treated, as brain damage can occur.

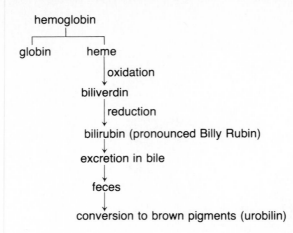

FIGURE 15.21 The Biotransformation of Hemoglobin to Bile Pigments.

A common treatment for neonatal jaundice involves shining a special fluorescent light onto the baby's skin (*bilirubin reduction light*). The energy in the light photolytically decomposes some of the bilirubin just beneath the surface of the skin, converting it to biliverdin or other colorless nontoxic products. The baby, with eyes protected, is usually under the light for several days, and is turned continuously to expose all skin surfaces. With this treatment, the baby's stools are often green.

CEREBROSPINAL FLUID

You have probably heard of a "spinal tap"—the withdrawal of fluid from the lumbar region of the spinal cord with a hypodermic syringe. The fluid that is withdrawn is *cerebrospinal fluid (CSF)*, and it exists to the extent of 100 to 150 mL in its own special compartment, quite separate from blood. CSF is not stagnant; it is used up and reformed each day.

Cerebrospinal fluid acts as a cushion for the brain and spinal cord, which can be thought of as "floating" in the CSF. In a blow to the head, it is the cushion of CSF that protects the brain.

CSF is generally similar in composition to plasma, with two major exceptions. There is 300 times more protein in plasma than in CSF, and 875 times more cholesterol.

However small the quantity of protein in CSF, it becomes very important in the diagnosis of disease. In syphilis of the brain, bacterial meningitis, and cerebral hemorrhage, the CSF contains elevated gamma-globulin levels.

Hydrocephalus, which the laity call "water on the brain," is an excessive accumulation of cerebrospinal fluid. In spina bifida patients, accumulation of cerebrospinal fluid is relieved by the permanent insertion of silicone rubber (polysiloxane) tubes and valves through which fluid drains to an outside bag.

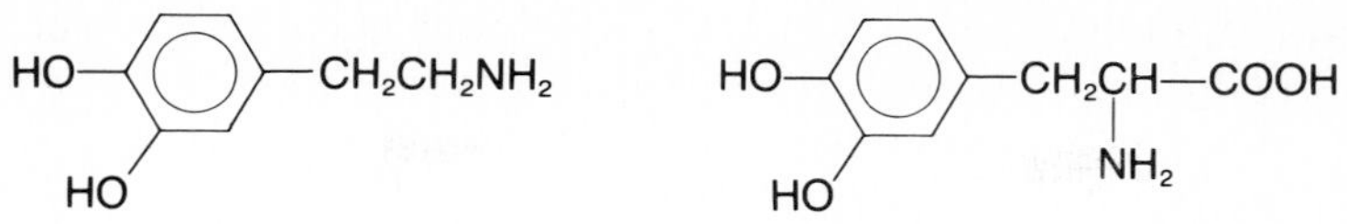

HO—⬡—CH$_2$CH$_2$NH$_2$ HO—⬡—CH$_2$CH—COOH
HO   HO   NH$_2$

dopamine   L-dopa (dihydroxyphenylalanine)

FIGURE 15.22 L-Dopa is the drug of choice in the treatment of Parkinson's disease. It can pass the blood-brain barrier far more easily than its metabolite, dopamine.

The concept of the **blood–brain barrier** has been created to explain the fact that many chemicals and drugs in the blood cannot readily pass over into the CSF (or into the extracellular fluid of the brain, which is essentially the same thing). Penicillin is a famous example. Oral or IV administration of penicillin may be essentially ineffective in the treatment of meningitis or other cerebrospinal infection. Although it is true that no substance is completely excluded from the brain, it is also a fact that only O_2, CO_2, and water pass the blood–brain barrier with ease.

It appears that the explanation for the existence of this barrier lies in the sheath of fat that surrounds capillaries in the brain. Water-soluble chemicals have a more difficult time passing through this sheath, and their rate of entrance into the CSF in thus slower than that of fat-soluble substances. Then, too, it appears that there are fewer pores in the capillaries of the brain than in other tissues.

The blood–brain barrier sometimes dictates the choice of drug in the treatment of a disease. Dopamine is effective in the treatment of Parkinson's disease, but dopamine's blood–brain barrier transfer rate is too low for it to be effective. However, the biochemical precursor to dopamine, L-dopa, passes the blood–brain barrier readily, and since it is biotransformed in vivo to dopamine, it is the drug of choice in Parkinson's disease (Figure 15.22).

SYNOVIAL FLUID

The yellow viscous fluid that lubricates the bony joints of the body is termed *synovial fluid*. The knee is a joint from which synovial fluid can be aspirated with a hypodermic syringe. *Gout* is a disease of the joints and other tissues. In gout, symptoms of arthritis appear, with the big toe (!) most often afflicted. The condition is very painful. We know that gout is associated with elevated blood and urine uric acid levels (formula given earlier), and that salts of uric acid deposit in the joints, presumably causing the pain.

In the body, uric acid is formed by the metabolism of purines. Any foods high in nucleoproteins, such as glandular organ foods, produce an increase in uric acid levels and thus exacerbate gout. Dairy products and eggs, on the other hand, are very low in purines. Colchicine USP, a drug obtained from *Colchicum autumnale,* a crocuslike plant, is used for the relief of acute attacks of gout, and in the prophylaxis of recurrent attacks.

Anti-inflammatory steroids are sometimes injected into bony joints, into the synovial fluid, for the treatment of chronic disease such as bursitis, or for the management of other acute inflammatory processes.

SUMMARY

The formed elements of the blood are red blood cells, white cells, and platelets. Extracellular fluid consists of the blood and the fluid bathing cells. The fluid inside cells is termed intracellular fluid.

The body is constantly striving to achieve homeostasis. Against the stresses of all external and internal forces, the body strives to maintain pH, O_2, CO_2, blood pressure, respiration, water volume, and all other factors at a physiologically normal level. We discussed mechanisms by which the body accomplishes homeostasis.

We defined plasma and serum. We specified the nature and function of three types of blood proteins: albumins, globulins, and fibrinogen. Plasma electrolytes consist mainly of NaCl with smaller amounts of K^+, Ca^{2+}, phosphate, Mg^{2+}, and HCO_3^-. The principal cation of plasma is sodium; that of intracellular fluid is K^+. The lymphatic system is a drainage system for proteins, fats, and other material; it drains interstitial fluid back to the blood.

Normal values for many blood constituents are given in Appendix IV. Too low a blood level of Na^+ or K^+ can have serious consequences. Hypokalemia can manifest itself in an irregular ECG.

The normal range of WBCs is 5000 to 10,000 per mm^3; that of RBCs, 4 million to 6 million per mm^3; in males. Red cells (erythrocytes) are flexible and easily distorted in shape. In sickle cell anemia, a genetic defect results in red cells that are sickled in shape and incapable of carrying a normal load of oxygen.

Hemopoiesis is red cell formation in the bone marrow. Hemolysis is rupture of red cells when placed in a hostile environment. The hematocrit is the percentage of blood that consists of formed elements; a hematocrit of 42 percent is normal for adult females. Platelets are an integral part of the clotting process.

Under blood functions, we discussed the blood gases O_2 and CO_2. The role of hemoglobin in O_2 and CO_2 transport was discussed. Gases in the blood are measured by their tension (their partial pressure in relation to the total pressure); the units are millimeters of mercury (Torr). Figure 15.13 is a summary of oxygen and CO_2 transport, binding with Hb, and tensions in the cardiovascular system.

CO_2 reacts readily with water to give H_2CO_3, which in turn ionizes to H^+ and bicarbonate. The reaction is reversible and is catalyzed by carbonic anhydrase. This reaction explains body acidosis when CO_2 accumulates. Carbon monoxide (CO) is highly toxic because the bond it forms with Hb is 240 times as strong as the bond O_2 forms with Hb. CO can thus effectively eliminate hemoglobin's oxygen carrying capacity.

There are two types of acidosis: respiratory (basis is in lung function) and metabolic (basis is in some physiologic process). We discussed the causes and clinical signs of acidosis and alkalosis and we learned (Table 15.4) how to diagnose acid–base status based on clinical blood gas data. We examined causes for metabolic acidosis and alkalosis and we learned how blood buffers work.

The blood acts to distribute heat to our body. Pyrexia (fever) is treated with antipyretics. Pyrogens are toxins or foreign proteins that induce fever.

The four primary blood types are A, B, AB, and O. People with type A blood have antibodies to type B blood and vice versa. Type O blood is the universal donor, it has no A or B antigens. Transfusions of the wrong type of blood can be fatal. Hemolytic diseases of the newborn can occur if an Rh− mother (previously sensitized during an earlier pregnancy) carries an Rh+ fetus. The mother's anti-Rh agglutinins can cross the placenta and destroy the baby's red cells.

In hemodialysis urea, uric acid, and creatinine are removed from the patient's blood by passing the blood through membranes permeable to these three chemicals but impermeable to formed elements. During hemodialysis, the membranes are bathed in dialysate fluid containing all the ions and molecules the physician does *not* want dialyzed out.

Next, we discussed the anatomy of the nephron and the formation of the urine and normal and abnormal constituents of urine. We discussed urine buffers, diuresis, and the antidiuretic hormone.

Normal breakdown of red cells releases Hb, which, in turn, is metabolized to the bile pigment bilirubin. This yellow-colored substance is normally excreted in the bile. But obstruction of bile flow or the presence of liver disease can result in a blood buildup of bilirubin high enough to cause a yellow color in the skin and sclera. These are cardinal signs of jaundice. Neonatal jaundice is common.

Cerebrospinal fluid cushions the brain and spinal cord. Its contents are used in diagnosis of illness. To get from the bloodstream to the cerebrospinal fluid, a drug must pass the blood–brain barrier.

The viscous fluid that lubricates the bony joints is synovial fluid. Drugs can be injected directly into synovial fluid. Gout is a disease of the joints; it is associated with elevated uric acid blood levels.

KEY TERMS

Check your understanding of this chapter. Can you explain what is meant by each of these terms?

albuminuria	jaundice
antibody	leukemia
antigen	lymph
bile pigment	metabolic acidosis
blood–brain barrier	metabolic alkalosis
blood buffer	occult blood
blood type	partial pressure
diuretic	plasma
erythrocyte	plasma protein
extracellular fluid	platelets
formed elements	pyrogen
hematocrit	RBC
hemodialysis	respiratory acidosis
hematuria	respiratory alkalosis
homeostasis	Rh factor
interstitial fluid	serum
intracellular fluid	WBC

STUDY QUESTIONS

1. **a.** Calculate the approximate number of liters of water there are in the body of a 55-kg woman.
   b. Do the same for a 70-kg man.

2. Name the three body fluid compartments and indicate which has the greatest percentage of body water and which has the least.

3. Name two circulating body fluids and four noncirculating body fluids.

4. Cite six ways or routes in which water can be lost from the body (normally or accidentally).

5. What is meant by the concept of homeostasis in the human? Give two examples of homeostatic mechanisms in the human.

6. Explain the difference between blood, plasma, and serum.

7. Name three categories of plasma proteins and give the general functions of each.

8. Examine Figure 15.3 and state the major difference in electrolyte composition between blood plasma and intracellular fluid.

9. Contrast white blood cells with red blood cells in terms of size, shape, number per cubic millimeter, place of synthesis, nuclei, and function(s) in the body.

10. What does the hematocrit value tell a physician about his patient?

11. What is the role of blood platelets in the body?

12. Write the equation that represents (a) the combination of oxygen with hemoglobin and (b) the combination of carbon dioxide with Hb.

13. True or false.
    a. Arterial blood had a low P_{O_2} and a high P_{CO_2}.
    b. The reaction $HbCO_2^- + H^+ \rightarrow HHb + CO_2$ takes place in the lungs.

14. Define the following:
    a. Phagocytosis
    b. PMNs
    c. Leukemia
    d. Hemopoiesis
    e. Hemolysis
    f. Formed elements of blood
    g. Anticoagulant
    h. Hypoxia
    i. P_{O_2}
    j. Buffer
    k. Febrile
    l. Pyrogen
    m. Antigen
    n. Blood–brain barrier

15. Most CO_2 gets to the lungs in a fashion that does not involve hemoglobin. Write two equations to show how CO_2 is transported in the blood. (*Hint:* The first equation shows the formation of an acid, the second, its ionization.)

16. Which compound forms a stronger bond to hemoglobin, O_2 or CO? And what is the implication of this in health and safety?

17. In a threatened metabolic acidosis, for example ketone body formation in diabetes mellitus, the patient can be observed to breathe with greater frequency. How would respiratory stimulation help restore homeostasis in metabolic acidosis?

18. Predict what will happen to arterial P_{CO_2} and arterial P_{O_2} after one has held his breath for a period of time.

19. Predict the effect on arterial P_{CO_2} after a period of hyperventilation.

20. Why are some patients instructed to breathe into (and out of) a paper sack for a period of time?

21. Does the system NH_3/NH_4^+ constitute a buffer? If so, write the balanced chemical equation for the addition of (a) HCl(aq) or (b) NaOH(aq).

22. An amino acid in its dipolar ion form, $R—CH—COO^-$, can act as a buffer. Write chemical equations to show this. $\underset{NH_3^+}{|}$

23. Some years ago, on a bitterly cold winter night, Chicago police discovered a woman alcoholic lying in the street. She had been there for some time. At the hospital, her rectal temperature was recorded as 69°F. What temperature is this on the Celsius scale? What is normal rectal Celsius temperature?

24. By what mechanism does alcohol rubbed on the skin lower body temperature?

25. What are the four types of blood? What type is of greatest frequency in the U.S. population? The least frequency?

26. Why is it dangerous to administer type B blood to a type A person? Explain in complete detail.

27. Examine Table 15.5. What blood type is correctly termed the universal donor? Explain why.

28. **a.** What chemical wastes are eliminated from the blood in hemodialysis?
    b. What prevents desirable electrolytes such as Na^+, K^+, Cl^-, and HCO_3^- from being lost?

29. Hemodialysis is used not only in kidney failure but also in barbiturate poisoning and other poisonings. Explain how it would work in barbiturate poisoning.

30. In the formation of urine, filtration of blood occurs in the glomerulus. This is followed by selective reabsorption. Where does reabsorption take place and what does it accomplish?

31. Name three end-products of nitrogen metabolism that are secreted in the urine.

32. Our text says that high protein foods with sulfur and phosphorus content are acid-residue foods. What does the sulfur and phosphorus content have to do with their potential for acidity?

33. What are diuretics? Give two medical conditions in which diuretics are used.

34. A hospital patient is asked to force fluids and so she drinks 2 L of water a day. Predict the effect this extra fluid intake will have on urea and creatinine excretion.

35. Students in graduate biochemistry laboratory courses sometimes take daily doses of bicarbonate to determine what effect it will have on the pH of their urine. What will the effect be, and what is the chemistry involved? Answer the same for orally administered NH_4Cl.

36. Give another name for hyperbilirubinemia. (*Hint:* Do not answer until you see the yellow of their eyes.)

37. Why is it reasonable to expect feces to be light in color in jaundice?

38. What is the origin of bile pigments in the human and what is their fate?

ADVANCED STUDY QUESTIONS

39. Describe the paths taken by, and the chemical modifications of, a molecule of O_2 from the time it enters the alveolar space until the time it leaves the body as either a molecule of water or as CO_2.

40. Why is bicarbonate ion considered a base in the body? Explain how and where it functions; mention any enzyme that is related to its functioning.

41. Describe and explain the cause of hemolytic diseases of the neonate.

42. In the kidney, hydrogen ions are actively secreted into tubular fluid. Write three equations to illustrate three buffering reactions that the kidney uses to keep the urine from becoming too acidic.

43. Explain why it is reasonable to expect a metabolic *acidosis* in diabetes mellitus.

44. Show how to distinguish a respiratory acidosis from a respiratory alkalosis on the basis of the following clinical data: P_{CO_2}, arterial blood pH, and the plasma level of bicarbonate ion.

45. How many milliequivalents of sodium are contained in 1 L of isotonic saline solution?

16 | Biotransformation of Chemicals, Drugs, and Foods

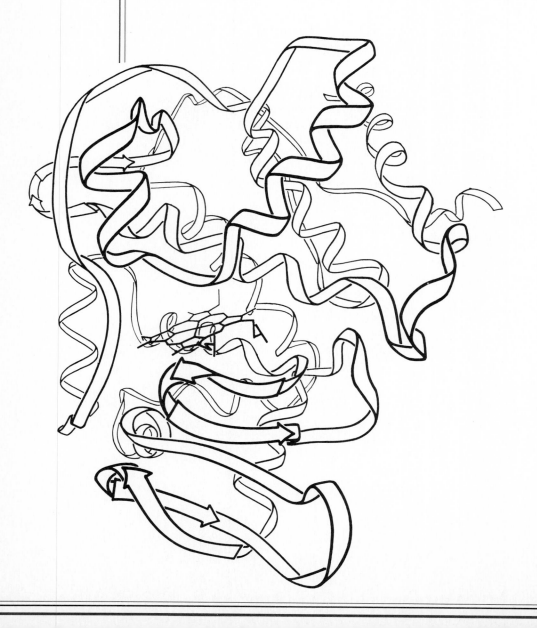

Psychiatrists use word-association tests to determine one's subconscious thoughts. If a biochemist were asked to associate a word with "biotransformation," he would probably answer "enzyme," for no biotransformation can occur at a sufficiently rapid rate in the body without the presence of an enzyme. (A biotransformation is any chemical change the body makes on a molecule to prepare it for further use or to eliminate it.)

We attach great importance to enzymes. They control a myriad of reactions that occur in living cells, including digestion of foods, transport of blood gases, detoxification of drugs, clotting of blood, and production of energy.

Enzymes are proteins. They are composed of amino acid molecules joined to form long chains. In our photo we see a drawing of the long chain that represents the backbone of the enzyme cytochrome C peroxidase. This large enzyme molecule has a complicatd three-dimensional structure, sometimes coiled, sometimes flattened out, yet fairly fixed and specific for the reactions it catalyzes. Located in one of the folds of the enzyme is one noncovalently bound heme molecule.

Cytochrome C peroxidase is a respiratory enzyme; it functions in the system of enzymes which transfer hydrogen to oxygen to form water. It is but one of the many examples of enzymes and their biotransformations that we shall study in this chapter. (Photo courtesy of Thomas L. Poulos, University of California, San Diego).

Knowledge of drug absorption, distribution, biotransformation and excretion are becoming essential to our ability to provide optimum patient care.

–Margaret Wallhagen, RN, MSN, California State University, Chico.

INTRODUCTION

This is another applications chapter. Here you will be able to apply the principles that you learned earlier in this book, and you will find the basis for answering the following questions:

1. How is the body prepared to handle (chemically accommodate) a wide range of normally ingested foods and dietary factors, ranging from cantaloupe to cornstarch and soybeans to sardines?
2. What happens to drugs such as aspirin, codeine, or barbiturates after we ingest them?

3. How does the drug-dependent person develop tolerance to ever larger and larger doses of a drug?
4. What are antimetabolites and how are they used in chemotherapy?
5. What are enzymes and coenzymes and what is their role in all of this?

It is true that the human body is prepared to handle chemically a remarkably great variety of normally and accidentally ingested substances. In addition to foods of all kinds, the body must somehow handle the chemicals in prescription drugs, alcohol, nicotine, marijuana, caffeine, and other recreational drugs, and the chemicals in agents in our environment—air and water pollutants, insecticide sprays, exhaust fumes, hair sprays, food additives, and chemicals in cosmetics.

Biotransformation is the modern term used to describe how our body handles endogenous substances as well as substances we ingest. You might be more familiar with other terms that mean approximately the same thing: **detoxification** and **metabolism.** When the body biotransforms a substance, it chemically alters it (by hydrolysis, oxidation, etc.) to yield one of the following:

1. A product useful in the synthesis of needed body chemicals.
2. A product that can be stored for future use.
3. A water-soluble substance that can be excreted in the urine.

The body can handle toxic materials (poisons) too, by biotransforming them. Rattlesnake venom is chemically degraded, albeit slowly, to harmless, inactive products. The chemical in an accidentally inhaled insecticide spray is degraded into nontoxic products. And in an overdose of a drug (such as a barbiturate), the offending chemical is biotransformed to a more urine-soluble form. Rarely, biotransformation produces a product *more* toxic than the original.

To summarize, whether it is carbohydrate, fat, or protein in foods, food additives, prescription or recreational drugs, or accidentally ingested toxicant, the body can usually handle the substance by the process termed biotransformation.

ENZYMES

The keys to understanding biotransformations are the **enzymes.** These are large protein molecules, manufactured in the liver and other tissues, that have an effect on the *rate* of specific types of chemical reactions. Enzymes are biological catalysts or biological regulators.

Enzymes speed up the rate of biochemical reactions. Without enzymes, the steps in the Krebs cycle might take weeks instead of milliseconds. Without enzymes, the half-life of the barbiturate Seconal might be 20 days instead of 20 hours. Without enzymes, the neurotransmitters at the ending of a nerve might not be destroyed, and the nerve might keep firing endlessly. Without its enzyme, phenylalanine cannot be metabolized; in children this leads to the serious disorder called PKU (phenylketonuria—see Chapter 18).

It is important to recognize that for a given biochemical reaction, an enzyme catalyzes the reverse reaction as well as the forward reaction. What the net chemical change will be depends on the starting point and the concentrations of reactants and products to begin with. An enzyme does no more than speed up the attainment of the equilibrium point that would be reached eventually even without the enzyme.

Enzymes, as biological catalysts, are not themselves used up in the reactions they catalyze. After they have done their job on one molecule, they remain intact to work on others. Enzymes, proteins themselves, catalyze the synthesis of other proteins essential for life. Hemoglobin, insulin, and FSH-releasing factor are all formed in enzymatically controlled processes.

If enzymes direct the synthesis of essential body proteins, where do the enzymes themselves come from? What directs their synthesis? The answer is that the DNA in genetic material directs the proper combination of amino acids to form the many hundreds of enzymes our body uses to catalyze syntheses and biotransformations (see Chapter 18). The design for human biochemistry, from beginning to end, is a grand one, and the more we learn about it, the grander it becomes.

Many enzymes in the body are synthesized in an *inactive* form and are stored in that form until the system requires them. Events then occur that convert the inactive enzyme, called the **proenzyme** or **zymogen,** into an active species. Examples of zymogens are

Prothrombin   the precursor to thrombin, a protease important in blood clotting.

Procarboxypeptidase A   the precursor to carboxypeptidase A, important in protein digestion.

Prophospholipase   the precursor to phospholipase, important in phopholipid digestion.

Pepsinogen   the precursor to pepsin, in gastric juice.

Trypsinogen   the precursor to trypsin, from the pancreas.

The wisdom of nature's plan for proenzymes is most evident in the case of pepsin, a proteolytic enzyme that would "eat" its way through any tissue in which it was stored!

Each enzyme has a temperature at which it catalyzes most effectively. The optimum temperature for most body enzymes is our normal body temperature, 37°C. At too high a temperature enzymes are denatured, and at too low a temperature they lose their catalytic effectiveness. Hence whole blood deteriorates more slowly if stored in the refrigerator, and seeds will not germinate in freezing weather. The pH, too, is very important in the regulation of enzyme activity. Most body enzymes catalyze best in the pH range of 7.0 to 7.5.

▬▬▬▬▬▬▬**EXAMPLE 16.1** ▬▬▬▬▬▬▬

Identify an enzyme that catalyzes best at a very acid pH (1 to 2) and one that catalyzes best in an alkaline medium.

Solution

What parts of the body normally are very acid or very alkaline? In strongly acidic stomach contents, the enzyme pepsin works optimally. The contents of the small intestine are alkaline (pH = 8), and pancreatic enzymes such as trypsin work best in this environment.

Enzymes as proteins are sensitive to and can be deactivated by heat and certain chemicals. At high temperatures, enzymes are simply denatured. Many of the chemicals known to be poisons in the human function by reacting chemically with functional groups in enzymes. Cyanide (CN^-), sulfide (S^{2-}), arsenic, and heavy metal ions (Hg^{2+}, Pb^{2+}) owe their toxic character to enzyme inhibition. Cyanide is extremely poisonous in mammals. It acts by binding tightly with metal ions in cytochrome oxidase (in the respiratory chain). The LDLo (lowest published lethal dose) of KCN in dogs is 3.8 mg per kilogram of body weight.

We recognize that enzymes are generally finicky molecules in that they are highly selective in the reactions they will catalyze. We term this their *specificity*. The size, shape, and electrical characteristics of an enzyme help determine its specificity. We learned in Chapter 12 that we have no enzyme that can catalyze the hydrolysis of cellulose and that glucose in cellulose is unavailable to us. Here enzyme specificity works against us. But specificity can be made to work for us in situations such as the use of antimetabolites in the treatment of disease (more later).

When it catalyzes a biological reaction, an enzyme merely is providing a surface for molecules to get together—a reaction site that helps reactants assume the correct configuration for interaction and product formation. (Think of the analogy of a student union acting as a catalytic site for many social activities.)

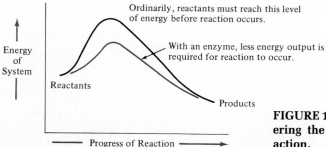

FIGURE 16.1   Enzymes work by lowering the activation energy of a reaction.

In a more sophisticated sense, a biological catalyst lowers the activation energy for a reaction, permitting it to take place at a lower temperature, that is, with less energy input (Figure 16.1).

Human steroid hormones (Chapter 14) act by stimulating processes that lead to increased synthesis of certain enzymes; this leads to increased protein synthesis. One of the hormones activated by steroids is the very well known substance cyclic-AMP (c-AMP).

DEFINITIONS AND RELATED TERMS

Before you learn the composition of enzymes and their roles in biotransformations, you should understand some terms commonly used in enzymology.

-ase the suffix customarily used to designate an enzyme; older names may not use it.

Substrate any food, drug, or other chemical substance on which the enzyme works to catalytically biotransform. Remember, enzymes catalyze anabolic as well as catabolic processes. The original substrate for cholesterol anabolism is acetyl.

Intracellular enzyme an enzyme that remains and functions in the cell in which it was produced; mitochondrial enzymes are intracellular.

Turnover number a measure of how active an enzyme is; defined as the number of moles of substrate biotransformed to product per mole of enzyme per minute; actual turnover numbers may be in the hundreds of thousands.

Activators Ions or molecules that convert a biochemically inactive form of an enzyme to a biochemically active form. Metal ions (Cu^{2+}, Fe^{2+}, Co^{2+} Zn^{2+}, Mg^{2+}), H^+ and enzymes themselves are activators of enzymes; some guns will not fire until the hammer is cocked: some enzymes will not catalyze until they are activated.

Inhibitor a drug, ion, molecule, or other chemical substance that combines with an enzyme making it incapable of carrying out its normal catalytic function; the inhibition can be reversible, that is, only temporary. A great many chemotherapeutic agents work by inhibiting enzyme systems; an example is Diamox, a carbonic anhydrase inhibitor. Antibiotics inhibit enzyme-catalyzed reactions in rapidly growing bacterial pathogens and in certain cancer cell systems.

THE COMPOSITION OF ENZYMES

Enzymes are proteins. Some enzymes are nothing but proteins and are active as such. They are termed *simple enzymes*.

Other enzymes require a cofactor before they can catalyze reactions. They are termed *conjugated enzymes*. The cofactor may be a metal ion activator such as the Zn^{2+} ion in carbonic anhydrase or the Fe^{2+} ion in cytochrome oxidase. But when the cofactor in an enzyme is an organic molecule, it is specifically called a **coenzyme**. Sometimes both a metal ion activator and a coenzyme are found in the same enzyme. Here is a brief summary of the composition of enzymes.

	Protein	Metal ion	Organic molecule
Simple enzyme	Yes	No	No
Conjugated enzyme	Yes	Yes and/or	Yes

Everybody has heard about the B vitamins. Until this chapter, we have not been able to specify their exact biochemical role. Now it is possible to state that vitamins B_1, B_2, and B_6 are part of the coenzymes of conjugated enzymes that are vital in the metabolism of carbohydrates, in the functioning of the Krebs cycle, and in the respiratory chain. Thiamin (B_1) (Figure 16.2) is the major part of the coenzyme thiamine pyrophosphate (TPP), the coenzyme in certain decarboxylases. Pyridoxine (vitamin B_6) appears in pyridoxal phosphate (Figure 16.2), a coenzyme for decarboxylation and transamination of amino acids.

Riboflavin (vitamin B_2) constitutes a large part of the coenzyme flavin adenine dinucleotide (FAD). It appears in enzymes that catalyze oxidation–reduction reactions in the respiratory chain.

Nicotinamide (niacinamide) is part of the coenzyme called NAD (nicotin-

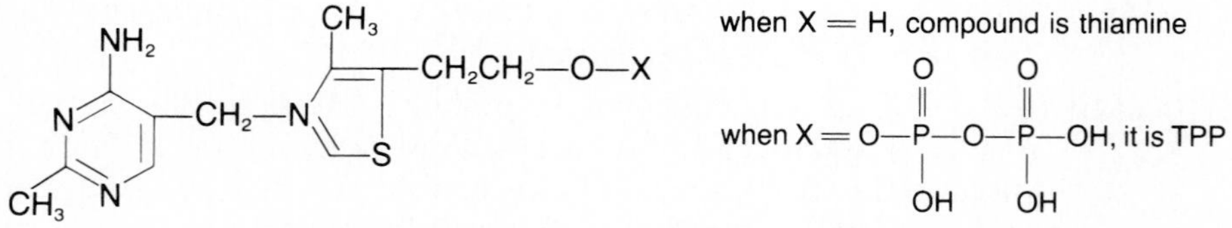

when X = H, compound is thiamine

when X = $O-P-O-P-OH$, it is TPP

thiamine is a major part of the coenzyme thiamine pyrophosphate (TPP)

when X′ = H, compound is pyridoxal

when X′ = $O-P-OH$, it is pyridoxal phosphate

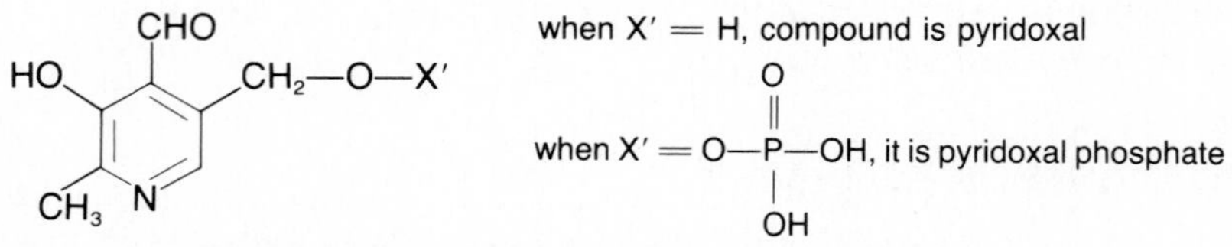

pyridoxine and other forms of vitamin B_6 appear in the coenzyme pyridoxal phosphate

FIGURE 16.2   The B vitamins serve as essential parts of the structures of coenzymes.

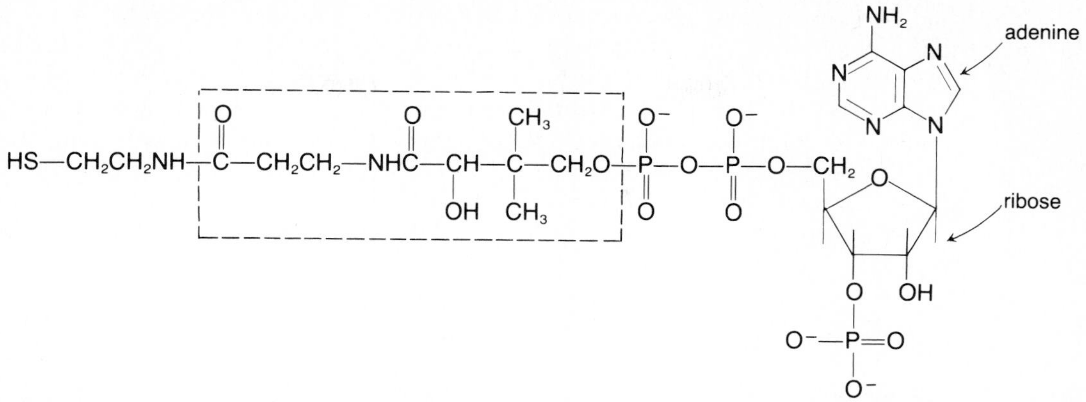

FIGURE 16.3   Coenzyme A (CoA) plays an important role in fatty acid metabolism, steroid synthesis, and carbohydrate metabolism. The dotted line outlines the pantothenic acid (vitamin) portion. The prefix *mercapto* designates an —SH group.

amide adenine dinucleotide; its older name is DPN) and it, too, functions in the respiratory chain.

In Chapters 12 and 13 we noted the action of Coenzyme A (CoA) in the Krebs cycle and in fatty acid metabolism. CoA (Figure 16.3) consists of adenine, D-ribose, three phosphate groups, pantothenic acid (considered to be a vitamin), and mercaptoethylamine. CoA plays a central role in metabolism by catalyzing the transfer of acyl groups, especially acetyl (CH_3CO).

There are many more examples of vitamins that function as a part of the coenzymes of conjugated enzymes. Together, they explain the vital role of vitamins in daily functioning and in health. See Chapter 17 for a discussion of vitamins and avitaminoses.

The protein portions of enzymes are made up of many amino acids joined in peptide bonds, just like any other polypeptide we have studied.

HOW ENZYMES WORK

Whether it is in the breakdown of a substance into smaller fragments or in the synthesis of a substance *from* smaller fragments, the mechanism of catalytic action of an enzyme is the same. In this chapter, with our emphasis on biotransformation of ingested substances, we will concentrate on the mechanism by which substances are catabolized.

A simplified explanation of enzyme action proposes the initial formation of an *enzyme–substrate complex:*

$$E + S \rightleftharpoons ES$$

enzyme + substrate → enzyme–substrate complex (held together by
noncovalent bonds such as dipole–dipole interaction)

The size, shape, and dipole characteristics of the **substrate** dictate whether or not a good fit will be made between E and S. Here we are again referring to the specificity of the enzyme. Specificity can vary greatly, with some enzymes being able to complex with numerous compounds. An example is a carboxypeptidase, which will complex with almost any compound containing the R—CO—NH—R link.

After the ES complex is formed, it is believed that the chemical hydrolysis, the oxidation, or whatever is to occur takes place while the substrate is yet bound:

$$\text{ES} \rightarrow \text{E—P}_1\text{—P}_2$$

After biotransformation is complete, the complex disintegrates, the products depart, and the enzyme is free to catalyze again:

$$\text{E—P}_1\text{—P}_2 \rightarrow \text{E} + \underbrace{\text{P}_1 + \text{P}_2}_{\text{products of reaction}}$$

A highly diagrammatic yet instructive picture of how enzymes work is given in Figure 16.4 (also consult Figure 11.5 in Chapter 11).

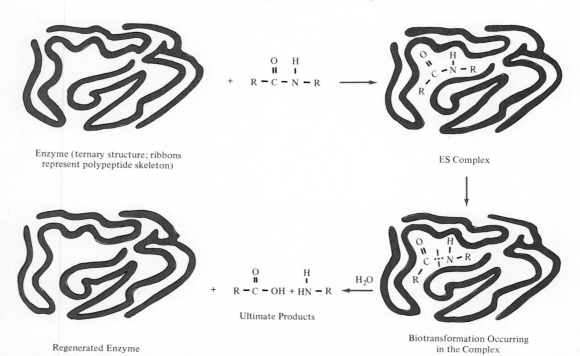

Enzyme (ternary structure; ribbons represent polypeptide skeleton)

ES Complex

Regenerated Enzyme

Ultimate Products

Biotransformation Occurring in the Complex

FIGURE 16.4   Highly diagrammatic representation of how an enzyme can catalyze a biotransformation (in this case the hydrolysis of a peptide bond).

Summary

In the mechanism of an enzyme-catalyzed reaction, an intermediate enzyme–substrate complex is formed. While in the complex configuration, the substrate is biochemically transformed. This can be by hydrolysis, oxidation, reduction, **decarboxylation,** rearrangement, or some other process. The complex then dissociates to release the product(s). The enzyme is freed to continue its catalytic role.

FIGHTING DISEASE BY INHIBITING ENZYMES

One of the ways medicinal chemists have learned to attack a disease is by interfering with the enzyme activity on which the progress of the disease depends. For example, in bacterial infection, the growth and proliferation of the bacteria depend on normal functioning of bacterial enzymes. Bacterial cells require certain substrates that they convert to components needed in growth; these substances we term *metabolites.*

Sulfa drugs (now much less used than formerly) are designed as **antimetabolites**—drugs that disrupt normal enzyme activity. They do this by tricking the bacterial enzymes into recognizing them as normal substrate, instead of the true substrate, para-aminobenzoic acid (PABA). Figure 16.5 shows how similar in chemical structure the sulfa drugs are to the normal substrate, PABA. If the circulating level of the sulfa drugs is high enough, it competes effectively with PABA for the enzyme,

$$E + S_{sulfa} \rightarrow ES_{sulfa} \text{ complex}$$
$$E + S_{PABA} \rightarrow ES_{PABA} \text{ complex}$$

thus tying up the enzyme and slowing down bacterial growth. This gives the body's normal defense mechanisms a better chance to overcome the infection.

Antimetabolites succeed because they are chemically similar enough to the true substrate to trick the enzyme into ES-type complexation, but chemically dissimilar enough to be useless in the overall metabolic scheme. It's a lot like installing pistonless engines in cars; they look alright after installation, but the cars are not going anywhere.

The antimetabolite concept has been applied quite successfully to the

PABA—the true substrate   typical sulfa drug structure
                                 (the antimetabolite)

FIGURE 16.5   Fighting Disease: Similarity in Chemical Structure between a Metabolite and Its Antimetabolite.

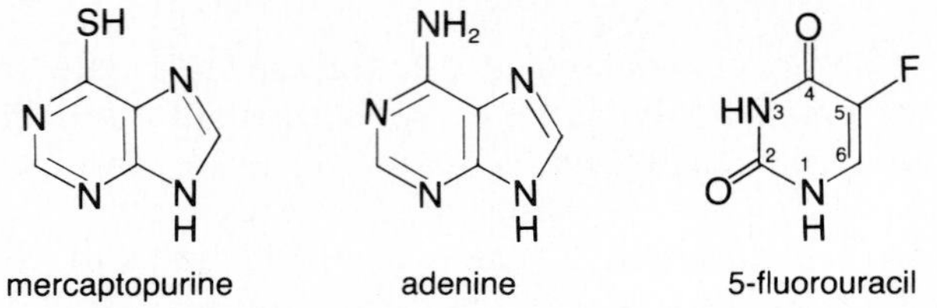

mercaptopurine   adenine   5-fluorouracil

FIGURE 16.6 **The concept of metabolites and anti-metabolites is used in the chemotherapy of cancer.**

chemotherapy[1] of cancer. Consider mercaptopurine (Figure 16.6), a well-known anticancer drug. This bogus molecule is so similar to the genuine metabolite adenine (see Figure 16.6) that rapidly growing cancer cells can mistake it for adenine, with consequent disruption of cancer cell proliferation. The use of S in place of N is possible because of their similar size and structure. Another example of an anticancer antimetabolite drug is 5-fluorouracil (Figure 16.6) in which a foreign fluorine atom has been introduced. Here again, the cancer cells incorporate the antimetabolite into their enzyme systems as though it were a useful substrate, but it only serves to foul up cell growth. 5-Fluorouracil is effective and is used in treating cancer of the breast, colon, stomach, ovary, bladder, and skin.

In some cancerous conditions, remarkable progress has been made in prolonging the life of the patient. With current chemotherapy, 50 percent of children with acute lymphoblastic leukemia can be expected to be free of the disease at 5 years; most will probably remain in remission and be considered cured of their disease (Figure 16.7). Antimetabolites are also used to treat viral infections of the eye. Antimetabolites to amino acids have been synthesized.

The idea of enzyme inhibition has been extended to the treatment of mental illness. Part of the reason we are mentally alert and not depressed most of the time is that our bodies secrete central nervous systems stimulants such as epinephrine (adrenalin) and norepinephrine (noradrenalin). These two are termed catecholamines. Ordinarily, these excitatory chemicals are destroyed in the body through the action of monoamine oxidase enzymes (MAO enzymes):

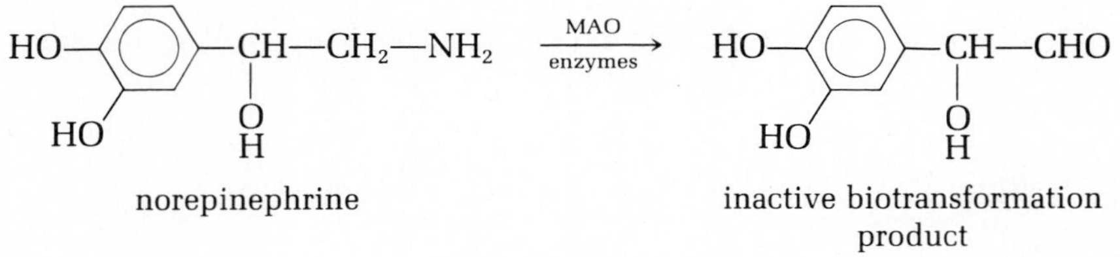

norepinephrine   inactive biotransformation
                                  product

[1] Chemotherapy is the treatment of a disease with a chemical specifically designed for that disease. It is expected that the chemical will attack the disease state but will not harm normal tissue or cause serious side effects.

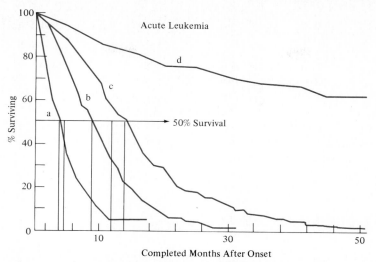

FIGURE 16.7   In certain cancers, great improvements in survival rates have been achieved with combined chemotherapy. (a) Thirty years ago, the median survival time for untreated children with acute lymphoblastic leukemia was between three and four months. (b) A report of 160 patients treated (January 1948 through January 1952) with antimetabolites and/or steroids showed some increase in survival time. (c) Further improvement was demonstrated with treatment of 205 cases (September 1959 through April 1965) using steroids, cyclophosphamide, vincristine, and an antimetabolite. (d) A more recent study involving 43 children (November 1969 through March 1972) used steroids, daunomycin, and vincristine as well as two antimetabolites and one alkylating agent. Of these patients, 50 percent can be expected to be free of the disease at 5 years and most will probably go on to be cured. (*Source:* R. E. LaFond, ed., *Cancer: The Outlaw Cell,* Washington, D.C.: American Chemical Society, p. 175. Used with permission.)

Phenelzine sulfate USP (Nardil) and tranylcypromine sulfate (Parnate) are two well-known antidepressant drugs that work by inhibiting the action of MAO enzymes. They are called monoamine oxidase inhibitors. Because they knock MAO out of action, they cause an increase in the concentration of epi, norepi, dopamine, and serotonin in storage sites throughout the nervous system. Presumably, this is the basis for their antidepressant activity. It is also an example of the chemotherapy of mental illness.

Acetylcholine, $CH_3COOCH_2CH_2N^+(CH_3)_3$, is the very important transmitter of nerve impulses in a large number of nerves in the human body. Acetylcholine, after it has assisted in getting the impulse from the end of one nerve to the beginning of the next, is ordinarily destroyed by the action of acetylcholinesterase:

$$CH_3COOCH_2CH_2\overset{\oplus}{N}(CH_3)_3 \xrightarrow[\text{H}_2\text{O}]{\text{acetylcholinesterase}} CH_3COOH + HOCH_2CH_2\overset{\oplus}{N}(CH_3)_3$$

Chemists have designed and synthesized a series of organophosphorus compounds that are exceedingly potent inhibitors of acetylcholinesterase. They are so effective, in fact, that their use causes a buildup of acetylcholine levels, with the result that nerves begin transmitting wildly and uncontrollably. Convulsions and death can follow. To our credit, we have applied this discovery to useful and effective insecticides; to mankind's discredit, this discovery has also been applied to the formulation of highly potent war gases (also called nerve gases).

INDUCTION OF LIVER ENZYMES: TOLERANCE

An insomniac receives a prescription for a barbiturate. For the first week, he takes the prescribed dose and it works well as a hypnotic (sleep-inducer). After several weeks, however, he finds that he must take a larger dose to get results. And as the weeks go by, he discovers that ever larger and larger doses of barbiturate are necessary for the desired hypnotic effect. Without realizing it, he has developed **tolerance** to the barbiturate.

Tolerance can develop to many drugs: alcohol, opiate pain relievers, amphetamines, nicotine, certain hallucinogens, and others.

We can explain tolerance in part by considering enzyme production and activity in the liver. Each drug that we have mentioned has an enzyme that catalyzes its biotransformation in the body. Many of these enzymes are found in the liver's cytochrome P-450 mitochondrial system. They are always present to some extent, but when we challenge the liver by ingesting a dose of a particular drug, we stimulate it to produce *more* enzymes to handle that drug. The process is termed **induction of liver enzymes**. As more enzymes are produced, the drug is more rapidly metabolized, and we discover that we must take a larger dose for the desired effect. But this only stimulates *greater* enzyme production, with consequent faster metabolism of the drug. A vicious cycle develops (Figure 16.8) the result of which is tolerance to the drug.

Tolerance is a dilemma for the patient. He has only two choices: stop the drug or take ever larger doses. For those who become physically dependent on the drug (addicted), the tolerance can become prodigious. A morphine addict was discovered to be taking 5000 mg a day (that's 500 times the therapeutic dose); a 22-year-old housewife on obesity control gradually built up to 250 mg of methamphetamine a day (that's fifty times the usual daily dose).

The lesson for us is, if a drug is not really necessary, do not begin taking it; try to find a nondrug modality. If the choice is made to use the drug, use it briefly in order to avoid the development of tolerance. Of course, these remarks apply only to elective drugs and not to drugs such as the antibiotics or an antihypertensive that must be taken for the duration of the illness.

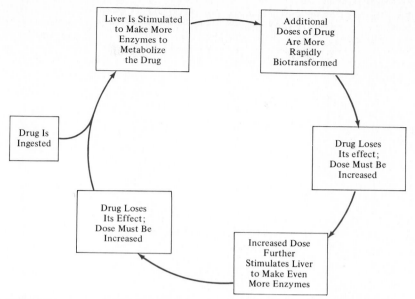

FIGURE 16.8 The Vicious Cycle that Leads to Development of Tolerance.

Cross tolerance can develop. Heavy alcohol users can handle barbiturates easily (when they are sober). This is because the same enzyme system that biotransforms alcohol biotransforms barbiturates. The development of cross tolerance between methadone, Demerol, and the opiate narcotics is well known.

ENZYMES USED IN DIAGNOSIS

More than fifty enzymes have been identified in human serum and about fifteen of these are used in clinical laboratories as aids in the evaluation of clinical disorders.

Physicians make use of certain transaminases, phosphatases, and dehydrogenases in the diagnosis of disease. Of the many transaminases found in human serum, glutamic oxalacetic transaminase (SGOT or GOT) is often used in diagnosis. SGOT catalyzes the transfer of an alpha-amino group from aspartic acid to alpha-ketoglutaric acid:

$$\text{aspartic acid} + \alpha\text{-ketoglutaric acid} \xrightarrow{\text{SGOT}} \text{oxalacetic acid} + \text{glumatic acid}$$

Phosphatases catalyze the hydrolysis of phosphate esters:

$$\text{R---O---PO}_3\text{H} + \text{H}_2\text{O} \xrightarrow{\text{phosphatase}} \text{ROH} + \text{H}_3\text{PO}_4$$

Table 16.1. Some Serum Enzymes and Their Diagnostic Value

Enzyme	Morbidity state	Diagnostic sign
LDH-1 (lactate dehydrogenase)	Myocardial infarct (MI)	Serum levels reach a maximum 4–5 days after a MI
SGOT (serum glutamic oxalacetate transaminase)	MI	Serum levels reach a maximum in 24 hours following an MI
CPK (creatine phosphokinase)	MI	Serum levels reach a maximum 24–36 hours following a MI
GPT (glutamate pyruvate transaminase)	Viral hepatitis	Elevated serum levels are seen in this disease
ACP (acid phosphatase)	Cancer of intestine or prostate	Elevated serum levels are seen in this disease
ALP (alkaline phosphatase)	Bone and liver diseases	ALP is elevated in active rickets, osteosclerosis, Paget's disease, and hyperparathyroidism

Lactic dehydrogenase (LDH) is an enzyme that catalyzes the following reversible reaction:

$$\text{pyruvic acid} + \text{NADH} \underset{}{\overset{\text{LDH}}{\rightleftharpoons}} \text{lactic acid} + \text{NAD}$$

Because blood levels of enzymes are either elevated or depressed in certain disease states, physicians can use the enzyme blood levels in their diagnosis of a disease state. After a heart attack (myocardial infarct, or MI), for example, serum levels of several enzymes rise markedly. This is because the enzymes leak out of the cells damaged in the attack. Cancerous growths cause an increase in the urine or serum levels of certain enzymes. In diseases of the liver, too, the blood level of an enzyme spikes. Table 16.1 lists some of these enzymes and the diseases to which they point.

A graphic summary of changes in enzyme values following a heart attack is given in Figure 16.9.

Isoenzymes (*isozymes*) are multiple molecular forms of the same enzyme, all having similar catalytic properties and the same cofactors but differing in the protein portion of the molecule. Their source in the body is also different; some arise in the heart, some in the brain, and some in skeletal muscle. To separate and distinguish isoenzymes, researchers use the technique of electrophoresis. CPK (see Table 16.1) is actually a group of several isoenzymes (designated CPK_1, CPK_2, CPK_3); when the clinician studies CPK values in the diagnosis of MI, he actually considers the serum levels of each CPK isozyme. The same is true for LDH.

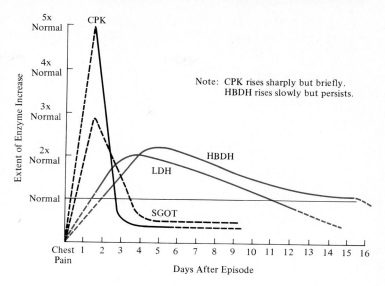

FIGURE 16.9 Typical Changes in Serum Enzyme Levels Following Myocardial Infarction. The changes in serum levels of certain enzymes after a heart attack can be used in diagnosis and prognosis. CPK = creatine phosphokinase, SGOT = serum glutamic oxalacetic transaminase, LDH = lactic dehydrogenase, HBDH = an isoenzyme of LDH.

Note: CPK rises sharply but briefly. HBDH rises slowly but persists.

A Case History

A 64-year-old man, a heavy smoker and coffee drinker, was well until a few weeks before admission, at which time he noted a viselike chest pain at rest that was aggravated by exertion. He remained at home for 2½ weeks, but then developed severe chest pain that required hospitalization. The ECG and laboratory data did not indicate MI during the first week in the hospital. While in the hospital, he again developed severe substernal nonradiating chest pain. This was felt to be similar to his previous anginal attacks, but to confirm serum was drawn for enzyme determinations. Later that evening he was found in bed, pulseless, apneic, with no measurable blood pressure, and in ventricular fibrillation. He was defibrillated and transferred to the coronary care unit (CCU). ECG studies now showed changes considered diagnostic of anteroseptal MI. Enzyme studies performed at the time of the chest pain and at 6, 12, 24, and 48 hours following the episode gave the following results.

	Episode	*6 hr*	*12 hr*	*24 hr*	*48 hr*	*Reference value*
Total CPK	7	197	328	1299	638	50 U/L
Total LDH	140	164	280	988	1168	200 U/L

These data plus isoenzyme data and sixteen other parameters were fed to a computer program. The computer-generated diagnostic possibilities were pernicious anemia, other macrocytic anemia, and myocardial infarction. The diagnosis: myocardial infarction.

CHEMICAL CHANGES IN BIOTRANSFORMATIONS

The body is remarkably well equipped with enzymes that are capable of handling a geat variety of ingested and naturally occurring substances. Biotransformations in the human can occur in the intestines, lungs, kidneys, skin, or liver. By far, the greatest number occur in the liver—the organ that receives and chemically processes almost every drug that enters the body. Enzyme systems in the body include the following.

1. *Digestive enzymes*, for carbohydrate, fat, and protein metabolism (see Table 16.2), and metabolic enzymes for glycolysis, Krebs cycle, respiratory chain, and beta-oxidation.
2. *Regulatory enzymes* for neurotransmitters; these include the monoamine oxidases (MAO) for regulating levels of catecholamines (epinephrine and norepinephrine) and for aminoid drugs such as the amphetamines; and acetylcholinesterase, for controlling levels of the neurotransmitter acetylcholine.
3. *Other enzymes* in great diversity for controlling many body processes such as kidney function, conversion of hemoglobin to bile salts for excretion, glycogen formation, fat deposition, nitrogen metabolism, and protein synthesis.
4. *Mixed-function oxidases*, the key enzyme of which is cytochrome P-450. As mentioned earlier, a cytochrome is a complex of heme and protein. As in hemoglobin, the heme in cytochrome serves to carry oxygen—in this case for the oxidation of a great variety of prescription drugs, street drugs, recreational drugs (alcohol, nicotine), and accidentally ingested organic compounds (like insecticides). The mixed-function oxidase system is highly inducible. It is located in the mitochondria of the liver. Neonates generally have a poorly developed oxidase system and a consequently poor capacity to handle drugs and other chemicals.

Digestion of foods involves enzymatic hydrolysis of three types of bonds. In carbohydrates, glycoside linkages are enzymatically broken with the insertion of a molecule of water:

polysaccharide

liberated monosaccharide

Table 16.2. Digestive Enzymes

Source	Enzyme	Substrate	Catalyc action
Saliva	α-amylase	Starch	Hydrolysis of starch to dextrins and maltose
Stomach	Pepsins	Proteins	Hydrolysis of peptide bonds adjacent to aromatic amino acids
Pancreas (see Figure 15.20)	Trypsin	Proteins	Cleaving of peptide links adjacent to arginine or lysine
Pancreas	Chymotrypsin	Proteins	Breaking of peptide links at phenylalanine and tyrosine to yield peptones, polypeptides, and amino acids
Pancreas	Carboxypeptidase A and B	Proteins	Hydrolysis of carboxy-terminal amino acids
Pancreas	Pancreatic lipase	Triacyl-glycerols	Cleaving of ester link between fatty acids and glycerol
Pancreas	Pancreatic α-amylase	Starch	See salivary amylase
	Phospholipase A	Lecithin	Hydrolysis to lysolecithin; also occurs in cobra venom
Intestinal mucosum	Aminopeptidases	Polypeptides	Cleaving N-terminal amino acids from peptides
Intestinal mucosum	Dipeptidases	Dipeptides	Hydrolysis
Intestinal mucosum	Lactase	Lactose	Hydrolysis to glucose and galactose
Intestinal mucosum	Maltase	Maltose and maltotriose	Hydrolysis to glucose
Intestinal mucosum	Sucrase	Sucrose	Hydrolysis to glucose and fructose

The complete metabolism of carbohydrates to CO_2 and H_2O is discussed in Chapter 12.

As we have learned, the fatty acid ester linkage in fats is enzymatically cleaved with the production of three moles of fatty acid and one of glycerin. Note that water is again involved, making this a hydrolysis reaction, too:

$$
\begin{array}{c}
\underset{|}{CH_2}-O-\overset{\displaystyle O}{\overset{\|}{C}}-R \\
\underset{|}{CH}-O-\overset{\displaystyle O}{\overset{\|}{C}}-R \\
CH_2-O-\overset{\displaystyle O}{\overset{\|}{C}}-R
\end{array}
\quad + \; 3\,H_2O \xrightarrow[\text{bile salts}]{\text{lipase and}}
\quad
\begin{array}{c}
CH_2OH \\
| \\
CHOH \\
| \\
CH_2OH
\end{array}
\quad + \; 3\,HOOC-R
$$

triacylglycerol　　　　　　　　　　　　　　　　glycerol　　fatty acids

Fatty acids are further oxidized by beta-oxidation, as described in Chapter 13.

Proteins are amides. Enzymatic hydrolysis frees the carboxylic acid and the amine that make up the peptide bond:

$$
\underset{\substack{| \\ NH_2}}{R'-CH}-\overset{\displaystyle O}{\overset{\|}{C}}-NH-\underset{\substack{| \\ R''}}{CH}-COOH
\;+\; H_2O \xrightarrow[\text{peptidases}]{\text{various}}
\underset{\substack{| \\ NH_2}}{R'-CH}-COOH \;+\; H_2N-\underset{\substack{| \\ R''}}{CH}-COOH
$$

However, protein digestion is not as simple as this. It begins with the action of pepsin on dietary protein in the stomach. Peptide bonds within the polypeptide chain are hydrolyzed by pepsin. Protein digestion continues in the small intestine. Trypsin and chymotrypsin also catalyze the hydrolysis of peptide bonds in the interior of the polypeptide chain. Finally, carboxypeptidases and aminopeptidases catalyze the stepwise cleavage of individual amino acids from the ends of the remaining peptide chains. The result is an intestine full of free amino acids. They pass across the intestinal membrane and enter the bloodstream. Figure 16.10 shows the possible fate of amino acid molecules in the blood. Figure 16.11 shows the formation of *urea*, the principal

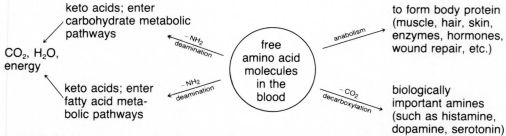

FIGURE 16.10 The Fate of Amino Acid Molecules in the Blood. **Alanine is one of the amino acids that enters the carbohydrate pathways; it is said to be *glucogenic*. Leucine, isoleucine, phenylaline, and tyrosine are amino acids that enter the fatty acid metabolic pathways; they are said to be *ketogenic*.**

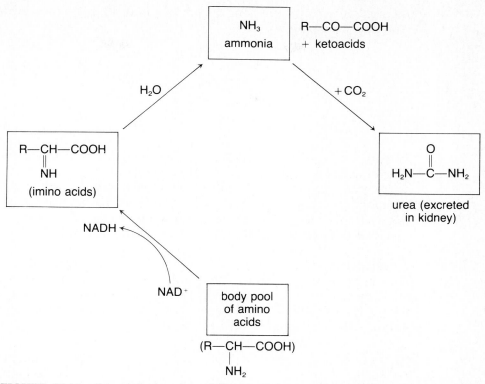

FIGURE 16.11   Ammonia is a by-product in the breakdown of amino acids in the human. In the liver, ammonia is converted to urea.

nitrogen waste made by the human body. The biotransformations in Figure 16.10 deserve explanation, as they involve very important body chemicals.

The word *decarboxylation* means loss of CO_2. When dihydroxyphenyl-alanine (L-dopa), for example, is enzymatically decarboxylated, CO_2 is lost, and the important brain amine dopamine is produced:

L-dopa →(a decarboxylase)→ CO_2 + dopamine

Dopamine imbalance is implicated in one current theory of the cause of schizophrenia (it is the dopamine receptors in the brain that are blocked by the major tranquilizers). By similar decarboxylations, the body makes histamine and serotonin; the precursors are the amino acids histidine and 5-hydroxy-tryptophan, respectively.

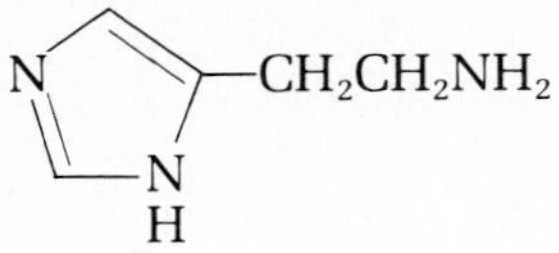

histamine 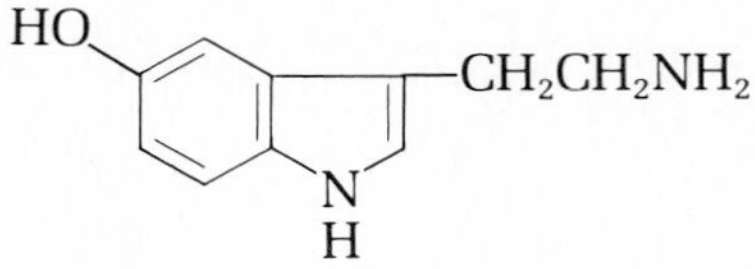   serotonin

Histamine and serotonin are of great importance in normal body functioning. Histamine is released from its storage sites in allergic reactions; it can cause vasodilation, swelling and redness of tissues, and constriction of bronchi. Serotonin occurs naturally in the brain and is important in regulating sleep/wake patterns and in modulating our response to stressful stimuli. It has been suggested that the hallucinogenic activity of LSD is due to its ability to block serotonin receptors in the brain.

Oxidative deamination, a widespread process in the human body, involves removal of an amino group with simultaneous oxidation of the carbon to an aldehyde or ketone. With amino acids it gives nitrogen-free products that can mix (or "pool") with similar molecules that arise in the course of carbohydrate or fat metabolism. Here is an example, using alanine:

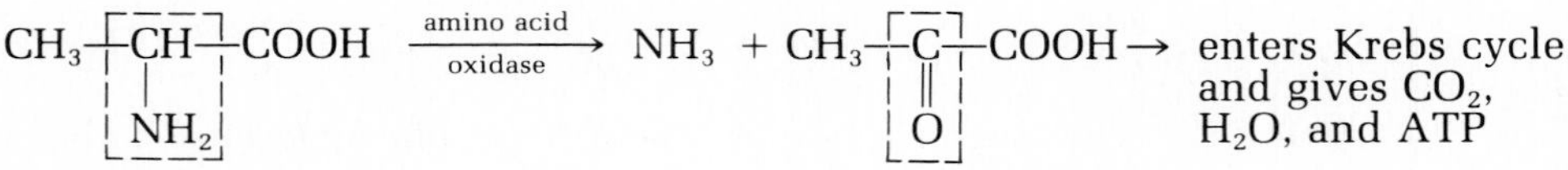

pyruvic acid

EXAMPLE 16.2

Serotonin is biotransformed into a biologically inactive product by oxidative deamination. Draw the structure of this product.

Solution

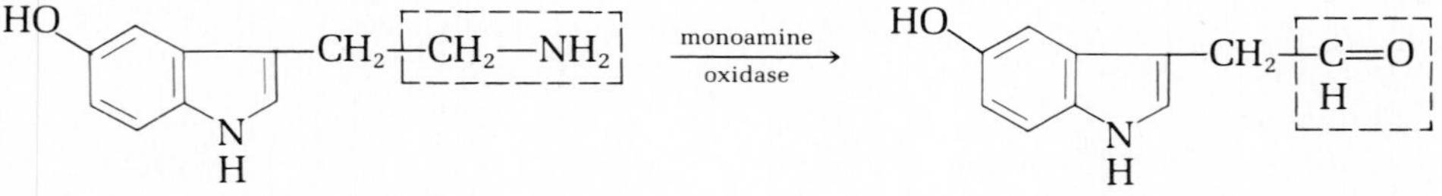

serotonin   aldehyde product
                                           (biologically inactive)

Summary

We have discussed chemical changes that occur when dietary carbohydrates, fats, and proteins are digested. All of these biotransformations are enzyme catalyzed. The monosaccharides, fatty acids, or amino acids that are produced are absorbed from the intestine, circulate in the blood, and eventually find their biochemical destination. Some intermingle with intermediates in other metabolic pathways. Some amino acids can be the source of important brain amines. Decarboxylation and oxidative deamination are two very important biotransformations that the body uses to deactivate biologically active molecules.

  Drugs are chemicals and must obey the rules for biotransformation much the same as foodstuffs. In fact, you may come across the phrase *metabolism of drugs*, and that really is not a misnomer. *Detoxification* is a similar term, except that it tends to emphasize the biotransformation of a *toxic* (poisonous) or very potent substance into nontoxic products. (In the world of drug dependence, to "detoxify" a person has another meaning; it means to get him off of his drug, to make him drug-free.)

  Isn't it remarkable that the human body just happens to have a wide array of enzymes for the biotransformation of narcotic analgesics, tranquilizers, high blood pressure drugs, alcohol, nicotine, marijuana, aspirin, cocaine, local anesthetics, amphetamines, caffeine, antispasmodics, muscle relaxants, vasoconstrictors, ganglionic blockers, MAO inhibitors, and all of the other types of prescription and over-the-counter (OTC) drugs Americans routinely put into their systems!

  Now it is true that some drugs are resistant to biotransformation; some are excreted mostly intact, with no chemical alteration. Others are very rapidly altered, and have short half-lives. But in general, the body strives to protect itself (to return to homeostasis) by quickly eliminating the foreign molecule and with it the altered physiology.

  The most prominent means of accomplishing this is to biotransform the drug molecule into a water-soluble degradation product that (a) has lost its pharmacological activity, and (b) is soluble in urine and can be quickly excreted. The body's ability to biotransform and eliminate drugs has implications for the health professional. It means that the dose of a drug and the frequency of dosing will have to be selected to maintan a blood level of drug that will accomplish the intended action. Further, the *initial* dose may have to be greater to "load" body fluids with the drug, whereas subsequent doses need be no more than is necessary to maintain effective blood levels. This is especially true with digitalis, where the initial digitalization dose is not the same as the maintenance dose.

  The major site of drug metabolism is the liver, specifically the mitochondria with their rich supply of enzymes. Figure 16.12 shows the liver, supplied abundantly with blood vessels that can absorb a drug from the gastrointestinal tract. Absorbed drugs enter the liver by way of the portal vein.

Some of the chemical formations that the body uses to handle drugs are the following.

1. Oxidation, in which polar groups are introduced into a molecule, increasing its solubility in the urine. Here are a few examples:

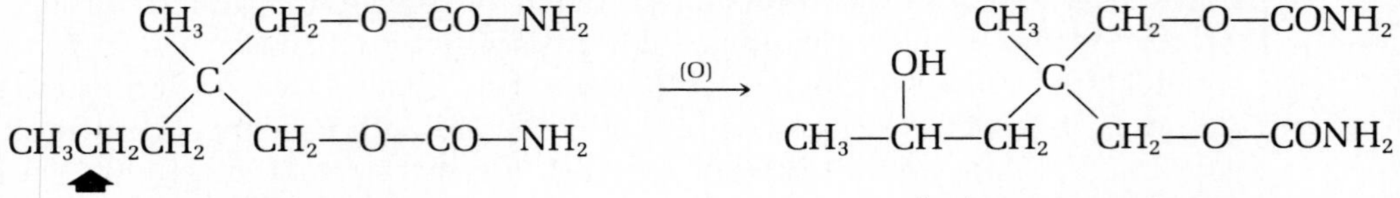

meprobamate

hydroxy meprobamate

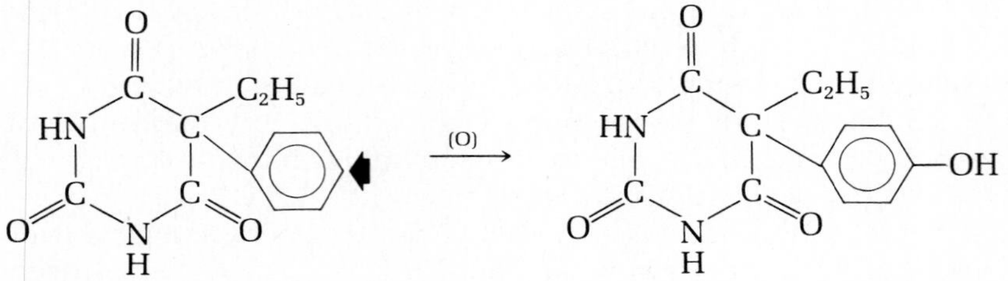

phenobarbital

hydroxylated phenobarbital

$$\text{HO—CH}_2\text{—CH}_2\text{—OH} \xrightarrow{\text{(O)}} \text{HOOC—COOH}$$

ethylene glycol

oxalic acid

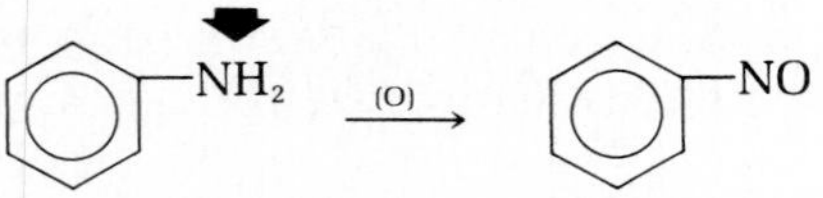

aniline

nitrosobenzene

2. **Conjugation** means "coming together" (as in wedded bliss). When, for example, the body wishes to excrete the foreign material benzoic acid, it conjugates it with glycine to form hippuric acid:

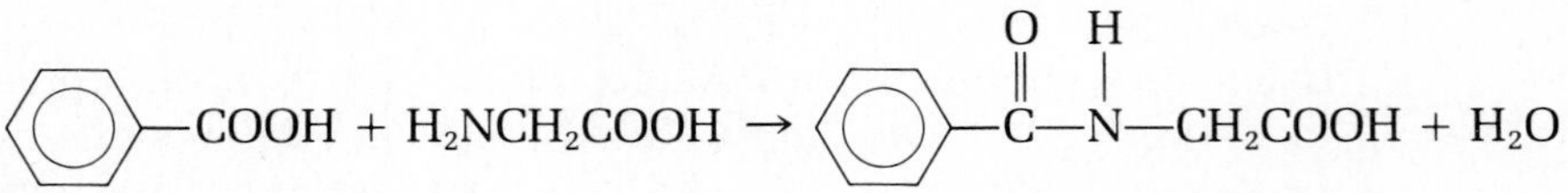

benzoic acid · · · · · glycine · · · · · hippuric acid, excreted in the urine

Hippuric acid is much more water soluble than benzoic acid. A supply of glycine is always available in the body for detoxifying ingested chemicals. The same is true for glucuronic acid, a substance very often

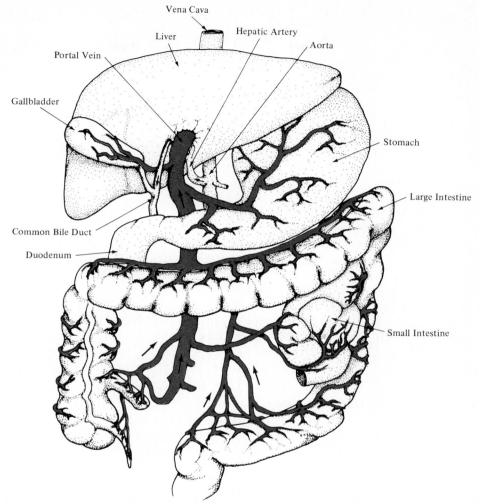

FIGURE 16.12   **The liver, as the primary site of drug metabolism, is supplied with many blood vessels that absorb the drug from the gastrointestinal tract and come together as the portal vein.**

used in the body to conjugate with and detoxify poisonous compounds having the OH functional group. Figure 16.13 shows how glucuronic acid can conjugate with phenol to give a urine-soluble product. Morphine and camphor react similarly.

3. *Ester hydrolysis* converts an ester group into the more polar alcohol and carboxylic acid groups:

$$H_2N-\bigcirc-COOCH_2CH_2N(C_2H_5)_2 \xrightarrow{\text{H}_2\text{O}} H_2N-\bigcirc-COOH + HOCH_2CH_2N(C_2H_5)_2$$

procaine   an acid   an alcohol

4. *O-Dealkylation* is the enzyme-catalyzed cleavage of an ether to an alcohol:

codeine   morphine

Note that in the metabolism of codeine, the product is a *more* active analgesic than the original drug.

The *metabolism of alcohol* deserves special attention because alcohol is so widely used in alcoholic beverages and as a solvent in OTC drug preparations. Of Americans over 21 years, 77 percent of males and 60 percent of females are drinkers. There are approximately 10 million alcoholics and problem drinkers in the United States.

The metabolism of alcohol can be summarized in the following steps:

$$C_2H_5OH \xrightarrow[\substack{\text{dehydrogenase} \\ \text{(liver)}}]{\text{alcohol}} CH_3CHO \xrightarrow[\text{dehydrogenase}]{\text{aldehyde}} CH_3\overset{\displaystyle O}{\overset{\|}{C}}-CoA \xrightarrow[\text{cycle}]{\text{Krebs}} CO_2 + H_2O + ATP$$

acetaldehyde   acetyl CoA

It is important to note that the biotransformation of alcohol occurs in the liver. Soon after a dose of alcohol, the liver is heavily perfused with this chemical. Note also that an aldehyde intermediate (acetaldehyde) is involved.

Antabuse (disulfiram) therapy in the treatment of the alcoholic is based on enzyme inhibition, which results in the interruption of alcohol metabolism at the acetaldehyde stage. Acetaldehyde thus builds up in the bloodstream of the drinker, producing very unpleasant symptoms: heart palpitation, throbbing headache, nausea, copious vomiting, sweating, and chest pain. It is possible, then, for a person earnestly desiring to stay away from alcohol to use Antabuse as a kind of chemical crutch. It works this way. The patient takes Antabuse under a physician's direction (along with supportive and psychotherapy treatment). He realizes that if he is tempted to drink, and does, the disulfiram in his system will make him seriously ill. This strong deterrent helps him through the temptation.

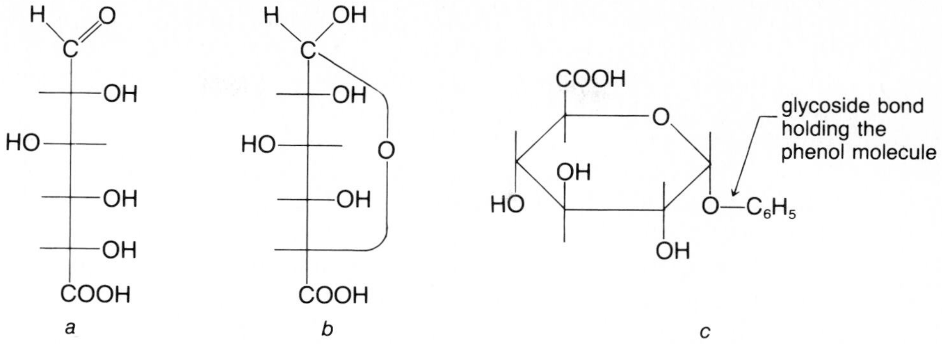

FIGURE 16.13 The body uses glucuronic acid to detoxify poisonous substances such as phenol. Glucuronic acid is shown in its free aldehyde form (*a*), as a cyclic hemiacetal (*b*), and conjugated with phenol (*c*).

Disulfiram therapy works and there are many physicians who use it extensively. It, by the way, is another example of the antimetabolite approach to chemotherapy.

SUMMARY

Biotransformation means the enzyme-catalyzed chemical change of a molecule (food, drug, or body substance) to a useful product, a storage form, or a product readily excretable in urine or feces.

Enzymes are the keys to biotransformations. Enzymes are proteins that affect the rate of a biotransformation by reducing the activation energy of the reaction. Some enzymes are stored until used in an inactive form called the proenzyme or zymogen.

Factors that can affect an enzyme's activity are temperature, pH, presence or absence of poisons, and activators such as metal ions. We discussed terms important in enzymology: specificity, turnover number, and inhibitors.

Simple enzymes are nothing but protein. Conjugated enzymes are protein plus a metal ion cofactor or an organic molecule. When an organic molecule is a cofactor, the enzyme is termed a coenzyme. Certain of the B-complex vitamins function as organic molecules in the coenzymes TPP, FAD, NAD, and CoA.

Enzymes work by forming an enzyme–substrate (ES) complex in which the biotransformation occurs. After reaction is complete, the products leave and the enzyme is free to catalyze again (and again).

Diseases can be fought by inhibiting the enzymes on which spread of the disease depends. We discussed the use of sulfa drugs as antimetabolites in treating bacterial infections, mercaptopurine in cancer, MAO inhibitors in mental illness, and acetylcholinesterase inhibitors as insecticides.

Tolerance to a drug occurs when certain enzymes in the liver which normally catalyze the biotransformation of the drug are induced to increase their numbers in response to the continued presence of quantities of the drug. With drugs like alcohol, barbiturates, and opiates, tolerance goes hand-in-hand with physical dependence (addiction).

Enzymes are used in the diagnosis of disease. This is because blood levels of certain enzymes are either elevated or depressed in certain diseases. Enzyme levels are used in the diagnosis of myocardial infarct.

Under the topic of chemical change in biotransformations, we studied the digestion of food. Digestion of carbohydrates involves cleavage of glycoside bonds to liberate monosaccharides. In fats, ester linkages are enzymatically cleaved to give glycerol and fatty acids. In proteins, peptide (amide) bonds are cleaved to yield amino acids.

Decarboxylations ($RCOOH \rightarrow RH + CO_2$) in the body give important compounds such as dopamine, histamine, and serotonin.

Oxidative deamination ($R—CH_2NH_2 \rightarrow RCHO + NH_3$) converts amino acids to keto acids. Drugs are biotransformed, too. We discussed oxidations ($Ar—H \rightarrow Ar—OH$), conjugations ($RCOOH + H_2NR \rightarrow RCONHR$), ester hydrolysis ($RCOOR' \rightarrow RCOOH + HOR'$) and O-dealkylation ($ROCH_3 \rightarrow ROH + CH_3OH$).

Finally, we discussed the body's biotransformation of ethyl alcohol (America's second most popular recreational drug). Acetaldehyde, the chemical intermediate in alcohol metabolism, is the key to understanding how Antabuse therapy works in the treatment of alcoholism.

KEY TERMS

Check your understanding of this chapter. Can you explain what is meant by each of these terms?

activator	induction of liver enzymes
antimetabolite	inhibitor
biotransformation	isoenzyme
coenzyme	metabolism
conjugation	proenzyme
decarboxylation	substrate
detoxification	tolerance
enzyme	zymogen

STUDY QUESTIONS

1. What is meant by the process of biotransformation?

2. What general types of ingested substances are biotransformed by the body?

3. True or false.
    a. Enzymes are biological catalysts.
    b. Enzymes are used up when they act to catalyze a reaction.
    c. Enzymes influence the rate of *attainment* of the equilibrium of a reaction.
    d. Enzymes influence the *position* of the equilibrium of a reaction.
    e. Enzymes are proteins.
    f. Enzyme activity is influenced by temperature but not by pH.
    g. Intracellular enzymes are those that are made inside of a cell but function outside of the cell.
    h. In a reversible reaction, an enzyme will affect only the rate of the forward reaction.

4. Match the term of the right with the statement on the left.

_____ The substance on which the enzyme acts.	**A.** Specificity
_____ A number indicating how active an enzyme is in catalyzing a reaction.	**B.** Turnover number
	C. Substrate
_____ An enzyme cofactor that is an organic molecule; often it is a vitamin.	**D.** Activators
	E. Coenzyme
_____ Ions such as Fe^{2+} and Zn^{2+} required before an enzyme can become active.	**F.** Simple enzyme
_____ An enzyme that is active without any cofactors.	
_____ Particular enzyme activity related to its three-dimensional shape, functional groups, and dipole characteristics.	

5. Some enzymes are made and stored in an inactive, *zymogen*, form; they are then released and activated. What special advantage does this approach have in the case of the peptidases?

6. Why is riboflavin (vitamin B_2) essential to our existence? Be specific.

7. In general terms, explain how an enzyme functions when it catalyzes a reaction. Use E for the enzyme, S for substrate, and P for product.

8. How can it be that only a few enzyme molecules can catalyze the biotransformation of a great number of its substrate molecules?

9. Explain the role of antimetabolites in fighting diseases such as infections and cancer.

10. On the left on page 474 you will find six structures representing normal metabolites for various biological processes. Match each with the compound on the right that might act as its antimetabolite.

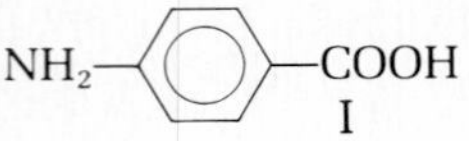
I

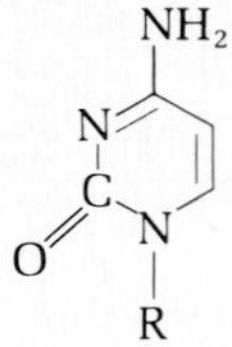
II

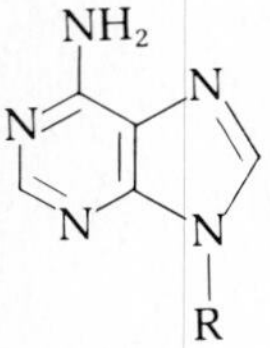
III

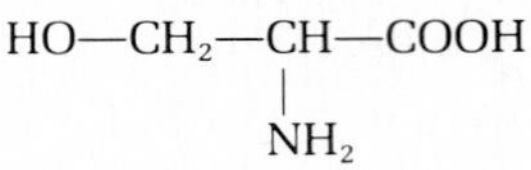
IV

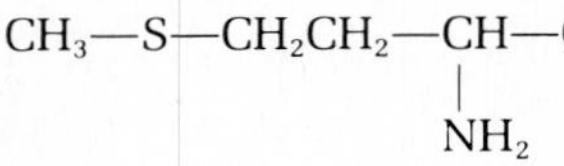
V

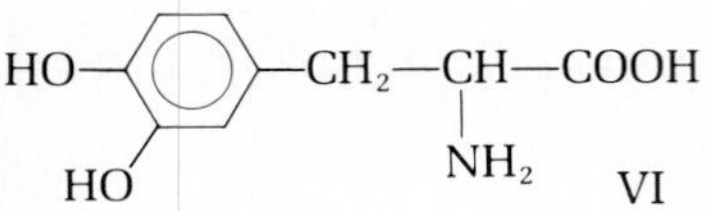
VI

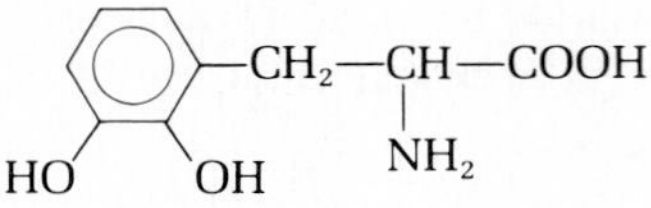
a.

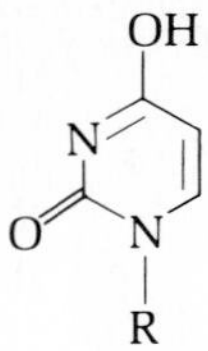
b.

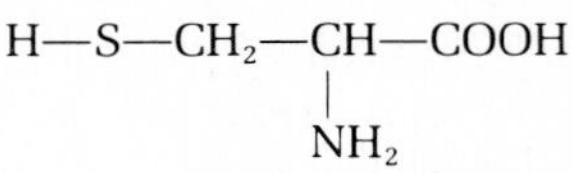
c.

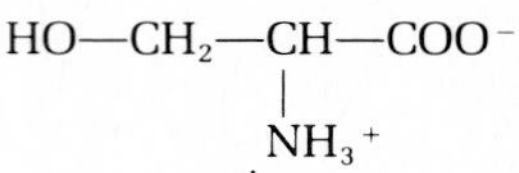
d.

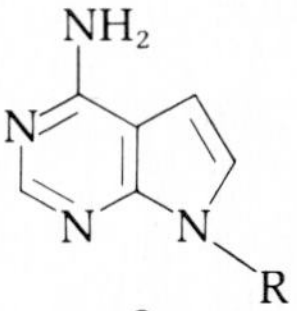
e.

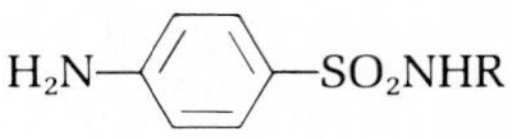
f.

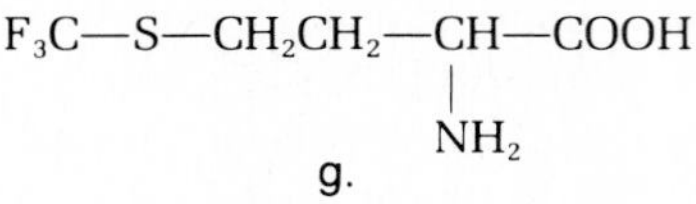
g.

11. Salivary amylase catalyzes the hydrolysis of starch in the mouth. Would you expect this enzyme to continue to function well in the stomach? Explain your answer.

12. Ascorbic acid oxidase requires Cu^{2+} for it to function. Is this enzyme a simple or conjugated enzyme?

13. If a carboxypeptidase catalyzes the hydrolysis of peptide bonds, what is the function of a (*a*) decarboxylase, (*b*) methyltransferase, (*c*) transaminase, (*d*) isomerase, (*e*) phosphorylase, (*f*) dehydrogenase?

14. What kind of substrate would be transformed by a (*a*) dipeptidase, (*b*) lipase, (*c*) sucrase, (*d*) oxidase, (*e*) salivary amylase?

15. A teenager smokes his first cigarette and gets so sick he thinks he is going to die. A couple of years later, he is smoking three packs a day. What has he developed biochemically, and how did he develop it?

16. Hospital doses of Demerol or morphine do not usually induce tolerance or drug dependence in the patient. Why not? What are the factors involved in the development of tolerance and dependence?

17. We learned from our text that histamine and serotonin arise by decarboxylation of amino acid precursors. Locate the structural formulas for

histamine and serotonin and then write the structural formulas of the amino acids that were decarboxylated.

18. List five organs or tissues in the body in which biotransformation of foods or drugs can occur. Which one of the five is the primary site for biotransformations?

19. Examine potential drugs in the following list and predict a *likely biotransformation* (oxidation, hydrolysis, etc.; see text) that the body would perform on it. Draw a heavy arrow to the bond(s) that would be broken or altered.

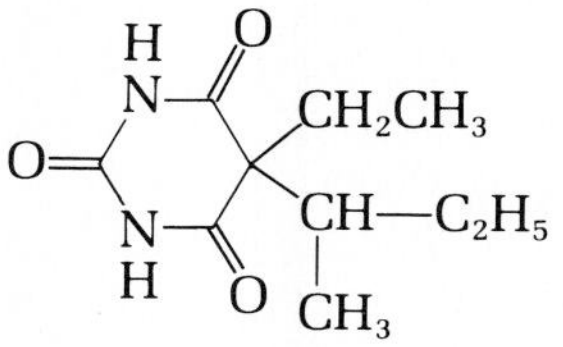

a barbiturate

$CH_3{-}CH_2{-}CH_2{-}OH$

propyl alcohol

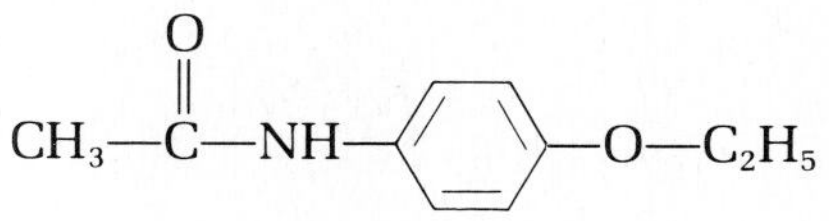

phenacetin

CH_3CHO

acetaldehyde

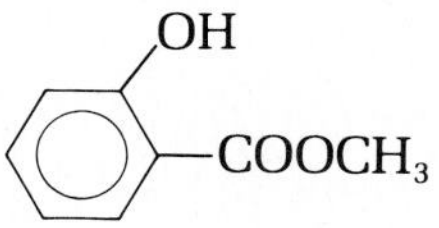

oil of wintergreen

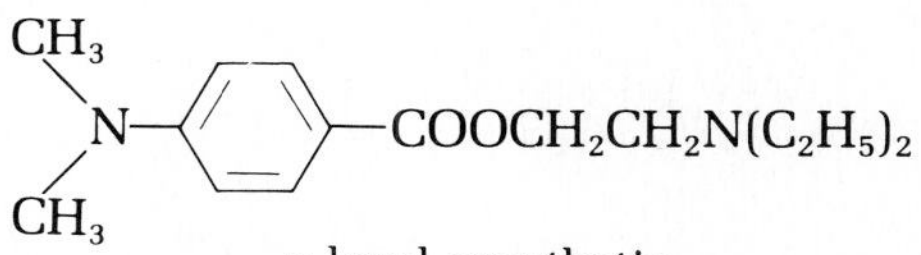

a local anesthetic

C_6H_5COOH

benzoic acid

$C_6H_5CH_3$

toluene

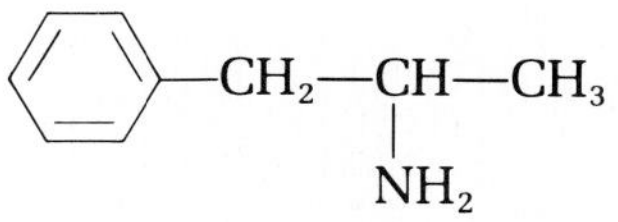

amphetamine

20. What is the rationale behind the use of Antabuse in the chemotherapy of alcoholism?

ADVANCED STUDY QUESTIONS

21. In the brain, dopamine is the immediate precursor of norepinephrine. Dopamine is inactivated partly by the enzyme catechol-O-methyltransferase. Predict the pharmacological effects of inhibiting the action of catechol-O-methyltransferase by administration of a suitable antimetabolite.

22. Assume that the growth of a pathogenic virus depends on the metabolism of the substrate *uracil*. Devise a chemotherapeutic approach to the treatment of such an infection, making use of the antimetabolite concept.

23. Using a textbook of pharmacology, report on the role of histamine, dopamine, and serotonin in natural body processes.

17

Diet and Body Chemistry

There is a Chinese proverb that says, "Give a man a fish and he will eat for a day; teach him how to fish and he will eat for the rest of his days." Our photo shows a grouper fish being carried to a market in India. The people who will eat the fish are lucky, for the meat is an excellent food and a good source of protein.

Not so lucky are millions of other people worldwide who suffer from malnutrition caused by lack of food or lack of quality food. The world's heavily populated and less developed regions will require increased production of cereals and animal products if their food standards are not to become lower than they are today.

But in affluent societies, problems also exist. We may have quality and variety of food available, but choose empty calories or processed foods instead. We may come up nutritionally short-changed by listening to the Madison Avenue hucksters, the food faddists, the diet planners, or those who stand to make a big profit selling their own ideas.

We cannot properly counsel patients on nutritional matters if we do not know the basics about diet and body chemistry. This chapter will introduce you to the essential food factors, the vitamins, the caloric values of foods, and the impact of minerals in our diet. (Photo courtesy World Health Organization.)

Adequate nutrition is essential to the maintenance of the morphophysiological correlates of the mind and therefore to normal mental and emotional behavior.

–Elizabeth M. Burns, RSM, Ph.D., Associate Professor of Nursing, University of Illinois, Chicago

INTRODUCTION

Perhaps more than we realize, our state of physical and mental health depends on what factors we consume in our diet. Although most people know that they should select foods from each of the four food groups—grains, vegetables, meats, and dairy products—not a lot of people pay attention to essental amino acids, essential fatty acids, complementary protein, excessive salt or sugar intake, or drug–food interactions. Nor are most Americans aware of the caloric value of different foods or of the presence of trace minerals in the diet.

In this chapter, we shall examine all of these topics and more. Our general goal is to relate diet to health. Our specific objectives are to accomplish the following:

1. Identify essential food factors.
2. Recognize symptoms of vitamin deficiencies and evaluate the need for vitamin supplementation.

3. Increase our awareness of caloric value of foods and be able to use calories for measuring energy consumption.
4. Understand hormonal control of basal metabolic rate.
5. Learn the importance of trace elements in the diet.
6. Identify food–drug interactions that are potentially dangerous to health.
7. Evaluate diet plans as they are promoted to the public.
8. Recognize the toxicity of heavy metal ions as they appear in foods.
9. Recognize the pervasiveness of sugar in the American diet.
10. Appreciate the importance of protein, and the problems associated with protein deficiency.

ESSENTIAL FOOD FACTORS

An **essential food factor** is any dietary substance required by the body for normal health and functioning but not produced at all in the body or produced at a rate too slow to meet body demands. Vitamins are essential factors, but they have their own category, and we do not include them in this immediate discussion. In the earlier literature, the word *indispensable* was used as synonymous with essential.

Essential food factors must, therefore, be included in the diet. They are of two types: essential amino acids, as found in protein, and essential fatty acids, as found in fats. We shall examine essential amino acids first.

Nine amino acids are essential for the *adult* human. They are as follows:

1. Isoleucine
2. Leucine
3. Lysine
4. Methionine
5. Phenylalanine
6. Threonine
7. Tryptophan
8. Valine
9. Histidine

One additional amino acid, arginine, is essential in growing children but not in adults. The remaining ten amino acids can be synthesized by the body from dietary precursors.

A protein food is said to be a **complete protein** if it contains all nine essential amino acids in the proper quantities. An *incomplete protein* either has one or more essential amino acids missing, or the quantities of one or more amino acids are too small to do much good. There is a fine point here that you should know about. It's called **complementary protein.** If a protein molecule (such as insulin or an enzyme) is to be synthesized in the body, and *one* of the amino acids required for its synthesis is missing, not only will the protein not be synthesized, but the other amino acids that are available will be deaminated, as though they were excess amino acid. In effect, they will be wasted. There must be enough of all the required amino acids for them to complement each other and join together in the synthesis of a protein.

Complete proteins, such as meats, fish, milk products, poultry, eggs, and soybean contain enough of all nine essential amino acids for complementation to occur. The protein in most cereals, nuts, and vegetables is incomplete; complementation will not always occur. To get around this, the incomplete protein can be supplemented with another protein food that contains the amino acids missing in the incomplete protein. For example, a cereal may be deficient in methionine or tryptophan. But when milk is added to the cereal, and consumed with it, complementation is accomplished. All of the amino acids in the cereal will then be utilized. This is one of the reasons nutrition experts urge us to *eat a variety of foods*. Beans and wheat are a complementary protein combination, for whereas beans are poor in the sulfur-containing amino acids and rich in lysine, wheat is rich in the sulfur-containing amino acids and poor in lysine. The high lysine content of seafood means that it can nicely complement grains and certain nuts that are low in lysine. There are many other complementary food combinations, including rice and legumes (bean, soy); wheat products with milk and cheese; cornmeal and beans; peanuts, milk, and wheat; and potatoes and milk (or cheese).

Of all of the foods we buy at the store, protein usually costs the most. And proteins are usually the most difficult and costly to produce (cows grow more slowly than corn). This is a reflection of the central and dominant role of protein in our metabolism. We must have the essential amino acids for protein synthesis, enzyme synthesis, muscle growth, and reproduction. In these areas, protein far surpasses carbohydrate and fat in dietary importance, and the nitrogen atom, in the form of the amino group, far surpasses C, H, and O atoms.

We know what happens when humans are deprived of complete protein. Certain Latin Americans, South Americans, Africans, and Asians live on diets that are deficient in complete protein. Yams, for example, are a staple in the diet of certain South Americans. Children growing up on yams can suffer from *kwashiorkor disease* (Figure 17.1), a protein deficiency condition that manifests itself in stunted physical and mental growth, malaise, anemia, scaly skin, diarrhea, discolored hair, and pot belly (from enlarged liver and spleen). The kwashiorkor child has a lowered resistance to infection, and if there is poor sanitation or living conditions, serious illness is almost certain.

Protein deficiencies such as kwashiorkor can be treated successfully if caught in time. Nonfat milk fortified with milk solids is given at first; whole milk and other complete proteins are given later.

General malnourishment in the human (Figure 17.2) often places the individual at the very serious disadvantage of trying to stand off an onslaught of childhood diseases, diarrheas, and parasites with the meager forces of resistance left in the body. Most of the protein burned during total starvation comes from the liver, spleen, and muscles. Fat is burned rapidly in starvation, a fact noted in hospitalized obese patients who were "starved" on water and vitamins. They lost about 1 kg of weight per day for the first 10 days.

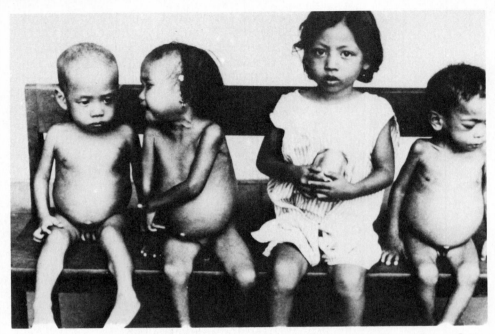

FIGURE 17.1 Children suffering from Kwashiorkor disease typically exhibit malaise, distended abdomen and poor hair growth. The protein deficiency also stunts their growth generally. (Photo courtesy World Health Organization.)

NITROGEN BALANCE

Are you in positive nitrogen balance this week? You are if you are taking in more nitrogen than you are excreting in your urine and feces. To be in positive **nitrogen balance** means that your body is actively incorporating nitrogen atoms into muscle growth, antibody synthesis, wound repair, and recovery from illness or in response to anabolic steroids. Growing children are normally in positive nitrogen balance (Figure 17.3).

The U.S. Department of Agriculture estimates that the average American eats from 10 to 12 percent more protein than his body can use as protein. The average Indian consumes 2.85 lb of meat and poultry in a year; the average Americans eats 212 lb, or seventy-five times as much. You may wonder what happens to the amino acids in all of the "extra" protein we Americans eat— the thick steaks, poultry, fish, soybean, dairy products, and eggs. The answer is that the amino acids obtained by digestion of all this unneeded protein are largely deaminated, as in the following equation:

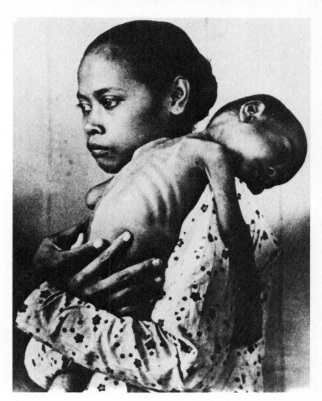

FIGURE 17.2 General malnourishment places a child at high risk from diseases, diarrhea or parasite. (Photo courtesy World Health Organization.)

$$CH_3-CH-COOH \xrightarrow[\text{deamination}]{\text{oxidative}} CH_3-\underset{O}{\overset{\|}{C}}-COOH + NH_3$$
$$|$$
$$NH_2$$

The ammonia is excreted as urea and the ketoacid mixes with the body's metabolic pool of pyruvate. In a sense, then, the protein is wasted. The expensive, hard-to-come-by aminoid nitrogen atoms are converted to ammonia and excreted. This is the reason we have some nutritionists and conservationists decrying our overdependence on protein as a source of food and calories. What a costly fuel for our bodies!

Recommended daily intake of protein for adult males is about 65 g and for females about 55 g (the rough rule is, for males multiply body mass in kilograms by the factor 0.9; for nonpregnant adult females multiply by 1.0). Pregnant women require about 30 g additional protein per day (second and third trimesters); an additional 20 g per day is required during lactation. All of these figures assume that a variety of protein food is being consumed (so

Balancing nitrogen intake and excretion.

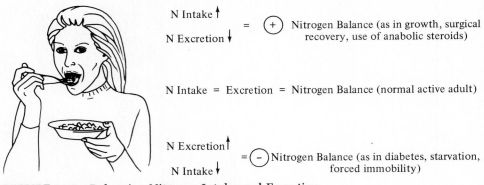

$$\begin{matrix} N \text{ Intake } \uparrow \\ N \text{ Excretion } \downarrow \end{matrix} = \left(+\right) \quad \text{Nitrogen Balance (as in growth, surgical recovery, use of anabolic steroids)}$$

N Intake = Excretion = Nitrogen Balance (normal active adult)

$$\begin{matrix} N \text{ Excretion} \uparrow \\ N \text{ Intake } \downarrow \end{matrix} = \left(-\right) \text{Nitrogen Balance (as in diabetes, starvation, forced immobility)}$$

FIGURE 17.3   Balancing Nitrogen Intake and Excretion.

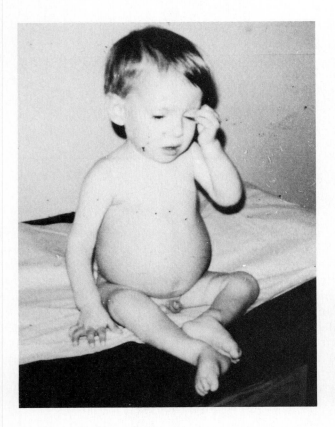

FIGURE 17.4   A 2½-yr-old Macrobiotic child, suffering from protein, mineral and vitamin deficiencies. He is unable to walk or crawl and has rickets. (Photo courtesy Cortez F. Enloe, Jr., M.D., *Nutrition Today*.)

that complementation is occurring and all of the protein consumed is available for protein use in the body).

How much protein food would a woman have to consume in 1 day to obtain 50 g of protein? Here are several possible dietary combinations:

1. 1 egg   7 g protein
    1 pt (473 mL) milk   16 g
    4 oz (113 g) meat or fish   <u>28 g</u>
                Total: 51 g protein

2. 5 oz (142 g) dry, raw soybean seeds   50 g protein

Three essential amino acids that are difficult to find generally in various incomplete protein foods are tryptophan, methionine, and lysine. Vegetarians and vegans must make special efforts to insure that their meatless diet offers adequate levels of these three amino acids. Nutritionists have reported on protein-deficiency disease among children in certain American communes where veganism is practiced (Figure 17.4). The children suffer from kwashiorkorlike debilitation. Many stores now offer bulk protein, such as hydrolyzed soy protein or hydrolyzed casein, that offer all or nearly all of the essential amino acids. Individual amino acids such as lysine can also be purchased, but they are expensive. Normally, persons consuming an adequate, varied diet will not need to supplement it with any of these products.

Table 17.1 lists some incomplete protein foods and identifies the insufficient amino acids with a circle.

ESSENTIAL FATTY ACIDS AND PROSTAGLANDINS

The principal dietary essential fatty acid in man is *linoleic acid* (*cis, cis*-9,12-octadecadienoic acid). It must be provided in our diet (in at least 1 percent of the caloric requirement) as it cannot be synthesized by animals or man. Linoleic acid is essential in man for cell membrane function, brain growth, and the prevention of atherosclerosis and thrombosis. When man or animals consume a diet completely devoid of fats, a dermatitis develops.

According to R. T. Holman, a recognized authority in the field of the biochemistry of fats and oils, *linolenic acid* is also a dietary essential fatty acid in the human. Holman discovered that, in one case, a surgical patient fed an essentially linolenic acid-free diet developed numbness, paresthesis, weakness, inability to walk, pain in the legs, and blurring of vision. Symptoms disappeared when linolenic acid was replaced in the diet. Long-chain fatty acids derived from linolenic acid are found in high proportions in nervous tissues, in the phospholipids of cerebral gray matter, and in the retina.

Table 17.1. Amino Acid Content of Some Incomplete Proteins

Food	Trp	Phe	Leu	Ile	Lys	Val	Met	Thr
Beans, common, dry, raw, 100g	(201)	1227	1918	1271	1650	1360	(223)	959
Lentils, whole dry, raw, 100 g	(222)	1136	1754	1309	1507	1334	(173)	864
Rice, milled, dry, 100 g	(82)	382	655	356	300	531	(137)	298
Wheat flour, enriched, 110 g	(139)	638	893	534	(267)	499	(151)	336
Corn flour, sifted, 110 g	(52)	387	1118	396	(249)	439	(163)	344
Coconut, fresh meat, 100 g	(33)	173	267	180	153	312	(73)	127
For comparison:								
Soy flour, full fat, 72 g	396	1426	2244	1558	1822	1531	396	1135
Egg whole, dried, 100 g	751	474	4039	3052	2959	3428	1456	2301
Milk, whole, powdered, 120 g	444	1520	3136	2032	2468	2188	792	1492

ª Values are given in milligrams. Deficiencies indicated by circles.

Holman considers linoleic and linolenic acids to be *dietary* essential fatty acids. He considers arachidonic and all other metabolically produced long-chain highly unsaturated fatty acids derived from linoleic and linolenic acids to be *functionally* essential fatty acids, that is, required for our biochemical functioning. We humans have a system of enzymes that can catalyze the conversion of linoleic and linolenic acids to a variety of polyunsaturated fatty acids including arachidonic.

Linoleic acid is abundant in several of our cooking and salad oils: corn, safflower, sunflower, soybean, and peanut, but not in coconut or olive oil. Linolenic acid is present in soybean oil (unhydrogenated). Hydrogenation reduces (obviously) the content of both linoleic and linolenic acids in the oil or fat, and what is more, the unusual fatty acids that result from the hydrogenations inhibit the body's metabolism of the polyunsaturated acids. It seems clear that food manufacturers should take a close look at the wisdom of partially hydrogenating oils just to offer a product that has more customer appeal.

The **prostaglandins** are a more recently discovered group of biochemically important compounds. The prostaglandins have been found to be involved in reproduction, respiration, pain, inflammation, and blood clotting. Structurally, they are twenty-carbon fatty acids (Figure 17.5) with varying degrees of unsaturation.

One of the most exciting theories about prostaglandins is that they are key factors in the perception of pain, and that pain-killing drugs such as aspirin, ibuprofen, mefenamic acid, and indomethacin work by inhibiting the body's synthesis of prostaglandins. Furthermore, research evidence indicates that excessive prostaglandin synthesis causes a wide variety of inflammatory, autoimmune, ulcerative, and degenerative bone diseases. Drugs that relieve symptoms of these diseases also act to inhibit the biosynthesis of prostaglandins. Menstrual pain (dysmenorrhea) is the result of release of prostaglandins in the uterus. It can be relieved by suppressing prostaglandin synthesis.

Prostaglandins are potent regulatory substances synthesized by every tissue in the body: lung, intestine, brain, heart, and kidney. They occur in large amounts in semen. There are many kinds of prostaglandins, and they have many kinds of pharmacological actions in the human:

- They increase the frequency and amplitude of uterine contractions.
- Generally, the PGF type contract bronchi and trachea, whereas the PGE type relax them.
- The PGE type inhibit gastric secretion.
- PGA, given IV to hypertensive patients, increases urine production and loss of Na and K.
- They induce luteolysis (regression of corpus luteum); in early pregnancy

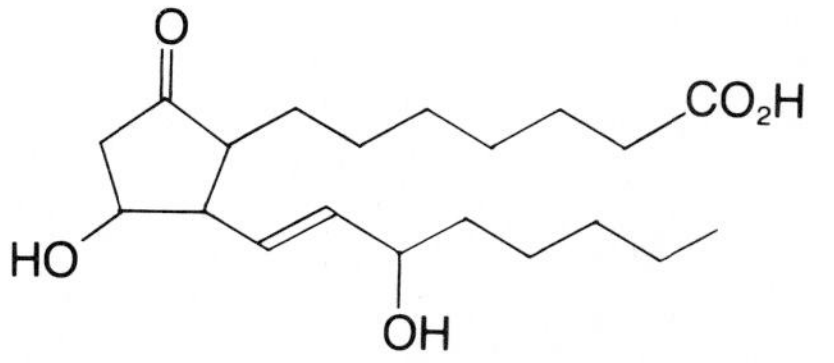

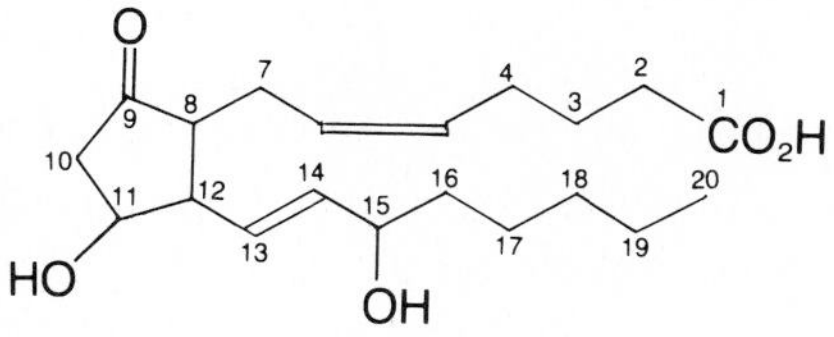

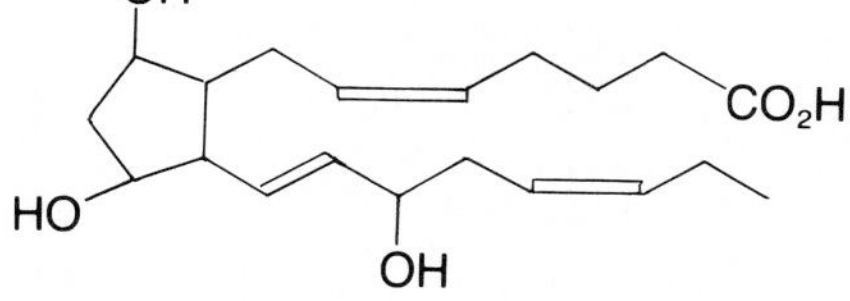

FIGURE 17.5   The three major types of prostaglandins are the PGE_1, the PGE_2, and the $PGF_{3\alpha}$. The cyclic five-membered ring determines the type (E, F, A, B, C, or D). All are C-20 fatty acids (see numbering) having different numbers and positions of double bonds in the side chains.

this produces an abortifacient effect. Thus in some circles the prostaglandins are being termed the new "morning after" pill.

- Given IV in 0.05 to 0.10 μg/kg/minute doses, the prostaglandins cause face flushing and pulsating headache (showing their great potency).

Proposals have been made to use prostaglandins to induce abortion, to induce labor at full term, to lower high blood pressure, to treat asthma and gastric hyperacidity, and to stimulate renin secretion. As yet, no prostaglandin product is on the market in the United States for general use.

ENERGY REQUIREMENTS AND CALORIC VALUE OF FOODS

An active adult can easily burn up as much as 2800 Cal of energy a day. When nutritionists and biochemists use the term **Calorie,** as above, they always mean the 1000-cal unit (kilocalorie), and it is spelled with a capital C. In the remainder of this chapter, the term will be used in this way.

Different types of foods show different caloric values when metabolized in the body.

Food	*Cal/g*
Carbohydrate	4.1
Protein	4.1
Fat	9.3

It can be seen, that gram for gram, fat can supply more than twice the calories of either carbohydrate or protein.

Exactly how many calories you and I need each day to break even on energy depends on several factors, the most important of which are (1) our body mass and (2) how active we are. A sedentary adult female weighing 58 kg might easily meet all of her daily energy needs with a 1600-Cal/day intake. An 80-kg basketball player, however, might consume 5000 Cal/day during the basketball season, and not gain a gram of body mass. Riding a bicycle for 16 hours has been measured as burning up 9000 Cal in 1 day. Table 17.2 provides some rough estimates to give you an idea of your daily energy consumption.

If you are overweight and desire to burn up some of that depot fat, you should be aware that for each pound of flesh lost (not water loss), approximately 3500 Cal of energy must be burned *in excess of intake.* Thus, if you are burning up 2500 Cal per day and are on a 2000-Cal diet, it will take you 7 days to lose 1 lb. If you are burning up 3000 Cal per day on the same diet, you will lose 2 lb in a week. Calculations like these should make us view with great skepticism

Table 17.2. Energy Burned Up by a 150-lb Person in Various Activities (Approximate)

Activity	Calories/hr	Activity	Calories/hr
Archery	160	Walking, $3\frac{3}{4}$ mph	300
Lying down or sleeping	80	Swimming, $\frac{1}{4}$ mph	300
Sitting	100	Square dancing, volleyball, roller skating	350
Driving an automobile	120		
Standing	140	Wood chopping, sawing	400
Domestic work	180	Tennis	420
Walking, $2\frac{1}{2}$ mph	210	Skiing, 10 mph	600
Bicycling, $5\frac{1}{2}$ mph	210	Squash and handball	600
Gardening	220	Bicycling, 13 mph	660
Golf, lawn mowing, power mower	250	Running, 10 mph	900
Golf, carrying clubs	360		
Skin diving	700		

diet schemes claiming weight losses of 10 lb or more per week. It just isn't likely. And if much weight is lost, it can often be attributed to water loss induced by diuretic drugs that are given as part of the diet scheme. When all of the advertising claims, drug crutches, and money-making schemes are swept aside, one fact remains: if you want to lose weight, eat less or exercise more.

Most Americans obtain most of their daily calories from carbohydrate. In fact, on the average, about half of our calories come from carbohydrate, with the remainder split up two-to-one between fat and protein.

There are no essential carbohydrate food factors. However, vegetables, with their content of carbohydrate and fiber, are an excellent food. Vegetables such as celery stalks, apple skins, carrots, broccoli, and asparagus provide indigestible cellulose fiber, add to the bulk in the intestine, and thus promote evacuation. Besides this, scientists have discovered that the incidence of diverticulitis is less in people who consume a high-fiber diet. For these reasons, cola drinks, sugary and starchy foods, processed snacks, powder puff bread, and potato chips are considered to be "junk foods" by nutritionists, who deplore their use as principal dietary factors by so many Americans (Figure 17.6, a and b).

THE BIOCHEMICAL ROLE OF THE THYROID GLAND

When we are sound asleep, our bodies are still engaged in metabolic activity, oxidizing carbohydrate and fat, and biotransforming protein. Let us take that a few steps further and consider our bodies' metabolism when we are also free of caffeine, nicotine, circulating adrenalin, and all other stimulant drugs and

(a)

(b)

FIGURE 17-6 *a* and *b*.   The rat in (a) was fed a standard laboratory diet and maintained a normal weight. The rat in (b) was permitted to consume all of the commercial "junk" food he wished and became obese. (Photos courtesy Drs. S. K. Gale and T. B. Van Itallie, Obesity Research Center, St. Luke's-Roosevelt Hosp. Center, N.Y. City.)

exciting circumstances, and when we have had no food for 12 to 14 hours. Under those conditions, we would have achieved a baseline metabolism, or a **basal metabolic rate** (*BMR*).

The thyroid gland, located at the base of the neck on either side of the voice box, plays a key role in determining the basal metabolic rate. In response to *thyroid-stimulating hormone (TSH)*, released from the pituitary, the thyroid gland secretes the thyroid hormones, the most important examples of which are *thyroxine* (T_4) and *triiodothyronine* (T_3) (Figure 17.7). We can see from Figure 17.7 that iodine is an essential part of the thyroid hormone molecules and must, therefore, be included in the diet.

The thyroid hormones have many, diverse and important effects.

1. They increase tissue metabolism, as indicated by oxygen consumption and increased BMR; this is termed their *calorigenic action*.
2. They help promote physical growth and maturation.
3. They help promote mental development.
4. They increase the rate of absorption of carbohydrates from the gut.
5. They influence lipid metabolism and lower circulating cholesterol levels.

Serveral decades ago, a patient's BMR was determined to aid in the diagnosis of thyroid disease (hyperthyroidism, hypothyroidism). But with the advent of modern clinical chemistry techniques, the BMR gave way to *protein bound iodine (PBI)* determinations. Most recently, diagnosis of thyroid disease has been made on the basis of the following tests.

1. *The T_4 test:* a measurement of total serum thyroxine by radioimmunoassay: used in the diagnosis of hyperthyroidism.
2. *The T_3 test* (or T_3 uptake): a measurement of the rate of uptake of radioiodinated triiodothyronine; high rates correlate to a hyperactive thyroid.
3. *The T_4 index test:* a comparison of T_4 and T_3 values.
4. *A test for circulating TSH:* in the hypothyroid patient, blood levels of TSH will be elevated. (Can you figure out why?)

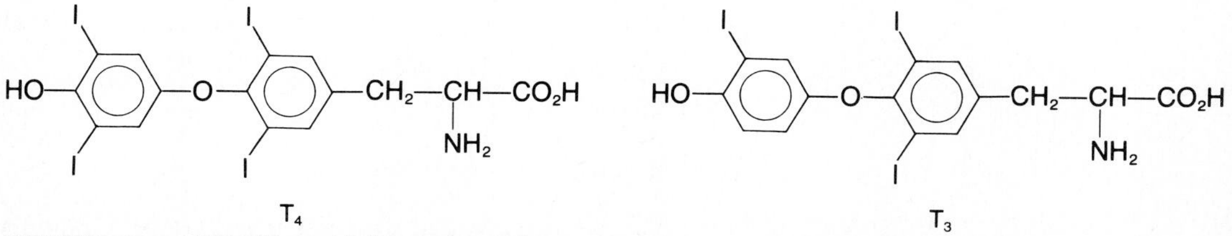

FIGURE 17.7   The Thyroid Hormones Thyroxine (T_4) and Triiodothyronine (T_3).

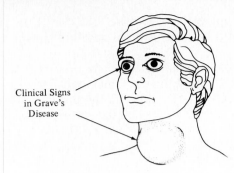

Clinical Signs
in Grave's
Disease

FIGURE 17-8 Exophthalmos and Goiter in Graves' Disease.

Goiter (enlargement of the thyroid gland) due to iodine deficiency is discussed under the subject of dietary minerals later in this chapter. Goiter due to Graves' disease is accompanied by exophthalmos—bulging of the eyeballs, a clinical sign (Figure 17.8).

Hypothyroidism (underactivity of the thyroid gland) in young children can lead to the disastrous condition called *cretinism;* in adults it can lead to *myxedema.* Both of these conditions are summarized in Table 17.3. We can easily conclude that thyroxine and the thyroid hormones play an immensely important role in human health and development.

Hyperthyroidism (overactivity of the thyroid gland) exists in all degrees of severity. Milder cases can be characteized by nervousness, weight loss, inability to tolerate heat, and inordinate hunger. All types of thyroid disease occur more commonly in women.

Table 17.3.   Forms of Thyroidism

Thyroid hormone levels	*Name of morbid state*	*Clinical picture*
Hypo- 1. In early childhood	Cretinism	Cretins are dwarf-sized, mentally retarded, and have pot bellies; with modern treatment, cretinism can be prevented if diagnosed early enough.
2. In adult	Myxedema	Patient cannot tolerate cold well; hair is coarse and sparse; husky voice, slow mentation (thinking), poor memory
Hyper- In adult	Graves' disease	Exophthalmos (bulging eyeballs), thyroid gland enlargement (goiter), warm skin, increased metabolic rate, systolic hypertension.

Drug preparations used in the treatment of thyroid diseases fall into one of two categories. For the treatment of hypothyroidism, **thyroid extract,** made from fresh, desiccated animal glands, is available. Synthetic thyroid hormones are also commercially available. Both substitute well for diminished natural thyroid hormone. *Antithyroid* drugs are designed to reduce thryoid functioning and diminish the secretion of thyroxine in cases of hyperthyroidism. Propyl-thiouracil inhibits the synthesis and therefore the secretion of thyroid hormones.

6-propyl-2-thiouracil—an antithyroid drug

It is quite effective in hyperthyroidism, but can produce serious side effects; users must be monitored closely. Hyperthyroidism is also treated surgically and with radioactive iodine.

MINERALS IN THE DIET

They may seem mundane, even boring, to some, but the inorganic ions Cr^{3+}, Mn^{2+}, Fe^{2+}, Co^{2+}, Cu^{2+}, Zn^{2+}, Mo^{2+}, Se^{2-}, and I^- play an essential role in human health. These elements are termed **trace elements** because our requirements for them are small; only traces need be in our daily diet.

Of course, the inorganic ions Na^+, K^+, Mg^{2+}, and Ca^{2+}, along with phosphorus, are essential in the human, but they occur in our diet and in our body in much more than trace amounts. They are termed **macronutrients,** whereas the Cr^{3+} and the other ions listed earlier are termed **micronutrients.** Table 17.4 lists some body minerals.

We rarely develop a deficiency of the trace elements because the tiny amounts we need daily are usually obtained in the ordinary diet of vegetables and other unrefined foods. When Americans do develop deficiencies, the shortages are most likely to be in iron and iodine; the latter occurs especially in parts of the midwest where the soil and water are unusually low in I^- content (the "goiter belt").

If you use a periodic chart of the elements to locate the micronutrients listed you will find that almost all of them are in the transition metal series. The three elements that have been given the most extensive biochemical study are iron, zinc, and iodine.

Iron, in the ferrous state, is an integral part of myoglobin, the cytochromes, and hemoglobin, where it plays a key role in oxygen transport (Chapter 15). Iron-rich foods include meats, generally, and heart and liver, specifically.

Table 17.4. Body Minerals and
Quantities

Mineral	Amount in a 60-kg human	
	Percent	Grams
Macronutrient		
Calcium	1.5–2.2	900–1320
Phosphorus	0.8–1.2	480–720
Potassium	0.35	210
Sulfur	0.25	150
Sodium	0.15	90
Chloride	0.15	90
Magnesium	0.05	30
Micronutrient		
Iron	0.004	2.4
Manganese	0.0003	0.18
Copper	0.00015	0.09
Iodine	0.00004	0.024

Spinach, lima beans, dates, clams, oysters, nuts, and dried fruits are good sources of iron. Milk is a *very poor* source of iron, and children fed an exclusive diet of milk—with no iron supplement—are in risk of becoming anemic.

Iron is very poorly absorbed from the GI tract, usually averaging between 5 and 10 percent of what is available. The Fe(II) form is more readily absorbed than the Fe(III) form, and that is why the iron supplements in medicinal use are *ferrous* sulfate and *ferrous* gluconate.

Lack of iron can result in iron-deficiency anemia. Adolescent girls are especially susceptible to this condition because of loss of iron in their menses (menstrual fluid) and because of the dietary indiosyncrasies associated with this age group.

The recommended dietary allowance **(RDA)**[1] of iron in the adult male is 10 mg; in the adult female it is 18 mg. Pregnant women in their last trimester and lactating mothers have a greater need for iron.

Multistate surveys have revealed that iron deficiency is common in America. But when a plan was proposed to force bakeries to incorporate extra iron in bread, it was rejected because *too much* iron is toxic to the human, and undoubtedly some people with adequate iron levels would have overdosed on iron. Symptoms of accidental poisoning with iron include pain and vomiting (due to irritation of the gastric mucosa) and liver damage. As few as fifteen

[1] RDAs are set by the Food and Nutrition Board of the National Research Council–National Academy of Sciences, and are generally accepted by the medical establishment. Some authorities are critical of some of the RDA values. MDRs (minimum daily requirements) are generally lower than RDAs.

0.3-g ferrous sulfate tablets have been lethal, but the ingestion of as many as seventy tablets has been followed by recovery. This shows the importance of *quick treatment.* The treatment of iron overdose may be started at home: induce vomiting immediately and then seek medical assistance.

Ferrous sulfate USP ($FeSO_4$) and ferrous gluconate NF (Fergon) are specific for the treatment of iron-deficiency anemia; the usual dose is 0.3 g tid (three times a day). Maximum reticulocyte response usually occurs 7 to 12 days after treatment begins.

The element *iodine* does not exist in nature as I_2 but as iodide ion (I^-) or as complex iodine ions (IO_3^-, iodate). In medicine, the words *iodine* and *iodide* ion are used interchangeably, but I^- is intended. An oral dose of iodide will wind up almost totally in the thyroid gland, where the iodine is used in the synthesis of T_3 and T_4 (see Figure 17.7). In this sense, the thyroid gland can be thought of as a scavenger of body iodine.

The RDA of iodine is 110 to 150 μg for men and 80 to 115 μg for women. When the daily intake of iodine consistently falls below 10 μg, synthesis of thyroxine is impaired and a condition called *goiter* can eventually develop. Goiter is the enlargement of the thyroid gland in its futile attempt to overcome its inability to manufacture enough thyroid hormone. The gland is said to hypertrophy. Unlike goiter in Graves' disease, iodine-deficiency goiter can easily be prevented by artificially increasing dietary iodine. People who season their food with iodized salt now seldom exhibit goiter, but in parts of the world goiter is still endemic. (Figure 17.9.)

Zinc is a trace mineral required in the synthesis of carbonic anhydrase, an enzyme we studied earlier. Zinc deficiency affects maintenance of skin, hair, taste, and wound healing. In zinc-deficient males, growth retardation and hypogonadism are seen. Galvanized water pipes are a nondiscretionary source of zinc in drinking water.

Copper is important in biochemistry because it is a cofactor in oxidative enzymes and becaue it is required in hemopoiesis—specifically the synthesis of hemoglobin. Copper deficiency, however, is rare in man.

Magnesium apparently is necessary for maintaining normal nerve conduction, for when a magnesium deficiency occurs, irritability of nerve tissue results in a condition called *tetany,* a syndrome characterized by intermittent spasms and possible convulsions. *Calcium tetany,* too, is well established; it results when Ca^{2+} plasma levels fall below their customary 10 mg % value. Calcium ions appear to play a near-universal role in body biochemistry. They are involved in blood clotting, teeth and bone formation, cellular adhesion and integrity, membrane stability, enzyme activity, control of certain aspects of cyclic nucleotide metabolism, mediation of some actions of prostaglandins, cell division, muscle contraction, glandular and other cellular secretory functions, nerve transmission, and probably many other body processes.

Calcium is important in the geriatric patient. For example, it is known that older women have the highest incidence of vertebral and hip fractures.

FIGURE 17-9 Goiter (enlargement of thyroid gland at base of neck) is endemic in areas of the world where iodine content of the diet is low. (Photo courtesy World Health Organization.)

(One-quarter of all white females over 60 years of age have spinal compression fractures due to osteoporosis.) The explanation lies in the fact that intestinal calcium absorption decreases with aging, contributing to postmenopausal osteoporosis (increased bone porosity). Estrogen and oral calcium supplements help prevent osteoporosis, preventing bone brittleness.

Too much calcium in the diet or in food supplments can be toxic as it can possibly lead to high Ca urine levels and thus to kidney stones in susceptible persons.

Fluoride (F-) has been established as essential in animals, but not necessarily in humans. However, in the prevention of dental caries, fluoride has assumed a very important role, even to the extent of being called the greatest advance in the decay-prevention field in the last 100 years.

Cavities are formed in teeth when lactic acid (from bacterial metabolism of sugar in the mouth) decomposes the *hydroxyapatite* in tooth enamel according to this equation:

$$Ca_{10}(PO_4)_6(OH)_2 + 8\ H^+ \rightarrow 10\ Ca^{2+} + 6\ HPO_4^{2-} + 2\ H_2O$$

hydroxyapatite   decalcified enamel

Element	Intake (mg/day)
Iron (males)	10
Iron (females)	18
Zinc	15
Manganese	2.5–5.0
Fluorine	1.5–4.0
Copper	2.0–3.0
Molybdenum	0.15–0.5
Chromium	0.05–0.2
Selenium	0.05–0.2
Iodine	0.15

FIGURE 17.10 Safe and Adequate Dietary Trace Element Intakes for Adults.

Fluoride reduces the incidence of cavities by converting a small part of hydroxyapatite to the more stable, less soluble fluoroapatite:

$$Ca_{10}(PO_4)_6(OH)_2 \;+\; 2\;F^- \rightarrow Ca_{10}(PO_4)_6F_2 \;+\; 2\;OH^-$$

  hydroxyapatite   fluoroapatite

Fluoroapatite is 100 times less soluble in lactic acid than hydroxyapatite.

About one-half of all Americans currently drink water that is fluoridated either naturally or artificially, and up to 12 million schoolchildren now take part in weekly fluoride mouthwashing programs in the nation's schools. In children whose teeth are calcifying and maturing, fluoride is especially helpful in preventing decay: 50 to 60 percent fewer cavities as compared to nonfluoride-water drinkers. About 85 percent of all toothpaste now sold in America is fluoridated; compounds used are NaF, sodium monofluorophosphate and SnF_2 (stannous fluoride). The fears for health expressed by some critics of fluoridation of municipal water supplies do not appear to be well founded.

Another trace element that has been identified as essential in humans is *selenium*. While Se deficiency is rare, when it does occur it appears to be associated with protein malnutrition disease (kwashiorkor) and multiple sclerosis. Public health officials have discovered an inverse relationship between blood selenium levels in adult males and cancer mortality in ten cities in the United States. Note that this relationship is statistical and does not constitute proof of anything.

Figure 17.10 provides a summary of safe and adequate daily intake of trace elements in adults, as recommended by the Food and Nutrition Board of the National Academy of Sciences.

Salt and the American Diet

Our emphasis in the previous section was on mineral *deficiencies*, but there is one dietary mineral that some Americans get far too much of: *NaCl* (sodium chloride, common table salt). Current estimates of the minimum adult daily requirement of NaCl are 0.5 g (at most). That's 200 mg of Na^+ since Na^+ is 39 percent by weight of NaCl. And what are we ingesting? Current estimates place the daily American intake at 10 to 12 g(!)—3 g from NaCl occurring naturally in the food eaten, plus 3 g added by the cook and at the table, plus approximately 4 to 6 g NaCl added during the commercial processing of the food. Of course, there is much individual variation in salt ingestion, but the figures still show that many of us daily ingest twenty times our nutritional requirement of salt. Furthermore, we Americans are taking in more and more sodium as we eat more processed and less fresh foods. Average per capita consumption of fresh vegetables in the United States fell by 22 percent from 1909 to 1978, and consumption of processed vegetables increased by 324 percent.

Some salty foods (more precisely Na^+-rich foods, for it is the Na^+ ion that is the biological monkey wrench) are given in Table 17.5. In 1978, over 1 million *tons* of salt was sold for use in food preparation in the United States.

Drinking water can contain much NaCl. Colorado River water, for example, has just under the level of NaCl that can be detected by taste.

Sodium ion in foods can come from leavening agents such as $NaHCO_3$, sodium acid pyrophosphates, and sodium aluminum phosphate. Medicines

Table 17.5. Some High-Sodium Foods

Food	Sodium content (mg/100 g food)
Pretzels	7800
Pickle, dill	4–5000
Bacon, Canadian	2555
Olives, green	2400
Cheese, Parmesan	1848
Soda crackers	1100
Ham, cured	860
Cheese, brick	557
Cheese, cream	294
Butter	224
Celery	126
Eggs	122
Ice cream	83
Codfish	70
Beef, ground	48
Carrots	47
Cauliflower	13
Cantaloupe	12
Peanuts, unsalted	5
Peanuts, salted	418

can be a source of Na^+, too. Certain antacids, laxatives, and analgesics are examples. Here it is important to *read the label*.

Why all this concern over salt intake? The answer lies in the statistical relationship between *hypertension* and dietary salt. This relationship has been studied for over 60 years; some of the conclusions are listed below.

- Hypertension afflicts over 20 percent of the world population; over 24 million cases are estimated in the United States.
- In most cases of high blood pressure, the actual cause cannot be determined (termed "essential" hypertension).
- When a hypertensive goes on a salt-restricted diet, his blood pressure goes down.
- When he returns to a salt diet, his blood pressure goes back up.
- Incidence of hypertension is high in Japan where salt intake commonly exceeds 20 to 25 g/day.
- In many primitive societies, blood pressure tends to be low and does *not* rise with age; prevalence of hypertension and heart disease approaches zero.
- When primitive tribes become acculturated, blood pressures rise; hypertension becomes a factor; this indicates environment is playing a role.
- Prevalence of hypertension may be as high as 20 percent in white Americans over 50 years of age, and up to 40 percent in black Americans.

To begin to understand the mechanism by which excessive Na^+ ion is believed to be involved in hypertension, remember that in the body Na^+ is always associated with water. The two go hand-in-hand. Excessive salt thus means excessive body water and a job for the kidneys to get rid of both. With an increased blood pressure, the kidneys are more effective in maintaining a high rate of excretion of water and Na^+ ion. Hence, the increase in blood pressure may simply be part of the body's homeostatic response to the stress of too much body fluid. An when the stress is removed the blood pressure need not be so high and will of itself fall toward normal.

In their 1980 *Guide for Healthful Diets*, the Food and Nutrition Board of the National Research Council suggests that a daily salt intake limited to 3 g may be helpful in the prevention of hypertension in susceptible individuals. This low intake would require the elimination of salt in cooking and at the table, since we receive about 3 g of NaCl daily in nondiscretionary salt intake in foods.

Potassium chloride (KCl) is becoming more widely used as a salt substitute. Recall that K^+ is the principal *intra*cellular cation and its ingestion will not add as much water to the extracellular compartment as will addition of Na^+. Although KCl does taste "salty," it is not an altogether satisfactory substitute for NaCl.

VITAMINS AND AVITAMINOSES

The word *vitamin* comes from the words *vital amine*, for it was originally thought that all of these compounds vital to sustaining life and health were aminoid. We now know, of course, that vitamins A, D, E, K, and C are nitrogen-free, but nonetheless vital.

Most of the vitamins were discovered in the great biochemical era of 1920–1940, which means that mankind survived many millenia without once trying a one-a-day vitamin supplement.

A *working definition of vitamin* is required because we must make personal and professional judgments about the true vitamin character of substances purported to be vitamins (laetrile is a current example). A vitamin is any organic, dietary substance required in very small quantities by the human body in its day-to-day functioning to promote growth or prevent the occurrence of debilitation or malfunctioning (exclusive of carbohydrates, fats, or proteins). Further, a substance can be classified as a vitamin only if deficiency symptoms occur in its absence *which can be relieved when the substance is readministered*. It is the latter half of this definition that eliminates a lot of compounds, such as laetrile (also known as amygdalin).

Some vitamins are synthesized inside the human body. Vitamin D, for example, is produced in the skin from provitamin D precursors. Intestinal bacteria are capable of synthesizing vitamin D, folic acid, thiamine, and biotin in the gut, but apparently only the folic acid so produced can be absorbed in any significant amounts into the general circulation.

Generally, vitamins have no metabolic activity by themselves, but act as coenzymes, that is, they only function as vitamins when combined with a protein (apoenzyme). It is now accepted that vitamin D also acts as a potent steroidal hormone.

The health professional should be an expert on the subject of vitamins. She or he should know the following facts:

1. The symbols, names, and synonyms of the vitamins.
2. Common food sources of each.
3. Symptoms of vitamin deficiencies (deficiencies are called **avitaminoses**).
4. The potential for toxicity of certain vitamins in overdosage.
5. The biological role played by vitamins.

In our studies so far we have discussed the biochemical role played by some of the B vitamins. In this section we will learn that vitamin C helps maintain the integrity of capillary membranes, that vitamin A prevents night blindness through its action on the retina, and that vitamin K is a key factor in preventing hemorrhage.

We indicated earlier the wisdom of eating a variety of foods, some from each of the four food groups: grains, vegetables, meats, and milk products. When it comes to obtaining vitamins, that is still good advice. Ordinarily, the average person should not have to take a vitamin supplement. Exceptions to this may occur in pregnancy, during lactation, or when recovering from illness or hospitalization. If fresh citrus fruits are unavailable, it is usually good practice to supplement with vitamin C (buy the cheapest brand you can get).

Tables 17.6 and 17.7 present names, synonyms, chemical structures, sources, and deficiency signs of all the known vitamins.

Four of the thirteen vitamins in Table 17.6 are fat soluble; the remainder are water soluble. Vitamins B_1, B_2, B_6, niacin, pantothenic acid, biotin, folic acid, and B_{12} constitute the B-complex, a group of water-soluble, nitrogen-containing vitamins that are poorly stored in the body and that must be replenished daily. Vitamin C is also water soluble; it should be included in our diet each day.

EXAMPLE 17.1

Examine the chemical structures of the thirteen vitamins in Table 17.6 and predict, on the basis of structures, which are fat soluble.

Solution
A, D, E, and K are fat soluble. Each has a structure in which relatively nonpolar hydrocarbon groups predominate. What few polar groups (such as OH) exist are insufficient to confer overall water solubility.

Occasionally, you will come across ads for other "vitamins," such as B_{17}. Remember, anybody can term their discovery a "vitamin" and give it a name, but if it does not fit our definition (see earlier), it is not a vitamin in the accepted, biomedical sense. Signs of vitamin deficiencies are shown for various animals in Figure 17.11.

The RDA for *ascorbic acid* is 45 mg (National Research Council). However, some authorities, especially Nobel Laureate Linus Pauling, believe that the average human body requires about 100 times that figure in order to prevent head colds and other viral infections and to maintain general good health. Daily doses of 2 to 4 g of vitamin C will be mostly excreted in the urine, but the proponents of such megavitamin therapy insist that large doses are required to *saturate* the tissues. Despite much debate, the matter remains unsettled.

Megavitamin therapy has been used to treat the mentally ill, in conjunction with conventional psychiatric treatment. Here the belief is that certain individuals *normally* require vastly more of the B vitamins than the average person in order to stay healthy. Vitamin E is also administered by some in very large doses. You should know that the concept of megavitamin therapy has not gained general acceptance by the medical establishment.

Table 17.6.   Symbols, Names, and Chemical Structures of the Vitamins

Symbol	Names	Chemical structure
A (A_1A_2)	Vitamin A group, retinols, dehydroretinol	CH_3 CH_3 ... $-CH=CH-C=CH-CH=CH-C=CH-CH_2OH$ (CH_3) (CH_3)
B_1	Thiamine, antiberiberi factor	NH_2 ... $-CH_2-N^+$ ... CH_3 N CH_3 CH_2CH_2OH (S)
B_2	Riboflavin	CH_3 ... CH_3 ... O, $N-H$, O, CH_2, $(HCOH)_3$, CH_2OH
B_6	Pyridoxine, pyridoxal	CH_2OH, HO, CH_2OH, CH_3 N
—	Nicotinic acid or niacin (nicotinic acid amide, niacinamide)	$-COOH$ N $\left(-CONH_2 \ N \right)$
—	Pantothenic acid (antigray hair factor)	CH_3 ... $HOCH_2-C-CH-CONH-CH_2CH_2COOH$ ... H_3C OH
—	Biotin (anti-egg-white-injury factor)	H, N, $O=$, S, N, H, $(CH_2)_4COOH$

Table 17.6. Symbols, Names, and Chemical Structures of the Vitamins *(Cont.)*

Symbol	Names	Chemical structure
	Folic acid derivatives, pteroylglutamic acid (PGA), folacin	
B_{12}	Cyanocobalamin, extrinsic factor	$C_{63}H_{88}O_{14}N_{14}PCo$
C	Ascorbic acid, antiscorbutic factor	
D ($D_2D_3D_4$)	Antirachitic factor, calciferol, irradiated ergosterol, activated ergosterol, ergocalciferol	
E	Vitamin E group, α-tocopherol, β-tocopherol, γ-tocopherol, antisterility factor (in animals)	 α-tocopherol
K (K_1K_2)	Menadione, vitamin K group, blood clotting factor, antihemorrhagic factor	 menadione—a synthetic compound with vitamin K activity

Table 17.7.   Vitamins: Food Sources and Deficiency Signs

Vitamin	Good food source	Deficiency signs
A	Fish liver oils, eggs, liver, milk and milk products, yellow vegetables. Green vegetables, squash, carrots, and tomatoes contain the precursor, beta-carotene	Night blindness, eye inflammation, dry skin
B_1	Liver, unrefined cereal grains, pork, beef, nuts	Beriberi (characterized by dermatitis, diarrhea, psychic disturbances, accumulation of body fluids, inflammation of nerve fibers)
B_2	Liver, milk, red meat, eggs, fish, whole wheat flour	Glossitis (tongue inflammation), cheilosis, dermatitis, anemia
B_6	Liver, yeast, meat, eggs, milk, wheat, corn, peas, beans	Hyperirritability, convulsions in infants, dermatitis, glossitis, increased susceptibility to infections
Niacin	Liver, red meat, yeast, eggs, peanuts, fish	Pellagra (characterized by dermatitis, psychic disturbances, diarrhea, inflammation of mucous membranes)

The RDA for *vitamin E* is 15 IU for adult males and 12 IU for females. Thousands of units of vitamin E daily have been recommended for heart conditions, sexual impotence, wound healing, and as an antoxidant for use in our polluted environment, but there is not enough firm evidence for any of these claims. We can say this about vitamin E. It is found in all body tissues and is important for proper tissue functioning. It helps maintain the life of the RBCs in the circulatory system; it aids the body in utilizing fully vitamin A; it is needed in increasing amounts as one's consumption of polyunsaturated fats increases. On the negative side, researchers at Baltimore Sinai Hospital and the Johns Hopkins University School of Medicine found that vitamin E may not be an innocuous substance, for in doses of 600 IU/day for 8 weeks, it increased the level of free cholesterol in low-density and very-low-density lipoprotein. The researchers concluded that vitamin E is an active drug, not a benign vitamin.

Table 17.7.   Vitamins: Food Sources and Deficiency Signs *(Cont.)*

Vitamin	*Good food source*	*Deficiency signs*
Pantothenic acid	Liver, milk, eggs, yeast, peas	Enteritis (GI disturbances), alopecia (falling out of hair), dermatitis, mental depression
Biotin	Liver, egg yolk, peanuts	Dermatitis, enteritis
Folic acid and folates	Green leafy vegetables, liver, wheat bran	Sprue, anemia, GI disturbances
B_{12}	Liver, meat, eggs, milk, oysters, kidney	Pernicious anemia, retarded growth, glossitis, spinal cord degeneration
C	Citrus fruits, tomatoes, berries, greens, cabbages, peppers	Scurvy (characterized by weakness, anemia, spongy gums, hemorrhage, fever)
D	Fish liver oils, vitamin-fortified milk; synthesized in skin with UV light from precursors	Rickets, osteomalacia (softening of bones)
E	Milk, eggs, vegetable oils, margarine, green leafy vegetables, wheat germ oil	Anemia in premature babies fed inadequate formula
K	Green, leafy vegetables, tomatoes	Spontaneous hemorrhage, increased clotting time in blood

Hypervitaminosis

Although little evidence exists to indicate that very large doses of vitamin C can actually be harmful, the evidence with vitamins A, D, E, and K is just the opposite. Cases of **hypervitaminosis** of A and D have been seen in recent years as some people overdo a good thing, prompted especially by advertising that is aimed more at making money than curing real deficiencies.

Too much vitamin A can result in liver damage, retarded growth in children, dry and cracked skin, headache, and bone pain. Hypervitaminosis A can further cause a lack of appetite, hair loss, and menstrual problems.

Too much vitamin D can result in calcification (Ca^{2+} deposit in tissues where we do not want it), kidney stones and kidney damage, retarded physical and mental growth in children, constipation, weakness, stiffness, and high blood pressure. Vitamin D is now considered a potent steroidal hormone that

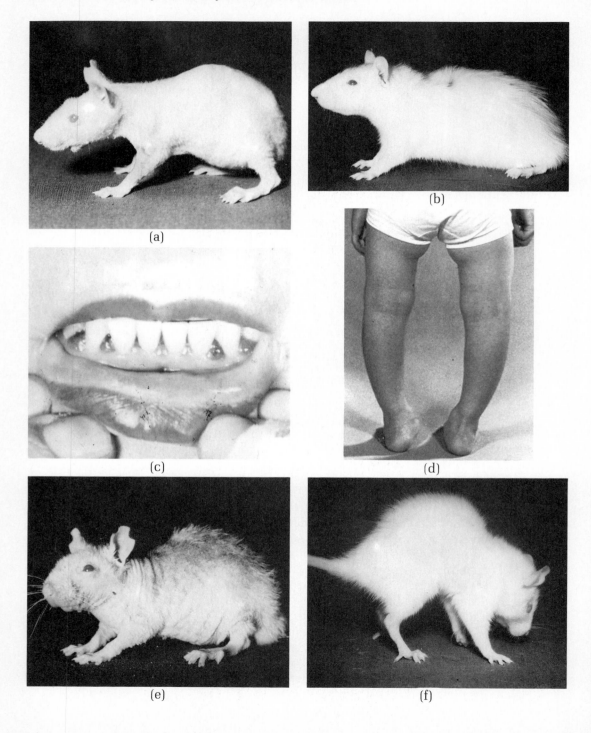

(a)

(b)

(c)

(d)

(e)

(f)

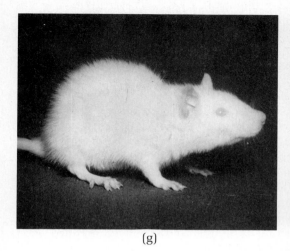
(g)

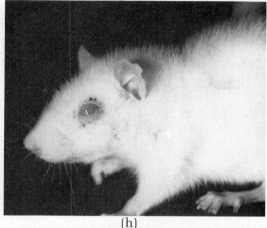
(h)

(i)

FIGURE 17.11 (a) Riboflavin-deficient rat; note dermatitis and inflammation of eye; (b) same rat as in (a) after two months' treatment with riboflavin; (c) gingivitis in adult scurvy; (d) in rickets, bowlegs curve laterally from the weight of standing; (e) biotin-deficient rat; dermatitis has progressed to alopecia; (f) thiamine-deficient rat; polyneuritis (generalized nerve inflammation) produces spastic gait and loss of balance; (g) same rat as in (f); 8 hours after receiving thiamin hydrochloride the rat has regained equilibrium and use of hind legs; (h) vitamin A deficient rat; late stages of eye damage shown here is incurable; (i) vitamin K deficient chick; spontaneous hemorrhaging has occurred under the skin. (Picture credits: a, b, c, e, f, g, h, i courtesy The Upjohn Company, Kalamazoo, MI 49001; d courtesy Cortez F. Enloe, Jr., M. M., *Nutrition Today*.)

clearly can be toxic in high doses. The RDA for vitamin D is 400 IU for all ages and both sexes. The RDA for vitamin A is 5000 IU for adult males and females.

■■■■■■■■**EXAMPLE 17.2**■■■■■■■■

How do the recommended dietary allowances given in the preceding paragraph compare with strengths of various A and D products available for sale in health food stores?

Solution

Presently, one can purchase vitamin D in 1000 IU doses and vitamin A in 25,000 IU doses.

Two Case Histories

In 1981, two cases of hypervitamosis were studied at UCLA hospital. A 4-year-old boy, suffering from increased intracranial pressure, edema, bone pain, cracked lips, enlarged liver, and blurred vision, had been treated with 15,000 IU of vitamin A daily on recommendation of the family chiropractor because of an increased frequency of upper respiratory infections. After 3 months, the dose of vitamin A was increased to 250,000 IU daily(!) along with 4000 IU of vitamin D. The boy's serum vitamin A level was 1626 IU/dL (normal 65–275 IU/dL). A second patient, a 2½-year-old boy with fever, rash, edema, vomiting, and irritability, showed a serum vitamin A level of 1812 IU/dL. He had been given one vitamin A tablet (25,000 units) daily or every other day, one to two multivitamin tablets containing 5000 IU of vitamin A, and multivitamin drops that also contained vitamin A. To treat the rash that developed, his mother applied (you guessed it) vitamin A cream.

Although the dangers of excessive vitamin D intake are well known, cases of hypervitaminosis D are still reported. Vitamin D has been used inappropriately in doses of up to 100,000 IU/day for cold hands and feet after thyroidectomy, for arthritis, and for psoriasis.

Excessive intake of vitamin K can affect the liver's ability to secrete bile; brain damage may be a consequence.

FOOD–DRUG INTERACTIONS

When over 1.5 billion prescriptions are written yearly for Americans, the strong likelihood arises that chemicals in the food we consume will interact in one way or another with the chemicals in the prescriptions to cause a health problem. Indeed, **food-drug interactions** are well known today. The health professional should know that in some cases these interactions may be of critical importance.

Consider the patient who is in mental depression and for whom a *monoamineoxidase inhibitor* (*MAO* inhibitor) has been prescribed. An MAO inhibitor is a drug that causes an accumulation of body amines (such as norepinephrine) with the result that the brain and spinal cord are stimulated and the depression is overcome. Now the patient eats some aged cheese, or Chianti wine, or chicken livers and suddenly gets a terrific reaction (severe headache, high blood pressure, brain hemorrhage, even death). It turns out that the foods mentioned all contain large amounts of another amine called *tyramine*, which is similar in action to norepi. The MAO inhibitor has made it impossible for the patient to metabolize his *endogenous* CNS-stimulating amines, and now he is getting an extra dose from an outside source. The combination is too much for him to handle, and a hypertensive crisis results. It is important to caution a patient taking an MAO inhibitor to avoid all aged or fermented foods, fermented sausages such as salami and pepperoni, sharp cheeses, yogurt, sour cream, beef and chicken livers, fava beans, canned figs, beer, Chianti wine, sherry, and other wines in large quantities.

Patients taking tetracycline antibiotics must avoid concurrent intake of milk or any product high in calcium. Calcium ions bind to the electron-rich groups in tetracyclines, reducing the solubility of the tetracycline and diminishing its bioavailability.

Oral contraceptives (the Pill) are known to lower blood levels of folic acid and vitamin B_6. Usually this depletion is not serious, but in women Pill users who have a preexisting diet-related deficiency of folic acid or B_6, oral contraceptive use could be critical. Such women should eat extra green, leafy vegetables, which are good sources of folic acid, and should seek nutritional guidance.

Other food–drug interactions include the following:

1. The action of the CNS-stimulant caffeine (in coffee, tea, and cola drinks) to counteract the sedative effects of a barbiturate or other sleep aid.
2. The ability of vitamin K from liver or green, leafy vegetables to hinder the effectiveness of anticoagulants.
3. The action of soybeans, brussel sprouts, cabbage, or kale to inhibit normal production of thyroid hormone in susceptible individuals (these vegetables contain substances identified as "goitrogens").
4. The ability of soda pop, acid fruit, or vegetable juices to produce enough acidity to cause a drug to dissolve in the stomach instead of in the intestine; poorer drug absorption into the blood can result.

Remember that drugs found in OTC preparations are also capable of causing undesirable interactions. Antacids are often high in calcium (see tetracycline discussion). Bulk-forming laxatives (e.g., cellulose derivatives) may interact and combine with salicylates and digitalis glycosides. Aspirin and other salicylates interfere with the action of probenecid and sulfinpyrazone used in the treatment of gout. Just as serious is the action of aspirin to prolong bleeding time and thus potentiate anticoagulant drugs to the point of inducing hemorrhage. The laity should be cautioned about indiscriminate use of drugs and about concurrent use of several drugs.

HEAVY METAL POISONS

Lead and mercury are widely distributed in nature and occur in our foods, as shown in Table 17.8.

Lead and mercury play no physiologic role in our nutrition or anywhere

Table 17.8. Lead Content
of Some Foods

Food	
Pineapple juice	0.26 μg/mL
Fresh orange juice	0.15 μg/mL
Tomato juice	0.3 μg/mL
Pears	0.32 μg/g
Prunes	0.32 μg/g
Potatoes	0.28 μg/g

in the body, but rather are toxic to human tissue. **Heavy metal poisons** can produce a **chronic toxicity,** that is, smaller doses can unknowingly be ingested over a longer period of time, producing toxic reactions not severe enough to kill immediately. When larger doses of a heavy metal poison are taken all at once, an **acute toxicity** can occur that may threaten life.

In a typical diet, the average American ingests about 300 μg of lead daily. Fortunately, only about 6 to 7 percent of the lead we ingest gets absorbed from the GI tract. Add to this the 15 μg of lead we inhale from the air (not all enters the circulation), and our total daily intake approximates 30 μg. Since we daily excrete about 30 μg of lead, there is usually no net accumulation of the metal.

Chronic exposure to higher lead levels is dangerous because lead is highly cumulative, with deposition as lead phosphate in the bone matrices being common. With the passage of time, absorbed lead is more deeply buried in the bone matrix, and a biological half-life of 2 to 3 years can be expected. About 3 months of steady lead ingestion are required before obvious symptoms of lead poisoning (plumbism) appear. If the ingestion of lead is stopped before symptoms appear, the actual disease does not occur. Symptoms of lead poisoning are often vague. Milder symptoms may include fatigue, loss of appetite, irritability, sleep disturbance, and sudden behavioral change. Acute symptoms include clumsiness, muscle weakness, abdominal pain, vomiting, constipation, and changes in conscious state.

Treatment of lead poisoning is accomplished by administering **chelating agents** that effectively *sequester* (bind and "hide") the metal ions, preventing their deposition on the surface of blood cells. Chelating agents that are frequently used to treat lead poisoning are *EDTA* (calcium disodium ethylenediamine tetraacetate), *BAL* (British antilewisite or 2,3-dimercaptopropanol) and *D-penicillamine.*

EXAMPLE 17.3

Structure *a*, below is BAL; *b* is EDTA. Chemically, which functional groups in BAL and EDTA help to bind and "hide" heavy metal ions like Pb^{2+} and Hg^{2+}?

$$
\begin{array}{ccc}
& \text{H} \quad \text{H} \quad \text{H} & \\
& |\quad\; |\quad\; | & \\
\text{H}- &\text{C}-\text{C}-\text{CH} & \\
& |\quad\; |\quad\; | & \\
& \text{OH SH SH} &
\end{array}
$$

a.

$$
\begin{array}{c}
{}^-\text{OOC}-\text{CH}_2 \qquad\qquad\qquad\qquad \text{CH}_2\text{COO}^- \\
\diagdown \qquad\qquad\qquad\qquad\qquad\qquad \diagup \\
\text{N}-\text{CH}_2\text{CH}_2-\text{N} \\
\diagup \qquad\qquad\qquad\qquad\qquad\qquad \diagdown \\
{}^-\text{OOC}-\text{CH}_2 \qquad\qquad\qquad\qquad \text{CH}_2\text{COO}^-
\end{array}
$$

b.

Solution

The groups $-COO^-$, $-OH$, $-SH$, and amine possess electron pairs for binding heavy metal ions.

Lead has appeared in foods when acidic fruit juices were served in pottery glazed with lead-containing compounds. Acids can dissolve some of the lead salts in the glaze, which are then ingested with the fruit drink.

Mercury has been detected in ocean fish. In nature, inorganic mercury is not as dangerous to man as organic mercury compounds. *Methyl mercury,* CH_3Hg^+, an organic compound, passes readily down the food chain, from smaller fish to larger fish to man. It is mercury in bioavailable form.

Symptoms of mercury poisoning include dizziness, headache, tremors, and loss of motor coordination—all the result of mercury's damage of the nervous system.

A Case History—Lead Poisoning

In Chicago in the summer of 1980 a 2-year-old child died of lead poisoning after a history of eating peeled paint chips in an old, rundown apartment building. Weak, listless, anemic, and nauseated at first, the child later convulsed and died of respiratory failure. At death, his blood level of lead was 160 µg/100 mL. A level of 40 µg/100 mL is considered normal in adults. Laws now prohibit the use of lead pigments in paints, but the metal may still be found in some older buildings. The craving in children for unnatural foods is termed *pica.* Children have been known to eat dirt, clay and plaster in addition to paint.

SUMMARY

Essential food factors are those substances required by the body for normal functioning but not produced by the body at all or in quantities large enough to meet body demands.

For the adult, there are nine essential amino acids. A protein is a complete protein if it supplies all nine essential amino acids in quantities large enough for complementation to occur, that is, for utilization of all the amino acids in complete protein synthesis. A food that is an incomplete protein source can be eaten with another food that supplies the deficient amino acids and thus complementation can occur.

Kwashiorkor disease is a protein-deficiency disease. The kwashiorkor child is in negative nitrogen balance, that is, he is excreting more nitrogen in his urine and feces than he is ingesting daily. Diabetics and immobile and starving people are in negative nitrogen balance. Growing children, surgical recovery patients, and those on anabolic steroids are typically in positive nitrogen balance.

The recommended daily protein intake for males is about 65 g; for females, about 50 g.

The principal essential fatty acid is linoleic acid. Linolenic acid is considered by some authorities to be essential, also. Prostaglandins are twenty-carbon unsaturated fatty acids. They have a wide variety of pharmacological effects including effects in reproduction, pain perception, respiration, inflammation, and blood clotting.

The Calorie used in nutrition is the kilocalorie. Gram for gram, fats supply more than twice the calories that carbohydrates or proteins supply. Table 17.2 shows the number of calories consumed per hour in various physical activities.

The thyroid gland secretes thyroxin (T_4) and triiodothyronine (T_3), two hormones that increase tissue metabolism, promote physical and mental growth, and influence carbohydrate and lipid metabolism. Goiter (enlargement of the thyroid gland) is typically due to a dietary deficiency of iodine. Hypothyroidism is underactivity of the thyroid gland. In children, this leads to cretinism, if severe.

Trace elements are inorganic ions essential to life but only in trace quantities. Another name for them is micronutrients. The macronutrients are found in larger quantities in our bodies. Fe(II) is required in the synthesis of myoglobin, the cytochromes, and hemoglobin. Fe(II) ions are more easily absorbed from the gut than Fe(III).

Zinc deficiency shows itself in lowered quality of skin, hair, taste, and wound healing. Copper is required in enzyme activity and in blood formation. Magnesium is needed to maintain proper nerve conduction. Calcium ions play a near-universal role in biochemistry. Our text lists ten physical and biochemical roles for the Ca^{2+} ion. Fluoride is of great value in the prevention of dental caries.

Americans consume too much salt—up to twenty times more than we need each day. Excess salt (NaCl) intake is statistically associated with hypertension. It is the Na^+ ion in the salt that is the troublemaker.

Vitamins are defined as any organic, dietary substance required in very small quantities to promote growth or prevent debilitation or malfunctioning (exclusive of carbohydrates, fats, proteins). Further, a substance can be classified as a vitamin only if deficiency symptoms occur in its absence that are relieved when the vitamin is readministered.

Table 17.6 presents symbols, names, and chemical structures for thirteen vitamins. Table 17.7 presents food sources and deficiency signs of these same vitamins. Some well-known avitaminoses are mentioned: night blindness, beriberi, pellagra, pernicious anemia, scurvy, rickets.

Megavitamin therapy was discussed in this chapter. Very large doses of vitamins such as C, E, and B complex, used in the treatment or prevention of disease, are still controversial. Megavitamin therapy has not gained wide medical acceptance.

Overdosing with vitamins A, D, E, and K can be harmful. This is especially true of A and D. Two case histories were presented to show this.

Food–drug interactions are well known. MAO inhibitors must not be mixed with foods containing tyramine. Avoid the consumption of milk or milk products with tetracycline antibiotics. Additional food–drug interactions are given.

Two important heavy metal poisons are lead and mercury. They can induce acute toxicity or chronic toxicity. Lead is highly cumulative and symptoms of its poisoning often are vague. Mercury can damage the nervous system. Methyl mercury (CH_3Hg^+) is mercury in its bioavailable form.

KEY TERMS

Check your understanding of this chapter. Can you explain what is meant by each of these terms?

acute toxicity	goiter
avitaminosis	heavy metal poison
basal metabolic rate	hypervitaminosis
Calorie	macronutrient
chelating agent	micronutrient
chronic toxicity	nitrogen balance
complementary protein	prostaglandin
complete protein	RDA
essential food factor	trace element
food–drug interaction	vitamin

STUDY QUESTIONS

1. From memory, name the nine amino acids essential in the adult human.

2. Examine the chemical structures of the nine essential amino acids. What chemical features do they have, generally, that the nonessential do not have?

3. If phenylalanine, with its aromatic ring, is an essential amino acid, why is not tyrosine also essential? It, too, has an aromatic ring.

4. Define the following:

   a. Complete protein

   c. BMR

   e. Trace element

   g. Fat-soluble vitamins

   i. MAO inhibitor

   b. Positive nitrogen balance

   d. Exophthalmos

   f. B complex vitamin

   h. Goitrogen

   j. Pica

5. What is meant by the term *complementary protein*. Give one example of foods complementing each other in regard to essential amino acid content.

6. How does protein deficiency disease arise, what are its signs, and what name has been given to it?

7. Diabetics are notoriously prone to infections and to wounds that will not heal. Diabetics normally show a negative nitrogen balance. Relate the diabetes and the negative N balance and explain the diabetic's susceptibility.

8. Name two essential fatty acids and give the number of double bonds contained in a molecule of each.

9. For a postoperative, debilitated patient, a physician prescribes an anabolic steroid. Predict the effect this type of drug will have on the patient's nitrogen balance.

10. **a.** What is the relationship of essential fatty acids to prostaglandins?

    b. What actions do the prostaglandins have on the uterus, on gastric secretion, and in early pregnancy?

11. How does a macronutrient differ from a micronutrient?

12. Theoretically, what is the maximum caloric value of a daily intake consisting of $\frac{3}{4}$-lb of carbohydrate, $\frac{1}{6}$-lb of fat, and $\frac{1}{4}$-lb of protein? (Remember, 1 lb = 454 g.)

13. If an obese person's daily energy consumption exceeds intake by 700 Cal, how long will it take him to lose 20 lb of flesh?

14. A person consumes 80 g of protein, 65 g of fat, and 400 g of carbohydrate in 1 day. He burns up 2350 Cal of energy that day. Did he gain or lose body mass that day?

15. Nutritionists evaluate quality foods from the standpoint of calorie content, vitamin and fiber content, essential amino acid and fatty acid content, and how much salt and sugar they contain. The potential for food–drug interactions is also considered. How do the so-called junk foods measure up to this yardstick?

16. How could anxiety over having one's BMR measured affect the results of the measurement?

17. Name and give the symbol for the two most important thyroid hormones.

18. What essential amino acid appears to be a likely precursor in the human to thyroxine? To serotonin? To norepinephrine? To dopamine?

19. How does cretinism differ from Graves' disease in terms of causes, signs and symptoms, and treatment?

20. If *phagocytosis* means "cell eating" (Chapter 15), what does hyper*phagia* mean? In what disease condition might hyperphagia be observed?

21. One of the heavily promoted diet plans of recent memory permitted the dieter all of the protein and fat he or she wanted, but no carbohydrate. The zero carbohydrate aspect was continued until ketone bodies were detectable in the urine. Biochemically, why is it reasonable to expect ketone bodies to appear in a zero carbohydrate diet? (*Hint:* What is now the source of energy?)

22. From what you have read in this chapter, predict the use of I^{131} in the treatment of human disease. (*Hint:* I^{131} is a beta-emitter and will destroy tissue in its immediate surroundings.)

23. Match the trace mineral with its biochemical role in the human:

    Fe _____   **A.** The synthesis of thyroxine.
    Co _____   **B.** A portion of vitamin B_{12}.
    Cu _____   **C.** A portion of carbonic anhydrase.
    Zn _____   **D.** Required for oxygen transport.
    I _____   **E.** Involved in hemopoiesis.
    Mg _____   **F.** Prevention of tetany.

24. Persons of what age and sex should be most conscious of their iron intake and iron blood levels, and why?

25. Cobalt is essential for the vitamin B_{12} molecule. Predict one effect of a deficiency of cobalt in the human.

26. How does fluoride act to protect teeth?

27. It is estimated that adults require 200 mg of Na^+ daily to replace losses. If one consumes $\frac{1}{10}$-lb of Canadian bacon, $\frac{1}{8}$-lb of cured ham, one egg (30 g), one olive (5 g), and $\frac{1}{2}$-oz of salted peanuts in 1 day, by how much would the 200-mg requirement be exceeded? (See Table 17.5 for values.)

28. Match the vitamin with each and all statements that apply to it. A statement may apply more than one time.

    Vitamin B_1 _____   **a.** Deficiency can result in spontaneous
    Vitamin B_2 _____   hemorrhage.
    Vitamin B_6 _____   **b.** Beriberi is the deficiency state.
    Vitamin C _____   **c.** Oranges and citrus fruits are a good source.

Vitamin A _____   d. Green, leafy vegetables are a good source.
Vitamin D _____   e. Pellagra is the deficiency state.
Vitamin E _____   f. Chemically, is a phenolic derivative.
Vitamin K _____   g. Scurvy is the deficiency disease.
Folic acid _____   h. Prevents night blindness.
Niacin _____   i. Is a ribose derivative.
                   j. Fish liver oils are good source.
                   k. Rickets is the deficiency condition.
                   l. Synonymous with retinol.
                   m. Megaloblastic anemia is the deficiency condition.
                   n. Contains no nitrogen.
                   o. Cyanocobalamin is a synonym.
                   p. alpha-tocopherol is an example.
                   q. Thiamine.

29. a. If a person swallows 0.100 g of PbO and 6 percent of it enters his circulation, how many micrograms of lead oxide has he absorbed?
    b. If all of the absorbed Pb^{2+} in part a remains in the blood, what concentration of Pb^{2+} in µg/100 mL would result? (Figure on an average 12 pt of blood.)

ADVANCED STUDY QUESTIONS

30. Describe the chemical differences and similarities among the three major types of prostaglandins; summarize the important pharmacological effects of prostaglandins in the human.

31. Compile a list of all factors mentioned in this chapter that are essential for our continued health and existence and which must come to us from external sources.

18

Chemistry of Heredity

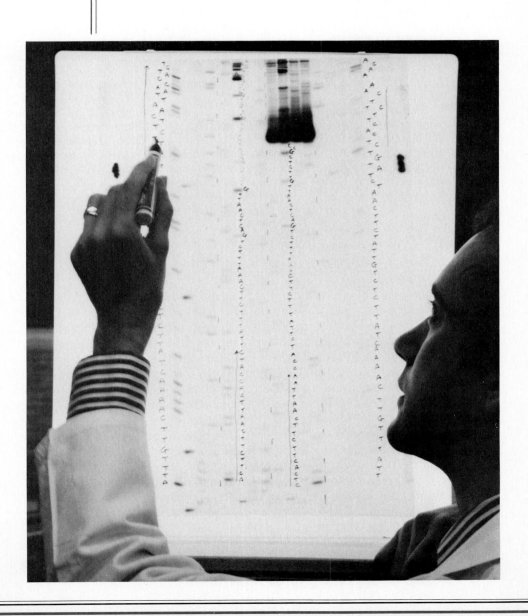

The researcher in our photo is a new breed of scientist—a genetic engineer. He is reading the sequence of letters in the genetic code identifying specific genes in DNA. His research could lead to great advances in biotechnology: weapons against disease; food crops that will grow in any soil; hormone supplies adequate to meet any needs; dairy cows as big as elephants, producing twenty tons of milk a year.

The age of genetic engineering has arrived! Scientists are now routinely capable of altering the DNA in an organism, that is, changing the hereditary material that dictates what the organism will look like and what it can do biochemically. One pharmaceutical house is now marketing human insulin made by a bacterium; the bacterium was given the human DNA that codes for insulin production. That human genetic material will now forever be a part of the bacterial genes.

As a health professional (and just as a private citizen) you will need to understand terms routinely used to describe progress in the chemistry of heredity. Be alert to terms like gene splicing, genetic code, DNA, and inborn errors of metabolism. Be aware of applications to health and disease: a ready supply of hormones, interferon, and insulin; the treatment of genetic disease; vaccines against disease. (Photo courtesy of G. D. Searle and Co.)

INTRODUCTION

Each of us is a hybrid, created from genetic material contributed by our parents. In this chapter we shall study the chemistry of heredity, that is, the chemical factors involved in the transfer of the staggering amount of information required to make an offspring from a set of parents. All of this information is "written" on long strands of remarkable substances called *nucleic acids*.

Nucleic acids are high-molecular-weight polymers that occur in all living matter, whether human, animal, or vegetable. There are two types of nucleic acid, *deoxyribonucleic acid, DNA,* and *ribonucleic acid, RNA.* DNA plays a vital part in heredity; it is capable of replicating itself and is the chief material in the chromosomes. In living cells that contain nuclei (eukaryotes), the DNA is found in the nucleus; in cells with no nuclei (such as bacteria) the DNA is found in the mitochondria, in the cytoplasm (Figure 18.1).

The monomer that goes together to make up giant polymeric nucleic acids is termed a **nucleotide.** Each nucleotide is composed of the following:

1. A sugar (deoxyribose in DNA; ribose in RNA).
2. Phosphoric acid.
3. Nitrogen-containing bases.

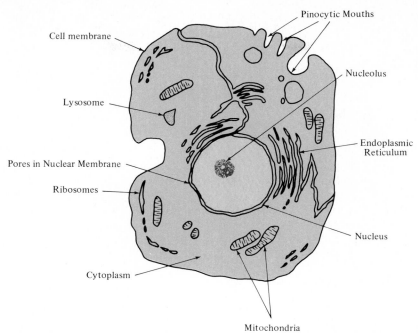

FIGURE 18.1   Pictorial representation of a typical animal cell. You will have occasion to refer back to this drawing as you read through the chapter.

Figure 18.2 shows how these chemical parts are arranged into nucleotide units and then into the long, polymeric chain we term **DNA**. Note that each nucleotide is a combination of one molecule each of nitrogen base, sugar, and phosphate. As we shall see, the sequence of bases in the nucleotides in DNA is extremely important.

THE BASES IN DNA

You should make a special effort to learn the names and structures of the very important bases that are a part of DNA. Chemically, they are of two types.

1. Bases that are derivatives of *pyrimidine*

pyrimidine (pie-rim′-eh-deen)

The three **pyrimidine bases** we encounter in nucleic acids are

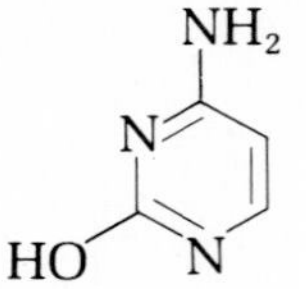

cytosine
(sy'-toe-seen)

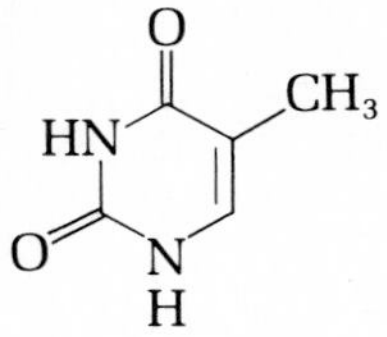

thymine
(thigh'-meen)

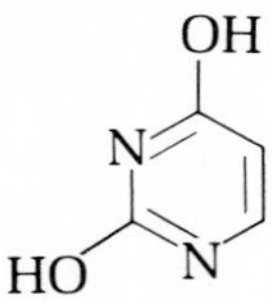

uracil (found in RNA)
(yur'-ah-sill)

2. Bases that are derivatives of *purine*

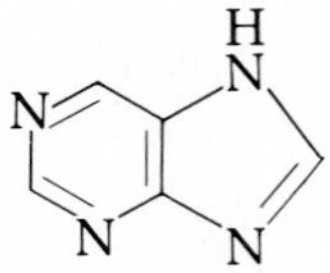

purine (pure'een)

The two **purine bases** we encounter in nucleic acids are

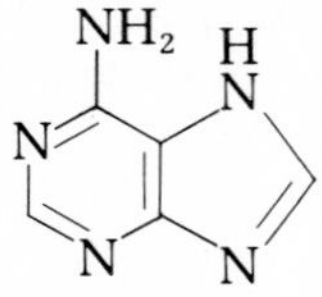

adenine
(add'-ah-neen)

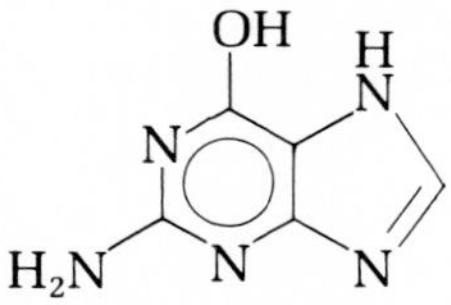

guanine
(g'wa-neen)

In biochemistry, the symbols for these bases are: *C, T, U, A,* and *G.*

Although DNA has been known for a long time (it was discovered in 1869 by Friedrich Miescher), it was not until 1953 that a tremendously important discovery was made concerning its geometry. J. D. Watson and F. H. C. Crick proposed that cellular DNA is not a single strand but a *double strand*, that the strands are coiled, and that the two interwoven, coiled strands form a **double helix** (Figure 18.3) held together by hydrogen bond bridges between the nitrogen base parts of the polymers. Typically, the strand revolves once every ten base pairs.

The DNA in Figure 18.3 is termed deoxyribonucleic acid because the sugar molecule in the polymer is *2-deoxyribose*, shown in Figure 18.4 in its Fischer projection and Haworth furanose forms. In RNA, which we discuss later, the sugar is D-*ribose*:

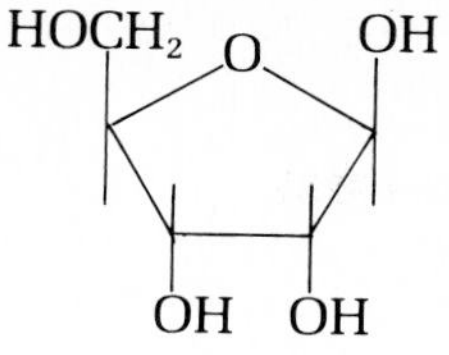

D-ribose

As shown in Figure 18.3, it is hydrogen bonding between base pairs that helps hold together the two DNA strands in their double helix. The base pairing is not random, but quite specific.

- Adenine . . . H-bonds . . . to thymine
- Guanine . . . H-bonds . . . to cytosine

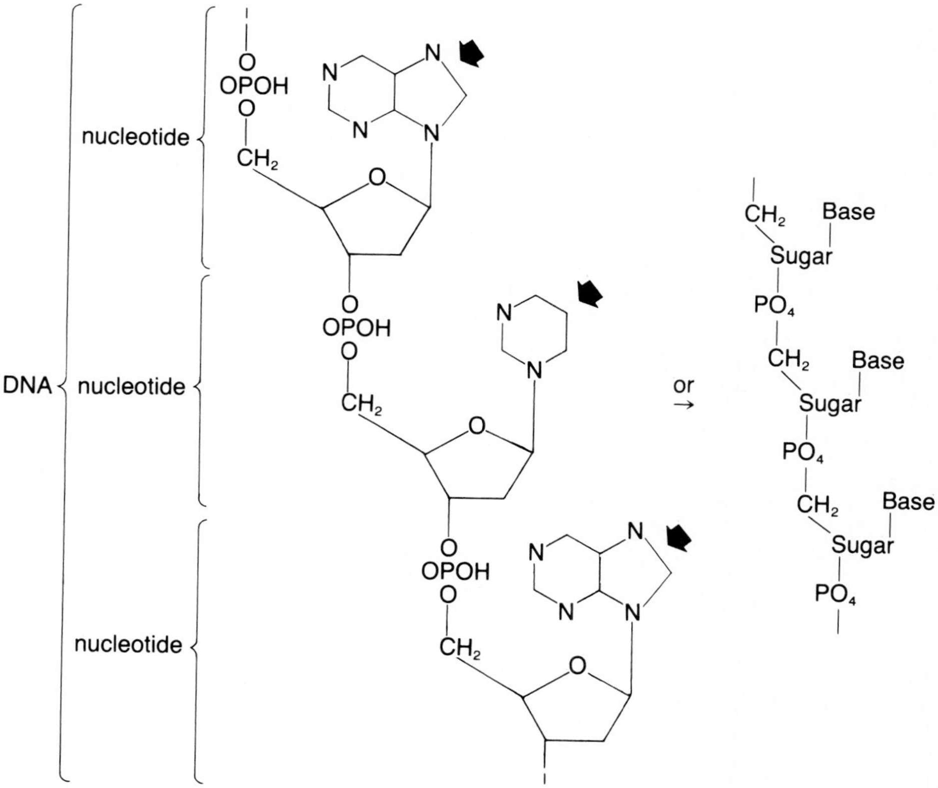

FIGURE 18.2 **Part of the DNA Polymer Chain. Three nucleotide units are shown. The arrows show the location of bases of either the purine or pyrimidine type. The bases (guanine, cytosine, adenine, or thymine) and their sequence determine the genetic code.**

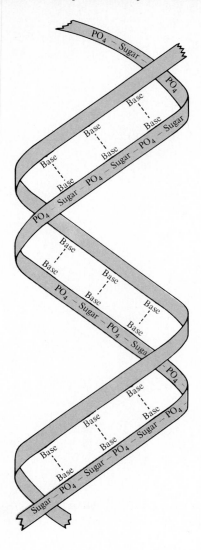

FIGURE 18.3   The Watson-Crick Model of DNA.
Two strands of DNA are coiled around each other
in a double helix and held together by hydrogen
bonding between purine and pyrimidine bases (the
broken lines represent hydrogen bonds).

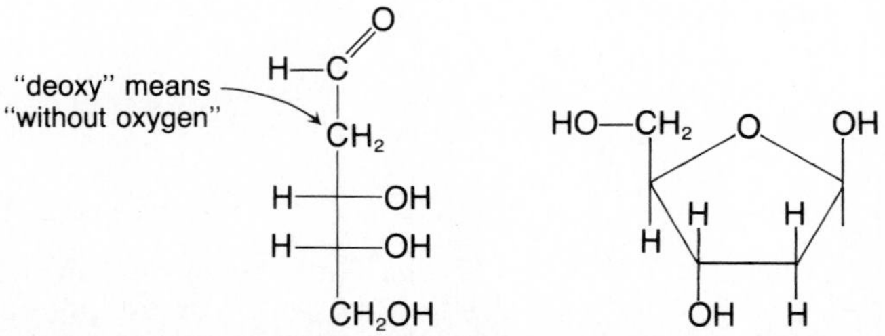

"deoxy" means
"without oxygen"

FIGURE 18.4   The sugar in DNA is 2-deoxy-D-ribose,
shown here in its Fischer projection and Haworth furanose
forms.

As a memory aid, remember A–T (all together) and G–C (go California). Note that uracil is found in RNA, not DNA; it base pairs with adenine (U–A or A–U).

We term the base pairs *complementary* pairs because they mutually *complete* the all-important bonding that holds the DNA strands together. Note that the sequences of bases in one chain automatically establish the sequence in the other chain because of the obligatory complementary base pairing.

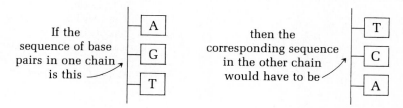

If the sequence of base pairs in one chain is this	A / G / T

then the corresponding sequence in the other chain would have to be — T / C / A

This is because the rules of complementation are A hydrogen bonds to T, G to C.

■■■■■■■■**EXAMPLE 18.1** ■■■■■■

Given the structural formulas of the DNA bases, predict the atoms involved in hydrogen bonding in the double helix.

Solution

adenine   thymine   guanine   cytosine

The double-strand concept of DNA is the usual case, but DNA in many situations can also occur as *rings*, consisting of either a single strand or two strands wound in a double helix. Both kinds of DNA ring can be converted to other three-dimensional configurations (rings can be tied in knots, interwoven, or "supercoiled"). Although the genetic information in DNA is independent of how the molecule is tangled, twisted, or knotted, the three-dimensional structure does play an important role in the events of DNA replication, transcription, and recombination. In man and bacteria, enzymes called *topo-isomerases* catalyze the conversion of one form of DNA to another.

The DNA that we are discussing is coiled up inside the nucleus of each body cell (in bacteria such as *Escherichia coli*, it is in the cytoplasm). If we

FIGURE 18.5 Diagrammatic representation of a cell that has been made to spill out its DNA. The magnification here is about 20,000 times actual size. The cell body is the dark spot in the center. Unraveled, the DNA of a typical bacterium would extend for 1 mm; in a human cell, it would extend for an incredible 6 ft!

could unravel it, it would stretch many thousands of times longer than the cell itself, showing how neatly the ultrathin DNA polymer is coiled and packed inside the cell. Figure 18.5 diagrammatically represents a cell that has been made to split open and spill out its DNA strands.

The DNA in the bacterium *E. coli* is in part circular. *E. coli* DNA has about 3 million base pairs (that is, the purine and pyrimidine bases that hydrogen bond to each other to hold the double helix together). Unravelled, all of the DNA in a single *E. coli* rod would extend for 1 mm! And all of this is packed into a cell 1 by 2 μm in size. The DNA, obviously, must be ultrathin.

DNA in man has 3 billion base pairs. But lest man boast, the newt has 30 billion base pairs. Even the lilly plant has fifty times as much DNA in its genetic makeup as man. The conclusion is obvious: the amount of DNA does not reflect the complexity of the organism.

Much of the DNA in higher organisms is "simple" DNA, that is, with the same two base pairs repeated over and over again. Up to 40 percent of higher organism's DNA can be of this "junk" or "selfish" type that is not used for directing the synthesis of protein. There are other types of DNA in the human. One type, found in very small quantities, has unique base-pairing sequencing. This type is used for protein synthesis. Humans have far more DNA than we need for protein synthesis. Why we have so much DNA, and especially the junk type, is not known. Writing in *Nature* (December 18, 1980), Francis Crick and L. E. Orgel said, selfish DNA "makes no specific contribution to an organism's phenotype [observable characteristics] and it spreads by making additional copies of itself within the genome [one set of all the DNA sequences within an organism]."

DNA, whether it is in a bacterial cell or in a human cell, is sensitive to damage by chemicals and by ionizing radiation (x-rays, gamma-rays, beta-rays, and even UV light). Damage to DNA consists of distortion of the double helix or interruption of base pairing. Fortunately, our cells have a highly developed enzyme repair system that usually can successfully repair damage to DNA.

GENES

Up until recently, geneticists thought they knew what a gene was, namely, a segment of the long DNA chain that contained the genetic information that coded for the biosynthesis of one complete protein. Molecular biologists believed there were up to 1 million different genes in each human cell (versus only 5000 in the fruit fly). But now, according to Nobel Laureate David Baltimore of Massachusetts Institute of Technology, we are much less certain about what a gene is. Genes do correspond to segments of DNA, but these segments can overlap, even "jump about." Genes are so complex that it is difficult even for molecular geneticists to say where they begin or end on the DNA strand. In fact, says Baltimore, the whole concept of *gene* has become so muddled that it is now difficult to define. For now, let's just say that it is too simplistic to think of genes as certain, fixed segments of the DNA chain.

CHROMOSOMES

In the nucleus of each animal and plant cell there exist threadlike strands consisting largely of DNA and protein. These structures are termed *chromatin* and they constitute what become the *chromosomes* at the time of cell division. Humans typically have forty-six chromosomes (twenty-three pairs) in most of their cells. When it comes time for a cell to divide, each chromatin strand contracts into two identical rods and then divides. Each new cell thus receives a set of chromosomes and of DNA exactly like that of the original cell. Hence the hereditary information contained in the DNA is passed on from mother cell to daughter cell. The term *replication* is used for the process in which the DNA makes a copy of itself for the daughter cell. Replication is an extraordinary process in which, through enzyme action, the original double strand of DNA opens, exposing two strands of DNA that serve as templates (patterns) on which new, complementary strands can build (Figure 18.6). As the new strands are attached, they naturally arrange themselves in a double helix identical with the original double helix. The bacterium *E. coli* can replicate (copy) the 3 million base pairs in its DNA in 20 minutes with an error of less than 1 base in 100 million.

Each of the billions of human cells contains a complete set of DNA, that is, each human cell is theoretically capable of synthesizing any of the 50,000 to 100,000 different proteins in the body. Why some cells get turned on to synthesize certain proteins (like insulin) and others do not remains a mystery.

DNA molecules contain from a few thousand to millions of nucleotide units. A chromosome contains many thousands of DNA molecules.

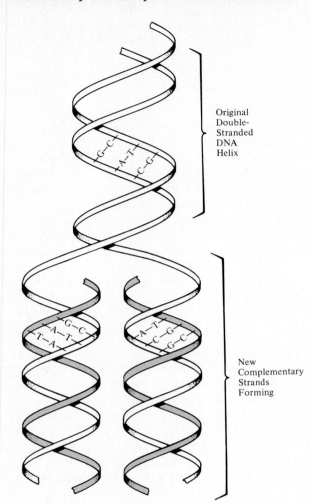

Original
Double-
Stranded
DNA
Helix

New
Complementary
Strands
Forming

FIGURE 18.6 Replication of DNA, as in Cell Division. **From one double helix, two new double helices are formed. Two exact copies result because, as the new complementary strands attach, there is mandatory base pairing of A with T and G with C.**

CELLS AS PROTEIN FACTORIES

A living cell is an efficient protein factory. Among the proteins it synthesizes are enzymes. The cell uses its enzymes to keep itself healthy and functioning and to catalyze the formation of other proteins such as hemoglobin or myoglobin.

Even in cells in simple organisms a great variety of proteins must be made. For example, in *E. coli,* there are about 3000 different proteins. In man, there are 50,000 to 100,000 at the most (Crick's estimate).

Different kinds of cells make different kinds of proteins. Beta cells in the islets of Langerhans in the pancreas make insulin; no other body cells do this.

Cells in the pituitary gland synthesize growth hormone; no other body cells do this. Liver cells synthesize enzymes used in the biotransformation of barbiturates. Some protein, as in muscle tissue, can be synthesized by cells in many parts of the body. It is difficult to overemphasize the importance of protein synthesis by cells. What, then, is the mechanism for selection and control of such protein synthesis?

THE ROLE OF DNA IN PROTEIN SYNTHESIS

Cells synthesize protein by **following a set of commands ordered by the arrangement of nucleotide units in a strand of DNA.** More specifically, the kinds of proteins synthesized depend on the *order of the bases* (adenine, thymine, guanine, cytosine, and uracil in the case of RNA) in the DNA polymer. *In sets of three* along the DNA double helix, these bases specify which proteins will be synthesized by specifying which amino acids will be employed. The integration between a set of three bases and the particular amino acids called forth (to make the protein) is termed the **genetic code.**

A set of three consecutive bases is DNA's way of passing on hereditary information in the form of enzymes and other proteins. Precisely how this is accomplished has been the subject of much research, and a mechanism has been elucidated. Following is a brief summary of how it works.

Transcription

It is not the DNA, as such, that selects the amino acids and joins them as peptides. The situation is more complicated than that. The DNA in the cell nucleus serves as a template (a pattern) for the assembling of a complementary strand of another nucleid acid, **messenger ribonucleic acid (mRNA).** The process is called *transcription.*

In **transcription,** the order of purine and pyrimidine bases is fixed by the order of bases in DNA according to the base-pairing rules we learned earlier:

1. Adenine always pairs with uracil (which, in RNA, replaces DNA's thymine).
2. Guanine always pairs with cytosine.

Figure 18.7 illustrates the process of transcription.

Thus the hereditary information originally contained in the DNA has now been passed on to a complementary strand of mRNA. Now each group of three bases along the mRNA can specify a particular amino acid, thus dictating what protein will be synthesized. **A set of three consecutive bases along the mRNA strand is termed a triplet or codon.**

The genetic code consists of sixty-four three-letter codons. The codons code for amino acids. That means that the codon orders up a particular amino

FIGURE 18.7 In the process of transcription, the genetic information stored in DNA is copied into another nucleic acid, mRNA, following the rules for base pairing. Transcription is similar to replication except that a strand of RNA is formed instead of two other strands of DNA. The difference lies in the enzymes the cell uses.

acid for incorporation into a protein. For example, the UUU codon codes for phenylalanine; ACU codes for threonine; and GGU for glycine (Table 18.1). Since there are twenty amino acids and sixty-four codons, it works out that all amino acids (except methionine and tryptophan) have more than one codon. In fact, some have as many as six. Thus GUU, GUC, GUA, and GUG all code for valine. Three codons, UAA, UAC, and UGA, do not code for amino acids. Instead, in their location in the nucleic acid chain, they act as "stop" signals, signifying that the genetic message is complete, that is, the protein has been synthesized.

Table 18.1. The Genetic Codons for mRNA

Amino acid	Codons
Alanine	GCA, GCC, GCG, GCU
Arginine	AGA, AGG, CGA, CGC, CGG, CGU
Asparagine	AAC, AAU
Aspartic acid	GAC, GAU
Cysteine	UGC, UGU
Glutamic acid	GAA, CAG
Glutamine	CAA, CAG
Glycine	GGA, GGC, GGG, GGU
Histidine	CAU, CAC
Isoleucine	AUA, AUC, AUU
Leucine	UUA, UUG, CUA, CUC, CUG, CUU
Lysine	AAA, AAG
Methionine	AUG (also chain initiation signal)
Phenylalanine	UUC, UUU
Proline	CCA, CCC, CCG, CCU
Serine	AGC, AGU, UCG, UCU, UCA, UCC
Threonine	ACA, ACC, ACG, ACU
Tryptophan	UGG
Tyrosine	UAC, UAU
Valine	GUG, GUA, GUC, GUU (GUU is sometimes used as a chain initiation signal)
Termination signals	UAA, UAG, UGA

ACAACCCCAGAAACAATCGAC—AUG—AGU—GAU—CUG—GUG—AGU—UAA
                                   ↓   ↓   ↓   ↓   ↓   ↓
                               Met . . . Ser . . . Asp . . . Leu . . . His . . . Ser   End

FIGURE 18.8   A portion of a hypothetical mRNA showing the sequence of base triplets that constitutes the genetic code. The genetic message does not begin until the AUG codon is reached; it stops when the UAA codon is encountered. In humans, there are sequences of DNA, like the "leader" sequence in this figure, that do not code for any protein synthesis. Intervening sequences of this type are called *introns*. The reason for their inclusion is unknown.

One codon, AUG, serves two purposes. It codes for the amino acid methionine, but at other times it indicates the beginning of a particular genetic message. Figure 18.8 shows a hypothetical base sequence in a nucleic acid and the amino acids it codes for. Note that the codon AUG initiates the peptide chain that results from the particular selection of amino acids.

The genetic code we are examining here has been worked out from research on *E. coli*. However, there is enough evidence now to conclude that the code is universally applicable. But although all cell nuclei follow the same genetic code, the special commands (such as begin, stop) are different in animal and bacterial cells.

Since each amino acid is encoded by a three-nucleotide codon, it is easy to calculate that the structural "gene" coding for an average protein of about 300 amino acids has a length of about 900 base pairs.

Messenger RNA molecules, once formed, carry the genetic information out of the nucleus and into the cytoplasm of the cell, to the **ribosomes** where complex reactions lead to protein synthesis. Ribosomes are submicroscopic, granular cell inclusions (250 Å or one-millionth of an inch in diameter), in which mRNA is bound and in which the amino acids necessary for protein synthesis are selected and hooked up in the correct order. We can think of mRNA as a temporary set of instructions. It is DNA's way of sending a signal to the cell: create this or that enzyme; prepare this or that protein. Note that the signal may have originated as the result of a drug or hormone action at some receptor site.

Messenger RNA does not hang around for very long. It does its job. It directs the synthesis of protein, and then it is metabolized. At any given time, there may be several different kinds of mRNA doing their job in the cell.

There is another kind of RNA that is involved in ribosomal protein synthesis. It is called **transfer RNA (tRNA),** and it is far smaller in size than either DNA or mRNA. It is the job of tRNAs to attach themselves to a particular amino acid and carry that amino acid to the ribosomes where protein synthesis is occurring. There are as many tRNAs as there are amino acids (twenty), and most amino acids have more than one tRNA. The tRNAs obtain their amino acids from the supply available to them in the cells following digestion of foods, plus *de novo* amino acid synthesis.

Translation

We come to the final step in our story of cellular protein synthesis (which is occurring in the ribosomes). The amino acids carried by the tRNAs to the site of protein synthesis must be lined up in the correct sequence (no easy task in the case of hemoglobin with 290 amino acids) and then chemically bonded to each other to make a polypeptide chain. The process is termed **translation.** Remember, the genetic code is now carried by mRNA. Because tRNA is able to recognize base triplets in the mRNA sequence, it can carry amino acids to their proper places (Figure 18.9).

In keeping with the scope of this book the description of the processes of transcription and translation has been brief. For the many additional details of the steps, consult an advanced textbook of biochemistry.

Summary

The information necessary for an organism to reproduce itelf and to synthesize protein is stored in its DNA. The genetic code is the triplet sequence of purine and pyrimidine bases in the DNA. The DNA code is read and the instructions executed by a second nucleic acid, messenger RNA, which obtains the genetic information in the nucleus (transcription) but which functions in the cytoplasm. In the process of translation, transfer RNA molecules pick up specific amino acids from the cell's chemical pool and line them up in an order dictated by

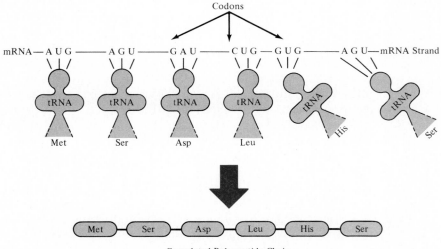

Completed Polypeptide Chain

FIGURE 18.9 The genetic code in mRNA is translated into polypeptide. Molecules of tRNA carry amino acids to their proper position for protein synthesis. They do this by recognizing base triplets in the mRNA. The amino acids have been activated previously to allow them to react head-to-tail, forming the polypeptide chain. When the final amino acid is in place, the polypeptide departs, permitting the synthesis to begin anew.

the order of bases in the mRNA strand. Thus the genetic code in DNA operates through mRNA, with the help of tRNA, to direct the synthesis of protein (enzymes, hormones, muscle, etc.).

THE AGE OF GENETICS

On June 16, 1980, the U.S. Supreme Court handed down a momentous decision on the most controversial patent case of the century. In a 5-to-4 decision, the court decreed that man-made life forms are patentable and that a patent be issued to General Electric Company for a laboratory-made bacterium that "eats" oil and helps clean up oil spills.

The Supreme Court decision will have great consequences, to put it conservatively. Many companies have sprung up to capitalize on the idea of using bacteria to manufacture human proteins, and the vistas seem limitless. It is anticipated that microorganisms can be made to solve problems in such areas as the following.

- Adequate supply of human growth hormone.
- Interferon for research purposes.
- Adequate supply of insulin in the future.
- Vaccines against hepatitis and hoof-and-mouth disease.
- Fuel technology.
- World hunger.
- Human genetic diseases.
- Adequacy of fertilizers for world food supplies.
- Agriculture.

Since the techniques of gene modification and cloning are relatively simple to learn, it is expected that an ever-increasing number of researchers will be applying them to a broad range of biological problems. Exciting success has already been achieved. In July 1980, Eli Lilly and Company announced that it had begun limited testing in humans of biosynthetic human insulin produced by the bacterium *E. coli*. Forty million dollars' worth of new plants are to be constructed for commercial-scale production of synthetic insulin, identical to human insulin. In 1981, the U.S. Department of Agriculture, working with Genentech Company of San Francisco, announced that an effective vaccine against hoof-and-mouth disease had been produced by recombinant DNA technology.

RECOMBINANT DNA TECHNOLOGY

The technique of taking a segment of animal DNA that codes for some desirable protein and successfully inserting it into the DNA of a bacterium such as *E. coli*, is termed **recombinant DNA technology** or, *gene splicing*. When the bacterium is then grown in large culturing tanks, it is possible to harvest significant amounts of the animal protein, manufactured by the bacterium. Microorganisms have, therefore, been called "living factories." (Except for the involvement of genes, the idea is not new: antibiotics, hormones, alkaloids, amino acids, and other drugs have been obtained that way for years.)

Here is a summary of the technique of gene splicing. Researchers extract the DNA out of the cells of the bacterium *E. coli.* Part of this DNA consists of rings called *plasmids.* Found in virtually all bacterial species, plasmids are circular molecules of double-stranded DNA that multiply independently within host cells in a regular manner as the host cells proliferate. Researchers use *restriction enzymes* to cut the DNA in the plasmid at specific points and insert the animal DNA (or gene) that they know codes for some specific protein (such as insulin, interferon, human growth hormone). They then close the plasmid ring with an annealing enzyme and put the plasmid back into the bacterium (a process called *transformation*). *Cloning* is then necessary to sort the new cells and isolate the one having the particular, desired gene. Next, with proper stimulants for growth, the *E. coli,* with its recombinant DNA, is made to replicate and multiply, all the while synthesizing the proteins coded for in its DNA, including the insulin, growth hormone, or whatever has been spliced in (Figure 18.10). The figure shows the process of gene splicing for one plasmid, however, actually millions are involved.

Although recombinant DNA research was begun only in 1973, it has already become almost routine in laboratories all over the world. Investigators have now shown that gene splicing can be used successfully to produce bacteria that will synthesize human insulin, growth hormone, interferon, and somatostatin. University of California researchers have used gene splicing to produce a vaccine for use against hepatitis B (the most dangerous type of hepatitis). They inserted human DNA into bacteria to enable the bacteria to produce hepatitis B surface antigen. When this antigen is isolated, purified, and injected into human patients, it stimulates the production of antibodies that fight off the hepatitis B virus.

One of the biggest problems in gene splicing is locating the segment of animal DNA that codes for the protein you are trying to splice in. After all, there are 3 billion base pairs in human DNA. One answer to the difficulty of locating and using the correct segment of human DNA is the *gene machine,* a comparatively cheap, desktop device that will automatically synthesize fragments of any gene whose genetic code a technician can type onto a keyboard. In 1 day the gene machine can make a synthetic version of a natural gene, or it could conceivably create a new gene at the notion of the researcher who could then insert the gene into a bacterium to see what would happen. All of this is possible because we now know what the genetic code is and how to duplicate it.

Much of the original concern that was expressed about the possible dangers inherent in the technique of gene splicing has dissipated. Although it is still

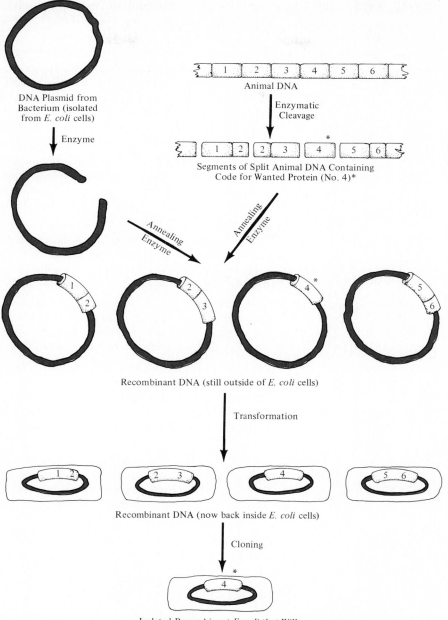

FIGURE 18.10 **The construction of a recombinant DNA molecule by insertion of a segment of animal DNA into the DNA of a bacterium.**

Table 18.2.   Some Recombinant DNA Technology Terms and Jargon

Term	Meaning
Cloning	Originally, the word meant obtaining a group of cells from one "parent" cell, making all of the cells genetically equal. In a more general sense, cloning means making many copies of a selected gene. The process involves moving genes into new host cells, sorting these cells, and selecting the cell having a particular gene from the cell mixture.
Eukaryote and Prokaryote	Eukaryotes are organisms whose cells contain nuclei; they include simple plants and animals such as fungi and protozoa, and complex species such as trees and humans. Prokaryotes have no nucleus; they are simple organisms, consisting of only single cells.
Gene Library	The DNA from an organism can be split into approximately gene-sized pieces by using enzymes that hydrolyze the large DNA molecules. The resulting collection of DNA fragments is available for splicing into recipient cells.
Host-vector System	The host is the organism, usually a bacterial species, into which a segment of DNA from another species is spliced. This "guest" gene is carried there by a vector, which can be a larger DNA molecule such as a plasmid.
Plasmid	A circular molecule of DNA found in a bacterium. Although outside of the chromosomes, it can reproduce itself. It can be moved outside its cell and then back into the cell. By inserting a desired gene into a plasmid, the circle can become a vector for moving the gene into a bacterium.
Promoter	A portion of a gene that turns on production of the protein specified by that gene. It can be attached chemically to the DNA sequence specifying a nonbacterial product, thus providing the control needed to increase the bacterium's synthesis of that product.
Restriction Enzymes	Enzymes that recognize and hydrolyze DNA molecules at specified nucleotide sequences. They can cut DNA into short, single-stranded regions that are called "sticky ends." These are biochemically eager to reform double strands.
Shotgun	The process of hydrolyzing a large amount of DNA and then splicing the separate and unidentified pieces into a host cell. The purpose is to clone a specific gene that cannot be identified conveniently before it functions in the host cell.

conceivable that a new, drug-resistant strain of organism could accidentally be produced that might elaborate a highly potent toxin, the chances seem remote. Besides, guidelines have been established that, if followed, greatly minimize the chance of accidental contamination. No one denies that great care must be exercised constantly in conducting this kind of research. After all, this is the creation of new gene combinations—of novel DNA.

In Table 18.2 we present a summary of some of the terms and jargon used by researchers in the field of recombinant DNA technology. The definition of cloning, as we have used it in this chapter, is given in the table.

VIRUSES

The proverbial question asked about **viruses** is, "Are they living organisms or just clumps of inanimate chemicals?" Most scientists would probably answer the latter, for viruses can reproduce themselves (replicate) only after they have entered a host cell. In the absence of a host, they appear to be lifeless, nonmetabolizing, crystalline chemicals.

Viruses are infectious agents. The agents in polio, herpes simplex, herpes zoster, hepatitis, regular measles (rubeola), German measles (rubella), mumps, rabies, smallpox, chicken pox, and influenza are all viruses. Viruses are exceedingly small, much smaller than bacteria, and are usually visible only under the electron microscope.

We mention viruses at this point because of their biochemical makeup. They contain a core of nucleic acid—either DNA or RNA, but never both—surrounded by a protein sheath (Figure 18.11). They may exist as spheres, threads, rods, particles with tails, or lengthy columns. A virus acts by penetrating a cell, hijacking the cellular machinery, and substituting its own genetic blueprint.

Viruses have been shown to cause cancer in animals, but the clear proof that links viruses and cancer in *humans* is lacking. Some authorities believe that there is sufficient evidence to link viral infection to human cancers of the cervix and lymphoid tissue (i.e., leukemias).

The herpes viruses are large, complex enveloped DNA viruses that in the human can cause herpes simplex (HSV), varicella-zoster, and infectious mononucleosis (Epstein-Barr virus). Genital herpes due to HSV is one of the major sexually transmitted diseases; infection can be asymptomatic in women. Acyclovir, a nucleoside of guanosine with potent antiviral activity against HSV, has received FDA approval for use in topical form.

MUTATIONS

Inherited (permanent) changes in genetic material are termed **mutations.** They may be induced by high-energy radiation (UV, x- or γ-rays) or by chemical mutagens (nitrous acid, nitrogen mustard), or they may occur without the

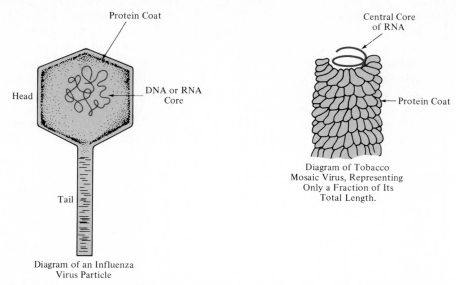

Diagram of an Influenza
Virus Particle

Diagram of Tobacco
Mosaic Virus, Representing
Only a Fraction of Its
Total Length.

FIGURE 18.11 **Viruses are composed essentially of a nucleic acid core surrounded by a protein sheath.**

intervention of any known external factors. In a mutation there may be a loss of genetic material (deletion), a gain (translocation), or an exchange (transduction), but in any event the original sequence of DNA bases is permanently altered. Since the order of bases in DNA is critical for protein synthesis, an alteration can have profound effects on the mutant. For example, sickle cell anemia is the result of a mutation that causes valine to replace glutamic acid in one of the two peptide chains of the hemoglobin molecule. Since the genetic code for glutamic acid is either G$\underline{A}$A or G$\underline{A}$G and the code for valine is either G$\underline{U}$A or G$\underline{U}$G, it would appear that the mutation that results in sickle cell hemoglobin may have been associated with the change in one base in a triplicate nucleotide.

Several thousand diseases in humans result from mutations; they have been given the name *inborn errors of metabolism.*

Inborn Errors of Metabolism

Some babies are born without the ability to produce certain key enzymes. The DNA they have inherited is imperfect. It cannot code for the synthesis of enzymes necessary for the proper biotransformation of some foods, for the synthesis of a hormone, or for the preparation of certain proteins. Biochemical inadequacies caused by defective DNA are termed inborn errors of metabolism, or **genetic diseases.** Some 12 million Americans suffer from genetic diseases.

Phenylketonuria (PKU), one of the most famous of the genetic diseases, was identified through the efforts of a mother of two mentally retarded children. She drew a connection between the peculiar odor of her children's urine and their mental retardation. It was discovered that her children lacked adequate amounts of phenylalanine hydroxylase, the enzyme that catalyzes one step in the metabolism of phenylalanine:

$$\text{C}_6\text{H}_5\text{—CH}_2\text{—}\underset{\underset{\text{NH}_2}{|}}{\text{CH}}\text{—COOH} \xrightarrow[\text{(normal metabolism)}]{\text{phenylalanine hydroxylase}} \text{HO—C}_6\text{H}_4\text{—CH}_2\text{—}\underset{\underset{\text{NH}_2}{|}}{\text{CH}}\text{—COOH}$$

Lacking this enzyme, her retarded children were biotransforming phenylalanine into phenylpyruvic acid:

$$\text{C}_6\text{H}_5\text{—CH}_2\text{—}\underset{\underset{\text{NH}_2}{|}}{\text{CH}}\text{—COOH} \xrightarrow[\text{(abnormal metabolism)}]{\text{alternate pathway}} \text{C}_6\text{H}_5\text{—CH}_2\text{—}\underset{\underset{\text{O}}{\|}}{\text{C}}\text{—COOH}$$

odor in urine

When dietary phenylalanine is improperly metabolized, high levels of it can accumulate in the blood and also in the brain. If diagnosis and treatment are not instituted in the first few weeks of life, these high levels can lead to permanent brain damage, resulting in a mentally retarded child. Since damage is irreversible, it is imperative to diagnose PKU very early and then carefully control the infant's diet so that just enough phenylalanine is given to sustain good health, but not enough to cause buildup in the brain with consequent brain damage. Milk, for example, contains significant amounts of phenylalanine, and must be replaced in the infant's diet.

Phenylketonuria can be diagnosed in the neonate using a simple test. A blood sample obtained by pricking the 3- to 14-day-old infant's heel is spotted on special paper. The paper is placed in a specially prepared growth medium and is incubated with a bacterium for 24 hours. If phenylalanine levels are high in the blood sample, bacterial growth will occur around the area of the paper, and a positive diagnosis of PKU is made. A home-test screening method for phenylketones (phenylpyruvic acid) is available commercially. A reagent-coated plastic strip is dipped into the baby's urine and a color comparison is made. The manufacturer warns that "positives" discovered through office or home testing should be confirmed with a laboratory blood test.

Presently, in all but four states (California, Hawaii, Mississippi, and Wyoming) and the District of Columbia, it is required by law that hospitals perform screening tests for PKU at birth. About 1 in every 15,000 babies born in the United States suffers from phenylketonuria. This makes PKU one of the most often observed congenital defects.

Diet modification is also used to treat neonates suffering from *galactosemia*,

Table 18.3.   Inborn Errors of Metabolism

Disease name	Missing enzyme or factor	Signs and symptoms of disease	Treatment
Albinism	Tyrosinase enzyme	Lack of melanin for skin pigmentation, white hair, pink iris	—
Dwarfism	Growth hormone	Insufficient physical growth	Administer growth hormone from human cadavers
Galactosemia	Galactose-1-phosphate uridyltransferase	Eye cataracts, liver and kidney malfunction, mental retardation	Dietary restrictions
Gaucher's disease	Glucocerebroside hydrolyzing enzyme	Anemia, spleen enlargement, skin pigmentation	Supportive only (e.g., blood transfusions)
Hemophilia	Any of a half-dozen factors	Bleeding or hemorrhage (spontaneous or from trauma)	Administer the missing factor (confers temporary protection)
Homocystinuria	Cystathionine synthetase	Mental retardation, shuffling gait, blood clots, increased levels of homocystine	Decrease methionine and increase cystine in the diet
Hurler's syndrome	Enzyme that catalyzes biotransformation of a muco-polysaccharide	Mental deficiency, deafness, saddle nose, heart disease	Provide cells with missing enzyme; can detect disease prenatally through amniocentesis
Maple sugar urine disease	A decarboxylase that catalyzes metabolism of leucine	Peculiar odor of urine, seizures, vomiting	Eliminate branched-chain amino acids from diet
Neonatal hypothyroidism	Thyroid gland fails to produce the hormone thyroxine	Lethargy, peculiar facial appearance	Administer exogenous thyroxine
Niemann-Pick disease	Sphingomyelin-hydrolyzing enzyme	Mental retardation, pigmentation, liver and spleen enlargement	Supportive only

Table 18.3.   Inborn Errors of Metabolism *(Cont.)*

Disease name	Missing enzyme or factor	Signs and symptoms of disease	Treatment
Phenylketonuria	Phenylalanine hydroxylase	Peculiar odor of urine, irritability, vomiting, mental retardation	Dietary restrictions
Tay-Sachs disease	Hexosaminidase	Mental retardation, muscular weakness, blindness, early death	Supportive only

an inborn error of metabolism in which the baby's DNA cannot code for the synthesis of galactose-1-phosphate uridyltransferase. Lacking this enzyme, the baby cannot properly metabolize galactose.

████████████ **EXAMPLE 18.2** ████████████

What are dietary sources of galactose, and with what should they be replaced to eliminate the galactose?

Solution

Human and cow's milk contain lactose—a disaccharide composed of glucose and galactose. Milk products such as cream, butter, cheese, and ice cream also contain lactose. Artificial infant formulas are made from soybean powder; they substitute well for natural milk.

██

Buildup of galactose in the body can result in damage to the liver, kidneys, and brain; death is possible if treatment is not initiated. Excessive body levels of galactose are associated with cataract development in the eye (opacity of the lens). About 1 in every 65,000 babies born in the United States suffers from galactosemia. Laws in twenty-two states require testing for galactosemia at birth.

Many more inborn errors of metabolism are recognized. Not all can be treated with diet modification as successfully as PKU and galactosemia; some cannot be treated at all. Table 18.3 gives a few of the more than 2000 known human genetic diseases.

SUMMARY

Nucleic acids carry all of the genetic information required to reproduce a species. Nucleic acids are of two types: DNA and RNA. They consist of very high molecular weight polymers made up of nucleotide units.

A nucleotide unit consists of a sugar, phosphoric acid (esterified), and a nitrogen base. The bases are of either the pyrimidine or purine type. The five common bases are cytosine, guanine, adenine, thymine, and uracil.

Much of our DNA consists of a double strand interwoven to form a double helix. Hydrogen bonding between complementary base pairs in the two strands holds the helix together (A pairs with T, G with C).

Replication is the process in which DNA makes a copy of itself for a daughter cell. In replication, the double strand of DNA opens so that each exposed strand can form its own double strand by complementary base pairing (again, A with T, G with C).

Organisms develop based on their ability to synthesize protein, especially enzymes. DNA has a most important role in protein synthesis. Cells synthesize protein by following a set of commands ordered by the arrangement of nucleotide units in a strand of DNA. More exactly, the order of purine or pyrimidine bases, taken in sets of three, specifies which protein will be synthesized. The integration between a set of three bases and the amino acids called forth (for protein synthesis) is termed the genetic code.

Transcription is the first step in protein synthesis. The DNA serves as a pattern for the synthesis of a complete strand of mRNA. A set of three consecutive bases in mRNA is called a codon. The genetic code consists of sixty-four three-letter codons, each corresponding to a particular amino acid or to a special command for protein synthesis.

The genetic information, now in the form of mRNA, is transported to the ribosomes where protein is synthesized in a process called translation. In translation, another type of RNA, called tRNA, acts to select the correct amino acids from the cell's pool of amino acids and bring them to the site of protein synthesis. There, they are joined to form the protein.

Recombinant DNA techniques, also called gene splicing, are used for inserting animal DNA into the DNA of a bacterium and then growing the bacterium in culture. By this technique, hard-to-obtain animal proteins are made available. Recombinant DNA technology has been successfully applied to the synthesis of insulin, human growth hormone, interferon, and somatostatin.

Viruses contain a core of nucleic acid, either DNA or RNA, but never both. Viruses are the infectious agents in many communicable disease.

Mutations are permanent accidental changes in a species' DNA. Mutations can be induced by high-energy radiation or by chemicals.

In an inborn error of metabolism, a child is born without the ability to synthesize a certain key enzyme. The child's DNA is defective; it cannot code for the synthesis of that protein. We discussed PKU, galactosemia, and other genetic diseases.

KEY TERMS

Check your understanding of this chapter. Can you explain what is meant by each of these terms?

codon	purine base
DNA	pyrimidine base
double helix	recombinant DNA technology
genetic code	ribosome
genetic disease	transcription
mRNA	translation
mutation	tRNA
nucleotide	virus
PKU	

STUDY QUESTIONS

1. What are nucleic acids, where do they occur, and what are their two types?

2. What is the relationship between a nucleic acid and a nucleotide?

3. Identify the three chemical parts that make up a nucleotide.

4. Structurally, what is the difference between (*a*) a purine base and (*b*) a pyrimidine base?

5. From memory, give the names of the five bases commonly found in nucleic acids. Their symbols are C, T, U, A, and G.

6. Explain the role of complementary base pairs in the double-stranded helix of DNA.

7. Which pyrimidine or purine base would hydrogen bond (i.e., base pair) with each of the following bases in DNA: (*a*) adenine, (*b*) guanine, (*c*) cytosine, (*d*) thymine?

8. **a.** Chemically, what is the difference between ribose and deoxyribose?
   b. Where does each occur in hereditary material?

9. In the broadest sense, what is the role of DNA in the body, and how does it accomplish its role?

10. What is meant by the genetic code?

11. Assume that the base sequence in a part of an mRNA molecule is

    —GUU—GGA—UUA—CCC—UGC—AAC—CAA—AUU—UAU—UGC

    What is the sequence of amino acids that is coded for by these codons? (*Hint:* GUU can function as a chain initiation signal.) See Chapter 11 for the name of this nonapeptide hormone.

12. Match the word on the left with the statement on the right:

    a. Transcription
    1. The formation of two new double helices from one double helix strand of DNA.

    b. mRNA
    2. The process in which DNA serves as a pattern for the assembly of a complementary strand of another nucleic acid (mRNA).

    c. Translation
    3. Threadlike strands of DNA and protein found in the nucleus of a cell; they become chromosomes at the time of cell division.

    d. Replication
    4. The final step in protein synthesis in which the amino acids are linked up in the correct order and then bonded into a polypeptide.

    e. Chromatin
    5. Nucleic acid molecules that carry genetic information from DNA to the ribosomes; after doing their job, they are metabolized.

13. What is the name of the variety of ribonucleic acid that is able to attach itself to a particular amino acid and carry it to the cell site where protein synthesis is occurring?

14. True or false. In recombinant DNA work to prepare insulin,
    a. DNA from human cells is incorporated into bacterial DNA.
    b. Insulin from gene splicing is *almost* identical to human insulin.
    c. The insulin is the only protein made by the *E. coli.*

15. What are plasmids?

16. In an inborn error of metabolism such as maple sugar urine disease, what is missing? What causes the disease?

17. If a child with Tay-Sachs disease has inadequate amounts of hexosaminidase, why not feed him the enzyme or inject it into his blood system?

ADVANCED STUDY QUESTIONS

18. Predict the structure of the *abnormal* metabolite of leucine that produces the signs and symptoms of maple sugar urine disease.

19. Explain the process by which Eli Lilly and Company is now preparing human insulin using recombinant DNA techniques.

19

Radiochemistry

This unusual-looking apparatus is part of a nuclear magnetic resonance (NMR) instrument being designed by biomedical engineers to detect internal body structures of infants. NMR is a technique that has been utilized by chemists for many years to analyze substances; it is just beginning to be applied in the medical field. The method is based not on x-rays but rather on the response of atomic nuclei to pulses in a magnetic field. Its advantages are that the technique is noninvasive and yet capable of very fine distinctions between normal and abnormal tissue. This illustrates just one of the many ways in which nuclear processes play a vital role in modern medicine.

You will focus in this chapter on the center of the atom, the nucleus. This is the place responsible for all the remarkable changes in the nature of atoms; it is also the source of energy changes in nuclear reactions. As a future health professional, you will want to understand these changes and how nuclear transformations can be useful in medical diagnosis and treatment. You will also find in this chapter several examples of using radioactive materials within the body to study the functioning of body systems.

As with all methods of diagnosis and therapy, the use of the special methods of radiochemistry must always be weighed against the potential for causing harm. The responsible health professional must always be sure the use of radioactive materials is justified for the particular case. This responsibility can only be carried out by understanding the processes used and the potential for interaction of the radioactive materials with living tissues. Even the promising method of NMR has limitations—it is not possible to examine a person with a pacemaker using NMR. (Photograph courtesy of the National Institutes of Health.)

INTRODUCTION

The rest of this book is devoted to chemical change. It was once thought that the rearrangement of atoms to form new compounds was the only type of change possible for atoms. In this chapter we will turn our attention to the more recent discovery of changes involving the nucleus of an atom. We will see that the result of nuclear change can be the emission of energy or particles, or both, from the atom. These emissions are called **radiation.** The isotopes emitting the radiation are referred to as **radioisotopes** or **radionuclides.** The branch of chemistry that studies these **nuclear transformations** is known as **radiochemistry.**

In the first part of this chapter we will study the types of emissions that are possible. Units to express the time necessary for atoms to undergo nuclear change will be presented. We will also look at units to express radiation damage.

The major part of this chapter deals with the interactions of radiation with matter. Radiation is used as a major tool for both the detection and the treatment of disease. Radioactive isotopes can be used to trace systems in living material and therefore to detect abnormalities. Some types of radiation are able to penetrate into tissues and shrink tumorous growth. Many examples of the usefulness of radiation in medicine will be studied.

The benefits of radiation do not come without risk. We will also study the methods used to detect radiation and explore the potential trade-offs of benefit versus risk.

Finally, some special methods involving radiochemistry are also presented in this chapter. These include procedures such as CAT and PETT scans and the methods of radioimmunoassay. These processes represent the newest ways that radiochemistry is used in the service of medicine.

NUCLEAR TRANSFORMATIONS

Many natural isotopes are stable. A nucleus of carbon-12, for example, does not spontaneously change. A nucleus of carbon-14, however, is not stable. It will spontaneously emit particles and energy from its nucleus in an attempt to become more stable. Of all the possible types of radiation, there are a few very common types that we will consider. These are shown in Table 19.1.

The common nuclear particles that are released differ in both their charge and their mass. The **alpha particle** is the heaviest and carries a positive charge. Because of its heavy mass, it does not easily penetrate any material. For this reason it is quite easy to shield against alpha radiation.

The **beta particle** and the positron each carry only a unit charge and have much lighter mass than the alpha particle. The beta particle and the positron can penetrate much better than alpha particle radiation but still can be fairly easily stopped by metals or concrete or water.

The most penetrating radiation is gamma radiation. **Gamma rays** have very high energy and are not easily stopped. Because gamma rays are not particles, they do not carry a charge or a mass. They are just a high-energy form of electromagnetic radiation. Visible light is also a form of electromagnetic radiation, just not as high in energy content as gamma rays. **X-rays** are another form of electromagnetic radiation. X-rays have lower energy than gamma rays but are still very penetrating.

The transformation that an isotope undergoes when releasing radiation can be represented by means of a **nuclear equation.** Earlier in this book we used nuclear symbols to represent the nuclei of atoms. These are the types of symbols used in nuclear equations. The spontaneous transformation that takes place when the nucleus of a carbon-14 isotope changes is represented in this manner.

$$^{14}_{6}C \rightarrow {}^{14}_{7}N + {}^{0}_{-1}\beta + \gamma$$

Table 19.1. Properties of the Major Types of Radiation

| Particle | Symbols | | Charge | Mass (u) | Penetrating power | |
	Nuclear symbol	Greek symbol			Relative	Numerical ratio (as determined through aluminum sheet)
Alpha	$^{4}_{2}\text{He}$	α	+2	4	Slowest moving and least penetrating Can be stopped by even clothing or paper Easily stopped by a thickness of most metals	1
Beta	$^{0}_{-1}\text{e}$	β^{-}	−1	0	More penetrating than alpha particle Stopped by several thicknesses of metals	100
Positron	$^{0}_{+1}\text{e}$	β^{+}	+1	0	Same penetrating power as a beta particle	100
Gamma rays	—	γ	—	—	Most penetrating Able to go through 10 in. of lead Stopped by enough lead or concrete Being electromagnetic radiation, travel at the speed of light (3×10^{8} m/s)	1000

Notice that the sums of the mass numbers on each side are equal. Also observe that the algebraic sums of the charge numbers on each side are equal. Emission of a beta particle has changed the identity of the isotope present, even though the mass is still 14. The atomic number has changed from 6 to 7, resulting in the production of an isotope of nitrogen from an isotope of carbon.

If an alpha particle is emitted, the nuclear equation still shows the same principles of balancing. For example, uranium occurs in several different isotopic forms. Uranium-235 will spontaneously emit alpha particles, forming an isotope of thorium in the process. The nuclear equation representing this change is

$$^{235}_{92}\text{U} \rightarrow {}^{4}_{2}\text{He} + {}^{231}_{90}\text{Th} + \gamma$$

Once again, notice that the sum of the mass numbers and the sum of the atomic numbers are the same on both sides.

Some spontaneous nuclear processes produce changes in the isotope present but maintain the identity of the element. For example, technetium can form an isotope that is widely used in cancer therapy. It is called **metastable** technetium-99. When it spontaneously undergoes nuclear change, it emits gamma radiation, leaving behind a more stable nucleus of the same isotope. The equation representing this transformation is

$$^{99\text{m}}_{43}\text{Tc} \rightarrow \gamma + {}^{99}_{43}\text{Tc}$$

In all of these changes, there has been a transformation of one isotope to another of the same or different element. These changes are known by the general name of **transmutations.** Transmutations often change the identity of the element present but, as we have seen, they may just result in a different isotope of the same element being formed. If the isotope produced is still radioactive, it still is classified as a radioisotope.

EXAMPLE 19.1

If iodine-131 decays by emitting a beta particle and gamma radiation, what isotope will be formed as the result of the decay? Write a nuclear equation for the change.

Solution

You will need the nuclear symbol for iodine-131. Finding iodine on the periodic table, you will see it has an atomic number of 53; the mass number 131 is specified in the question.

Next you will need to be sure of the nuclear symbol for the beta particle. It is an electron, with a charge number of -1 and an apparent mass of zero. The gamma radiation will appear in the equation but will not influence the balancing.

The nuclear equation for the transformation is

$$^{131}_{53}I \rightarrow \, ^{0}_{-1}e + \, ^{131}_{54}Xe + \gamma$$

The nuclear equation shows that the isotope formed is xenon-131.

HALF-LIFE

Radioactive atoms differ greatly in how quickly they will emit their radiation. The concept that expresses the time necessary for decay is called the half-life. Imagine a radioactive sample of a particular isotope is isolated at this very moment. The average time interval necessary for that radioactive isotope to decay so that only half the number of atoms of that isotope remain is called the **half-life.** Its symbol is $t_{\frac{1}{2}}$. Figure 19.1 summarizes this concept in both a table and a graph.

Take note particularly of the shape of the decay curve in Figure 19.1. Although the absolute amount of time for half the sample to decay varies from fractions of seconds to millions of years, the shape of this decay curve remains the same. Fifty percent of the number of atoms, and therefore of the mass of the isotope, decays during the first half-life. During the next half-life, another 50 percent of what remains will undergo transmutation. This pattern continues until all of the atoms have undergone change.

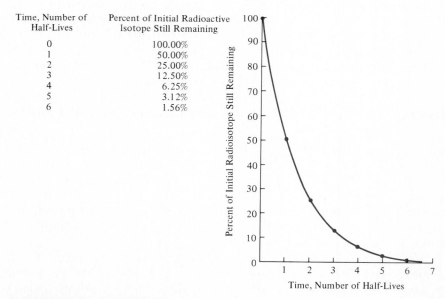

Time, Number of Half-Lives	Percent of Initial Radioactive Isotope Still Remaining
0	100.00%
1	50.00%
2	25.00%
3	12.50%
4	6.25%
5	3.12%
6	1.56%

FIGURE 19.1   Radioactive Decay.

Another characteristic of half-life is that none of the usual means that alter the rate of a chemical reaction—heat, pressure, acid, or catalysts—has any effect on nuclear reactions. The transmutation process, originating in the nucleus, is also independent of the chemical life of the atom. Whether carbon-14 is combined in carbon dioxide or is free as carbon particles, it still will have its typical half-life, 5730 years.

The half-life of a particular radioisotope is one of the major factors in determining the medical suitability of that radioisotope. In general, isotopes with short half-lives are preferred for diagnostic tests so that no long-term residual radiation remains in the body of the person being tested. If radiation is to be used for certain types of therapy, such as implanting isotopes into a tumor, a longer lived isotope may be preferred. We will study many examples of the medical applications of radioisotopes later in this chapter.

The half-life of a radioisotope also is a determining factor in the proper disposal of that isotope. If the isotope has a short half-life, it can just be stored surrounded by the proper shielding until it has decayed. If the half-life is much longer, storage is a very different problem. Plutonium-239, an alpha particle emitter with a half-life of 2.436×10^4 years, presents special challenges if it is to be safely stored. This particular isotope is of concern because of its production in certain types of nuclear reactors.

EXAMPLE 19.2

Iodine-131 has a half-life of 8 days. Assume that 10.0 μg of iodine-131 is delivered to the clinical laboratory. How many days will go by before you can no longer be sure of having at least 2.5 μg of iodine-131.?

Solution
For every half-life that elapsed, the starting mass of the radioisotope will be cut in half. At the end of 8 days, the 10.0 μg of I-131 will be reduced to 5.0 μg of I-131, the rest having undergone transmutation. At the end of 8 more days, only 2.5 μg of the I-131 will remain. Past that time, you cannot be guaranteed of having more than 2.5 μg.

Therefore, after 16 days (two half-lives) have gone by, you can no longer be sure of having at least 2.5 μg of I-131. This illustrates the practical planning that must take place for fresh samples of radioisotopes to be delivered as they are needed.

UNITS OF RADIATION

The units used to measure radiation are of two basic types. One type of unit focuses on the intensity of the source itself. The second type of unit describes the interaction of that radiation with the absorbing material, such as air or tissue. This second type of unit is directly related to the potential biological impact of radiation.

The common unit of radiation intensity is the **curie** (Ci). It is named after Marie Curie (1867–1934), the Polish scientist who, while working in France, discovered radium. The curie is defined as an activity of 3.7×10^{10} disintegrations per second. This number was chosen because it is the number of disintegrations occurring in 1 g of radium in 1 second. The unit curie does not differentiate among types of radiation emitted or the nature of the source. As long as the source provides 3.7×10^{10} disintegrations per second, it is said to have an activity of 1 Ci. If smaller measures of activity are needed, all the standard prefixes can be combined with this unit. Therefore, you may see sources rated in millicuries (10^{-3} Ci), microcuries (10^{-6} Ci), or picocuries (10^{-12} Ci).

The SI unit of activity is the *becquerel* (Bq). It is named after the French physicist A. H. Becquerel (1852–1908), credited with the discovery of radioactivity. A becquerel is defined as an activity of one disintegration per second. This unit is not in common use in the United States at the present time.

The second type of unit measures radiation in terms of its potential to interact with an absorbing material. The **roentgen** (R) is a measure of the ionizing ability of x-rays and gamma rays. Specifically it is defined as the amount of x-rays or gamma rays required to produce ions carrying 1×10^9 units of electrical charge in 1 cc of dry air at standard conditions of pressure and temperature. This unit was named after the German physicist Wilhelm Roentgen (1845–1923), who discovered x-rays in 1895. Because the roentgen measures energy lost by the passage of x-rays and gamma radiation in air, it is not specific to tissue absorption.

The rad and the rem are the major units used to measure the interaction of radiation with tissue. The **rad** (radiation *a*bsorbed *d*ose) is a measure of absorption of a standard amount of energy (10^{-5} J) per gram of absorbing tissue. The abbreviation for a rad is D. With moderate energy gamma rays, an exposure of 1 R will produce an absorbed dose in muscle tissue of 0.97 D. For some cases, therefore, it is a fairly good approximation to say that the dosage in roentgens equals the absorption in rads. For exact comparison, the type of absorbing tissue and the nature of the radiation itself must both be considered.

The SI unit of absorption is the *gray* (Gy). It is named after a British radiologist, Louis Harold Gray (b. 1905). A gray is equal to an absorption of 1 J of energy per kilogram of tissue. One rad is equal to 10^{-2} Gy. Again, this SI unit has not achieved widespread use in the United States at this time.

We have seen that the different types of radioactive particles and waves vary in mass, charge, and penetrating ability. These factors will affect the potential biological impact of the radiation. The **rem** (*r*oentgen *e*quivalent *m*an) is the quantity of radiation that causes damage equivalent to absorption of 1 R. The rem is most easily calculated from the product of the absorbed dosage in rads and the relative biological effectiveness (RBE) of the incoming radiation.

$$rems = rads \times RBE$$

The RBE is a value that is selected to take into account the relative ability of the different types of radiation to interact with tissue. The heavy and highly charged alpha particle can form more ions than the lighter and singly charged beta particle. On the other hand, because it forms fewer ions, a gamma ray has a greater penetrating power than an alpha particle. The RBE is set at 1 rem/D for gamma rays and beta particles, at 10 rems/D for alpha particles, and at 20 rems/D or more for some of the heavier, charged nuclear fragments that might be present in a nuclear disintegration. All of these values can also be expressed in millirems per millirads. Multiplying both numerator and denominator of a fraction by the same conversion ratio, 10^3, does not change the value of the RBE.

The advantage of using the unit rem or any of its common subunits is that the weighting factor is already included. This makes it a useful unit for expressing and comparing overall biological effects of exposure due to all forms of radiation. It also means that dosages expressed in rems are numerically additive, making total exposure easier to evaluate. Table 19.2 gives you some ideas of current exposure levels in the United States.

■■■■■■■■**EXAMPLE 19.3**■■■■■■■■

If you were exposed to 55 mD of gamma radiation and 55 mD of alpha particle radiation, would the biological effect be the same? Express each of these exposures in millirems.

Solution
The effect of each of these exposures would be quite different, because of the different types of radiation present. Alpha particles are charged and heavy, whereas gamma radiation is very penetrating. The RBE for gamma radiation is equal to 1 and for alpha particles, it is equal to 10.

$$rems = rads \times RBE$$

or, multiplying both sides of this equation by 10^3,

$$millirems = millirads \times RBE$$

For the gamma rays,

$$millirems = 55 \; \cancel{mD} \times \frac{1 \; mrem}{\cancel{mD}}$$

$$= 55 \; mrems$$

For the alpha particles,

$$\text{millirems} = 55 \,\cancel{mD} \times \frac{10 \text{ mrems}}{\cancel{mD}}$$

$$= 550 \text{ mrems}$$

Notice that the difference in the dose when expressed in millirems clearly shows the big difference in potential biological effect between these two exposures.

One final unit that you may encounter being used to describe radiation is the *electron volt* (eV). This is a non-SI energy unit that is used to describe the energy emitted by x-ray generators or by radioactive sources emitting gamma rays. One electron volt is very small indeed; $1 \text{ eV} = 1.602 \times 10^{-19}$ J. The magnitude of the energy of x-rays will be on the order of 0.10 million electron volts (MeV). The gamma radiations used in cancer treatment have energies of about 1.2 MeV.

BIOLOGICAL EFFECTS OF RADIATION

The basis for most damage caused by radiation lies in the ionizing ability of the incoming radiation. As particles and rays from radioactive sources strike atoms in the absorbing material, the electrons from the atom struck are torn away from the electron cloud region. This leaves ions behind. The ionizing power of radiation may break apart the chemical bonds between atoms in a complex molecule, altering the expected chemical characteristics. Cells cannot carry on their usual chemical activities if they are ionized.

Cells may also be damaged if the interaction with radiation produces free radicals, which are highly reactive uncharged species. These atoms or groups of atoms having at least one unpaired electron can interact with other of the complex molecules in the body, once again preventing the usual chemical functions associated with those molecules.

Radiation has the potential to kill or to damage cells. If enough cells die, the entire organism will die. In humans, a dosage of 600 rems occuring in a period of not more than a day would probably be lethal to 100 percent of the exposed population. In the range from 100 to 200 rems, radiation sickness will occur, characterized by such symptoms as decrease in white blood cell count, nausea, diarrhea, general weakness, loss of hair, and skin reactions.

For low levels of radiation, cell death is not as prevalent as cell damage. Cell damage is of particular concern for potential long-range effects. Although dead cells may be replaced, damaged cells may replicate and carry a defective

Table 19.2. Radiation Exposure Levels[a]

Sources of radiation dose to persons in the United States.
(All numbers involve a range of values depending on location.)

Natural

Terrestrial radiation; earth's crust, principally U, Th	55 mrems/year
Internal sources; principally ^{40}K, but also ^{87}Rb, ^{14}C, ^{226}Rn	25 mrems/year
Cosmic radiation (add 1 mrem for each 100 ft elevation)	40 mrems/year
(range: 100–150)	120 mrems/year

Man-made

Medical exposures	
Average chest x-ray	100–200 mrems/exposure
Average GI tract exam	200–500 mrems/exposure
Dental x-rays	20 mrems/exposure
Per capita dose for diagnostic x-rays in the United States	55 mrems/year
Television viewing:	
Black and white (Multiply number of hours of viewing per day by 1 mrem to get yearly dose.)	
Color (Multiply number of hours of viewing per day by 2 mrem to get yearly dose.)	
Typical operating nuclear power reactor at boundary	1–5 mrems/year
All aspects of nuclear industry	2–10 mrems/year
Nuclear weapons testing, 1954–61	29 mrems/year
Nuclear weapons testing, present levels of fallout	<10 mrems/year

Radiation protection standards: Doses above natural background

Occupational exposures to individuals employed in the nuclear industry[b]	
Whole body (30 year maximum)	5,000 mrems/year
Gonads and red bone marrow	5,000 mrems/year
Skin, thyroid, bone	30,000 mrems/year
Hands, forearms, feet, ankles	75,000 mrems/year[c]
General public	170 mrems/year[d]

[a] Table taken from L. Pryde, *Environmental Chemistry*, Cummings Publishing Company, 1973, p. 48.

[b] The occupational exposure limits are set higher than those for the general public as the workers are all healthy adults who are less susceptible to radiation damage than children. It is also assumed that the hazards can be foreseen and controlled.

[c] The different allowable exposures reflect the different susceptibility of these organs to radiation-induced cancers. Leukemia is the most radiogenic of all human cancers (most easily induced by radiation) and therefore allowed exposures are lower for red bone marrow than for other organs.

[d] This number approximately coincides with the national average of radiation exposure from natural and medical sources. Therefore, the philosophy behind this standard is that no person shall receive more radiation from other artificial sources than from the combination of natural and medical exposure.

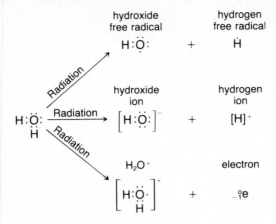

FIGURE 19.2 Possible Interactions of Radiation with Water Molecules.

message. Thus, radiation may also cause damage by altering genetic codes or by changing the DNA molecules in chromosomes, resulting in biological mutations or in cancer. Such long-term changes are particularly difficult to predict and measure. Certain forms of leukemia may have a latent period of 30 years or more before symptoms occur. The incidence of leukemia in residents of Hiroshima and Nagasaki peaked many years after the population was exposed to radiation from the explosion of atomic bombs in the 1940s. The same population showed a similar delay in developing cataracts.

The cells most susceptible to damage are those undergoing rapid growth. Bone marrow, blood-forming tissues, lymph nodes, and embryonic tissue represent examples of particularly susceptible cell systems. Ironically, this damaging of rapidly reproducing tissue also explains the success of radiation in treating some types of cancer. The fast-multiplying cancerous cells are more easily killed by radiation than are the slower growing, healthy cells surrounding the cancer.

Because cells are approximately 80 percent water, the effect of ionizing radiation on the water molecule is of particular interest. In each of the mechanisms indicated in Figure 19.2, the net effect is the same. Neutral water molecules no longer exist and therefore cannot carry out their expected functions.

ISOTOPES IN MEDICINE

Despite the known hazards of radiation, radioisotopes and other forms of radiation are indispensable in modern medicine. The uses are broadly classified as either diagnostic or therapeutic. Diagnostic uses include the use of radioactive tracers and of x-rays. Therapeutic uses may involve the implantation or other direct application of radioactive isotopes with the intention of damaging or killing rapidly developing cells. Therapy may also be carried out by directing a beam of high-energy gamma rays toward the area under treatment.

Radioisotopes are used successfully in diagnostic tracer studies because the radioactive form of an element will behave *chemically* just as the stable form. Therefore, the location of the tracer in the body can be predicted if the chemistry of that element in the body is known. For example, ingested iodine naturally has a high affinity for the thyroid gland. This gland uses the iodine in the synthesis of thyroxin. Knowing this, we can predict that iodine-131 can be used to study thyroid function because the isotope will naturally find its way to that organ and be concentrated there. The thyroid gland cannot distinguish between non-radioactive iodine and its radioactive isotope, I-131. The only difference between a radioactive isotope and a stable isotope of the same element is that the radioisotope will be emitting radiation and can therefore be detected. When a patient is given a radioisotope and then the emitted radiation is detected and measured, the process is referred to as a **scan.**

The radioisotope selected for any scanning study should be one with a short half-life and a known mechanism for elimination so that the radioactive material does not remain in the body for any longer time than necessary. For use in diagnosis, it is desirable that the selected radioisotope decay by emitting gamma radiation. If alpha or beta particles were emitted, they could do more biological damage in the area of concentration. Although this may be part of the plan if the material is to be used for therapy, the goal during diagnosis is to minimize any harm while maximizing the information received.

A particularly important tracer is the artificially produced technetium-99m mentioned earlier in this chapter. It has a short half-life (6.0 hours) and the product of its decay has a very low level of activity. Tc-99m decays entirely by emitting gamma radiation. It is usually administered in the form of the pertechnitate ion, TcO_4^-. This ion behaves in the body in a fashion similar to the chloride ion. Its radioactivity may be detected as a "hot spot" if a tumor incorporates the ion. It may also show up as a "cold spot" if there is an obstruction preventing its circulation, such as a blood clot. Although Tc-99m has now been replaced in some specific applications by CAT scans (see the last section of this chapter), Tc-99m still has a wide variety of applications in diagnosis of abnormalities of the spleen, kidneys, lungs, and liver. Some of the other radioisotopes useful in diagnosis are shown in Table 19.3.

When radioisotopes are used for therapy, the choice of the material is changed. Here the intent is to give a highly concentrated dose of radiation directly to the tumor while sparing the healthy surrounding tissue as much as possible. High-energy beams of gamma radiation have traditionally been employed to accomplish therapy. Such methods are referred to as **teletherapy,** which means therapy at a distance. It is difficult to use such external energy sources on deep tumors, particularly without causing high whole-body exposures that cause the symptoms of radiation sickness.

Table 19.3. Radioisotopes Used in Diagnosis

Isotope	Half-life	Application
Arsenic-74	18 days	Concentrates in brain tissue; useful for the detection of brain tumors.
Barium-131	11.6 days	Detection of bone tumors and finding the sites of rheumatoid arthritis.
Chromium-51	27.8 days	Attaches to red blood cells. Useful in determining blood volume, red blood cell survival times. Administered as sodium chromate.
Cobalt-57	270 days	Diagnosis of abnormalities in absorption and utilization of vitamin B_{12}.
Cobalt-58	71 days	Also used for tracing vitamin B_{12}.
Copper-64	12.8 hr	Concentrates in the brain. Useful for detection of brain tumors. Aids in diagnosis of copper storage abnormalities in the body. May be administered as copper acetate.
Fluorine-16	109.7 min	Bone scans. Administered as sodium fluoride.
Gallium-67	77.9 hr	Concentrates in chest and deep abdominal tumors. Also useful for diagnosis of lymphomas and Hodgkin's disease. Administered as gallium citrate.
Gold-198	2.693 days	Used for scanning the liver to evaluate its function. Administered as colloidal gold.
Indium-111	2.81 days	Tumor-seeking agent used for lung, liver, and spleen scans. Administered as a colloid with iron(III) hydroxide.
Indium-113m	100 min	Brain scans.
Iodine-123	13.3 hr	Evaluation of thyroid function.
Iodine-125	60 days	Evaluation of the thyroid, blood, and liver function.
Iodine-131	8.07 days	Measurement of thyroid activity, used for evaluating the concentration of certain hormones in the blood.

Newer methods focus the radioactive material by the use of implanted seeds contained in hollow needles or tubes. This type of internal radiation therapy is referred to as **brachytherapy,** meaning therapy at short range. The radioactive material is contained within applicators made of nonreactive materials such as alloy of gold or platinum. The shapes of these containers are designed to bring the radioactive material in close contact with the tumor. The seeds may be left in place only momentarily, a few hours, or permanently, depending on the details of the therapy planned. While the implants are in place, the family and nursing staff need to be educated as to the proper precautions to avoid unnecessary exposure to radioactivity. Table 19.4 lists some of the common isotopes employed for brachytherapy.

Radiopharmaceutical therapy means giving radioactive material orally or intravenously in order to achieve a therapeutic goal. The material may also be flooded directly into a cavity if the cancer being treated is located on the walls of the stomach, for example. This branch of therapy is part of the broader area

Table 19.3. Radioisotopes Used in Diagnosis *(Cont.)*

Isotope	*Half-life*	*Application*
Iron-59	45.1 days	Used for study of transfer and metabolism of iron in red blood cells. May be used to determine blood volume, rates of iron metabolism. Administered as iron(II) sulfate, citrate, or chloride.
Krypton-79	34.9 hr	(Gas) Used for studying blood flow in the cardiovascular system.
Krypton-85	10.76 yr	(Gas) Study of cardiac abnormalities, diagnosis of tumors in the brain, lung, liver, kidney, pancreas. Also used to study fat absorption in the pancreas.
Mercury-197	65 hr	Brain scans, spleen function evaluation, renal studies.
Phosphorus-32	14.3 days	Blood studies, brain tumors, and treatment of breast carcinoma.
Potassium-42	12.4 hr	Detection of brain tumors. Determination of intercellular spaces in fluids. Administered as potassium carbonate.
Rubidium-82	75 s	Evaluation of the heart.
Selenium-75	120.4 days	Pancreatic function, size and shape of the pancreas.
Sodium-24	15.0 hr	Study of sodium metabolism, circulation studies, particularly cerebrospinal fluid and peripheral vascular disease. Administered as sodium chloride.
Strontium-85	64 days	Bone scans. Strontium behaves chemically like calcium in the body. Administered as sodium nitrate.
Xenon-133	5.27 days	Lung tumors and evaluation of lung function. Also used to detect cardiovascular abnormalities and to evaluate blood flow in skeletal muscles. Administered as a gas or as a gas in saline solution.

of **chemotherapy.** Drugs used for chemotherapy, radioactive or not, are designed to treat disease without seriously harming the patient. Chemotherapy may be combined with hyperbaric oxygen therapy or chemical therapy designed to add oxygen to the cells under attack. If the cells are more fully oxygenated, chemotherapy appears to be more effective in many applications.

DETECTION OF RADIATION

The earliest method used to detect the presence of radiation was photographic. The fact that the natural salts of uranium caused a "fogging" of photographic plates was discovered quite accidentally. We now know it was due to the gamma radiation emitted by naturally unstable uranium nuclei. Each type of radiation produces a characteristic track on a photographic plate. The use of film badges (see Figure 19.3) is a common and relatively inexpensive method

Table 19.4.   Radioisotopes Used in Therapy

Isotope	Half-life	Application
Cesium-137	30.2 yr	Beta, gamma emitter. Used for implantation therapy.
Cobalt-60	5.26 yr	Gamma emitter. Used for cancer treatment both by teletherapy and brachytherapy.
Gold-198	2.693 days	Beta, gamma emitter. Used for implantation therapy.
Iodine-131	8.07 days	Beta, Gamma emitter. Treatment of hyperthyroidism, cancer of the thyroid. Also used for some types of heart disease.
Iodine-136	83 s	Beta, gamma emitter. Treatment of hyperthyroidism, cancer of the thyroid.
Iridium-192	74 days	Beta, positron, gamma emitter. Used for implantation therapy.
Phosphorus-32	14.3 days	Beta, gamma emitter. Used for treatment of some types of leukemia, lymphomas, and widespread carcinomas. Administered as disodium hydrogen phosphate.
Plutonium-238	86 yr	Alpha, gamma emitter. Energy generated from decay is converted to electricity by means of a nuclear battery. Heart pacemakers operate on this principle. Will need to be replaced every few years.
Radium-226	1620 yr	Alpha, gamma emitter used for cancer therapy. First radioisotope used for therapy. Now used in implantation therapy.
Radon-222	3.823 days	Alpha, gamma emitter. Used to treat uterine, cervical, oral, and bladder cancers. May be used for skin cancer treatment.
Strontium-90	28.1 yr	Beta, gamma emitter used for implantation therapy.
Yttrium-90	64 hr	Beta, gamma emitter. Yttrium oxide ceramic beads used for direct implantation.

of monitoring radiation today. Such badges contain sensitive film that is shielded from light but not from radiation.

Radiation can also be detected by taking advantage of the property that certain types of materials will emit light when struck by radiation. Zinc sulfide is one such material that exhibits this **fluorescence.** The very simplest type of fluorescence detection unit is called a **spinthariscope** (see Figure 19.4). It was used for many years as the only tool for counting individual radioactive events by counting the flashes on the zinc sulfide screen. Modern adaptations of the spinthariscope include the necessary electronics to count the flashes emitted by different types of crystals or liquids. Sodium iodide and thallium iodide are the chemicals commonly used to interact with the radiation. These modern instruments are called **scintillation counters.** Often their usefulness is increased by combining them with computers, helping to analyze the types and intensities of radiation received.

FIGURE 19.3 Film Badges to Monitor Radiation Exposure. **This chemist is checking samples in a gamma counter. To monitor personal exposure to radiation, she is wearing a small film badge on her collar. (Photograph courtesy of National Institutes of Health.)**

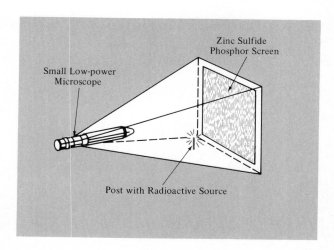

FIGURE 19.4 Spinthari-scope for Detecting Radiation.

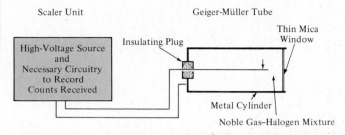

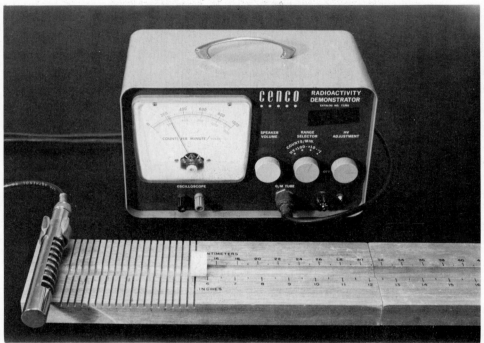

FIGURE 19.5   Geiger Counter and its Schematic Representation.

The **Geiger counter** is a very useful tool for detecting radiation, particularly beta radiation. The heart of this unit is the Geiger-Müller tube, a two-electrode, gas-filled tube. One electrode is a thin metal wire. The other is a thin-walled metal cylinder. The tube is evacuated to low pressure and then filled with a noble gas and halogen mixture to a pressure of only a few millimeters of mercury. When radiation passes through the tube, the ions created cause the noble gas in the tube to become an electrical conductor. The resulting surge in current is amplified and used to activate any number of signaling or recording devices, such as a "click," a flashing light, or, more important, a recording unit. The schematic of a typical Geiger counter is shown in Figure 19.5.

SPECIAL METHODS

X-rays have been used for many years as a diagnostic tool. They are successful because they have enough energy to penetrate soft tissue but do not penetrate the harder material such as teeth or bone. The result of an x-ray is a **radiograph,** which, like a photograph, is a two-dimensional image. Although the routine use of x-rays for screening purposes has been greatly reduced in recent years, there are still many applications. The use of x-rays can be expanded by incorporating chemicals temporarily suspended in a system that provide a block to the passage of x-rays. These materials are called **radiopaque** if they cannot be penetrated by x-rays. For example, barium sulfate may be administered orally if the purpose of the x-ray is to evaluate the gastrointestinal system. Special dyes containing iodine can be injected into the body to outline organs. Both the iodine dye and the barium sulfate appear radiographically dense and so help to visualize the system under study.

One problem with x-rays has always been that they do produce just two-dimensional images. This can create serious problems in locating abnormalities. If you were to have a chest x-ray taken while wearing a gold chain, that chain could appear to be *in* your chest when viewed on a two-dimensional radiograph. A second x-ray taken at right angles would help to clarify the chain's position on the surface of the body. This is the principle behind **CAT** (computerized axial tomography) scans. These scans can help to locate the hidden structures of the body by combining a special x-ray machine with a computer. A series of x-rays can be taken, making it possible to obtain a series of radiographic cross sections.

One of the newest and fastest growing diagnostic tools is **PETT** (positron emission transaxial tomography). This is another method that depends on radiochemicals being administered and then scanned after the chemicals have assumed their appropriate function and location in the system. For example, a compound containing carbon-11 is administered by inhalation or injection. Carbon-11 is a positron (β^+) emitter, but the emitted positron can only travel a short distance in the body before meeting up with an electron (β^- or e^-). As these two particles meet, two gamma rays are produced and sent off in nearly opposite directions. By placing the detection units on a rotating assembly around the body, these gamma rays are detected. By computer analysis, a cross-sectional "image" of the system under study can be constructed. Furthermore, you can not only locate a tumor precisely, you can use PETT to study metabolism within the tumor.

Table 19.5 lists some of the short-lived, positron-emitting isotopes that have been utilized for PETT studies in the short time since the development of the method. One of the most difficult parts of this technique has been to synthesize the biologically appropriate compounds to carry the positron-emitting radioisotope. One of the earliest radiolabeled materials used in PETT was glucose. Abnormal glucose metabolism in the brain is a key indicator of

Table 19.5. Radioisotopes Used in PETT Studies

Isotope	Half-life (min)	Application
Carbon-11	20.3	Study of glucose metabolism, nerve damage from multiple sclerosis, amino acid transport, heart muscle function, pancreas, and brain tumor visualization.
Fluorine-18	109.7	Study of dopamine metabolism, brain receptor studies, proposed as a tracer for nerve damage and neurological disorders. Also used for estrogen receptor studies for breast tumors.
Oxygen-15	2.07	Study of blood flow, blood volume, oxygen metabolism.
Nitrogen-13	9.97	Labeling of nitrogen-containing compounds for metabolism studies.

the biochemical basis of mental illness and senility and may also be important in brain tumor development. Current research is centering on the production of closely related glucose derivatives that can be labeled with the radioisotope fluorine-18.

EXAMPLE 19.4

Carbon-11, used for the preparation of the organic compounds necessary for PETT studies, is artificially produced by bombarding nitrogen-14 nuclei with protons. Write a nuclear equation to represent this process.

Solution

A nuclear equation must be balanced from the standpoint of nuclear charge and atomic mass. The reactants are nitrogen-14 with an atomic number of 7 and a proton with a charge of $+1$ and a mass number of 1 unit. Therefore, the products must have a total charge of $(7 + 1) = 8$ and must have a total mass of $(14 + 1) = 15$ units. One of the products is specified to be carbon-11, with atomic number of 6. Therefore, the other product must have a charge of $(8 - 6) = 2$ units and a mass of $(15 - 11) = 4$ units. An alpha particle meets this description.

All of this information is represented in this nuclear equation:

$$^{14}_{7}N + ^{1}_{1}H \rightarrow ^{11}_{6}C + ^{4}_{2}He + \gamma$$

A final special method that depends on radioactivity is known as radioimmunoassay (RIA). This is a testing method used for both qualitative and quantitative determinations of drugs present in a person's urine. RIA is based on the use of a radioactive isotope, typically iodine-125 with a half-life of 60 days. This radioisotope is used to "tag" an **antigen,** which is an enzyme, toxin, or other substance of high molecular weight to which the body reacts by producing antibodies. The tagged antigen then enters into competition with untagged antigens for binding sites on specific molecules being tested for in the urine.

In an RIA test for morphine, for example, a morphine and protein mixture is injected into goats in whom it acts as an antigen and induces the production of specific antibodies. The antibodies are then collected from the blood of the goat. Some of the original morphine-protein mixture is made radioactive with iodine-125. This material is capable of attaching itself to binding sites on the antibodies. At the time of the test, urine is combined with the radioactive morphine-protein and with the antibodies previously collected from the blood of the injected goats. If there is no morphine or its derivatives present in the urine, the radiolabeled morphine will stay attached to the antibodies and the test will not show any activity in the rest of the solution. If the urine has morphine-protein derivatives present, these compounds will try to attach themselves to specific binding sites on the antibodies. The more morphine present in the urine, the more it will displace the I-125 labeled morphine from its binding sites on the antibodies. The result will be that the analyst performing the test will see a higher radioactivity count in the nonbonded part of the sample for those urine samples with morphine present. The amount of released I-125 activity is related to the concentration of the morphine present, making identification both qualitative and quantitative if the proper test protocols have been followed. The flow chart for this method is shown in Figure 19.6.

In addition to morphine, RIA can be used to screen for such drugs as cocaine, phencyclidine (PCP), methaqualone, barbiturates, and amphetamines.

SUMMARY

In this chapter we surveyed the topic of radiation and its use to the health professional. We found that there are several major types of radiation emitted by radioisotopes. These types include alpha, beta, and positron particles, as well as gamma rays. Each of these emissions results in a nuclear transmutation for the radioisotope involved.

The time necessary for half of the atoms of a radioactive isotope to decay is known as the half-life. This value can be used to compare rates of decay for different radioactive isotopes.

The units most commonly used for the measurement of radiation are the rem and the rad. These are both units of dosage. Intensity of radiation is measured in curies. Radiation can be detected by the use of photographic devices, Geiger counters, or scintillation counters. Most medical personnel wear film badges or other dose-measuring devices in order to monitor their own exposure if their jobs bring them in contact with ionizing radiation.

Radiation interacts with the complex molecules of the body, disrupting

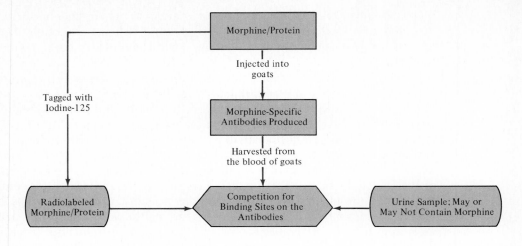

Case I. No Morphine or Its Derivatives in the Urine

Result: Low residual radioactivity in the rest of the solution because the radiolabeled morphine is all attached to the antibodies.

Case II. Morphine (Or Its Derivatives) Present in the Urine Sample

Result: Not all the radiolabeled morphine is able to attach to the antibodies. The concentration of residual radioactivity in the rest of the solution is proportional to the amount of morphine in the urine.

FIGURE 19.6   RIA for Morphine Detection.

the normal functioning of those molecules. This may cause damage, but it is also the basis for radiation therapy as the rapidly dividing cancerous cells are more susceptible to the ionizing effects of radiation than normal cells.

There are many radioisotopes that are useful for diagnosis and treatment. One of the earliest applications was utilization of a beam of high-energy gamma rays directed at the region under treatment. Radioisotopes may be administered by implantation or by actually circulating the radioactive material in the system. The use of radiation as weapon to diagnose and treat disease must always be weighed against the hazards created.

Newest methods utilizing radiochemistry in medicine are CAT and PETT scanners as detection methods. RIA is used for the detection and measurement of drugs.

KEY TERMS

Check your understanding of this chapter. Can you explain what is meant by each of these terms?

alpha particle	radiation
antigen	radiochemistry
beta particle	radiograph
brachytherapy	radioisotope
CAT	radionuclide
chemotherapy	radiopaque
curie	radiopharmaceutical therapy
fluorescence	RBE
gamma rays	rem
Geiger counter	roentgen
half-life	scan
metastable	scintillation counter
NMR	spinthariscope
nuclear equation	teletherapy
nuclear transformation	transmutation
PETT	x-ray
rad	

STUDY QUESTIONS

1. What is the charge and mass for each of these particles?
   a. Alpha
   b. Beta
   c. Positron

2. Thorium-232 decays by giving off an alpha particle as well as gamma radiation. Write the nuclear equation for this transmutation.

3. Cesium-137 decays by giving off a beta particle as well as gamma radiation. Write the nuclear equation for this transmutation.

4. Each of the following isotopes decays by emitting an alpha particle. What isotope will be formed as the result of the transmutation?
   a. Bismuth-213

   b. Radon-222
   c. Actinium-225
   d. Radium-226

5. Each of the following isotopes decays by emitting a beta particle. What isotope will be formed as a result of the transmutation?
   a. Gold-198
   b. Iodine-136
   c. Strontium-90
   d. Yttrium-90

6. Iridium-192 emits both a beta particle and a positron, as well as gamma radiation, when it decays. Represent this information in a nuclear equation.

7. Fluorine-16 has a half-life of approximately 110 minutes. If you isolated 10 g of this pure isotope, what time would elapse before you were no longer sure of having 2.5 g of the pure isotope?

8. If you start with 20 g of a radioisotope and find that after 2 hours, you have only 1.25 g of the original substance, what must be the half-life of that material?

9. Potassium-42 has a half-life of 12.4 hours. Assuming that you start with 6 trillion atoms, how many of these atoms would you expect to have after 2 days?

10. Which of the following conditions, if any, will affect the rate of decay of a radioactive isotope?
    a. Applying pressure.
    b. Subjecting the material to a strong magnetic field.
    c. Forming a compound.

11. How many disintegrations per second are there in a microcurie?

12. What is the difference between radiation units of intensity and units of dosage?

13. A 6000 mile jet airplane trip exposes you to approximately 4 mrem of radiation. How does this exposure compare to the average dental x-ray?

14. How many hours of color TV viewing are radiologically equivalent to the dose received from one 6000-mile jet airplane trip? Even though these two exposures are numerically equivalent, what are the differences between the two types of exposure?

15. List the factors that contributed to your total radiation exposure over the last year.

16. The average exposure to fallout from nuclear weapons testing is estimated to be 4 mrems per year on the average for a resident of the United States.
    a. If all of this exposure were in the form of alpha particles, what would be the dose in rads?

b. If all of this exposure were in the form of beta particles, what would be the dose in rads?

c. If all of this exposure were in the form of gamma rays, what would be the dose in rads?

17. If an x-ray machine has a capacity of 0.15 million eV, how many joules of energy can be produced by this machine?

18. Describe the principle of operation of each of these detection methods.
    a. Film badge
    b. Spinthariscope
    c. Geiger counter

19. Why is it impossible for a Geiger counter to count alpha particle radiation?

20. What types of cells are most susceptible to radiation damage?

21. One of the components of the fallout from atmospheric testing of nuclear weapons was strontium-90. There has been a great deal of concern about this particular radioisotope, especially its impact on children. What is the basis for this specific concern about strontium-90? (*Hints:* What chemical group does strontium belong to? What is the half-life of this isotope?)

22. Both Marie Curie and her daughter Irene died of leukemia. Find out enough about the work of these two women to be able to suggest the probable cause of the leukemias.

23. If your job requires handling radioactive materials, what precautions can be taken to minimize exposure?

24. If your patient is undergoing radioactive brachytherapy for the treatment of a thyroid tumor, what measures can you take to protect yourself from unnecessary radiation exposure?

25. Why is implantation of radioactive seeds preferable to the use of gamma radiation treatment for some types of tumors?

26. If your doctor suggests the use of radioisotopes to perform a diagnostic test, what questions would you want answered concerning the use of these materials?

27. How would the criteria for choosing a radioisotope for diagnosis differ from the criteria for choosing a radioisotope for therapy?

28. Write the nuclear equation for the production of fluorine-18 used in PETT studies. This is done by bombarding neon-20 with deuterium.

29. Write the nuclear equation for the production of oxygen-15 used in PETT studies. This is done by bombarding nitrogen-14 with deuterium.

30. Write the nuclear equation for the production of nitrogen-13 used in PETT studies. This is done by bombarding oxygen-16 with a proton.

Appendices

I. Glossary

II. Table of Common Ions

III. A Glossary of Solution Terminology

IV. Normal Reference Laboratory Values of Blood and Urine Constituents

V. A Brief Classification of Prescription and OTC Drugs Based on Pharmacological Action

VI. Answers to Odd-Numbered Questions

Absolute alcohol One hundred percent (i.e., water-free) ethanol; highly hygroscopic.

Absolute zero The temperature at which molecular motion stops and the volume of an ideal gas becomes zero. Equal to $-273.15°C$, which is defined to be zero on the Kelvin temperature scale.

Acid **a.** Arrhenius theory: Any substance capable of liberating the hydrogen ion, H^+, in aqueous solution.
b. Bronsted-Lowry theory: Any substance that can donate a proton, H^+.

Acidic solution A solution in which the concentration of the hydrogen ion is greater than the concentration of the hydroxide ion.

Activation energy The minimum amount of energy that a chemical system must receive before it will undergo a chemical reaction.

Activator A substance that combines with an inactive form of an enzyme to change it into an active enzyme.

Acute Sharp, relatively severe, occurring over a brief course.

Adipose tissue Body fat.

Aerobic In the presence of oxygen.

Agonist The drug that binds to the receptor site and brings about the desired pharmacological action; opposite of antagonist.

Aliphatic The alkanes and their derivatives (as distinguished from aromatics).

Alkali metal Any element of the IA group of the periodic table. The alkali metals are Li, Na, K, Rb, and Cs.

Alkaline earth metal Any element of the IIA group of the periodic table. The alkaline earth metals are Be, Mg, Ca, Sr, and Ba.

Alkaloid An alkaline, aminoid plant product often having significant pharmacological activity.

Alkalosis A pathological condition arising from an accumulation of excess body base.

Alkane   A hydrocarbon of the general formula C_nH_{2n+2}.

Alkene   A hydrocarbon containing a C=C double bond.

Alkyne   A highly unsaturated hydrocarbon; contains a C≡C triple bond.

Alloy   A solid solution of two or more metals.

Alopecia   Falling out of hair.

Alpha particle   A particle made up of two protons and two neutrons identical to the nucleus of a helium atom. Symbols: He-4, $_2^4$He, α. Alpha particles are emitted when certain unstable nuclei decay.

Alveoli (lung)   Tiny air sacs in the lungs through which gas exchange occurs.

Amphoteric   Capable of either donating or accepting a proton.

Anabolic steroid   A synthetic steroid capable of stimulating protein synthesis or promoting anabolism generally.

Anabolism   The building up or construction of needed body chemicals from simpler substrates.

Anaerobic   In the absence of oxygen.

Analgesic   A pain reliever.

Androgens   Male sex hormones.

Anesthetic   A drug or substance used to abolish the sensation of pain; can be local, general, or inhalational.

Angstrom   A unit of wavelength equal to 10^{-7} mm or 10^{-8} cm.

Anion   A negatively charged ion.

Antabuse   A chemical employed to produce an aversion to ethyl alcohol.

Antagonist   A substance that prevents a drug from acting by preferentially occupying the drug's receptor site.

Anticholinergic   Any drug that counteracts the effects of acetylcholine or of parasympathetic stimulation.

Antigen   Any substance that can induce a specific immune response and that can react with the antibodies produced in that response.

Antimetabolite   A chemical so structurally similar to one required for normal biosynthesis that it can interfere with the utilization of the normal substrate; concept is used in the chemotherapy of disease.

Antipyretic   A drug that lowers fever.

Antirachitic activity   Prevention of the occurrence of rickets (a vitamin D deficiency disease).

Antisialogogue   A drug that reduces the flow of saliva.

Apnea   Cessation of breathing.

Apoenzyme   The protein portion of an enzyme; requires a coenzyme to function.

Aqueous Water-based; prepared with water.

Aromatic A type of organic compound characterized by six pi electrons, a 5- or 6-membered ring, planarity, and unusual stability.

Ascorbic acid Vitamin C.

Asepsis Freedom from infection or infectious organisms.

Association Another name for hydrogen bonding.

Astringent An agent that causes the skin to contract or close up.

Atherosclerosis Hardening of the arteries; damage to inner walls of arteries with plaque formation.

Atmosphere Unit of pressure in the English system of measurement.

Atom Smallest unit of an element. Composed of a dense nucleus that contains neutrons and protons, and electrons outside of the nucleus. Note: the simplest hydrogen atom consists of a single proton and a single electron.

Atomic mass number A number that indicates the total number of neutrons and protons in the nucleus of an atom.

Atomic mass unit, unified Unit of mass based on one-twelfth the mass of an atom of carbon-12. One atomic mass unit is equivalent to 1.66×10^{-27} kg.

Atomic number The number of protons in the nucleus of an atom. This value is specific for each element and does not vary for that element. If the atom is neutral, the atomic number also gives the number of electrons possessed by that atom.

Atomic weight The average relative mass of all naturally occurring isotopes of an element, expressed in unified atomic mass units.

Avitaminosis Hypovitaminosis, that is, the lack or deficiency of a certain vitamin.

Avogadro's law The volume of a constant mass of dry gas is directly proportional to the number of moles present, temperature and pressure being constant.

Avogadro's number The constant number of formula units in a mole of any substance. For any element existing in monoatomic form, the formula unit is a single atom; for a covalently bonded compound, it is a molecule. For an ionically bonded substance the formula unit is the simplest neutral group of oppositely charged ions. The value is 6.022×10^{23}.

Balanced equation An equation in which all the atoms of each type in the reactants are accounted for the products.

Bar Unit of pressure. Recently accepted as the standard international unit to be used for pressure.

Base
- **a.** Arrhenius theory: Any substance capable of liberating the hydroxide ion, OH$^-$, in aqueous solution.
- **b.** Bronsted-Lowry theory: Any substance that will act as a proton receptor.

Basic solution   A solution in which the concentration of the hydroxide ion is greater than the concentration of the hydrogen ion.

Beta-oxidation   The process by which fatty acid molecules are metabolized, two-carbon segments at a time.

Beta particle   A particle emitted when certain unstable nuclei decay. Beta particles are identical with electrons but come from the nucleus. Symbols are $_{-1}^0e$, β.

Bile salts   Salts of glycocholic and taurocholic acids, released by the gallbladder to facilitate the digestion of triacylglycerols.

Binary compound   A compound composed of just two different elements. Examples are NaCl, AlBr$_3$, K$_2$S.

Bioavailability   The degree to which a drug or other substance becomes available to or usable by the target tissue after administration.

Biotransformation   The chemical alteration of a biological molecule, substrate, food, or drug occurring within the organism.

Blood-brain barrier   The barrier separating the blood from the fluid of the brain.

Boyle's law   At constant temperature, the volume of a constant mass of dry gas varies inversely with the pressure.

Brachytherapy   Therapy at short range.

Bronsted-Lowry theory   An acid–base theory in which an acid is defined as a proton donor and a base is defined as a proton acceptor.

Buffer   A chemical system that tends to resist threatened changes in pH.

Calorie
- **a.** Spelled calorie (with a small case c), the amount of heat necessary to raise the temperature of 1 g of water by 1°C. Equal to 4.184 J.
- **b.** Spelled Calorie (with a capital C), equal to 1000 cal, hence a kilocalorie. Common unit of heat energy for food metabolism.

Carbolic acid   Phenol, C$_6$H$_5$OH.

Carbonmonoxyhemoglobin   A compound formed between CO and Hb; the bond is stronger than that between oxygen and Hb.

Carboxamide   An amide of a carboxylic acid, RCONH$_2$.

Carcinogen   A compound capable of causing cancer.

CAT   Computerized Axial Tomography. A type of scanning technique.

Catabolism Biotransformations in which complex molecules are broken down to simpler molecules.

Catalyst A substance that changes the rate of a chemical reaction but is itself chemically unchanged at the end of the reaction.

Catecholamine An amine having the 3,4-dihydroxyphenethylamine structure: $(HO)_2C_6H_3$—$CH_2CH_2NH_2$.

Cathartic A drug that induces evacuation of the bowels.

Cation A positively charged ion.

Celsius A metric system scale for reporting the measurement of temperature. Also refers to the degree itself, which is the same size as the degree Kelvin.

Charles' law At constant pressure, the volume of a constant mass of dry gas varies directly with the Kelvin temperature.

Chelating agent A substance that can combine with a metal ion to form a unique complex in which the metal ion is "hidden" from chemical reaction.

Chemical bond A force between atoms, holding them together.

Chemical change A change that involves an alteration of the chemical identity of a substance. New substances with new properties are produced as the result of chemical change. The process is often accompanied by changes in energy.

Chemical equation (See Equation, chemical.)

Chemical formula A set of symbols showing the elements present and their simplest ratio when combined.

Chemical property Any characteristic of a substance that serves to describe the substance's chemical (not physical) nature. Examples are ability to combine with oxygen, ability to react with metals or nonmetals, and reaction with acids or bases.

Chemistry The study of matter and the changes it undergoes.

Chemotherapy Using drugs to kill diseased cells while minimizing harm to normal cells.

Chiral center Usually a carbon atom, substituted by four different atoms or groups, that makes a molecule asymmetric and therefore optically active.

Chirality The absence of symmetry in a molecule leading to "handedness," that is, right-handed and left-handed molecules.

Chronic Occurring or persisting over a long period of time (months or years).

Cloning In the more general sense, cloning means making many copies of a specific selected gene by moving genes into new "host cells," sorting these cells, and isolating the wanted cells from the cell mixture.

CNS depressant A substance that dulls the activity of the brain and spinal cord.

Codon The peculiar sequence of three adjacent nitrogen bases in an mRNA strand that calls forth a specific amino acid in protein synthesis.

Coefficient Number that appears in front of a formula in a balanced chemical equation.

Colligative properties Characteristics of behavior of solutions that depend on the *number* of solute particles present rather than on their nature.

Colloid A fluid containing dispersed particles of a substance ranging in size from 1 to 100 nm; particles are too small to filter out or settle out.

Combination A type of chemical reaction characterized by two or more reactants combining to form one new compound. Also called composition or synthesis.

Combined general gas law The volume of a constant mass of dry gas varies inversely with the pressure and directly with the Kelvin temperature.

Combustion Rapid oxidation. Burning accompanied by heat and light production.

Complete protein One containing all nine adult essential amino acids in proper quantities.

Composition (See Combination.)

Compound A pure substance containing two or more elements chemically united in a fixed combination by mass. May be decomposed by chemical means into elements.

Conjugate acid In Bronsted-Lowry theory, the product formed from the combination of a base and a proton.

Conjugate base In Bronsted-Lowry theory, the product formed by the acid after it has lost its proton.

Conjugate pair In Bronsted-Lowry theory, either the acid and its conjugate base or the base and its conjugate acid.

Conjugated protein A complex molecule consisting of a simple protein bound to a nonprotein substance (called the prosthetic group).

Conjugation The chemical combination of a body chemical with a toxic agent to produce a new compound, less toxic and more urine soluble.

Conservation of energy, law of Energy cannot be created or destroyed by chemical means. It may only be transformed from one type to another. This statement excludes nuclear reactions.

Conservation of mass, law of Matter cannot be created or destroyed by chemical means. This statement excludes nuclear reactions.

Conversion factor A ratio that relates the size of one unit to another.

Coordinate covalent bond A covalent bond in which both electrons in the shared pair were donated by one of the atoms involved. Once formed, the coordinate covalent bond cannot be distinguished from any other covalent bond.

Corupus luteum   ("yellow body") Ovarian tissue that develops at the site of ovulation. Secretes progesterone to maintain a pregnancy.

Covalent bond   A chemical bond in which electrons are shared between atoms. May be polar (unequal sharing of electrons) or nonpolar (equal sharing of electrons).

Covalent bond length   The distance between two nuclei in a covalent bond. The bond length maximizes forces of attraction and minimizes forces of repulsion between bonded atoms.

Crenation   The notched, shrunken appearance of an RBC owing to its shrinkage after suspension in a hypertonic medium.

Cubic centimeter   A unit of volume, abbreviated cm^3 or cc. One cubic centimeter is approximately equivalent to 1 mL.

Curie   Unit of radiation activity. Defined as the amount of a radioactive substance necessary to give 3.7×10^{10} disintegrations per second.

Cycloalkane   A ringed aliphatic hydrocarbon of the general formula C_nH_{2n}.

Dalton's law   The total pressure exerted by a mixture of gases in a closed container is equal to the sum of the individual pressures that each of the gases would exert if they were alone in the container.

Decomposition   A type of chemical reaction in which one reactant compound breaks down into two or more products.

Delta   A sign, Δ, used to represent heat in a chemical reaction; it is placed over the arrow in a chemical equation. Also used to indicate the position of a C=C double bond, as in Δ^4.

Denaturant   A poison deliberately added to ethanol to render it unfit for human consumption. In protein chemistry, a substance that alters the tertiary structure of a protein.

Denaturation (of proteins)   The process of drastically altering the secondary, tertiary, or quaternary structure of a protein without breaking any peptide bonds.

De novo   Prepared from the basic elements (as opposed to being obtained intact from some other source).

Density   Physical property defined as the ratio of mass to unit volume. In symbols, $d = m/v$.

Desiccated   Dried; with water removed.

Dextrans   Linear polysaccharides composed of D-glucose units; found in various organisms.

Dextrins   Degradation fragments formed in the hydrolysis of starch.

Dextrorotatory   Rotating plane-polarized light to the right (as the viewer looks at the light source).

Diatomic molecules   Molecules composed of two identical atoms, such as Cl_2, H_2.

Diffusion   The spontaneous movement of atoms or molecules in order to reach a random, uniform concentration throughout a system; membrane passage may or may not be involved.

Dipole   A covalently bonded molecule that acts with the equivalent of two oppositely charged electrical poles.

Displacement   (*See* Replacement reactions.)

Diuretic   An agent that increases the flow of urine.

Diverticulitis   Inflammation of the tiny pouches or sacs in the lining of the intestines.

Double bond   A covalent bond in which four electrons (two electron pairs) are shared between nuclei.

Double replacement reaction   (*See* Replacement reactions.)

Drug   Any absorbed substance that changes or enhances any physical or psychological function in the body.

Drug depot   Any reservoir or tissue in the body (fat, blood proteins) that attracts and holds a drug for a period of time.

Edema   Abnormal accumulation of extracellular fluid.

Electrolyte   Any substance whose aqueous solution will conduct an electrical current. The term may also refer to the solution itself.

Electron   A subatomic particle found outside the nucleus of an atom. It has an extremely small mass and carries a negative electrical charge. In some cases an electron may be totally separated from an atom by heat, light, electrical energy, or radiation.

Electron dot structure   A chemical symbol or set of symbols showing elements or compounds with their valence electrons.

Electronegativity   The relative measure of the electron-attracting power of a bonded atom. There are no units to the electronegativity scale.

Electronic configuration   The arrangement of electrons in an atom, ion, or molecule. Lists orbitals and shows how many electrons are in each orbital.

Element   A pure substance that cannot be decomposed by ordinary chemical means. Made up of atoms of only one atomic number. Because of the existence of isotopes, the atoms in any element may have small differences in their masses.

Emulsification   Dispersal and stabilization of tiny globules of oil in water (or water in oil).

Enantiomers   Stereoisomers that are nonsuperimposable mirror images of each other.

Endergonic Describes a change accompanied by the absorption of energy. The energy does not just initiate the change (see activation energy) but is necessary to sustain it.

Endogenous Developing or originating from within the organism.

Endothermic Describes a change accompanied by the absorption of heat. The heat energy does not just initiate the change (see activation energy) but is necessary to sustain it.

End point The end of a titration. In an acid-base titration, the point at which one detects, usually by using a pH meter or a colored indicator, that the equivalence point has been reached.

Energy A capacity to cause a change that can, in principle, be harnessed for useful work. Some types of energy are electromagnetic, heat, electrical, mechanical, and chemical.

Energy level The energy states of an electron in an atom. In general, there is a greater probability of finding electrons of higher energy levels farther from the nucleus.

Entropy A measure of the disorderliness of a system; entropy increases in all spontaneous and irreversible processes.

Enzyme A protein produced in a cell and having a catalytic action by which it greatly accelerates the biotransformation of a substance (the substrate).

Epiphyses The ends of the long bones. The target tissue for growth hormone.

Equation, chemical The shorthand representation of a chemical reaction, employing the correct formulas for the reactants and products. An equation is balanced when all the atoms in the reactants are accounted for in the products.

Equation, nuclear The shorthand representation of a nuclear reaction, employing nuclear symbols for the reactants and products. A nuclear equation is balanced when the sum of the atomic numbers of the reactants equals that of the products and the sum of atomic mass numbers of the reactants equals that of the products.

Equilibrium The situation in which two opposing events, a forward reaction and its reverse, are occurring at the same rates, resulting in no net change with time.

Equivalence point In an acid–base titration, the equivalence point is reached when the number of moles of hydrogen ion equals the number of moles of hydroxide ion.

Equivalent weight One mole of an ionized substance divided by its valence.

Erythro A prefix meaning red.

Essential food factor A dietary substance, required by the body, but not made at all by the body or made in quantities too small to meet body demands.

Estrogens The female sex hormones.

Etiology   The causes or origins of disease.

Exergonic   Describes a change accompanied by the release of some form of energy.

Exogenous   Developed or originating outside the organism.

Exothermic   Describes a change accompanied by the release of heat energy.

Exponential notation   Expresses any number as the product of two factors. The first factor is not restricted to a value between 1 and 10 (*see* scientific notation). The second factor is a power of 10.

Fahrenheit   An English system scale for reporting the measurement of temperature. Also the degree itself, which is only five-ninths the size of a degree Celsius.

Family   Vertical column of elements in the periodic table. Also known as a group.

Fat   A fatty acid ester of glycerol; an ester formed between glycerol and three acids such as oleic, palmitic, or stearic.

Fatty acid   Any of the long-chain (C_4—C_{20}) carboxylic acids found esterified to glycerol in fats and oils.

Feedback, negative   The control of the output of a biological system by the inhibitory effect on a key step in the system, caused by a product of that system.

FFA   Free fatty acid (long-chain alkyl-COOH).

Fluid   Any substance that will assume the shape of its container and flow under pressure. Gases and liquids are fluids.

Fluorescence   Phenomenon in which a substance absorbs energy and then gives off visible light.

Food-drug interaction   A chemical or medical incompatibility between a food substance and a drug in which the drug's absorption or actions are undesirably altered.

Foreign protein   A protein exogenous to the organism and capable of eliciting an antigenic response.

Formed elements   The particles of the blood, namely, red blood cells, white cells, platelets.

Formula weight   The sum of the atomic weight of all the atoms represented in the chemical formula of a compound. Expressed in unified atomic mass units.

Free base   An alkaloid or other amine in the nonsalt form, that is, having an *unshared* pair of electrons on the N atom.

FSH   Follicle-stimulating hormone, released from the pituitary gland.

Functional group   A chemical group in a molecule that confers characteristic chemical, physical, or pharmacological properties.

Gamma rays   High-energy electromagnetic radiation emitted when unstable nuclei decay. Similar to x-rays but more penetrating and shorter in wavelength. Travel at the speed of light.

Gas   A state of matter in which the substance has no definite shape or volume (unless confined).

Gay-Lussac's law   The pressure of a constant mass of a dry gas is directly proportional to Kelvin temperature, volume and number of moles being constant.

Geiger counter   Instrument for detecting radiation.

Gene splicing   The insertion of a desirable segment of human DNA into the DNA of a bacterium so that the bacterium can grow and synthesize a wanted human protein. Also called recombinant DNA technology.

Genetic code   A code written in groups of three nitrogen bases in a nucleic acid strand; the code translates into amino acids for protein synthesis.

Genetic disease   (*See* Inborn error of metabolism.)

Geriatric   Pertaining to the aged.

GI tract   Gastrointestinal tract.

Glycogen   The human storage form of reserve carbohydrate; corresponds to starch in plants.

Glycogenolysis   The breakdown of glycogen into glucose.

Glycolipid   A lipid containing carbohydrate (galactose, glucose, inositol).

Glycolysis   The anaerobic, enzymatic conversion of glucose to lactate or pyruvate, with the production of energy-rich ATP.

Glycosides   Products formed by chemical combination between sugars and alcohols by the loss of a molecule of water.

Goiter   Enlargement of the thyroid gland, frequently due to a deficiency of dietary iodine.

Gout   A hereditary form of arthritis characterized by an excess of blood uric acid and the deposition of monosodium urate in joints.

Gram   Unit of mass in both SI and metric systems.

Gram atomic weight   The atomic weight of an element expressed in the unit grams. Also the weight in grams of 1 mol of an element. May be referred to as a gram atom.

Gram formula weight   The formula weight of a substance expressed in the unit grams. Also the weight in grams of 1 mol of that substance.

Gram molecular weight The molecular weight of a covalently bonded substance expressed in the unit grams. Also the weight of 1 mol of that substance.

Group Vertical column of elements in the periodic table. Also known as family.

Half-life
  a. Nuclear chemistry: Time needed for half the atoms of a radioactive isotope to decay into a different atomic form.
  b. Drug: Time required for the body to rid itself of half of an ingested dose of a drug.

Halogen Family of elements consisting of fluorine, chlorine, bromine, iodine, and astatine. Group VIIA elements.

Halogenation The substitution of a halogen atom into a hydrocarbon.

HCG Human chorionic gonadotropin; a hormone of pregnancy.

HDL High-density lipoprotein; considered by some authorities to play a key role in mobilizing cholesterol to the liver for metabolism.

Heavy metals High atomic weight metals important in human biochemistry. Some are useful, some are poisonous. Examples: Pb, Hg, Au, Bi.

Hemo A prefix meaning "related to the blood."

Hemolysis Splitting open of an RBC (red blood cell) with liberation of its Hb.

Hemopoiesis The formation and development of blood cells.

Henry's law The solubility of a gas in a liquid is directly related to the partial pressure of the gas above the liquid, temperature being held constant.

Heterocyclic A ringed organic compound having at least one noncarbon atom in the ring.

Heterogeneous matter Matter that is nonuniform in composition.

Homeostasis A tendency toward stability in the normal body environment, achieved by a system of control mechanisms.

Homogeneous matter Matter that is uniform in composition.

Hormone A highly active chemical substance that is produced in the body and that, in tiny amounts, exerts a specific regulatory action on the activity of a body organ or tissue.

Humor Anatomically, a fluid or semifluid substance. Example: aqueous humor of the eye.

Hydrocarbon A compound containing only carbon and hydrogen.

Hydrogen bonding Bonding between an electropositive part of a polar covalent molecule and an electronegative part of another polar molecule. May also be between polar regions of the same large polar molecule. Because of size limitations of such bonds, hydrogen bonding usually only involves atoms of oxygen, nitrogen, and fluorine with hydrogen. Although weaker than covalent or ionic bonding, hydrogen bonding is relatively strong compared with other types of intermolecular forces.

Hydrolysis
   a. A reaction with water, in which a bond is broken.
   b. A reaction in which the concentration of hydronium and hydroxide ions in an aqueous solution is changed because of water interacting with either a positively charged ion (other than H_3O^+) or a negatively charged ion (other than OH^-).

Hydronium ion   Ion with formula H_3O^+ that forms when a water molecule accepts a proton. Often the simplification H^+ is used when the hydronium ion is the true species present.

Hydroxide ion   Ion with formula OH^-. Characteristic of a classic base.

Hygroscopic   Having a tendency to absorb water from the surroundings.

Hyperbaric   Refers to pressures greater than normal atmospheric pressure.

Hypercholesterolemia   Elevated serum cholesterol.

Hypertension   A too-high blood pressure.

Hypertonic   Denotes a solution having greater osmotic effects than blood.

Hypervitaminosis   An abnormally high, sometimes dangerous, body level of a vitamin.

Hypoglycemia   A too-low blood sugar level; has serious effects in some people.

Hypotonic   Denotes a solution having lesser osmotic effects than blood.

Ideal gas   A gas that obeys the ideal gas law. An ideal gas has no interactions between particles, perfectly elastic collisions between particles, and the particles themselves do not occupy significant volume.

Ideal gas law $PV = nRT$.

Inborn error of metabolism   The inability to metabolize a certain substrate owing to the absence of the enzyme required for that substrate because of defective DNA; may have serious consequences.

Indicator   A substance that shows a change in the physical or chemical nature of a system by a characteristic change in color. Usually means a pH-sensitive substance.

Inositol   A sugarlike vitamin of the B-complex.

Interstitial fluid   The aqueous body fluid bathing the cells, excluding blood.

Intubation   Insertion of a tube into a body orifice.

In vitro   In the test tube.

In vivo   In the living organism.

Ion   An electrically charged atom. A covalently bonded group of atoms carrying an electrical charge may also be called an ion, or a polyatomic ion.

Ionic bonding   A chemical bond in which electrons are transferred between atoms, resulting in oppositely charged ions that are attracted to each other. Also called electrovalent or electrostatic bonding.

Ionic bond length   The distance between two atomic nuclei in an ionic bond.

Irradiation   The passage of high-energy radiation through or onto a substance or organ in order to bring about a chemical or biologic change.

Isoelectric point   The pH of an aqueous solution of an amino acid or protein at which the number of negative charges on the substance equal the number of positive charges (that is, maximum charge exists on the molecule).

Isomerism   The existence of two or more different compounds having the same number and kinds of atoms; types include structural, geometric, optical.

Isotonic   Denotes a solution having approximately the same osmotic effects as blood.

Isotopes   Atoms having the same atomic number but different atomic mass. They are atoms of the same element but differ in the number of neutrons in the nucleus.

IUPAC   International Union of Pure and Applied Chemistry. Chemists around the world agree to abide by their rules for naming chemicals.

Jaundice   Hyperbilirubinemia, that is, a too-high blood level of bilirubin with consequent yellow color of skin, membranes, and sclera.

Kelvin   An SI temperature scale. Also the degree itself, which is the same size as a degree Celsius.

Ketoacidosis   Excess acid in the body accompanied by the accumulation of ketone bodies.

Kilogram   Base unit of mass in the SI. Equal to 1000 g. The kilogram is the base unit for mass in the SI.

Kinetic energy   Energy of motion.

Kinetic molecular theory   A set of hypotheses about the movement of particles in a gas. The particles are assumed to be always in motion and have perfectly elastic collisions.

Krebs cycle   The cyclic, metabolic mechanism by which the complete oxidation of the acetyl group in acetyl Coenzyme A is accomplished.

Le Chatelier's principle When a change in condition is imposed on a system at equilibrium, the concentrations of reactants or products will shift in such a way as to partially relieve the imposed change. This reestablishes an equilibrium with new concentrations.

Leukocyte A white blood cell or corpuscle.

Levorotatory Rotating plane-polarized light to the left (as the viewer looks at the light source).

Lewis structure A symbol or set of symbols showing the number of valence electrons. May be simplified by the use of a dash to represent a bonded electron pair.

Lipid A fat or fatlike substance characterized by water insolubility but solubility in most nonpolar solvents.

Lipophilic Having an affinity for or solubility in fats or nonpolar solvents.

Liquid A state of matter in which the substance has no definite shape (except that of the container) but does have definite volume.

Liter An unit of liquid volume in the metric system.

Litmus An organic substance derived from lichen; indicates a change in pH by changing color.

Lymph The fluid that drains from the interstitial spaces back to the blood via the complex system of ducts and nodes called the lymphatic system.

Lysis Cleavage or splitting.

Mass Measure of the tendency of an object to stay at rest if it is stationary or to stay in motion if it is already moving. Compares the quality of matter present with a standard set of masses. The value of mass for an object is not dependent on gravitational force.

Mass number The total number of protons and neutrons in the nucleus of an atom.

Matter Anythng that has mass and occupies space.

Metal An element with a tendency to lose electrons in its attempt to attain noble gas electron configuration. Metallic character is more pronounced on the left-hand side of the periodic table.

Metalloids Elements with properties intermediate between those of metals and nonmetals. Examples are antimony and arsenic. Metalloids are found on the periodic table along a diagonal zigzag line separating metals from nonmetals. Newer name is semimetals.

Metastable A radioisotope that undergoes a transmutation but maintains the identity of the element. Only energy is emitted.

Meter Unit of length in the SI and metric systems. It is the base unit for length in the SI.

Metric system System of reference units used for scientific work and by most people of the world as the common system for measurement (*see* SI). The base units are the kilogram or gram for mass, the meter for length, the second for time, and the Kelvin or degree Celsius for temperature.

MI Myocardial infarct. Heart attack.

Micelle A colloidal-sized particle, as of fat in the gut.

Micro A prefix meaning one one-millionth.

Microliter Unit of liquid volume equivalent to 1×10^{-6} L or 1/1000th mL.

Milliliter Unit of liquid volume equivalent to 1/1000 L. Approximately equal to 1 cc.

Miscible Capable of being mixed, as two solvents, to form a homogeneous mixture.

Mitochondria Spherical or rod-shaped organelles found in the cytoplasm of cells; principal sites of energy generation; contain the enzymes of the Krebs and fatty acid cycles and respiratory chain.

Mixture Sample of matter composed of two or more substances present in indefinite proportions by weight. May be either homogeneous or heterogeneous. Separation is possible by physical methods.

Molar volume The volume that 1 mol of any gas will occupy at standard contions. Equal to 22.4 L at 273 K and 1.00 atm.

Mole The amount of a substance containing Avogadro's number of particles. Equal to 1 gram-formula weight for any substance.

Molecule Smallest unit of a compound that still retains all significant properties of that compound. It is an electrically neutral (but may be polar) unit in which the atoms are held together by covalent bonding.

Monomers Small chemical units that add to each other to form giant molecules called polymers.

Morbidity Illness or the diseased state.

mRNA Messenger RNA, a ribonucleic acid fraction of intermediate molecular weight that transmits genetic information from DNA to the protein-synthetic system of the cell.

Neo A prefix meaning new.

Neurotransmitter A chemical agent released from the end of a nerve into the synapse to either excite or inhibit the target cell (another neuron or an effector organ).

Neutralization The process in which an acid reacts with a base to produce a salt and water.

Neutral solution A solution in which the concentration of the hydronium ion (hydrogen ion) is equal to the concentration of the hydroxide ion.

Neutron A subatomic particle found in the nucleus. It is electrically neutral and approximately equal in mass to a proton.

NF The National Formulary, a legally recognized compendium for United States' drug standards.

NMR Nuclear Magnetic Resonance. A type of scanning technique.

Noble gas Any member of the family of elements consisting of neon, argon, krypton, xenon, and radon. Also called the rare gas family; known previously as the inert gases.

Noncovalent Not involving shared electron pair bonds. Examples of noncovalent bonds are hydrogen bonds, ion-dipole bonds, and van der Waal's forces.

Nonmetal An element that tends to gain electrons in its attempt to attain a noble gas electron configuration. Such elements are commonly found in Groups IVA, VA, VIA, and VIIA of the periodic table.

Nonpolar covalent bond A covalent bond in which there is no significant difference in the relative electron-attracting abilities of the bonded atoms. Considered nonpolar if the electronegativity difference between the two bonded atoms is less than 0.5.

Nuclear symbol An elemental symbol shown with its atomic number and mass number. Gives enough information to determine the number of protons and neutrons in the nucleus of the atom represented.

Nuclear transformation Changes in the nucleus of an atom, resulting in a different radioisotope.

Nucleoside A combination of a sugar with a purine or pyrimidine base.

Nucleotide The repeating unit in nucleic acids; consists of a nitrogen base, a sugar, and a phosphate group.

Nucleus Dense central region of an atom. Mass and positive charge are both concentrated in the nucleus.

Octet rule Explains many chemical changes in which elements attempt to gain a stable "octet" of electrons as present in the stable noble gases. The eight electrons occur in full s and p orbitals at any one energy level.

Olefin Synonymous with alkene.

Ophthalmic Pertaining to the eye.

Optically active Capable of rotating plane-polarized light.

Orbit A Bohr theory "path" that an electron follows around the nucleus.

Orbital Describes a volume in space around a nucleus where an electron has a high probability of being found. Term is used when describing electron arrangement by use of a quantum mechanical model of atomic structure.

Organic chemistry Branch of chemistry dealing with carbon compounds.

(Excludes simple oxides of carbon, carbonic acid and carbonate salts, and hydrocyanic acid and its salts.)

Osmole The formula weight of a substance divided by the number of ions, atoms, or molecules it liberates in solution. The standard unit of osmotic pressure.

Osmosis The passage of solvent through a semipermeable membrane from a solution of lesser to one of greater solute concentration.

OTC Over-the-counter (as in the sale of drugs).

Oxidant A substance that gains electrons and thus acts as an oxidizing agent.

Oxidation
 a. Chemically combining with oxygen.
 b. Occurs if electrons are removed, which implies an increase in oxidation number for the element oxidized.

Oxidation number An assigned number representing the charge an element would have if the bonds formed were totally ionic. Any free element has an oxidation number of zero. May also be called *valence*.

Oxidizing agent A substance that gains electrons in a chemical reaction. It is, itself, reduced. An oxidizing agent is also called an oxidant.

PABA *Para*-aminobenzoic acid.

Paraffin A solid mixture of hydrocarbons of C_{24}—C_{30} length, obtained from crude petroleum.

Parenteral solution A solution to be given by injection, rather than by oral or other routes of administration.

Paresthesis An abnormal sensation such as burning or tingling.

Partial pressure The vapor pressure exerted by an individual gaseous component of a mixture of gases. (*See* Dalton's law.)

Pascal Unit of pressure in the SI.

Peptide Any member of a class of compounds of low molecular weight that yield two or more amino acids upon hydrolysis.

Peptide bond The amide-type bond formed between two alpha-amino acids.

Period A horizontal row or sequence of elements in the periodic table.

Periodic law Generalization stating that the properties of the elements are periodic functions of their atomic number. There are regular repeating patterns of properties, based on position on the periodic table.

Periodic table Table that shows all elements arranged in order of increasing atomic number and also grouped according to the regular recurrence of their properties. Elements in the same families have the identical number of electrons in outer energy levels.

Peripheral   At the edges of; situated away from the center.

Peritoneum   The membrane lining the abdominal cavity and covering the viscera.

Petrolatum   A semisolid hydrocarbon mixture obtained from crude petroleum; used as an ointment base.

PETT   Positron Emission Transaxial Tomography. A type of scanning technique.

pH   A scale of values used to represent the relative acidity or alkalinity of a solution. A pH of 7 represents a neutral solution, from 0 to 7 is an acidic solution, and from 7 to 14 is an alkaline solution.

Phenotype   The complete biochemical, physical, and physiological makeup of an individual as determined both genetically and environmentally.

Phospholipid   Any lipid containing phosphorus, including those with a glycerol or sphingosine backbone.

Physical change   A change that involves no alteration of chemical identity of a substance but may involve change of state, color, form, or other physical properties.

Physical property   Any characteristic of a substance that serves to describe the substance's physical, not chemical, nature. Examples are color, density, melting, and boiling temperatures.

Plane-polarized light   Light that has passed through a Polaroid sheet or a Nicol prism and thus is vibrating all in the same plane.

Plasma   The fluid portion of the blood in which the formed elements are suspended.

Plasmid   In bacteria, plasmids are circular DNA molecules separate from chromosomal DNA. Used in gene splicing.

pOH   A scale of values used to represent the relative acidity or alkalinity of a solution. A pOH of 7 represents a neutral solution, from 0 to 7 is a basic solution, and from 7 to 14 is an acidic solution at 25°C.

Polar
   a. A polar covalent bond is one in which electrons are not shared equally between two atoms.
   b. A polar molecule is one in which polar covalent bonds are present in a nonsymmetrical arrangement, resulting in an overall electrical polarity.

Polyatomic compound   A compound composed of three or more types of elements.

Polyatomic ion   A group of two or more atoms covalently bonded, carrying a charge.

Polymer   A very large molecule made by joining many small chemically similar monomer units. Also called a macromolecule.

Polymerization A chemical reaction in which a high molecular weight compound is formed by the combination of many simpler molecules.

Polyol An organic compound containing more than two OH groups.

Polyuria Frequency of urination.

Postprandial After a meal

Potential energy Stored energy.

Precipitate An insoluble solid that forms from the combination of components in aqueous solution.

Pressor agent A substance that raises blood pressure.

Pressure Force per unit area.

Product Any of the substances formed in a chemical change. Products appear on the right-hand side of a chemical equation.

Progestin A synthetic progesteronelike steroid, used in birth control pills.

Prophylactic dose Dose given to prevent disease, as contrasted to a therapeutic dose.

Prostaglandins Naturally occurring long-chain C-20 cyclic fatty acids that stimulate uterine and other smooth muscles, affect blood pressure, regulate gastric secretion, body temperature, and platelet aggregation, and influence the inflammatory process.

Prosthetic group Nonpolypeptide molecules joined to protein to form conjugated protein.

Proteins Complex organic molecules of high molecular weight, widely distributed in plants and animals, and consisting essentially of combinations of amino acids in peptide linkages.

Proton A subatomic particle found in the nucleus. It is electrically positive and approximately equal in mass to a neutron.

Pure substance Homogeneous substance composed of just one characteristic material. Has a constant and invariable chemical composition by weight.

Pyrogen A fever-producing substance.

Quantum mechanics A view of the structure of atoms in which electrons are treated as matter waves and energy is quantized. Probability regions in space are used to describe electron arrangement within the atom.

Quaternary amine salt A tetra-substituted amine of the type formula $R_4N^+ \cdot X^-$. The nitrogen atom always bears a positive charge.

Racemic mixture An optically inactive 50–50 mixture of enantiomers.

Rad A unit of radiation dosage.

Radiation   Emission of energy or particles or both from the nucleus of an atom.

Radioactivity   A phenomenon in which there is disintegration of an unstable atomic nucleus, resulting in the emission of radiation. This may occur naturally or be stimulated by bombarding a nucleus with particles.

Radiochemistry   That branch of chemistry dealing with the nuclear transformations of atoms.

Radiograph   Two-dimensional image resulting from an exposure to x-rays.

Radioisotope   A radioactive isotope; an isotope emitting radiation.

Radionuclide   A radioactive isotope; an isotope emitting radiation.

Radiopaque   Materials capable of blocking the passage of x-rays.

Radiopharmaceutical Therapy   Giving radioactive material orally or intraveneously to achieve a therapeutic goal.

RBE   Relative Biological Effectiveness

RDA   Recommended dietary allowance (daily), set by the Food and Nutrition Board of the National Research Council (not necessarily the same as U.S. Recommended Daily Allowance), as satisfactory to meet the nutritional needs of most healthy persons.

Reactant   Any of the substances started with in a chemical reaction. Reactants appear on the left-hand side of a chemical equation.

Receptor site   Specialized cells in many body locations to which chemicals bind because of distinctive size, shape, and electrical characteristics.

Redox   Abbreviation for *reduction–oxidation* reaction. A redox reaction is a type of chemical change in which electrons are transferred from one reactant to another.

Reducing agent   A substance that loses electrons in a chemical change. It is itself oxidized. A reducing agent is also called a reductant.

Reducing sugar   A carbohydrate that can chemically reduce Tollens', Benedict's, or Fehling's solutions.

Reduction
   a. Chemically removing oxygen.
   b. Occurs if electrons are gained, which implies a decrease in oxidation number for the element reduced.

Releasing factors   Proteins, made in the hypothalamus, that stimulate the release of pituitary hormones.

Rem   A unit of radiation dosage.

Replacement reactions   Any reaction in which one or more ions, atoms, or groups of atoms are replaced by a different ion, atom, or group of atoms. May be single replacement with a general pattern of $A + BC \rightarrow AC + B$, or may be double replacement with a general pattern of $AB + CD \rightarrow AD + CB$.

Replication   The process in which an exact (duplicate) copy of a polynucle-otide strand of DNA or RNA is made.

Representative elements   Groups on the periodic table with an "A" designation.

Restriction enzyme   Enzyme that catalyzes specific cleavage of DNA at various points along the strand.

Reticulocyte   A young red blood cell.

Ribosomes   Ribonucleoprotein particles inside cells on the surface of which protein synthesis occurs.

Roentgen   A unit of radiation dosage.

Salt   A crystalline substance consisting of an orderly arangement of oppositely charged ions (other than H^+ and OH^-) occurring in a definite ratio. A salt can be produced in a neutralization reaction from the positive ion of the base and the negative ion of the acid.

Saturated solution   One that is ordinarily holding its maximum quantity of solute at the specified temperature.

Scabies   A contagious skin disease caused by the itch mite.

Scan   The image of an organ that is created when a radioactive isotope accumulates in that organ.

Scientific notation   Expresses any number as a product of two factors. The first factor is between 1 and 10 and will have just one digit before the decimal point. The second factor is a power of 10.

Scintillation counter   Fluorescence detection unit, usually combined with a computerized information system.

Self-ionization   The ability of a compound to react with itself to produce ions.

Semimetal   (See Metalloid.)

Sequestering agent   (See Chelating agent.)

Serum   The clear portion of the blood after formed elements and the clot are separated.

SI   The abbreviation for the system of weights and measures known as the International System of Units (in French, Système Internationale d'Unites). The SI is very similar to the metric system but differs in some reference standards.

Significant figure   A figure having scientific meaning due to the measurement process.

Single replacement reaction   (See Replacement reactions.)

Soap   The water soluble K or Na salt of a fatty acid; capable of emulsifying oils and greases.

Solid A state of matter in which the substance has both definite shape and volume.

Solute The substance that is dissolved by the solvent to form a solution.

Solution The dispersal of one substance in another to yield a homogeneous single-phase preparation.

Solvation The formation of ion–dipole bonds between a solvent and its solute. Stabilization of the ion (the solute) results.

Sphingolipid A lipid containing the long-chain amino acid sphingosine; bound to it are a fatty acid and choline. They occur especially in brain and nerve tissue.

Spinthariscope Simple fluorescence detection unit.

Standard conditions A fixed set of temperature and pressure conditions. For gases, standard pressure is usually 0°C (273 K) and standard pressure is 1.00 atm (760 mm Hg).

Stereoisomers Isomers having the same number and kinds of atoms, the same structure, but different arrangements in space. Examples are optical isomers and geometric isomers.

Steroids A group of lipids containing a characteristic tetracyclic nucleus. Are nonsaponifiable.

Sterol Chemically, an alcohol derivative of a steroid.

Strong (acid, base, elecrolyte)
  a. Describes a high percent ionization or dissociation.
  b. In Bronsted-Lowry theory, a strong acid is any acid that holds its proton weakly and therefore readily donates it; a strong base will bind an accepted proton strongly.

Structural formula Represents the nucleus-to-nucleus sequence within an individual molecule. Simplified from a Lewis electron dot structure.

Subatomic particle Any of the particles that comprise an atom. The three common subatomic particles are the electron, proton, and neutron.

Subscript A lowered number following a symbol in a formula. Indicates how many of that unit are present in one complete formula unit.

Substitution reaction (*See* Replacement reactions.)

Substrate A substance on which an enzyme acts to biotransform.

Supersaturated Refers to an unstable situation in which a solution is temporarily holding more solute than it can ordinarily hold at equilibrium conditions.

Surface tension A kind of barrier across the surface of a liquid that resists (however slightly) penetration of the liquid.

Surfactant A chemical that reduces the surface tension of a liquid.

Suspending agent A substance used to thicken a solution so that suspended particles will not settle out so quickly.

Suspension   A preparation of a finely divided substance suspended in a fluid. The particle size is greater than about 100 nm; particles can be filtered out.

Syndrome   All of the signs and symptoms associated with a disease.

Synergism   The mutual potentiating action of two drugs on each other to give a superadditive effect.

Synthesis reaction   (See Combination.)

Tachycardia   A faster-than-normal heart rate.

Teletherapy   Therapy at a distance.

Temperature   A measure of the hotness or coldness of an object. It is a measure of average kinetic energy of the molecules.

Tension   The partial pressure of a gas in a fluid, for example, oxygen dissolved in blood.

Teratogen   A substance or a factor that damages the embryo, producing a birth defect.

Ternary compound   A compound composed of just three different elements. Examples: H_2SO_4, $Ca(NO_3)_2$, NaOH.

Tetany   A condition manifested by sharp muscle spasms, muscle twitching, cramps, and convulsions. May be due to abnormal calcium metabolism.

Tincture   A solution of a medication prepared using alcohol or alcohol and water as the solvent.

Titration   A laboratory procedure in which a standard solution is caused to react with an unknown until an end point is determined. Frequently used to determine quantitatively the concentration of an acid, base, or some other solute in solution.

Tocopherols   A group of compounds having vitamin E activity.

Tolerance   The state of acquired resistance to a drug such that ever larger doses must be taken to obtain desired results.

Torr   A unit of gas pressure. A torr is equivalent to a millimeter of mercury. Standard pressure is 1 atm, which is 760 mm Hg.

Trace element   Essential inorganic (mineral) dietary factor required by the body in only tiny (trace) amounts.

Transcription   The process in which the genetic information contained in DNA is transferred to a complementary sequence of bases in mRNA.

Transdermally   Through the skin.

Transition element   An element found in the center block of the periodic table between the representative s and p block elements. Such elements have electrons filling d orbitals.

Translation   The formation of a polypeptide chain, the amino acid sequence of which is directed by the genetic information carried in mRNA.

Transmutation   Transformation of one isotope to another by means of nuclear change. Often alters the identity of the element.

Triacylglycerol   (*See* Fat.)

Triple bond   A covalent bond in which six electrons (three pairs of electrons) are shared between two nuclei.

tRNA (transfer RNA)   An RNA fraction of low molecular weight that combines with one amino acid, transferring it from activating enzyme to ribosome for the synthesis of a polypeptide.

Turnover number   The number of moles of substrate biotransformed to product per mole of enzyme per minute.

Tyndall effect   Name given to the phenomenon in which colloidal-sized particles suspended in a fluid make visible a transverse beam of light passing through the fluid.

Unified atomic mass unit   Unit of mass based on one-twelfth the mass of an atom of carbon-12. Equal to 1.66×10^{-27} kg.

Unsaturated (hydrocarbon)   Containing one or more double or triple carbon-to-carbon bonds.

USP   The U.S. Pharmacopoeia—the official compendium for U.S. drug standards.

Valence   Chemical combining ability of an element or radical in a compound. (*See* Oxidation number.)

Vasodilator   A substance that expands the inside diameter of a blood vessel.

Vegan   An extreme vegetarian; excludes *all* animal protein from diet.

Virus   A submicroscopic infectious agent that can reproduce only within living host cells; consists of nucleic acid with a protein coat.

Viscosity   The thickness of a liquid (dependent on the friction of component molecules as they slide by one another).

Wax   An ester formed between a fatty acid and a long-chain monohydric alcohol.

Weak   (as applied to acid, base, electrolyte)
   a. Describes a low percent ionization or dissocation in water.
   b. In Bronsted-Lowry theory, a weak acid is any acid that holds its proton strongly and therefore does not donate it readily. A weak base will bind an accepted proton weakly.

Weight   A measure of the gravitational pull on an object. Dependent on position in relation to the gravitational field.

Word equation   An equation that uses words rather than chemical formulas to represent chemical change.

X-ray   Part of the electromagnetic spectrum. Penetrating, high-energy waves much like gamma rays. May be produced when electrons change to lower energy levels within an atom.

Zymogen   An inactive precursor that is converted to an active enzyme. Also termed a proenzyme.

Appendix II Table of Common Ions

Positive ions		Negative ions	
Name	*Formula*	*Name*	*Formula*
Aluminum	Al^{3+}	Acetate	$C_2H_3O_2{}^-$
Ammonium	$NH_4{}^+$	Bromate	$BrO_3{}^-$
Barium	Ba^{2+}	Bromide	Br^-
Bismuth	Bi^{3+}	Carbonate	$CO_3{}^{2-}$
Cadmium	Cd^{2+}	Chlorate	$ClO_3{}^-$
Calcium	Ca^{2+}	Chloride	Cl^-
Chromium(II) (chromous)	Cr^{2+}	Chlorite	$ClO_2{}^-$
Chromium(III) (chromic)	Cr^{3+}	Chromate	$CrO_4{}^{2-}$
Cobalt(II) (cobaltous)	Co^{2+}	Cyanide	CN^-
Cobalt(III) (cobaltic)	Co^{3+}	Dichromate	$Cr_2O_7{}^{2-}$
Copper(I) (cuprous)	Cu^+	Dihydrogen phosphate	$H_2PO_4{}^-$
Copper(II) (cupric)	Cu^{2+}	Fluoride	F^-
Hydrogen	H^+	Hydrogen carbonate (bicarbonate)	$HCO_3{}^-$
Hydronium	H_3O^+		
Iron(II) (ferrous)	Fe^{2+}	Hydrogen sulfate (bisulfate)	$HSO_4{}^-$
Iron(III) (ferric)	Fe^{3+}	Hydrogen sulfide ion (bisulfide ion)	HS^-
Lead(II) (plumbous)	Pb^{2+}		
Lead(IV) (plumbic)	Pb^{4+}	Hydrogen sulfite (bisulfite)	$HSO_3{}^-$
Lithium	Li^+	Hydroxide	OH^-
Manganese(II) (manganous)	Mn^{2+}	Hypochlorite	ClO^-
Manganese(III) (manganic)	Mn^{3+}	Iodide	I^-
Magnesium	Mg^{2+}	(Mono)hydrogen phosphate	$HPO_4{}^{2-}$

Positive ions		Negative ions	
Name	Formula	Name	Formula
Mercury(I) (mercurous)	Hg_2^{2+}	Nitrate	NO_3^-
Mercury(II) (mercuric)	Hg^{2+}	Nitrite	NO_2^-
Nickel(II) (nickelous)	Ni^{2+}	Oxalate	$C_2O_4^{2-}$
Nickel(III) (nickleic)	Ni^{3+}	Oxide	O^{2-}
Potassium	K^+	Perchlorate	ClO_4^-
Silver	Ag^+	Permanganate	MnO_4^-
Sodium	Na^+	Peroxide	O_2^{2-}
Strontium	Sr^{2+}	Peroxydisulfate	$S_2O_8^{2-}$
Tin(II) (stannous)	Sn^{2+}	Phosphate	PO_4^{3-}
Tin(IV) (stannic)	Sn^{4+}	Phosphite	PO_3^{3-}
Zinc	Zn^{2+}	Sulphate	SO_4^{2-}
		Sulfide	S^{2-}
		Sulfite	SO_3^{2-}
		Tetraborate	$B_4O_7^{2-}$
		Thiosulfate	$S_2O_3^{2-}$

Appendix III A Glossary of Solution Terminology

This glossary lists some words and phrases associated with solutions used in the health science field.

Astringent   An aqueous solution of an aluminum, zinc, or other metal salt that checks secretions of mucous membranes or constricts blood vessels.

Bactericidal solution   A preparation that is capable of killing bacteria.

Buffered solution   A solution that has been prepared specifically to resist changes in pH.

Carbonated solution   An aqueous solution of carbon dioxide.

Concentrated solution   Typically, a solution having greater than 10 percent of solute; its concentration is usually expressed as percent weight to weight (% w/w).

Dilute solution   Typically, a solution having less than 10 percent of solute. Usually expressed as percent weight to volume (% w/v).

Elixir   A clear, sweetened hydroalcoholic (i.e., water and alcohol) solution intended for oral use.

Intravenous   A sterile, pyrogen-free, isotonic solution of a substance in water suitable for injection directly into a vein.

Irrigating solution   A sterile, isotonic solution of a substance in water used for treating wounds or as an artificial organ bath. Not suitable for injection.

Isotonic solution   An aqueous solution of an ionized or unionized solute having the same osmotic pressure as blood.

Keratolytic solution   A solution of a chemical such as resorcinol or salicylic acid that is capable of dissolving away the horny layer of skin.

Limewater   The alkaline solution of calcium hydroxide in water; used as an antacid and astringent.

Neutral solution   A solution having a pH equal to 7.

Nonaqueous solution   A solute dissolved in a solvent other than water; typical nonaqueous solvents are glycerin, ether, petrolatum, ligroin; not generally intended for internal use.

Normal saline solution   An isotonic NaCl solution (0.9% w/v).

Normal solution   An ambiguous term that could refer to acid–base concentrations, to milliequivalents of electrolytes, or to physiologically normal saline solution.

Ophthalmic solution   A sterile, isotonic solution free of solid matter, suitable for use in the eye.

Otic solution   A solution for use in the ear.

Parenteral solution   A solution that is to be administered by injection.

Plasma   The complex solution that remains after the formed elements (RBCs, WBCs, platelets) have been removed from the blood.

Proof solution   A 100-proof solution contains 50% v/v of ethyl alcohol in water.

Pyrogen-free solution   An aqueous solution, intended for hypodermic use, from which has been removed all foreign protein matter which, if present, would induce fever in the recipient; such foreign proteins are termed *pyrogens*.

Saline solution   Generally, a solution that contains one or more mineral salts, such as NaCl, Na_2SO_4, or $MgSO_4$. In specific use, saline solution refers to aqueous NaCl.

Saturated solution   A solution that contains the maximum quantity of dissolved solute at a given temperature.

Serum   The watery portion of the blood after coagulation; hence serum is plasma with the clotting factors removed; it is still a solution.

Smelling salts   A solution of ammonium carbonate, $(NH_4)_2CO_3$, in ammonia water; causes reflex stimulation upon inhalation of NH_3 fumes.

Spirit   An alcoholic or hydroalcoholic solution of a volatile substance; an example is aromatic ammonia spirit.

Sterile solution   A solution that contains no viable forms of bacteria, fungi, yeasts, or viruses; an antiseptic solution.

Stock solution   A more concentrated solution, usually not used itself as a medicine, from which dilutions are made.

Supersaturated solution   An unstable solution, temporarily containing more solute than is ordinarily soluble at the given temperature; seeding a supersaturated solution induces precipitation of the "excess" solute.

Tincture   A solution of a medicinal substance in alcohol.

Topical solution   A solution intended for application to external parts of the body.

Unsaturated solution   A solution that is capable of dissolving more solute than is presently dissolved, at a given temperature.

Appendix IV Normal Reference Laboratory Values for Blood and Urine Constituents

Constituent	Fluid	Reference range	
		Conventional	SI
Ammonia	Blood	80–110 μg/100 mL	47–65 μmol/L
Ascorbic acid	Blood	0.4–1.5 mg/100 mL	23–85 μmol/L
Bilirubin	Serum	0.2–1.0 mg/dL	up to 7 μmol/L
Calcium	Serum	8.5–10.5 mg/100 mL	2.1–2.6 mmol/L
Calcium	Urine	150 mg/day or less	3.8 or less mmol/day
Carbon dioxide content	Serum	24–30 mEq/L	24–30 mmol/L
Chloride	Serum	100–106 mEq/L	100–106 mmol/L
Creatine phospho-kinase (CPK)	Serum	Female 5–35 mU/mL Male 5–55 mU/mL	0.08–0.58 μmol·s^{-1}/L
Fructose	Urine	0	
Glucose	Plasma	Fasting: 70–110 mg per 100 mL	3.9–5.6 mmol/L
Glucose	Urine	0	
Hematocrit	Blood	Male: 45–52%	Male: 0.42–0.52
Hemoglobin	Blood	Male: 13–18 g/100 mL Female: 12–16 g/100 mL	Male: 8.1–11.2 mmol/L Female: 7.4–9.9 mmol/L
Iron	Serum	50–150 μg/100 ml	9.0–26.9 μmol/L
Lactic acid	Blood	0.6–1.8 mEq/L	0.6–1.8 mmol/L
Lactic dehydrogenase	Serum	60–120 U/mL	1.00–2.00 μmol·s^{-1}L^{-1}
Lead	Urine	0.08 μg/mL	0.39 μmol/L or less

Constituent	Fluid	Reference range	
		Conventional	SI
Lipids			
Cholesterol	Serum	120–220 mg/100 mL	
Cholesterol esters	Serum	60–75% of cholesterol	
Phospholipids		9–16 mg/100 mL as lipid phosporus	2.9–5.2 mmol/L
Total fatty acids		190–420 mg/100 mL	1.9–4.2 g/L
Total lipids		450–1000 mg/100 mL	4.5–10.0 g/L
Triglycerides (triacyl-glycerols)		40–150 mg/100 mL	0.4–1.5 g/L
P_{CO_2}	Blood	35–45 mm Hg	4.7–6.0 kPa
pH	Blood	7.35–7.45	Same
P_{O_2}	Blood	75–100 mm Hg (depends on age)	10.0–13.3 kPa
Phosphatase, acid	Serum	Male, total: 0.13–0.63 Sigma U/mL	36–175 nmol·s^{-1}/L
Phosphatase, alkaline	Serum	13–39 IU/L	0.22–0.65 μmol·s^{-1}/L
Phosphorus (inorganic)	Serum	3.0–4.5 mg/100 mL	1.0–1.5 mmol/L
Phosphorus (inorganic)	Urine	average 1 g/day	32 mmol/day
Potassium	Serum	3.5–5.0 mEq/L	3.5–5.0 mmol/L
Protein, total	Serum	6.0–8.4 g/100 mL	60–84 g/L
Albumin	Serum	3.5–5.0 g/100 mL	35–50 g/L
Globulin	Serum	2.3–3.5 g/100 mL	23–35 g/L
Pyruvic acid	Blood	0–0.11 mEq/L	0–0.11 mmol/L
Sodium	Serum	135–145 mEq/L	135–145 mmol/L
Stool nitrogen	—	less than 2 g/day or 10% of urinary N	less than 2 g/day
Sulfate	Serum	0.5–1.5 mg/100 mL	0.05–1.2 mmol/L
Total T_3 by RIA	Serum	70–190 ng/100 mL	1.08–2.92 nmol/L
Total T_4 by RIA	Serum	4–12 μg/100 mL	52–154 nmol/L
Transaminase (SGOT)	Serum	10–40 U/mL	0.08–0.32 μmol·s^{-1}/L
Urea nitrogen (BUN)	Serum, blood	8–25 mg/100 mL	2.9–8.9 mmol/L
Uric acid	Serum	3.0–7.0 mg/100 mL	0.18–0.42 mmol/L
Vitamin A	Serum	0.15–0.6 μg/mL	0.5–2.1 μmol/L
Titratable acidity	Urine	20–40 mEq/day	20–40 mmol/day
Vanillylmandelic acid (VMA)	Urine	up to 9 mg/24 hr	up to 45 μmol/day

Appendix V   A Brief Classification of Prescription and OTC Drugs Based on Pharmacological Action

Adrenergic Blocking Agents (block the action of catecholamines)
   a. propranolol (Inderal)
   b. ergot alkaloids
Analgesics (pain relievers)
   a. OTC: aspirin and other salicylates; acetaminophen
   b. narcotic analgesics: morphine, codeine, pentazocine
Anesthetics, inhalational
Antacids
Anticancer (antineoplastic) drugs
Anticoagulants (prevent or delay clotting)
Anti-emetic drugs (used to prevent motion sickness)
Antihistamines
Antihypertensives (used to treat high blood pressure)
Anti-infectives
   a. antibiotics
   b. fungicides
   c. antituberculars
   d. sulfonamides
   e. urinary germicides
Anti-inflammatory agents (used in arthritis)
   a. gold compounds
   b. enzymes
   c. phenylbutazones
   d. salicylates
   e. steroids
Antimalarials
Anti-Parkinsonism drugs
Antipyretics (fever reducers)
Antispasmodics
   a. belladonna alkaloids

Cardiovascular drugs
 a. vasodilators
 b. vasoconstrictors
 c. heart stimulants (Digitalis)
 d. anti-arrhythmics (Quinidine, Inderal, Ca slow-channel blockers)
CNS Depressants
 a. sedatives & hypnotics (the barbiturates, chloral hydrate)
 b. ethyl alcohol
 c. anticonvulsants (anti-epileptics such as Dilantin)
CNS Stimulants
 a. amphetamines, cocaine, caffeine, strychnine
 b. antidepressants (tricyclic)
Digestants (including enzymes)
Diuretics (increase flow of urine)
Expectorants
Heavy Metal Antagonists (used in Hg-, Pb- and As-poisonings)
Hormones
 a. estrogens and androgens (the sex hormones)
 b. adrenal cortex hormones
 c. insulin
Laxatives and Cathartics
Local Anesthetics
Narcotic Antagonists (naloxone or Narcan)
Ophthalmic (eye) Drugs
 a. miotics and mydriatics
Oral Contraceptives
Oxytocics (uterine stimulants)
Parasympathomimetics (mimic the effects of acetylcholine)
Plasma Volume Expanders
Skeletal Muscle Relaxants
 a. curare
Sympathomimetics
 a. epinephrine and norepi
 b. isoproterenol; tyramine
 c. ephedrine
 d. amphetamines
Tranquilizers
 a. minor: Valium, Serax, the meprobamates
 b. major: Compazine, Haldol, Thorazine
Uricosuric Agents (used in treatment of gout)
 a. Probenecid
 b. colchicine
Vaccines
Vitamins

Chapter 1

1. a. pure substance   b. mixture
   c. mixture   d. pure substance
   e. mixture   f. mixture
   g. pure substance   h. mixture
   i. mixture   j. mixture

3. a. Salt is more soluble in water than is pepper so you could add water. This would dissolve the salt and you could then filter out the pepper. By evaporating the water, you would have the salt back.
   b. Alcohol and water have different boiling points so you might try distilling the mixture. Some chemicals absorb water but not alcohol so that might provide an alternate way to separate.
   c. Iron is magnetic but sulfur is not. A good magnet should separate out the iron. As an alternative, some solvents may dissolve the sulfur but not the iron.
   d. With two different salts, you may have to make use of their difference in solubility. A measured amount of water can be added, allowing one salt to dissolve but not the other. Temperature may play a role in the separation as well for salts differ in the solubility at different temperatures.

5. Evaporation of the water will leave the salt behind from a homogeneous mixture of salt and water.

7. Salt and sugar are both white, crystalline solids. They will be quite different in such physical properties as melting point, boiling point, density, and of course, taste!

601

9. a. physical change   b. chemical change
   c. physical change   d. chemical change
   e. physical change   f. chemical change
   g. physical change   h. chemical change
   i. chemical change

11. Direct observations are useful but not sufficiently detailed or reliable. The senses are easily fooled. Measurements can extend the senses and give a more quantified way of identifying substances.

13. a. $675 \text{ dm} \times \dfrac{1 \text{ m}}{10 \text{ dm}} = 67.5 \text{ m}$

   b. $92 \text{ dam} \times \dfrac{10 \text{ m}}{1 \text{ dam}} = 920 \text{ dam}$

   c. $5{,}600{,}000 \text{ μm} \times \dfrac{1 \text{ m}}{10^6 \text{ μm}} = 5.6 \text{ m}$

   d. $800 \text{ mm} \times \dfrac{1 \text{ m}}{1000 \text{ mm}} = 0.8 \text{ m}$

   e. $25 \text{ km} \times \dfrac{1000 \text{ m}}{1 \text{ km}} = 25{,}000 \text{ m (or } 2.5 \times 10^4 \text{ m)}$

15. a. $6.7 \text{ L} \times \dfrac{10^3 \text{ mL}}{\text{L}} = 6.7 \times 10^3 \text{ mL or 6,700 mL}$

   b. $900 \text{ kL} \times \dfrac{10^3 \text{ L}}{\text{kL}} \times \dfrac{10^3 \text{ mL}}{\text{L}} = 9 \times 10^8 \text{ mL}$

   c. $6 \times 10^{-4} \text{ L} \times \dfrac{10^3 \text{ mL}}{\text{L}} = 6 \times 10^{-1} \text{ mL or 0.6 mL}$

   d. $0.0056 \text{ nL} \times \dfrac{1 \text{ L}}{10^9 \text{ nL}} \times \dfrac{10^3 \text{ mL}}{\text{L}} = 5.6 \times 10^{-9} \text{ mL}$

   e. $0.045 \text{ L} \times \dfrac{10^3 \text{ mL}}{\text{L}} = 4.5 \times 10^1 \text{ mL}$

17. a. $750 \text{ g} \times \dfrac{1 \text{ kg}}{10^3 \text{ g}} = 7.5 \times 10^{-1} \text{ kg or 0.75 kg}$

   b. $5{,}000{,}000 \text{ dag} \times \dfrac{10 \text{ g}}{1 \text{ dag}} \times \dfrac{1 \text{ kg}}{10^3 \text{ g}} = 5 \times 10^4 \text{ kg or 50,000 kg}$

   c. $4 \times 10^5 \text{ g} \times \dfrac{1 \text{ kg}}{10^3 \text{ g}} = 4 \times 10^2 \text{ kg}$

   d. $3.4 \times 10^{-7} \text{ mg} \times \dfrac{1 \text{ g}}{10^3 \text{ mg}} \times \dfrac{1 \text{ kg}}{10^3 \text{ g}} = 3.4 \times 10^{-13} \text{ kg}$

   e. $7.124 \times 10^4 \text{ cg} \times \dfrac{1 \text{ g}}{10^2 \text{ cg}} \times \dfrac{1 \text{ kg}}{10^3 \text{ g}} = 7.124 \times 10^{-1} \text{ kg}$

19. a. 2.469×10^1; 4 significant figures
   b. 2.1×10^{-4}; 2 significant figures

c. 7.4×10^4; 2 significant figures

d. 4.1329×10^0; 5 significant figures

e. 1.2×10^4; 2 significant figures

Note that the number of significant figures does not change just because the measurement is now expressed in scientific notation.

$$200 \text{ m} \times \frac{39.4 \text{ inches}}{\text{m}} \times \frac{1 \text{ yard}}{36 \text{ inches}} = 219 \text{ yards}$$

21. Therefore, since 200 meters has been calculated to be 219 yards, a distance of 220 yards is greater than 200 meters.

$$90 \text{ kg} \times \frac{10^3 \text{ g}}{1 \text{ kg}} \times \frac{1 \text{ lb}}{454 \text{ g}} = 198 \text{ lbs}$$

23. If your usual weight is 180 lbs, you should start a diet.

25. $25°C + 273° = 298 \text{ K}$

and

$1.8(25°) + 32° = 77°F$

27. a. $1.8(100°) + 32° = 212°F$

b. $\dfrac{(65° - 32°)}{1.8} = 18°C$

c. $\dfrac{(100° - 32°)}{1.8} = 38°C$

d. $5° + 273° = 278 \text{ K}$

e. $\dfrac{(50. - 32)}{1.8} + 273 = 283 \text{ K}$

29. $\dfrac{(102° - 32°)}{1.8} = 39°C$ This temperature is not yet above 40°C so you will not need to call the doctor in this case.

31. $1.003 \dfrac{g}{mL} \times \dfrac{10^3 \text{ mg}}{g} \times \dfrac{10^3 \text{ mL}}{L} \times \dfrac{1 L}{10 \text{ dL}} = 1.003 \times 10^5 \dfrac{mg}{dL}$

33. A: $\dfrac{55 \text{ g}}{15 \text{ cc}} = 3.7 \text{ g/cc}$

B: $\dfrac{295 \text{ g}}{120 \text{ cc}} = 2.5 \text{ g/cc}$ Therefore, object A is more dense as there is a larger number of grams per unit volume compared with object B.

Chapter 2

1. Not all atoms of one particular element are the same in mass due to the existence of isotopes. Atoms themselves are no longer believed to be indestructible because we now know about nuclear reactions. It is still true, however, that atoms are indestructible during *chemical* change. The rest of Dalton's ideas are also considered correct today.

3. $3.28 \times 10^{-23} \not{g} \times \dfrac{u}{1.66 \times 10^{-24} \not{g}} = 1.98 \times 10^{1}$ u or 19.8 u

$$\overbrace{\hspace{10cm}}^{\text{In the nucleus}}$$

5. a. $^{31}_{15}\text{P}$ 15 protons, 16 neutrons, 15 electrons

b. $^{40}_{20}\text{Ca}$ 20 protons, 20 neutrons, 20 electrons

c. $^{233}_{92}\text{U}$ 92 protons, 141 neutrons, 92 electrons

d. $^{33}_{16}\text{S}$ 16 protons, 17 neutrons, 16 electrons

7.

Element	Atomic number	Mass number	#of protons	#of neutrons
$^{4}_{2}\text{He}$	2	4	2	2
$^{56}_{26}\text{Fe}$	26	56	26	30
$^{238}_{92}\text{U}$	92	238	92	146
$^{75}_{33}\text{As}$	33	75	33	42

9.

Element	Atomic number	Mass number	# of protons	# of neutrons	# of electrons
Sr	38	90	38	52	38
C-13	6	13	6	7	6
Al	13	27	13	14	13

11. 19.99 u $\times$ 0.9092 = 18.17 u
    20.99 u $\times$ 0.0026 = 0.05 u
    21.99 u $\times$ 0.0882 = $\underline{\;1.94\;\text{u}}$
                20.16 u

13. The elements carbon, silicon, germanium, tin, and lead make up the IVA subgroup. The electron configurations of this subgroup all end with ns^2np^2, where n = the number of the period in which the element is found.

15. An atom of chlorine should be larger than an atom of fluorine. This is because chlorine has more electrons than fluorine and the electrons of chlorine are at higher energy levels, further, on the average, from the nucleus.

17. The Bohr theory predicted all circular paths, so the overall positions of the electron could be found on the surface of a sphere. The shape of a p orbital in quantum theory is not at all spherical. The boundary surface enclosing regions of probability of finding the electron in a p orbital is rather "dumbbell" in shape. See Figure 2.3.

19. Sodium's electron configuration of $1s^22s^22p^63s^1$ gives the distribution and arrangement of electrons within an atom of sodium. On the first energy level, there are two electrons in a spherically shaped s orbital. On the second energy level, there are two electrons in a spherically shaped s orbital and 6 electrons found in the three orientations of dumbbell-shaped p orbitals. On the third energy level there is one electron in a s orbital.

21. The IIIb subgroup contains:
    Sc $[Ar]4s^23d^1$
    Y $[Kr]5s^24d^1$
    La $[Xe]6s^25d^1$
    Ac $[Rn]7s^26d^1$

23. The maximum number of electrons associated with all orbitals within the second energy level is 8. This is the same number as the number of elements in the second period of the periodic table.

25. Element 107, when discovered, will fall under rhenium, the element with atomic number 75 on the periodic table. This will make element 107 a member of group VIIB.

27. Any element can be toxic if present in concentrations beyond the normal tolerances. Below these tolerances the effects may not be obvious or there may not be any effect.

Chapter 3

1. The noble gas family members each have electron configurations ending with complete s and p sublevels. (Helium, being a very light element has only $1s^2$ electrons.) The members of the halogen family can attain noble gas electron structures by sharing electrons in diatomic molecules.

3. a. Of the common diatomic molecules listed, O_2 contains a double bond and N_2 contains a triple bond.
   b. The gaseous diatomic molecules at room temperature are H_2, N_2, O_2, F_2, and Cl_2.

5. HF would have more polar bonds because the difference in electronegativity between hydrogen and fluorine is greater than is the case for HCl.

7. a. Each element obeys the "octet rule" as best it can. Hydrogen, being a very light element, can only attain a share in two electrons. Carbon has a share in 8 electrons.
   b. Yes. Hydrogen has the noble gas electronic structure of helium and carbon has the noble gas electronic structure of neon.

9. Water is an excellent solvent for dissolving salts and polar covalent substances because water is itself a polar molecule. The generalization is that "like dissolves like."

11. Noble gases do not commonly form bonds so they cannot be ranked in a comparative scale based on the relative ability to attract bonded electrons.

13. Oxygen, having a higher electronegativity than hydrogen, will have the greatest share of the bonding electrons.

15.

Electronegativity		Electronegativity difference	Bond type
a. K = 0.8	O = 3.5	2.7	Ionic
b. H = 2.1	O = 3.5	1.4	Polar covalent
c. C = 2.5	O = 3.5	1.0	Polar covalent
d. Ca = 1.0	Cl = 3.0	2.0	Ionic
e. Na = 0.9	H = 2.1	1.2	Polar covalent
f. N = 3.0	H = 2.1	0.9	Polar covalent

17. a. Sodium has a greater electronegativity than potassium.
    b. Sodium is a smaller atom than potassium. Any bonded electron will be closer to the sodium nucleus than to the potassium nucleus, making the electron held more strongly to sodium.

19. Aluminum has electron structure $[Ne]3s^23p^1$. It tends to lose three electrons, making its most usual valence $3+$. The formula of its chloride is $AlCl_3$, its sulfide Al_2S_3, and its nitride AlN.

21. $$\left[\begin{array}{c} H \\ H\!:\!\ddot{N}\!:\!H \\ \ddot{H} \end{array} \right]^+$$ is the Lewis structure for the ammonium ion.

One of these bonds is a coordinate covalent bond. The nitrogen atom brings five valence electrons. Each hydrogen normally brings one valence electron but the charge on the ion is $+$, meaning one hydrogen came without its electron.

23. a. ZnO b. BaS
    c. SO_3 d. H_2O
    e. Na_3P f. P_2O_5
    g. SnF_2 h. $HgCl_2$

25. a. $Ba(MnO_4)_2$ b. $Al_2(CO_3)_3$
    c. $AgClO_3$ d. $Na_2Cr_2O_7$
    e. Li_2SO_3 f. $Sn(OH)_2$
    g. $PbSO_4$ h. $(NH_4)_3PO_4$

Chapter 4

1. a. 3 atoms of sodium, 1 atom of phosphorus, 4 atoms of oxygen.
   b. 1 atom of calcium, 4 atoms of carbon, 6 atoms of hydrogen, 4 atoms of oxygen

   c. 2 atoms of hydrogen, 1 atom of sulfur, 4 atoms of oxygen

   d. 3 atoms of nitrogen, 12 atoms of hydrogen, 1 atom of arsenic, 4 atoms of oxygen

   e. 2 atoms of carbon, 6 atoms of hydrogen

   f. 12 atoms of carbon, 22 atoms of hydrogen, 11 atoms of oxygen

3. a. 18.0 g   b. 159.6 g
   c. 241.8 g   d. 60.0 g
   e. 183.1   f. 153.1

5. $1 \text{ g Na} \times \dfrac{1 \text{ mol Na}}{23.0 \text{ g Na}} \times \dfrac{6.02 \times 10^{23} \text{ atoms Na}}{1 \text{ mol Na}} = 3 \times 10^{22} \text{ atoms Na}$

   $1 \text{ g He} \times \dfrac{1 \text{ mol He}}{4.0 \text{ g He}} \times \dfrac{6.02 \times 10^{23} \text{ atoms He}}{1 \text{ mol He}} = 2 \times 10^{23} \text{ atoms He}$

There is a larger number of atoms in 1 g of He than in 1 g of Na.

7. $38 \text{ g SO}_2 \times \dfrac{1 \text{ mol SO}_2}{6.41 \text{ g SO}_2} \times \dfrac{6.02 \times 10^{23} \text{ molecules SO}_2}{1 \text{ mol SO}_2} = 3.6 \times 10^{23} \text{ molecules}$

9. a. $5.5 \text{ g Ca(OH)}_2 \times \dfrac{1 \text{ mol Ca(OH)}_2}{74.1 \text{ g Ca(OH)}_2} = 7.4 \times 10^{-2} \text{ mol Ca(OH)}_2$ (0.074 mol)

   b. $825 \text{ g C}_{12}\text{H}_{22}\text{O}_{11} \times \dfrac{1 \text{ mol C}_{12}\text{H}_{22}\text{O}_{11}}{342 \text{ g C}_{12}\text{H}_{22}\text{O}_{11}} = 2.41 \text{ mol C}_{12}\text{H}_{22}\text{O}_{11}$

   c. $0.44 \text{ g SO}_2 \times \dfrac{1 \text{ mol SO}_2}{64.1 \text{ g SO}_2} = 6.9 \times 10^{-3} \text{ mol SO}_2$ (0.0069 mol)

   d. $2.25 \text{ g Ca(HCO}_3)_2 \times \dfrac{1 \text{ mol Ca(HCO}_3)_2}{162.1 \text{ g Ca(HCO}_3)_2} = 1.39 \times 10^{-2} \text{ mol Ca(HCO}_3)_2$
                                             (0.0139 mol)

   e. $4.0 \times 10^{14} \text{ g NaOH} \times \dfrac{1 \text{ mol HaOH}}{40.0 \text{ g NaOH}} = 1.0 \times 10^{13} \text{ mol NaOH}$

   f. $7.7 \times 10^{-3} \text{ g H}_2\text{SO}_4 \times \dfrac{1 \text{ mol H}_2\text{SO}_4}{98.1 \text{ g H}_2\text{SO}_4} = 7.8 \times 10^{-5} \text{ mol H}_2\text{SO}_4$

11. a. $1.00 \text{ g NaOH} \times \dfrac{1 \text{ mol NaOH}}{40.01 \text{ g NaOH}} \times \dfrac{6.022 \times 10^{23} \text{ formula units NaOH}}{1 \text{ mol NaOH}} =$
      1.51×10^{22} formula units NaOH

   b. $3 \text{ mol NaOH} \times \dfrac{6.02 \times 10^{23} \text{ formula units NaOH}}{1 \text{ mol NaOH}} = 2 \times 10^{24}$ formula
                                              units NaOH

   c. $155 \text{ kg NaOH} \times \dfrac{10^3 \text{ g}}{1 \text{ kg}} \times \dfrac{1 \text{ mol NaOH}}{40.01 \text{ g NaOH}} \times \dfrac{6.022 \times 10^{23} \text{ f.u. NaOH}}{1 \text{ mol NaOH}}$
                    $= 2.33 \times 10^{27}$ formula units NaOH

   d. $5.6 \times 10^2 \text{ mg NaOH} \times \dfrac{1 \text{ g}}{10^3 \text{ mg}} \times \dfrac{1 \text{ mol NaOH}}{40.0 \text{ g NaOH}} \times \dfrac{6.02 \times 10^{23} \text{ f.u. NaOH}}{1 \text{ mol NaOH}}$
                          $= 8.4 \times 10^{21}$ formula units NaOH

13. A formula equation shows not only the qualitative nature of the reactants and products, it also gives quantitative information. Correct formulas as well as correct mole ratios are shown in the balanced chemical equation.

15. a. $2 \text{ Al(s)} + 6 \text{ HCl(g)} \rightarrow 2 \text{ AlCl}_3\text{(s)} + 3 \text{ H}_2\text{(g)}$
    b. $2 \text{ Na(s)} + 2 \text{ H}_2\text{O} \rightarrow \text{H}_2\text{(g)} + 2 \text{ NaOH(aq)}$
    c. $2 \text{ SO}_2\text{(g)} + \text{O}_2\text{(g)} \rightarrow 2 \text{ SO}_3\text{(g)}$
    d. $\text{CuCO}_3\text{(s)} \rightarrow \text{CuO(s)} + \text{CO}_2\text{(g)}$
    e. $\text{H}_3\text{PO}_4 + 3 \text{ NH}_4\text{OH} \rightarrow 3 \text{ H}_2\text{O} + (\text{NH}_4)_3\text{PO}_4$
    f. $\text{Na}_2\text{SO}_3 + 2 \text{ HCl} \rightarrow 2 \text{ NaCl} + \text{H}_2\text{O} + \text{SO}_2$

17. a. $\underline{2} \text{ PbO}_2 \rightarrow \underline{2} \text{ PbO} + \text{O}_2$
    b. $\underline{2} \text{ Al} + \underline{3} \text{ Br}_2 \rightarrow \underline{2} \text{ AlBr}_3$
    c. $\text{Cr}_2\text{O}_3 + \underline{3} \text{ H}_2 \rightarrow \underline{2} \text{ Cr} + \underline{3} \text{ H}_2\text{O}$
    d. $\underline{2} \text{ KClO}_3 \rightarrow \underline{2} \text{ KCl} + \underline{3} \text{ O}_2$
    e. $\underline{2} \text{ C}_4\text{H}_{10} + \underline{13} \text{ O}_2 \rightarrow \underline{8} \text{ CO}_2 + \underline{10} \text{ H}_2\text{O}$

19. $2 \text{ FeCl}_3\text{(aq)} + 3 \text{ Ba(OH)}_2\text{(aq)} \rightarrow 3 \text{ BaCl}_2\text{(aq)} + 2 \text{ Fe(OH)}_3\text{(s)}$

21. $2 \text{ Na} + \text{Cl}_2 \rightarrow 2 \text{ NaCl}$

$$1.55 \text{ g Na} \times \frac{1 \text{ mol Na}}{22.99 \text{ g Na}} \times \frac{2 \text{ mol NaCl}}{2 \text{ mol Na}} \times \frac{58.45 \text{ g NaCl}}{1 \text{ mol NaCl}} = 3.94 \text{ g NaCl}$$

23. $2 \text{ AgNO}_3\text{(aq)} + \text{Cu(s)} \rightarrow \text{Cu(NO}_3)_2\text{(aq)} + 2 \text{ Ag(s)}$

$$5.00 \text{ g Ag} \times \frac{1 \text{ mol Ag}}{107.9 \text{ g Ag}} \times \frac{1 \text{ mol Cu}}{2 \text{ mol Ag}} \times \frac{6.355 \text{ g Cu}}{1 \text{ mol Cu}} = 1.47 \text{ g Cu}$$

25. $3 \text{ Cu} + 8 \text{ HNO}_3 \rightarrow 3 \text{ Cu(NO}_3)_2 + 2 \text{ NO} + 4 \text{ H}_2\text{O}$

$$25 \text{ g NO} \times \frac{1 \text{ mol NO}}{30.0 \text{ g NO}} \times \frac{3 \text{ mol Cu}}{2 \text{ mol NO}} \times \frac{63.5 \text{ g Cu}}{1 \text{ mol Cu}} = 79 \text{ g Cu}$$

Chapter 5

1. a. O_2
   b. Fe_2O_3
   c. N_2
   d. O_2

3. a. decomposition, redox
   b. synthesis, redox
   c. single replacement, redox
   d. decomposition, redox
   e. redox (this is also a combustion reaction)

5. Nothing will happen chemically if hydrogen gas is bubbled into a zinc chloride solution. Since it is given that zinc is more active than hydrogen, hydrogen gas will be unable to replace the zinc from zinc chloride. In addition, hydrogen gas is not very soluble and so will just escape from solution.

7. a. $\underline{3}$ $AgNO_3$ + $H_3PO_4 \rightarrow Ag_3PO_4(s)$ + $\underline{3}$ HNO_3
   double replacement reaction
   b. $\underline{2}$ $AgNO_3$ + $Mg \rightarrow Mg(NO_3)_2$ + $\underline{2}$ Ag
   single replacement reaction

9. $CaCO_3(s)$ + $H_2SO_4(aq) \rightarrow CaSO_4(s)$ + $[H_2CO_3](aq)$
$$\downarrow H_2O(l) + CO_2(g)$$

11. An increase in randomness means an increase in entropy. Physical changes:
    a. If water changes from liquid to gaseous state, the molecules will be much less orderly, far more random in the gaseous state. There will be an increase in entropy expected.
    b. Solid water will contain more orderly molecular arrangement than will liquid water and so a decrease in entropy is expected here.
    Chemical changes:
    a. If solid wood burns, several gaseous products are formed. This will mean an increase in entropy.
    b. Iron rusting is a synthesis reaction. Iron and oxygen combine to form an oxide of iron, a more complex, orderly arrangement of the atoms. A decrease in entropy is expected.

13. $2 AgCl(s) \rightarrow 2 Ag(s) + Cl_2(g)$
    $2 AgI(s) \rightarrow 2 Ag(s) + I_2(s)$
    A greater entropy change is expected in the first decomposition. A gas is formed, Cl_2. The increased randomness associated with a gas forming will be greater than for the solid I_2.

15. The erm "endergonic" is a more general expression for the requirement that energy is required for a reaction to take place. "Endothermic" specifies that the energy will be provided in the form of heat.

17. $250 \cancel{g} \times \dfrac{20. \cancel{kJ}}{\cancel{g}} \times \dfrac{1 \text{ kcal}}{4.184 \cancel{kJ}} = 1.2 \times 10^3 \text{ kcal}$

19.

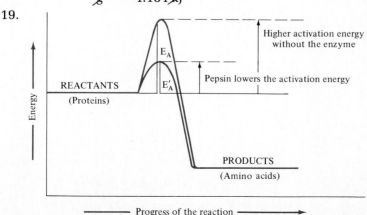

Chapter 6

1. Gases are easily compressed because they are largely empty space. The molecules making up the gas are very small and very far apart, making compression far easier for a gas than for either a liquid or solid.

3. "Absolute zero" means the theoretical limit to molecular motion. If there is no molecular motion, the temperature is at absolute zero and the volume of a gas becomes zero.

5. a. Charles' law
   b. $p \propto n$ (no specific name)
   c. Gay-Lussac's law
   d. Boyle's law
   e. Charles' law
   f. Henry's law
   g. Henry's law

7. If the temperature remains constant, pressure will have to increase to cause the volume to decrease. As this happens, a *real* gas will start to condense when the pressure becomes great enough for the real molecules to show enough intermolecular attractions to cause the condensation.

9. a. $2.45 \text{ atm} \times \dfrac{760. \text{ mm Hg}}{1.00 \text{ atm}} = 1.86 \times 10_3 \text{ mm Hg}$

   b. $1.00 \times 10^6 \text{ Pa} \times \dfrac{760. \text{ mm Hg}}{1.01 \times 10^5 \text{ Pa}} = 7.52 \times 10^3 \text{ mm Hg}$ which is greater than 780 mm Hg.

   c. $\dfrac{55 \text{ lbs}}{\text{sq inch}} \times \dfrac{1.01 \times 10^5 \text{ Pa}}{14.7 \text{ lbs/sq inch}} \times \dfrac{1 \text{ kPa}}{10^3 \text{ Pa}} = 3.78 \times 10^2 \text{ kPa}$ so yes.

   d. $125 \text{ mm Hg} + 760. \text{ mm Hg} = 885 \text{ mm Hg}$
      $82 \text{ mm Hg} + 760. \text{ mm Hg} = 842 \text{ mm Hg}$
      The blood pressure would be reported in atmospheres as:

   $$\dfrac{885 \text{ mm Hg} \times \dfrac{1 \text{ atm}}{760. \text{ mm Hg}}}{842 \text{ mm Hg} \times \dfrac{1 \text{ atm}}{760. \text{ mm Hg}}} = \dfrac{1.16 \text{ atm}}{1.11 \text{ atm}} = \dfrac{1.16}{1.11}$$

11. $25 \text{ L} \times \dfrac{760 \text{ mm Hg}}{1500 \text{ mm Hg}} = 13 \text{ L}$

13. The volume will decrease. The new volume will be:

   $5.0 \text{ L} \times \dfrac{1.0 \text{ atm}}{15 \text{ atm}} = 0.33 \text{ L}$

15. $25°C + 273 = 298 \text{ K}$

   $298 \text{ K} \times \dfrac{9.0 \text{ L}}{18 \text{ L}} = 149 \text{ K}$ which is $-124°C$.

17. $\dfrac{(93°F - 32°)}{1.8} = \dfrac{(61)}{1.8} = 34°C$ which is cooler than $38°C$.
   It is warmer in Tijuana than in San Diego on that day.

19. $50.\text{ mL } \times \dfrac{323\ \cancel{K}}{293\ \cancel{K}} \times \dfrac{500.\ \cancel{\text{mm Hg}}}{700.\ \cancel{\text{mmHg}}} = 39\text{ mL}$

21. $1.8(34°C) + 32 = 93°F$ This is a very low temperature for a human patient and so you should be concerned.

23. $P_T = P_{N_2} + P_{O_2} + P_{H_2}$

    $P_T = 27\text{ mm Hg} + 300.\text{ mm Hg} + 65\text{ mm Hg}$

    $P_T = 392\text{ mm Hg}$

25. The major component of our atmosphere is nitrogen gas. It makes up 78.084% by volume of our atmosphere. Because the partial pressures of gases such as O_2 and CO_2 are different in exhaled breath from inhaled breath, the partial pressure of nitrogen gas is probably different also.

27. The partial pressure of oxygen is higher in the lungs than it is in the active cells in tissues throughout the body. Arterial blood transports oxygen and distributes it throughout the body.

29. Compressed air used by divers must contain a lower percentage of oxygen than normal air. This is because the increased pressure experienced by the divers will maintain a "normal" partial pressure of oxygen, 159 mm Hg. The rest of the compressed air is principally helium.

31. The concentration of a gas dissolved in a fluid increases with pressure. If the partial pressure of the helium decreases, its solubility in the blood would also decrease.

33. "Artificial blood" is a synthetic material capable of carrying oxygen to tissues. It is an organic mixture of perfluorocarbons. At the present time, it is capable only of the function of carrying oxygen and cannot be considered a true replacement for all of the crucial functions of blood.

Chapter 7

1. A *solution* is a dispersal of one substance into another to yield a homogeneous, single-phase preparation. The substance that is dispersed is termed the *solute*. In *nonaqueous* solutions, liquids other than water are used as solvents ($CHCl_3$, CCl_4, alcohol, ether, toluene, liquid ammonia, etc.) *Otic* solutions are intended for use in the ear. *Polar solvents* have high dipole moments (see Table 7.1). Miscibility is the measure of one liquid's ability to mix with another, forming a single-phase mixture. *Cationic surfactants* are surface-active agents carrying a positive charge.

3. In both *spirits* and *elixirs*, alcohol is the primary solvent. But spirits are solutions of *volatile* substances, usually not the case with elixirs. All *syrups* are aqueous solutions containing sugar, plus active ingredient (except Simple Syrup).

5. Polar: a, b, d, g.

7. Blood gases would come out of solution and be exhaled—as the external pressure drops.

9. Soluble in water: a, b, c, g.

11. All the statements are descriptive.

13. Mass out 9 g of NaCl. Dissolve in some water and dilute with mixing to 1 liter.

15. 500 ml.

17. All 5% w/v solutions contain 5 g of solute per 100 ml of soln; the nature of the solute is not a factor.

19. a. 1/120   b. 3/2
    c. 1:100   d. 1:2
    e. $\frac{1}{3}$%   f. 35 ml of 0.12 M
    g. 0.25%   h. 1:7500

21. 35:1000.

23. 0.01%.

25. Use 0.4 ml of the 1:1000 epinephrine soln.

27. 1:5000 = 0.02%. Thus dilute it 600 times. Ans: Enough water to make 6 liters.

29. Enough to dilute the 400 ml of ethanol to a total volume of 1000 ml. Exact volume of water cannot be specified since water and alcohol contract.

31. Use 20 ml (to two significant figures). Use Equation 7-2, and solve for vol.

33. (a) False it is once, $\frac{1}{2}$, $\frac{1}{3}$, etc.; (b) true; (c) true.

35. Hypertonic.

37. The solution, at 1.0% w/v, is hypertonic with blood.

39. An osmotic membrane allows passage of water only. A dialyzing membrane allows passage of electrolytes generally plus water (but not colloids or large molecular weight compounds).

41. Zephiran, Bactine, etc., lower the surface tension of water, increasing its ability to wet (permeate) objects.

43. Cationic.

45. Carbon dioxide is added to oxygen in order to stimulate CO_2 detectors in the brain and thus stimulate respiration.

47. (a) conserve; (b) decrease.

49. By osmosis, water was drawn out of the body, dehydrating and preserving it.

Chapter 8

1. a. acid   b. neither   c. base   d. acid   e. neither

3. $Zn + 2\,HCl \rightarrow ZnCl_2 + H_2(g)$
   $Zn + 2\,NaOH \rightarrow Na_2ZnO_2 + H_2(g)$

5. a. non-electrolyte   b. non-electrolyte
   c. strong electrolyte   d. weak electrolyte
   e. strong electrolyte   f. strong electrolyte

7. a. A classic acid produces hydrogen ions in water solution. An example is hydrochloric acid.
   b. A Bronsted-Lowry acid is a proton donor. An example is hydrogen chloride gas if placed into water. The proton from HCl will be donated to the water which in this case acts as a Bronsted-Lowry base.
   c. Water can act as a Bronsted-Lowry acid under the right circumstances. For example, water can donate a proton to ammonia gas.

9. The term "acidic" is an adjective used to describe all solutions with properties associated with any acid. "Acetic" is the name of a particular acid, acetic acid, $HC_2H_3O_2$.

11. $MgCO_3(s) + 2\,HCl(aq) \rightarrow MgCl_2(aq) + H_2O(aq) + CO_2(g)$
    The magnesium carbonate in *Tums* reacts with "excess stomach acid," forming magnesium chloride in solution, water, and carbon dioxide.

13. a. [0.001] HCl means $[H^+] = 1 \times 10^{-3}$ M because HCl is a strong acid.
    $$[OH^-] = \frac{1 \times 10^{-14}}{1 \times 10^{-3}} = 1 \times 10^{-11} \text{ M}$$
    b. Acetic acid is a weak acid and therefore even though its molar concentration is 0.001 M, it does not release all of its hydrogen ions into solution. If you do not know the hydrogen ion concentration, you cannot find the hydroxide ion concentration.

15. In order of increasing acidity, the solutions are milk of magnesia, blood, lemon juice, and stomach acid (gastric juice).

17. If the pH of a sample of orange juice is 4, then the hydrogen ion concentration is 1×10^{-4} M. This means the hydroxide ion concentration is 1×10^{-10} M.

19. A solution with pH = 5.0 has a hydrogen ion concentration of 1.0×10^{-5} M. This is not as high a concentration as 2.5×10^{-5} M, so the second solution is more acidic.

21. $[H^+] = 1.0 \times 10^{-4}$ M means a pH of 4.0. The colors expected in the different indicators are:

a. green   b. yellow   c. colorless

23. Moles of hydrogen ion = moles of hydroxide ion

$$\left(X \frac{\text{moles}}{\text{L}} \text{HCl}\right)(0.0100 \text{ L}) = \left(0.0250 \frac{\text{moles}}{\text{L}} \text{NaOH}\right)(0.0256 \text{ L})$$

$$X = \frac{(0.0250)(0.0256)}{(0.0100)}$$

$$X = 0.0640 \frac{\text{moles HCl}}{\text{L}}$$

The HCl is 0.0640 M.

25. $HC_2H_3O_2 + NaOH \rightarrow NaC_2H_3O_2 + H_2O$

$$0.500 \frac{\text{mol NaOH}}{\cancel{L}} \times 0.0200 \cancel{L} \times \frac{1 \text{ mol } HC_2H_3O_2}{1 \text{ mol NaOH}}$$

$$\times \frac{1}{0.0100 \text{ L } HC_2H_3O_2} = 1.00 \text{ M } HC_2H_3O_2$$

If the acetic acid is 1.00 M, to find the percent (by weight) in vinegar, you will need to convert to grams of vinegar in comparison to the total number of grams of solution present. Assume the approximate density of the solution is 1 gram per ml.

$$\frac{1.00 \text{ mol } HC_2H_3O_2}{\text{L solution}} \times \frac{60.05 \text{ g } HC_2H_3O_2}{\text{mol } HC_2H_3O_2} \times \frac{1 \text{ L solution}}{10^3 \text{ grams solution}} \times 100$$

$$= 6\% \text{ acetic acid}$$

27a.
$$\overset{\displaystyle \quad \text{conjugate pair} \quad}{CO_3^{2-}(aq) + H_2O(l) \rightleftarrows HCO_3^-(aq) + OH^-(aq)}$$
     base   acid   acid   base

conjugate pair

27b.
$$\overset{\displaystyle \quad \text{conjugate pair} \quad}{NH_3(l) + NH_3(l) \rightleftarrows NH_4^+(l) + NH_2^-(l)}$$
     acid   base   acid   base

conjugate pair

Note that $NH_3(l)$ acts as both an acid and a base in this reaction. The NH_3 is amphoteric in liquid state. (This is not the case in aqueous solution.)

29. a. neutral   b. basic   c. acidic   d. basic

31. The purpose of a buffer solution is to hold the pH of a solution constant

or nearly constant, no matter what acidic or basic stress is placed on the solution.

33. Breathing out into a confined volume such as a paper bag would increase the concentration of carbon dioxide in the bag. When the patient inhaled, the partial pressure of carbon dioxide coming into the system would be higher than normal. Increasing the partial pressure of CO_2 increases dissolved CO_2 in the blood, forming additional H_2CO_3 (carbonic acid) to combat the condition of alkalosis.

Chapter 9

1. Organic compounds are characterized by lower melting point, volatility, tendency to form covalent compounds, capacity for absorbing radiant energy, and the potential for complexity in structure.

3. In a tetrahedral carbon atom, the four sp^3 hybrid bond orbitals point to the four corners of a regular tetrahedron; bond angles are regular and are 109°28′.

5. In n-pentane: 6-1°, 6-2°. In isobutane: 9-1°, 1-3°. In isopropyl bromide: 6-1°, 1-2°. In t-butyl bormide: 9-1°.

7. The general formula for alkenes is: C_nH_{2n}.

9. The same compound written twice.

11. Mineral oil is a mixture of liquid C_{18}–C_{24} hydrocarbons obtained from petroleum.

13. Four: the 1,1-, 1,2-, 1,3-, and 2,2-.

15. sec-Butyl alcohol, 1-chloro-2-methylbutane. (Demerol, no, because it shows a plane of symmetry.)

17. C_2H_2, cyclopentene, oleic acid, C_6H_6.

19. a. $CH_2{=}CH_2 + HCl \rightarrow CH_3{-}CH_2Cl$

b.
$$\begin{array}{c} CH_3 \\ \diagdown \\ \end{array} \begin{array}{c} CH_3 \\ \diagup \\ \end{array}$$
$$\underset{H}{\overset{}{}}\,C{=}C\,\underset{H}{\overset{}{}} + H_2 \xrightarrow{\text{catalyst}} CH_3CH_2CH_2CH_3$$

c. $x\ CH_2{=}CHCl \xrightarrow{\text{catalyst}} {+}CH_2{-}CHCl{+}_x$

21. Aromaticity requires planarity, a 5- or 6-membered ring, and six pi electrons.

23. Anisole, toluene, aspirin, acetaminophen, DDT, acetophenone, Valium, Demerol, diphenylmethane, a detergent.

25. a. [structure: furan ring with O]   b. [structure: pyran ring with O]   c. $CH_2{=}CH_2$ d. $CH_3CH{=}CH_2$

e. $\underset{CH_3}{\overset{H}{}}C{=}C\underset{H}{\overset{C_2H_5}{}}$ f. $CH_3C{\equiv}C{-}C_3H_7$ g. [benzene ring]   h. $(CH_3)_2CH{=}CH_2$

Chapter 10

1. a. ether, b. alcohol, carboxylic acid, c. amine, d. ketone, e. ester, f. aldehyde, g. alcohol, h. carboxylic acid, ester, i. ketone, j. carboxylic acid, k. amine salt, l. alcohol, ether, amine, phenol, m. phenol, n. carboxylic acid, ester, o. carboxylic acid.

3. A-12, B1,6, C-10, D-2, E-2,4, F-5, G-8, H-4,9, I-4,6, J-4,7, K-8, L-11, M-3,11.

5. Fermentation, and hydration of ethylene. See text for details.

7. a. $2\ C_2H_5OH \xrightarrow{\text{heat, } H_2SO_4} C_2H_5{-}O{-}C_2H_5 + H_2O$
   b. $C_2H_5OH \xrightarrow{K_2Cr_2O_7} CH_3CHO$
   c. $C_2H_5OH \xrightarrow{KMnO_4} CH_3COOH$

9. a. 4-methyl-2-hexanol, b. 7-chloroheptanoic acid, c. potassium propionate, d. 2,4-dimethyl-3-pentanol, e. methyl butanoate, f. cyclohexanol

11. o-Hydroxybenzoic acid (salicylic acid)

13.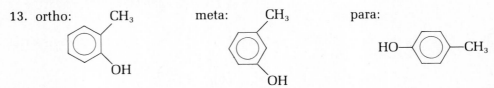

ortho: (2-methylphenol, CH_3 / OH)   meta: (3-methylphenol, CH_3 / OH)   para: ($HO{-}$benzene$-CH_3$)

15. All aldehydes have a hydrogen atom on the carbonyl carbon: R—CHO. Ketones, in contrast, have two Carbon-containing groups: e.g., Ar—CO—R. Both have a carbonyl group.

17. a. $CH_3CH_2COOH + NaHCO_{3(aq)} \rightarrow CH_3CH_2COONa_{(aq)} + H_2O + CO_2$
   b. $CH_3COOH + CH_3OH - CH_3COOH_3 + H_2O$
   c. $CH_3CHOHC_2H_5 + \frac{1}{2}O_2 \rightarrow CH_3COC_2H_5 + H_2O$
   d. $CH_3CH_2CH_2CH_2OH + \frac{1}{2}O_2 \rightarrow C_3H_7CHO + H_2O$
   e. $(CH_3)_3N + H_2O \rightarrow (CH_3)_3NHOH$
   f. $(CH_3)_3N + HNO_3 \rightarrow (CH_3)_3NH^+ + NO_3^-$
   g. $CH_3(CH_2)_8COOCH_3 + H_2O \rightarrow CH_3(CH_2)_8COOH + CH_3OH$

19. a. Both have a tertiary amine, b. methadone is also a ketone, but heroin is a phenolic ester, an alcoholic ester, and an ether, c. because cross tolerance

develops between methadone and heroin. Methadone can prevent withdrawal symptoms from heroin.

21. a. the hydrochloride salt is

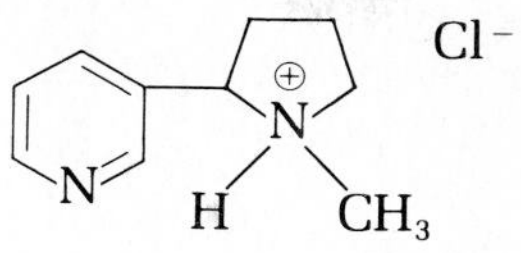

   b. the quaternary salt is

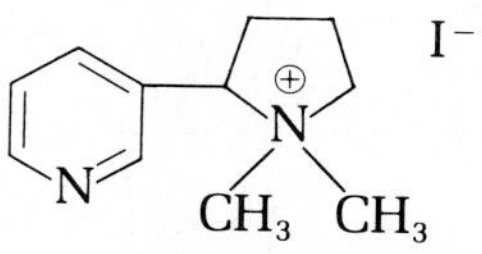

   c. the free base is the non-protonated amine form in which the N atom retains sole control of its unshared pair of electrons.

23. Being so similar structurally to histamine, the antihistamines preferentially occupy the histamine receptor sites. This prevents histamine from exerting its pharmacological effects.

25. $CH_2{=}CH_2 + H_2O \xrightarrow{H_2SO_4} CH_3CH_2OH \xrightarrow{KMnO_4} CH_3COOH$

   Then, $CH_3COOH + CH_3CH_2OH \xrightarrow[\text{heat}]{\text{acid}} CH_3COOC_2H_5 + H_2O$

27. Isoflurane is a new general inhalation anesthetic, much like several discussed in this chapter.

Chapter 11

1. a. $CH_3CH_2CONH_2$
   b. $C_6H_5CONH_2$
   c.

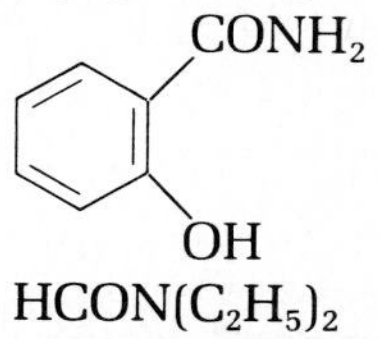

   d. $HCON(C_2H_5)_2$
   e. $H_2NCH_2CONHCH_2COOH$
   f. $CH_3(CH_2)_{16}CONH_2$

3. Polymers are high molecular weight, long chain macromolecules that form when many molecules join together in a head-to-tail fashion through addition reactions or condensation reactions. Polymers are made up of chemical groupings that recur over and over again (the "repeating unit"). Monomers are the molecules that join together to form the polymers.

5. Hexamethylenediamine is unambiguous, for there is only *one* way that six —CH_2— groups and two amino groups can be joined and still follow the rules of organic chemistry.

7.

$$\text{C}_6\text{H}_5-\text{CH}_2-\underset{\underset{\text{NH}_2}{|}}{\text{CH}}-\boxed{\text{CONH}}-\underset{\underset{\underset{\underset{\text{CH}_3\ \ \text{CH}_3}{}}{\text{CH}}}{\text{CH}_2}}{\text{CH}}-\text{COOH} \quad \text{and} \quad \text{H}_2\text{N}-\text{CH}-\boxed{\text{CONH}}-\text{CH}-\text{COOH}$$

with side chains $-\text{CH}_2-\text{CH}(\text{CH}_3)_2$ and $-\text{CH}_2-\text{C}_6\text{H}_5$

9. Trp-Lys-Ala-Glu-Leu.

11. Tryptophan, as found in proteins, is optically active. It and all 20 amino acids (except glycine) have a chiral center.

13. (a) Valine, leucine, isoleucine, (b) phenylalanine, tryptophan, tyrosine, histidine, (c) lysine, arginine, histidine, (d) aspartic acid, glutamic acid, (e) isoleucine, threonine.

15. Muscle tissue. The primary function of protein is body building and maintenance.

17. The R group can confer either hydrophobic or hydrophilic properties; it can have a profound effect upon hydrogen bonding (with consequent regulation of 3-D geometry); it can link peptide chains via disulfide bridges; it can regulate electrical charge on protein (via $-\text{NH}_3^+$ and/or $-\text{COO}^-$ groups); it can modify chemical behavior in acid/base, redox, or other types of reactions; it can act as a source of methyl groups. Non-polar R groups can provide a site for solubilizing non-polar drugs or agents (like dissolves like).

19. $$\text{CH}_3-\underset{\underset{\text{NH}_2}{|}}{\text{CH}}-\text{COO}^- \xleftarrow{\ +\ \text{OH}^-\ } \text{CH}_3-\underset{\underset{^+\text{NH}_3}{|}}{\text{CH}}-\text{COO}^- \xrightarrow{\ +\ \text{H}^+\ } \text{CH}_3-\underset{\underset{^+\text{NH}_3}{|}}{\text{CH}}-\text{COOH}$$

21. (a) Essential amino acids are those that the body cannot synthesize at all or at a rate fast enough to meet body demands, (b) the isoelectric point is the pH at which the amino acid exists in its dipolar ion form (i.e., maximum charge exists on the molecule, (c) in denaturation of a protein, its ternary structure is altered, (d) the epiphyses are the ends of the long bones, (e) HDL stands for high density lipoprotein: a good mobilizer of cholesterol, (f) enkephalins are small peptides secreted naturally in the brain; they bind to the same receptor sites to which opiates bind, affording pain relief, (g) proteins are nitrogenous organic compounds of high molecular weight, synthesized by plants and animals, that upon hydrolysis yield amino acids.

23. The alpha-helix represents the secondary structure; pleated sheets represent secondary structures. Receptor sites are ternary structures; hemoglobin represents quaternary structure. Insulin's amino acid sequence = 1° structure.

25. Hydrogen bonding in α-helix has its origin in $\diagdown$C=O and —NH— groups.

Hydrogen atoms form the bridge to carbonyl groups, to oxygen atoms, and to nitrogen atoms:

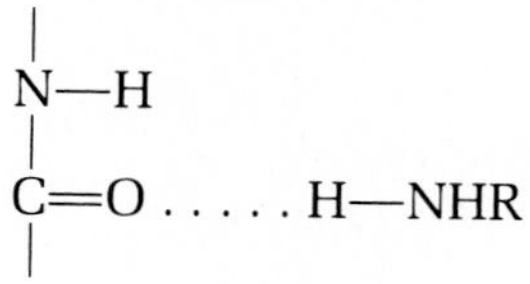

$$
\begin{array}{l}
\mid \\
\text{N—H} \\
\mid \\
\text{C=O} \ldots\ldots \text{H—NHR} \\
\mid
\end{array}
$$

27. (a) Contraction of the uterus, milk ejection, (b) retention of body water (anti-diuresis), (c) increased blood pressure, (d) establishes the anti-viral state in tissues, (e) luteinizing hormone releasing factor promotes release of LH, (f) insulin promotes tissue uptake and metabolism of glucose.

29. $1 \times 10^{-3}\, g \times \dfrac{1\ mole}{59 \times 10^{6}\, g} \times \dfrac{6.0 \times 10^{23}}{1\ mole}$ molecules $= 1 \times 10^{13}$ molecules

31. 60,000 ($0.0032x = 32$, $x = 10{,}000$, the minimum molecular weight if one atom of sulfur existed per molecule of protein. Since we know that six sulfur atoms actually exist per molecule, the molec. wt. must equal 60,000.)

Chapter 12

1. (a) Carbohydrates are polydroxy aldehydes or polyhydroxy ketones, or compounds that can be hydrolyzed to them, (b) monosaccharides are carbohydrates that cannot be hydrolyzed to simpler molecules, (c) disaccharides are carbohydrates that upon hydrolysis yield two monosaccharides, (d) polysaccharides are carbohydrates that upon hydrolysis yield many monosaccharides.

3. One is a five-carbon aldehyde, the other a six-carbon ketone.

5. (a) A ketopentose, (b) an aldotetrose, (c) an aldohexose, (d) aldotriose.

7. 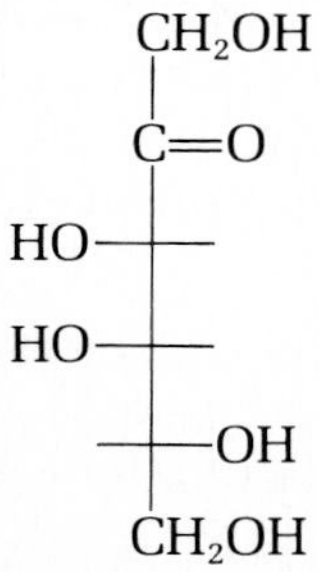
$$
\begin{array}{l}
\text{CH}_2\text{OH} \\
\mid \\
\text{C=O} \\
\text{HO——} \\
\text{HO——} \\
\text{——OH} \\
\text{CH}_2\text{OH}
\end{array}
$$

9.

CH_2OH

HO   OH

OH   OH

11. b and e

13. Reducing sugars are those sugars capable of reducing Ag^+ in Tollens' solution or Cu^{2+} ion in Benedict's solution. They have a free or potentially free aldehyde group. Yes, the sugar in question 12 is a reducing sugar.

15. Someone told Fred that the cellulose in plants is made up of D-glucose units. But they forgot to tell Fred that humans don't have the enzymes that catalyze the hydrolysis of cellulose. In the diet, however, cellulose provides needed bulk for the G–I tract.

17. The hormone insulin, released to the blood stream in the pancreas, promotes tissue uptake and metabolism of glucose. Hence insulin acts to lower blood sugar levels. Glucagon is the pancreatic hormone that promotes glycogen breakdown (glucogenolysis) with resultant increase in blood sugar levels.

19. Thyroxine is a general metabolic stimulant. It increases oxygen consumption of many body tissues. Naturally, the body will require more glucose under such conditions, and glycogenolysis provides it.

21. Normal blood sugar levels are: 70–100 mg%. In typical mild hypoglycemia, blood levels are 50–60 mg%. Normal fasting venous blood levels of glucose are 60–80 mg%. Fasting arterial blood levels are about 80–100 mg%.

23. Diabetes mellitus is characterized by abnormal carbohydrate and fat metabolism, by excretion of large volumes of urine, by presence of glucose in the urine, polydipsia, wounds that are slow to heal, weakness, loss of weight and hunger.

25. Ketone bodies—acetone, acetoacetic acid, beta-hydroxybutyric acid—occur in the blood and sometimes on the breath of severe diabetics. They are always present to some extent, but they greatly multiply from greatly increased fat metabolism in the liver caused by tissue failure to take up and metabolize glucose.

27. No. Oral hypoglycemic drugs act to stimulate the pancreas to secrete more insulin. That is why they work only in diabetics who still retain integrity of pancreatic beta cells. These patients usually fall into the maturity-onset category.

29. The citric acid cycle (Krebs cycle) operates aerobically in the mitochondria. A net of 30 ATPs are gained per glucose molecule.

31. The sources of the electrons are the covalent bonds to H atoms in pyruvate (and acetate). These electrons are transported through a series of enzymes (that can exist reversibly in reduced and oxidized forms) to molecular oxygen, to form water.

33. Better diagnosis, more thorough reporting; medical advances are saving the lives of diabetics so that they can live and beget (sometimes diabetic) children; intermarriage, spreading the deficient DNA in the population. Further, Americans are living longer, permitting maturity-onset diabetes to manifest itself. Obesity and stress contribute to the etiology of the disease. If we are more obese and more stressed than formerly, we can expect to see more diabetes.

35. Injected insulin must be given a chance to be absorbed and distributed to body tissues so that when glucose levels rise after a meal, the hormonal mechanism is there to handle it. Apparently, the time required for washing and dressing coincides nicely with the time required for insulin distribution, for the researchers found neither too-high post-meal glucose levels nor too-low pre-meal glucose levels. In other words, the injected insulin did not act too soon nor too slowly.

Chapter 13

1. Fatty acids are alkanoic acids, typically of the C_4–C_{18} length. Fats and oils chemically are very similar, both being fatty acid esters of glycerol. Oils (liquid at room temperature), however, generally have a higher degree of unsaturation.

3.

5. Vegetable product highest in saturated fat is coconut oil. Animal product highest in saturated fat is butter. Safflower oil has highest percentage of linoleic acid.

7.

9. At body temperature (ca. 38°C) one desires a suppository to melt and release its active ingredient.

11. Facilitating slow absorption: c, d, g

13. Soaps possess a long, non-polar "tail" and a polar carboxylate anion "head." Fat micelles are stabilized (that is, prevented from coalescing) when the non-polar soap "tail" dissolves in the fat leaving the polar head outside of the micelle. This effectively gives each micelle a multinegative surface charge; these like-charged micelles repel each other, preventing coalescence.

15. Lipids in the blood (triacylglycerols, FFA, phospholipids and sterols such as cholesterol) are bound to blood proteins (albumin, alpha-, and beta-globulins). These lipid-protein complexes, called lipoproteins, vary in their content of lipid: those containing the most lipid and the least protein are least dense, and *vice versa*. Based on density, we have VLDL, LDL, and high density lipoprotein. Triacylglycerols are bound to VLDL, while HDL is best able to bind and transport cholesterol.

17. Normal blood cholesterol levels are in the range 120–220 mg%. An Arizona physician, conducting research in childhood hypercholesterolemia, told us he was working with a ten-year-old child with a 350 mg% cholesterol level.

19. In the dietary approach, one should avoid high cholesterol foods (brains, caviar, egg yolk, liver, kidney, oysters, butter, shrimp), and avoid saturated fats (coconut oil, lard).

21. a. CH_2—O—COR′

    CH—O—COR″

    CH_2—O—PO_3R‴

    R′ = R″ = fatty acid residue
    R‴ = an amino alcohol

    b. CH_2—O—COR′

    CH—O—COR″

    CH_2—O—$PO_3CH_2CH_2\overset{+}{N}(CH_3)_3$

    R′ = R″ = fatty acid residue

23. The scheme for this answer would show the enzyme-(lipase) catalyzed hydrolysis of dietary fat, occurring in the small intestine. It would also show transport of fat across intestinal membranes into the lymph system for distribution to the blood, and lipoprotein-directed transport of fat to the liver or to fat depots.

Chapter 14

1.

3. The two types of adrenal cortex hormones are (a) the glucocorticoids, and (b) the mineralocorticoids. For names, structures and actions, see Table 14-1.

5. Some of the corticosteroids (the glucocorticoids, especially) have powerful anti-inflammatory activity in acute and chronic allergy conditions (hay fever, serum sickness, urticaria, dermatitis, drug reactions, and bee stings), where they suppress asthma, bronchospasm, and other symptoms. Normally, corticosteroids are secreted as protection against stress caused by chemical, mechanical, radiant, infectious, or immunological agents.

7. Major differences: A C-18 aldehyde exists in Aldosterone, but not in hydrocortisone, and a C-17 OH exists in hydrocortisone but not in aldosterone. Both have the *trans* B/C ring fusion. Metabolically, aldosterone, with an aldehyde group, is much more highly susceptible to oxidation to a carboxylic acid than is hydrocortisone.

9. Estradiol, estriol, estrone; all are steroids. Ring A is aromatic and carries a phenolic OH group. All possess an oxygen functional group at C-17, and an angular methyl group (C-18).

11. The key change in DES would be cyclization to form rings C and D. Secondly, the pattern of aromaticity would have to be altered to bring it closer to that in the estrogens.

13. Male sex hormones have a suppressant, masculinizing effect on all female tissue; they cause the breast to atrophy (shrink). Thus any tumor of the breast would shrink, too.

15. A — 2   B — 6   C — 5
    D — 1   E — 8   F — 7
    G — 4   H — 8   I — 8

17. Prophylaxis means prevention of disease or morbidity by proper administration of drugs or treatment *before* disease strikes; a therapeutic dose of a drug is part of the therapy or treatment of an *existing* disease. Therapeutic doses are usually greater in magnitude than prophylactic doses (an ounce of prevention is worth a pound of cure).

19. Do not permit UV radiation to impinge upon retina. Do not expose skin

to too-long a radiation period. True UV is invisible and therefore potentially insidious.

21. Progestins are synthetic progesterone-like steroids that help fool a woman's pituitary into behaving as though the woman were pregnant; hence pituitary release of FSH and LH is inhibited; ovulation ceases; conception is impossible.

23. Progesterone, testosterone, estradiol, estriol, estrone (plus bile acids, and adrenocorticosteroids).

25. Eight chiral carbons can give rise to 256 optically active isomers ($2^n = 2^8$).

Chapter 15

1. a. Approx. 30 liters; b. approx. 38 liters.

3. Circulating: blood, cerebrospinal fluid, lymph. Noncirculating: intracellular fluid, aqueous (or vitreous) humor of the eye, tears, seminal fluid, amnionic fluid.

5. In homeostasis, the body reacts to stress or to situations that would tend to alter body pH or fluid or electrolyte balance. The body reacts in such a way as to minimize the impact of the stress and thus keep body parameters within a normal, physiological range. The three most important areas of control are: blood pH, fluid balance, electrolyte balance. Examples: (1) dehydration due to lack of water intake, to excessive perspiration, or vomiting, is handled by conservation of water and sodium ion in the kidney (active tubular reabsorption); (2) threatened acidosis, due to diabetes, loss of HCO_3^- from the intestine, excessive acid production, respiratory failure, or impaired renal function, is combated by increased urinary excretion of acid, increased breathing rates and carbon dioxide exhalation, the action of blood buffers (e.g., $H^+ + HPO_4^{-2} \rightarrow H_2PO_4^-$; $H^+ + HCO_3^- \rightarrow H_2CO_3$).

7. Albumins contribute to osmotic pressure; they buffer, and they bind drugs and other molecules. Globulins comprise the antibodies and are involved in blood lipid transport. Fibrinogen is the precursor to fibrin, the material that forms the matrix of a blood clot.

9.

	Size	Shape	#/mm³	Site of syn.	Nuclei	Function
White cells	13μ diameter	spherical	5–10,000	bone marrow thymus spleen	variety of	fight against infections
Red cells	7μ × 1μ	biconcave disc	5×10^6	bone marrow	absent	transport O_2 & CO_2

11. Platelets help initiate the blood clotting process. Inadequate platelet levels can lead to excessive bleeding or hemorrhaging.

13. a. False. Just the opposite; b. true.

15. $CO_2 + H_2O \rightarrow H_2CO_3 \qquad H_2CO_3 + H_2O \rightarrow H_3O^+ + HCO_3^-$

17. A greater breathing rate would mean increased loss of CO_2. Since CO_2 is the anhydride of carbonic acid, that would mean less acid formation and reduced systemic acidity.

19. During voluntary hyperventilation, arterial P_{CO_2} falls from 40 torr to as low as 15 torr.

21. Yes, an effective buffer.
    a. $HCl_{(aq)} + NH_3 \rightarrow NH_4Cl_{(aq)}$
    b. $NaOH_{(aq)} + NH_4^+ \rightarrow NH_3 + H_2O + Na^+$

23. $9\,C = 5\,F - 160$. Therefore, $69°F = 20.6°C$ ($37.6°C$ is normal rectal temp.).

25. Blood types: A, B, AB, and O. Most frequent is type O, least is type AB.

27. Type O blood is safely donated to any recipient because it has no A or B antigens. Hence there can be no antibody reaction to it. Furthermore, even though type O blood has antibodies to both A and B blood, its small volume (when transfused) is easily assimilated into the recipient's larger blood pool.

29. To remove barbiturate molecules from the blood, the blood can be passed through dialyzing membranes. The dialysate fluid has a zero barbiturate concentration so that barbiturate will dialyze from fluid of high concentration to fluid of low concentration. This method of barbiturate detoxification is faster than waiting for the liver to biotransform barbiturate. Used in conjunction with a respirator, it can maintain life during the critical period of barbiturate-induced respiratory depression.

31. Urea, uric acid, creatinine.

33. Diuretics are substances (hormones, polypeptides, drugs) that cause an increase in the production of urine. They are used (a) in congestive heart failure—in which body fluid accumulates because of inadequate pumping action, and (b) in hypertension where loss of sodium and water helps reduce the too-high blood pressure.

35. Bicarbonate acts as a base in the body. Daily doses of it can increase the pH of the urine to greater than 8.0, as the body's homeostatic mechanisms attempt to resist the effect of added base by excreting base. With NH_4Cl, the urine becomes acidic because ammonium ion is converted in the liver to neutral urea with the release of HCl.

37. Since the brown color of fecal matter is due to the end products of bilirubin metabolism, it follows that blockage of bilirubin metabolism (as in jaundice) will result in absence of such brown color (Fig. 15-21).

39. In capillary beds in the alveoli, oxygen combines with hemoglobin: $O_2 + HHb \rightarrow HbO_2^- + H^+$. This oxygen-rich blood is pumped from the heart to body tissues where in capillary beds it diffuses across membranes and into mitochondria of cells. Here it accepts hydrogen ion and electrons to form water: $4\,H^+ + 4\,e^- + O_2 \rightarrow 2\,H_2O$. This water is pooled with other body water and is excreted as required. Oxygen is also consumed in the metabolism of carbohydrate; $C_6H_{12}O_6 + 6\,O_2 \rightarrow 6\,CO_2 + 6\,H_2O$. Specifically, CO_2 arises from the decarboxylation of polycarboxylic acids in the Krebs cycle. More specifically, oxygen enters at the point of oxidative decarboxylation of pyruvate. CO_2 is transported from tissues to the lungs as (a) a simple solution in the blood, where a small amount of it combines with water to form carbonic acid: $CO_2 + H_2O \rightarrow H_2CO_3 \rightleftarrows H^+ + HCO_3^-$, and (b) as a chemical combination with hemoglobin: $HHb + CO_2 \rightarrow HbCO_2^- + H^+$. In the lung, the last two chemical reactions are reversed to yield carbon dioxide, which is exhaled since the P_{CO_2} in the alveolar space is greater than ambient P_{CO_2}.

41. If an Rh− woman becomes pregnant with an Rh+ fetus and, at birth, some of the baby's blood somehow enters her circulation (as via the episiotomy), the woman can develop antibodies to the Rh factor. If, in a *subsequent* pregnancy, the woman carries another Rh+ baby, the anti-Rh agglutinins so acquired can cross the placenta and agglutinate fetal blood. If the fetus dies, the disease is termed erythroblastosis fetalis.

43. In diabetes mellitus, lack of insulin means that carbohydrate metabolism is diminished and fat metabolism becomes the dominant source of body energy. Acetoacetic acid and betahydroxybutyric acid are byproducts of fat metabolism and are produced in excessive amounts in diabetes. Because they are *acids* they can cause a marked decrease in blood pH. The extent of the acidosis depends on the extent of the diabetes mellitus.

45. 154 mEq (remember, isotonic saline is 0.9% w/v NaCl and only 39.3% of NaCl is sodium).

Chapter 16

1. A biotransformation is any body process in which an endogenous or exogenous chemical, foodstuff, hormone or drug is enzymatically broken down into simpler, usually nonbiologically active products. We do not consider anabolic reactions to be biotransformations in the same sense as detoxifications or metabolism. However, in the broadest sense, anabolic reactions would be biotransformations, too.

3. a. true, b. false, c. true, d. false, e. true, f. false, g. false, h. false.

5. Storage of the body's peptidases in an inactive zymogen form prevents them from catalyzing the digestion of protein constituting the lining of the digestive tract. The peptidases become active only when released by stimulation due to the presence of food.

7. Enzyme (E) binds to substrate (S) to form the enzyme-substrate complex (E–S). In the complex form, biotransformation of substrate occurs. Subsequently, the products of the reaction desorb from the enzyme, freeing it to catalyze again. Enzymes lower the activation energy for a reaction, permitting it to occur at a lower temperature.

9. Antimetabolites are molecules so similar in size, shape and chemico-electrical characteristics to the genuine metabolite, that they fool the enzyme into accepting them as the real metabolite. Thus antimetabolites can bind to enzymes, forming the E–S complex. However, enough chemical difference exists in the antimetabolite to make it useless in normal cell metabolism. Cancer cells take up mercaptopurine (Fig. 16-6), but cannot make anything out of it. This blunts their growth, giving body defenses a better chance to destroy the cancerous cells.

11. Stomach pH is vastly more acidic than mouth pH. Salivary amylase's optimum pH is between 6 and 8, and can thus be expected to function very poorly if at all in the stomach contents. Pancreatic amylase functions in the intestine, where alkaline conditions once again obtain.

13. a. Catalyze decarboxylation ($—CO_2$) reactions, b. transfer of methyl group, as from donors such as methionine, c. transfer of amino groups, d. isomerization, i.e., relocation of a double bond or functional group to yield an isomer of original compound, e. esterification with phosphate, f. removal of the elements of H_2.

15. The heavy, chronic tobacco smoker has developed a tolerance to nicotine and other tobacco chemicals. His tolerance consists in part in the development of higher levels of a liver enzyme that catalyzes the biotransformation of nicotine. That is, the liver has been induced to synthesize enzymes in response to ingested nicotine. And as smaller amounts of nicotine no longer satisfy the smoker, he smokes more—which only serves to increase his tolerance even more (vicious cycle).

17. For histamine:

histidine   histamine

For serotonin:

5-hydroxytryptophan   serotonin

19. Barbiturates are oxidized in the alkyl side chain to give hydroxy derivatives or carboxylic acids; propyl alcohol is oxidized to propionic acid *via* propionaldehyde; phenacetin is hydrolyzed to the free phenol (which is then conjugated with glucoronic acid and excreted in the urine); acetaldehyde is oxidized to acetic acid; methyl salicylate is hydrolyzed to salicylic acid; local anesthetics that are esters of PABA derivatives are hydrolyzed to the corresponding carboxylic acid together with the aminoalcohol; in man, benzoic acid is conjugated with glycine to form hippuric acid ($C_6H_5CONHCH_2COOH$); toluene is oxidized to benzoic acid; amphetamine is oxidatively deaminated to $C_6H_5CH_2COCH_3$.

21. Inhibiting the enzyme that helps account for the destruction of dopamine would have the result of increasing tissue levels of dopamine. We could expect this to manifest itself in increased norepi synthesis—and increased norepi pharmacological activity (such as CNS stimulation).

23. For a complete answer, consult modern pharmacology and/or physiology texts. Histamine increases capillary permeability; it may be a synaptic mediator in the brain; it is a vasodilator; it contracts smooth muscle in the bronchioles; it causes increased gastric secretion; it stimulates lacrimal and nasal secretions. Dopamine is the immediate precursor of norepi; it is recognized as a synaptic transmitter in the brain; it is central to one theory of the etiology of schizophrenia; dopamine is involved in the control of prolactin secretion. Serotonin receptors exist in the brain; they are blocked by LSD. Serotonin is involved in sleep/wake patterns in animals and man; its presence helps us to modulate our responses to powerful stimuli. Serotonin probably plays an excitatory role in the regulation of the secretion of prolactin.

Chapter 17

1. Val, Leu, Ile, Lys, Met, Phe, Thr, Trp, His.

3. The human body, once it has a supply of Phe, is able enzymatically to hydroxylate the aromatic ring in the *para* position. Actually, since we can synthesize catecholamines like epi and norepi, it is apparent that we can 3,4-dihydroxylate phenyl rings.

5. When a protein food or mixture of foods has enough of all nine essential amino acids so that all nine are taken up in the body and utilized in protein synthesis, the food is said to contain complementary protein. If even one essential amino acid is missing, complementation cannot occur, and all of the amino acids may be metabolized for energy and thus "lost" for protein synthesis. Cereal and milk are complementary.

7. In diabetes (i.e., the absence of insulin) protein synthesis in muscle is diminished, blood amino acid levels rise, and more amino acids are metabolized to glucose in the liver (increased protein catabolism). This is because insulin stimulates protein formation, a fact illustrated in diabetic children who fail to grow normally. Thus diabetics show a marked negative nitrogen balance, protein depletion and wasting. The excess of sugar in the diabetic's body fluids provides a good culture medium for the growth of pathogenic microorganisms.

9. Anabolic steroids promote tissue synthesis of proteins; this will tend to put the patient into positive nitrogen balance.

11. Both are essential, but micronutrients are required only in micro amounts in the diet (mg or mcg). The RDAs for macronutrients are in the order of grams.

13. $20 \text{ lb} \times \dfrac{3500 \text{ Cal}}{1 \text{ lb}} \times \dfrac{1 \text{ day}}{700 \text{ Cal}} = 100 \text{ days}$

15. By and large, so-called "junk foods" are higher in calories, sometimes vitamin fortified, almost zero in fiber content, deficient in essential amino acids and essential fatty acids, and high in salt and sugar. Their potential for food-drug interactions is not high.

17. Thyroxine (T_4) and triiodothyronine (T_3).

19.

	Cause	Signs & symptoms	Treatment
cretinism	too little thyroid secretion in childhood	dwarf size, mental retardation, pot belly	early administration of exogenous thyroxine
Graves' disease	excessive thyroid levels	exophtholmos, ↑ BMR, goiter, warm skin	suppress synthesis of excess thyroxine

21. In zero carbohydate diets (e.g., Atkins'), as soon as body stores of glycogen are depleted, fat becomes the main source of energy. Now the pathways for ketone bodies (which are always active to some degree) begin turning

out large quantities of ketone bodies, which become detectable in the urine.

23. Fe–D, E; Co–B; Cu–E; Zn–C; I–A; Mg–F.

25. Tendency towards pernicious anemia (megaloblastic anemia).

27. Bacon, 1160 mg; ham, 504 mg; egg, 37 mg; olive, 120 mg; peanuts, 59 mg; for a total of 1880 mg, or 1680 mg over needs.

29. a. 6000 mcg, b. 106 mcg/100 ml.

31. Nine essential amino acids (Leu, Ile, Lys, Met, Phe, Thr, His, Trp, Val); two essential fatty acids (linoleic, linolenic); trace elements (Cr, Mn, Fe, Co, Cu, Zn, Mo, Se, I), macronutrients (Na, K, Mg, Ca), and the vitamins (see Table 17-6). Some bulk (fiber) might be included in the list.

Chapter 18

1. Nucleic acids are very high molecular weight polymers occurring in the nucleus or cytoplasm of a cell; they are of the DNA or RNA type. They carry all of genetic information required to reproduce the species.

3. Each nucleotide consists of a sugar, phosphoric acid and a nitrogen base.

5. cytosine, thymine, uracil, adenine and guanine.

7. a. Thymine, b. cytosine, c. guanine, d. adenine

9. DNA directs cell synthesis of proteins, including enzymes. It is the arrangement and sequence of nucleotide units (actually, nitrogen bases) in the DNA strand that directs cell synthesis of protein. Enzymes, once available in the cell, can catalyze the synthesis of other compounds vital for growth, maintenance, defense, etc. Thus the link of DNA to cell proteins is paramount.

11. Gly-Leu-Pro-Cys-Asn-Gln-Ile-Tyr-Cys. (Oxytocin)

13. Transfer RNA

15. Plasmids are inclusions in bacterial cells that consist of circular molecules of double-stranded DNA. They are used in recombinant DNA work.

17. Enzymes are proteins and will be digested if administered orally. Upon injection, the enzyme is likely to be bound to blood proteins or otherwise deactivated. Little will reach the desired tissue goal. Besides, this approach would require frequent hypodermic injections for the remainder of the patient's life.

19. Write the Public Relations Department, Eli Lilly and Company, Indianapolis, Indiana, for information about insulin made by their recombinant DNA process.

Chapter 19

1. particle   charge   mass
   a. alpha   2 +   4 u
   b. beta   1 −   0 u
   c. positron   1 +   0 u

3. $^{137}_{55}\text{Cs} \rightarrow \,^{0}_{-1}\text{e} + \gamma + \,^{137}_{56}\text{Ba}$

5. a. $^{198}_{80}\text{Hg}$ b. $^{136}_{54}\text{Xe}$ c. $^{90}_{39}\text{Y}$ d. $^{90}_{40}\text{Zr}$

    10. g $\xrightarrow[\text{min}]{110}$ 5.0 g $\xrightarrow[\text{min}]{110}$ 2.5 g

7. After 220 minutes, you could no longer be sure of having 2.5 g of pure fluorine-16.

9. 2 days $\times \dfrac{24 \text{ hours}}{\text{day}} \times \dfrac{1 \text{ half-life}}{12.4 \text{ hours}}$ = 4 half-lives have elapsed

   6 trillion atoms $\xrightarrow[\text{1/2 life}]{\text{one}}$ 3 trillion $\xrightarrow[\text{1/2 life}]{\text{2nd}}$ 1.5 trillion

   $\xrightarrow[\text{1/2 life}]{\text{3rd}}$ 0.75 trillion $\xrightarrow[\text{1/2 life}]{\text{4th}}$ 0.375 trillion ($3.75 \times 10^9 \approx 4 \times 10^8$)

   At the end of 2 days, you would expect approximately 4×10^8 atoms of potassium-42 to remain.

11. 1 $\mu\text{Ci} \times \dfrac{1 \text{ curie}}{10^6 \, \mu\text{Ci}} \times \dfrac{3.7 \times 10^{10} \text{ disintegrations per second}}{1 \text{ curie}}$ = 3.7×10^4 dps

13. A 6000 mile jet airplane trip exposes you to 4 millirems of radiation. This is only one-fifth of the average dental x-ray, which gives 20 millirems per exposure.

15. See Table 19.2. The list will vary by occupation and medical exposure, but should include radiation from the earth's crust, internal sources and cosmic radiation adjusted for elevation.

17. 0.15×10^6 e.v. $\times \dfrac{1.602 \times 10^{-19} \text{ J}}{\text{e.v.}}$ = 2.4×10^{-14} J

19. Alpha particles are not very penetrating. They cannot go through the thin window of a Geiger counter.

21. Strontium-90 can mimic the behavior of its chemical family member calcium. This means there is a risk of Sr-90 becoming a part of such essential foods as milk, entering the body, and then becoming incorporated into the bones. Sr-90 has a half-life of 28 years so its presence in the fallout from atmospheric testing would be a long-range problem.

23. Minimizing exposure can be accomplished by maximizing the distance between yourself and the radioactive source, by maximizing the amount

of dense shielding material between you and the source, and by minimizing exposure time.

25. Radioactive implants can reach precisely the optimum location for effective therapy. The seeds also can continue the therapy over a longer period of time, at a lower and continuous dose. High doses of gamma radiation may damage more surrounding tissue and still not reach the target efficiently unless that target is on the surface.

27. A radioisotope used for diagnosis typically has a short half-life and a known method of elimination from the body. If used for therapy, the radioisotope is often one with a very high energy radioactive emission. These radioisotopes may have much longer half-lives if you want to treat by means of implanting a radioactive seed directly into a tumor.

29. $^{14}_{7}\text{N} + ^{2}_{1}\text{H} \rightarrow ^{15}_{8}\text{O} + ^{1}_{0}\text{n} + \gamma$

Index

Atomic Numbers and Atomic Weights of the Elements

Based on $^{12}_{6}C$. Numbers in parentheses are the mass numbers of the most stable isotopes of radioactive elements.

Element	Symbol	Atomic Number	Atomic Weight	Element	Symbol	Atomic Number	Atomic Weight
Actinium	Ac	89	227.0278	Europium	Eu	63	151.96
Aluminum	Al	13	26.98154	Fermium	Fm	100	(257)
Americium	Am	95	(243)	Fluorine	F	9	18.998403
Antimony	Sb	51	121.75	Francium	Fr	87	(223)
Argon	Ar	18	39.948	Gadolinium	Gd	64	157.25
Arsenic	As	33	74.9216	Gallium	Ga	31	69.72
Astatine	At	85	(210)	Germanium	Ge	32	72.59
Barium	Ba	56	137.33	Gold	Au	79	196.9665
Berkelium	Bk	97	(247)	Hafnium	Hf	72	178.49
Beryllium	Be	4	9.01218	Helium	He	2	4.00260
Bismuth	Bi	83	208.9804	Holmium	Ho	67	164.9304
Boron	B	5	10.81	Hydrogen	H	1	1.0079
Bromine	Br	35	79.904	Indium	In	49	114.82
Cadmium	Cd	48	112.41	Iodine	I	53	126.9045
Calcium	Ca	20	40.08	Iridium	Ir	77	192.22
Californium	Cf	98	(251)	Iron	Fe	26	55.847
Carbon	C	6	12.011	Krypton	Kr	36	83.80
Cerium	Ce	58	140.12	Lanthanum	La	57	138.9055
Cesium	Cs	55	132.9054	Lawrencium	Lr	103	(260)
Chlorine	Cl	17	35.453	Lead	Pb	82	207.2
Chromium	Cr	24	51.996	Lithium	Li	3	6.941
Cobalt	Co	27	58.9332	Lutetium	Lu	71	174.967
Copper	Cu	29	63.546	Magnesium	Mg	12	24.305
Curium	Cm	96	(247)	Manganese	Mn	25	54.9380
Dysprosium	Dy	66	162.50	Mendelevium	Md	101	(258)
Einstenium	Es	99	(252)				
Erbium	Er	68	167.26				